S0-BSK-919

Springer Monographs in Mathematics

To my Riemann zeros:
Odile, Julie and Michaël, my muse and offspring

Michel L. Lapidus

To Jena, David and Samuel

Machiel van Frankenhuijsen

Michel L. Lapidus Machiel van Frankenhuijsen

Fractal Geometry, Complex Dimensions and Zeta Functions

Geometry and Spectra of Fractal Strings

With 53 Illustrations

Michel L. Lapidus
University of California
Department of Mathematics
231 Surge Building
Riverside, CA 92521-0135
USA
lapidus@math.ucr.edu

Machiel van Frankenhuijsen
Utah Valley State College
Department of Mathematics
800 West University Parkway
Orem, Utah 84058-5999
USA
vanframa@uvsc.edu

The front cover shows a tubular neighborhood of the Devil's staircase (Figure 12.2, page 335) and the quasiperiodic pattern of the complex dimensions of a nonlattice self-similar string (Figure 3.7, page 86).

Mathematics Subject Classification (2000): Primary–11M26, 11M41, 28A75, 28A80, 35P20, 58G25
Secondary–11J70, 11M06, 11N05, 28A12, 30D35, 81Q20

Library of Congress Control Number: 2006929212

ISBN-10: 0-387-33285-5 e-ISBN: 0-387-35208-2
ISBN-13: 978-0-387-33285-7

Printed on acid-free paper.

Printed in the United States of America. (EB)

9 8 7 6 5 4 3 2 1

springer.com

Contents

Preface

This book is a self-contained unit which encompasses a broad range of topics that connect many areas of mathematics, including fractal geometry, number theory, spectral geometry, dynamical systems, complex analysis, distribution theory and mathematical physics.

The material in our earlier book, *Fractal Geometry and Number Theory* ([Lap-vF5], Birkhäuser, January 2000), has been recast and greatly expanded, taking into account the latest research developments in the field. Compared with that foundational monograph, the present book is longer by almost 200 pages and has twice as many illustrations. Further, it contains two new chapters, Chapter 3 about the complex dimensions of nonlattice self-similar strings, and Chapter 7 about flows, and a new appendix, on an application of Nevanlinna theory, as well as many new sections within the original chapters of [Lap-vF5]. It also provides a number of new examples, comments and theorems, many of which have not previously been published in the mathematical literature.

Chapter 1 provides a gentle introduction to some of the main topics. We extend the definition of self-similar string in Chapter 2 to include self-similar strings with more than one gap, following [Fra1, 2]. These self-similar strings correspond to (standard) self-similar sets in $\mathbb{R}$. Moreover, their geometric zeta functions may have both poles (i.e., complex dimensions) and zeros, so that, as shown in the new Section 2.3.3, cancellations may occur and some of the potential complex dimensions may disappear as a result.

Since the publication of [Lap-vF5], the material on Diophantine approximation of complex dimensions of nonlattice strings, which was included in [Lap-vF5, Sections 2.4–2.6], has grown into a new chapter, Chapter 3. In that chapter, we give a large amount of new information on the fine structure of the complex dimensions of nonlattice self-similar strings. In particular, both qualitatively and quantitatively, we obtain a much better understanding of the quasiperiodic patterns of the complex dimensions. Our new theoretical and numerical results are illustrated by a variety of diagrams throughout Chapter 3, and several conjectures and open problems are proposed. Further, dimension-free regions are obtained that are used to derive good (and sometimes sharp) error estimates in much of the remainder of the book. In addition, we introduce Dirichlet polynomials and we slightly change the definition of generic nonlattice. We use a new approach to computing the density of the complex dimensions of a nonlattice self-similar string, improving upon the density results of Chapter 2. We have kept the argument of [Lap-vF5] in Chapter 2, and the relevant parts of Nevanlinna theory are presented in the new Appendix C. These results were first partly presented in the paper [Lap-vF7] (see also [Lap-vF9]).

We have added a new and previously unpublished discussion of the Euler product of the spectral operator in Section 6.3.2. The interest of this construction is that it provides an operator-valued Euler factorization of the spectral operator (associated with the spectral counting function) that is valid in the critical strip $0 < \operatorname{Re} s < 1$.

Chapter 7 has been largely expanded from Section 2.1.1 of [Lap-vF5]. It contains material about the periodic orbits of self-similar flows, leading to an Euler product representation of the geometric zeta function of a self-similar fractal string. We obtain an explicit formula (expressed in terms of the underlying dynamical complex dimensions) for the prime orbit counting function of a suspended flow, and thereby deduce a Prime Orbit Theorem in this context. Our results are most precise in the special case of self-similar flows for which we determine (via Diophantine approximation) dimension-free regions, from which we deduce a Prime Orbit Theorem with (often sharp) error term. This chapter is partly based on the paper [Lap-vF6].

We found a more efficient way to study the tubular neighborhoods of fractal strings (Chapter 6 of [Lap-vF5]), which led us to establish a pointwise tube formula in Section 8.1.1, Theorem 8.7. Under suitable somewhat stronger hypotheses, this theorem complements and improves upon the conclusion of Theorem 8.1 (the distributional tube formula). The latter theorem is central to our book (and played a key role in [Lap-vF5] as well). In fact, upon the request of some of our readers, we have provided more details for the proof of Theorem 8.1 as well as of several other results in Chapter 8. Thanks to Erin Pearse, we were also able to include Figure 8.1, illustrating the structure of the proof of Theorem 8.15 and the interdependence of many of the explicit formulas and other results in Chapters 5 and 8. In Sections 8.4.2 and 8.4.4, which discuss the important class of self-similar strings, we have provided a significantly more detailed discussion of the lattice case and of the nonlattice case. Furthermore, our earlier statements are extended to general self-similar strings (i.e., lattice and nonlattice strings with multiple gaps). In Section 8.4.3, we define and compute the average Minkowski content of an arbitrary lattice string.

The geometry and the spectrum of Cantor strings and truncated Cantor strings is presented in Chapter 10 (Chapter 8 of [Lap-vF5]). The material on truncated Cantor strings is new, and is applied in Section 11.1.1.

We include in Chapter 11 (Chapter 9 of [Lap-vF5]) an exposition of the work of Mark Watkins on shifted arithmetic progressions [Watk, vFWatk]. We thank him for allowing us to include this result. We also include in Section 11.1.1 an exposition of [vF3], on finite arithmetic progressions of zeros of the Riemann zeta function. These works build upon and further develop the earlier work in [Lap-vF5, Chapter 9] on infinite arithmetic progressions of zeros. They also provide additional tools to attempt to solve some of the problems and conjectures proposed in [Lap-vF5, Section 10.1] (see Section 12.1).

Chapter 12 contains a summary of the recent results of Erin Pearse and the first author [LapPe1] on the complex dimensions and the volume of the tubular neighborhoods of the von Koch snowflake curve (see Section 12.3.1), which is pursued in a somewhat different direction in [LapPe2–4] (see Section 12.3.2). The latter work can be viewed as a step towards the long-term goal of developing a higher-dimensional theory of complex dimensions of fractals. We also discuss (in Section 12.4.1) recent results of Ben Hambly and the first author [HamLap] on the complex dimensions of random fractal strings, including random self-similar strings and the zero set of Brownian motion. Furthermore, in Section 12.4.2, we give a short introduction to the theory of fractal membranes (quantized fractal strings), proposed by the first author in the forthcoming book [Lap10] (see also [Lap9]) and further developed by Ryszard Nest and the first author in the papers in preparation [LapNes1–3]. The last section of the book, Section 12.7, corresponding to [Lap-vF5, Chapter 10.5], has been expanded, elaborating our proposed theory of complex cohomology.

In Appendix A, we have added a brief introduction to the two-variable zeta functions of Pellikaan [Pel], Schoof and van der Geer [SchoG], and Lagarias and Rains [LagR].

Several mistakes and misprints were pointed out to us by a number of people and have been corrected. We want to thank those people for their helpful comments. Without a doubt, new mistakes have been added, for which we take full responsibility. As was the case for the first book [Lap-vF5], we welcome comments from our readers.

Some, but by no means all of the main results of this book appeared in [Lap-vF1–7, 9]. Our earlier book [Lap-vF5] combined and superseded our two IHES preprints [Lap-vF1–2] and was announced in part in the paper [Lap-vF4], which was a slightly expanded version of the IHES preprint M/97/85. Most of the material in [Lap-vF5] was entirely new. The interested reader may wish to consult [Lap-vF4], as well as the research expository article [Lap-vF9]—in conjunction with the introduction and Chapter 1—to have an accessible overview of some of the main aspects of this work.

Michel L. Lapidus and Machiel van Frankenhuijsen

May 2006

List of Figures

List of Tables

Overview

In this book, we develop a theory of complex dimensions of fractal strings (i.e., one-dimensional drums with fractal boundary). These complex dimensions are defined as the poles of the corresponding (geometric or spectral) zeta function. They describe the oscillations in the geometry or the frequency spectrum of a fractal string by means of an explicit formula. Such oscillations are not observed in smooth geometries.

A long-term objective of this work is to merge aspects of fractal, spectral, and arithmetic geometries. From this perspective, the theory presented in this book enables us to put the theory of Dirichlet series (and of other zeta functions) in the geometric setting of fractal strings. It also allows us to view certain fractal geometries as arithmetic objects by applying number-theoretic methods to the study of the geometry and the spectrum of fractal strings.

In Chapter 1, we first give an introduction to fractal strings and their spectrum, and we precisely define the notion of complex dimension. We then make in Chapter 2 an extensive study of the complex dimensions of self-similar fractal strings. This study provides a large class of examples to which our theory can be applied fruitfully. In particular, we show in Chapter 3 that self-similar strings always have infinitely many complex dimensions with positive real part, and that their complex dimensions are quasiperiodically distributed. This is established by proving that the lattice strings—the complex dimensions of which are shown to be periodically distributed along finitely many vertical lines—are dense (in a suitable sense) in the set of all self-similar strings. We present the theory of Chapter 3—in which we analyze in detail the quasiperiodic pattern of complex dimensions, via Diophantine approximation—by using the more general notion of Dirichlet polynomial.

In Chapter 4, we extend the notion of fractal string to include (possibly virtual) geometries that are needed later on in our work. Then, in Chapter 5, we establish pointwise and distributional explicit formulas (explicit in the sense of Riemann's original formula [Rie1], but more general), which should be considered as the basic tools of our theory. In Chapter 6, we apply our explicit formulas to construct the spectral operator, which expresses the spectrum in terms of the geometry of a fractal string. This operator has an Euler product that is convergent, in a suitable sense to be explained in Section 6.3.2, in the critical strip $0 < \operatorname{Re} s < 1$ of the Riemann zeta function. We also illustrate our formulas by studying a number of geometric and direct spectral problems associated with fractal strings.

In Chapter 7, we use the theory of Chapters 3 and 5 to study a class of suspended flows and define the associated dynamical complex dimensions. In particular, we establish an explicit formula for the periodic orbit counting function of such flows and deduce from it a prime orbit theorem with sharp error term for self-similar flows, thereby extending in this context the work of [PaPol1, 2]. We also obtain an Euler product for the zeta function of a self-similar fractal string (or flow).

In Chapter 8, we derive an explicit formula for the volume of the tubular neighborhoods of the boundary of a fractal string. We deduce a new criterion for the Minkowski measurability of a fractal string, in terms of its complex dimensions, extending the earlier criterion obtained by the first author and C. Pomerance (see [LapPo2]). This formula suggests analogies with aspects of Riemannian geometry, thereby giving substance to a geometric interpretation of the complex dimensions.

In the later chapters of this book, Chapters 9–11, we analyze the connections between oscillations in the geometry and the spectrum of fractal strings. Thus we place the spectral reformulation of the Riemann hypothesis, obtained by the first author and H. Maier [LapMa2], in a broader and more conceptual framework, which applies to a large class of zeta functions, including all those for which one expects the generalized Riemann hypothesis to hold. We also reprove—and extend to a large subclass of the aforementioned class—Putnam's theorem according to which the Riemann zeta function does not have an infinite sequence of critical zeros in arithmetic progression. This work is supplemented in Section 11.1.1 with an upper bound for the possible length of an arithmetic progression of zeros, and in Section 11.4.1, where we present Mark Watkins' work on the finiteness of shifted arithmetic progressions of zeros of L-series.

In the final Chapter 12, we propose as a new definition of fractality the presence of nonreal complex dimensions with positive real part. We also make several suggestions for future research in this area. In particular, we summarize the recent results of [LapPe1] on the complex dimensions and the volume of the tubular neighborhoods of the von Koch snowflake curve, which provides a first example of a higher-dimensional theory of complex dimensions of fractals.

Le plus court chemin entre deux vérités dans le domaine réel passe par le domaine complexe.

[The shortest path between two truths in the real domain passes through the complex domain.]

Jacques HADAMARD

Introduction

A fractal drum is a bounded open subset of $\mathbb{R}^m$ with a fractal boundary. A difficult problem is to describe the relationship between the shape (geometry) of the drum and its sound (its spectrum). In this book, we restrict ourselves to the one-dimensional case of fractal strings, and their higher-dimensional analogues, fractal sprays. We develop a theory of complex dimensions of fractal strings, and we study how these complex dimensions relate the geometry and the spectrum of fractal strings. See the notes to Chapter 1 in Section 1.5 for references to the literature.

In Chapter 1, we define the basic object of our research, *fractal strings*. A standard fractal string is a bounded open subset of the real line. Such a set is a disjoint union of open intervals, the lengths of which form a sequence

$$\mathcal{L} = l_1,\ l_2,\ l_3, \dots,$$

which we typically assume to be infinite. Important information about the geometry of $\mathcal{L}$ is contained in its *geometric zeta function*

$$\zeta_{\mathcal{L}}(s) = \sum_{j=1}^{\infty} l_j^s.$$

We assume throughout that this function has a suitable meromorphic extension. The central notion of this book, the *complex dimensions* of a fractal string $\mathcal{L}$, is defined as the poles of the meromorphic extension of $\zeta_{\mathcal{L}}$. These can also be referred to as the complex fractal dimensions of $\mathcal{L}$.

The spectrum of a fractal string consists of the sequence of frequencies[1]

$$f = k \cdot l_j^{-1} \qquad (k, j = 1, 2, 3, \dots).$$

The *spectral zeta function* of $\mathcal{L}$ is defined as

$$\zeta_\nu(s) = \sum_f f^{-s}.$$

The geometry and the spectrum of $\mathcal{L}$ are connected by the following formula [Lap2]:

$$\zeta_\nu(s) = \zeta_{\mathcal{L}}(s)\zeta(s), \tag{$*$}$$

where $\zeta(s) = 1 + 2^{-s} + 3^{-s} + \dots$ is the classical Riemann zeta function, which in this context can be viewed as the spectral zeta function of the unit interval.

We also define a natural higher-dimensional analogue of fractal strings, *fractal sprays*, the spectra of which are described by more general zeta functions than $\zeta(s)$ [LapPo3]. The counterpart of $(*)$ still holds for fractal sprays and can be used to study their spectrum. We refer the interested reader to Appendix B for a brief review of aspects of spectral geometry—including spectral zeta functions and spectral asymptotics—in the classical case of smooth manifolds.

We illustrate these notions throughout Chapter 1 by working out the example of the Cantor string. In this example, we see that the various notions that we have introduced are described by the complex dimensions of the Cantor string. In higher dimensions, a similar example is provided by the Cantor sprays.

This theory of complex dimensions sheds new light on, and is partly motivated by, the earlier work of the first author in collaboration with C. Pomerance and H. Maier (see [LapPo2] and [LapMa2]). In particular, the heuristic notion of complex dimension suggested by the methods and results of [Lap1–3, LapPo1–3, LapMa1–2, HeLap1–2] is now precisely defined and turned into a useful tool.

In Chapters 2 and 3, we make an extensive study of the complex dimensions of self-similar strings, which form an important subclass of fractal strings. This amounts to studying the zeros of the function

$$f(s) = 1 - r_1^s - r_2^s - \dots - r_N^s \qquad (s \in \mathbb{C}),$$

for a given set of real numbers $r_j \in (0,1)$, $j = 1, \dots, N$, $N \geq 2$. We introduce the subclass of lattice self-similar strings, and find a remarkable difference between the complex dimensions of lattice and nonlattice self-similar

[1] The eigenvalues of the Dirichlet Laplacian $-d^2/dx^2$ on this set are the numbers $\lambda = \pi^2 k^2 l_j^{-2}$ $(k, j \in \mathbb{N}^*)$. The (normalized) frequencies of $\mathcal{L}$ are the numbers $\sqrt{\lambda}/\pi$.

strings. In the lattice case, each number r_j is a positive integral power of one fixed real number $r \in (0, 1)$. Then f is a polynomial in r^s, and its zeros lie periodically on finitely many vertical lines. The Cantor string is the simplest example of a lattice self-similar string, and we refer to Section 2.3 for additional examples. In contrast, the complex dimensions of a nonlattice string are apparently randomly distributed in a vertical strip. In Chapter 3, however, we show that these complex dimensions are approximated by those of a sequence of lattice strings. Hence, they exhibit a quasiperiodic behavior (see Theorems 2.17 and 3.6, along with Section 3.4). On page 52, Figure 2.12, the reader finds a diagram of the complex dimensions of the golden string, one of the simplest nonlattice self-similar strings. This and other examples are discussed in Section 2.3, and many other such examples are discussed in more detail in Chapter 3. In fact, much of Chapter 3 is devoted to the careful study (via Diophantine approximation techniques) of the beautiful and intriguing quasiperiodic patterns of the complex dimensions of nonlattice strings, both rigorously and computationally. We also obtain in that chapter specific dimension-free regions of the complex plane for nonlattice self-similar strings (see Section 3.6). When combined with our explicit formulas from later chapters, this will enable us, in particular, to give good estimates for the error term in the resulting asymptotic formulas, depending on the Diophantine properties of the scaling ratios.

Chapters 4, 5 and 6 are devoted to the development of the technical tools needed to extract geometric and spectral information from the complex dimensions of a fractal string. In Chapter 4, we introduce the framework in which we will formulate our results, that of *generalized fractal strings*. These do not in general correspond to a geometric object. Nevertheless, they are not just a gratuitous generalization since they enable us, in particular, to deal with virtual geometries and their associated spectra—suitably defined by means of their zeta functions—as though they arose from actual fractal geometries. In Chapters 9, 10 and 11, the extra flexibility of this framework allows us to study the zeros of several classes of zeta functions.

The original explicit formula was given by Riemann [Rie1] in 1858 as an analytical tool to understand the distribution of primes. It was later extended by von Mangoldt [vM1–2] and led in 1896 to the first rigorous proof of the Prime Number Theorem, independently by Hadamard [Had2] and de la Vallée Poussin [dV1] (as described in [Edw]). Writing $f(x)$ for the function $= \sum_{p^n \leq x} \frac{1}{n}$ that counts prime powers $p^n \leq x$ with a weight $1/n$, the explicit formula of Riemann is (see [Edw, p. 304 and Section 1.16])

$$f(x) = \mathrm{Li}(x) - \sum_{\rho} \mathrm{Li}(x^{\rho}) + \int_x^{\infty} \frac{1}{t^2 - 1} \frac{dt}{t \log t} - \log 2,$$

where the sum is over all zeros ρ of the Riemann zeta function, taken in order of increasing absolute value, and $\mathrm{Li}(x)$ is the logarithmic integral $\int_0^x dt/(\log t)$ (see (5.78)). For numerical purposes, the left-hand side of

Riemann's explicit formula is easy to compute. For theoretical purposes, however, the right-hand side is more useful. For example, the Prime Number Theorem $f(x) = \mathrm{Li}(x)(1 + o(1))$ as $x \to \infty$,[2] which was the primary motivation for Riemann's investigations, follows if all zeros satisfy $\mathrm{Re}\,\rho < 1$, that is, the function $\log \zeta(s)$ has a singularity at $s = 1$ but no other singularities on the line $\mathrm{Re}\, s = 1$.

An example of an explicit formula in our theory is formula (**) below, which expresses the volume of the tubular neighborhoods $V(\varepsilon)$ of a fractal string as an infinite sum of oscillatory terms $\varepsilon^{1-\omega}$, where ω runs over the complex dimensions of the fractal string (and hence $1 - \omega$ runs over the complex codimensions). Much like Riemann, we use that formula to show that $V(\varepsilon) = \varepsilon^{1-D}(v + o(1))$ as $\varepsilon \to 0^+$, for some constant v which is related to the D-dimensional volume of the fractal, if and only if there are no nonreal complex dimensions on the line $\mathrm{Re}\,\omega = D$. Here, D denotes the Minkowski dimension of the fractal boundary of the string.

In Chapter 5, we state and prove our explicit formulas, which are the basic tools for obtaining asymptotic expansions of geometric, spectral or dynamical quantities associated with fractals. Our first explicit formula, which expresses the counting function of the lengths as a sum of oscillatory terms and an error term of smaller order, is only applicable under fairly restrictive assumptions. To obtain a more widely applicable theory, we show in Section 5.4 that this same function, interpreted as a distribution, is given by the same formula, now interpreted distributionally. The resulting distributional formula with error term is applicable under mild assumptions on the analytic continuation of the geometric zeta function. We also obtain a pointwise and a distributional formula without error term, which exists only for the geometry of a smaller class of fractal strings, including the self-similar strings and the so-called prime string. In Section 5.5, we use this analysis of the prime string to give a proof of the Prime Number Theorem.

We note that our explicit formulas are close relatives of—but are also significantly more general than—the usual explicit formulas encountered in number theory [Edw, In, Pat]. (See the end of Section 5.1 and the notes in Section 5.6 for further discussion and additional references.)

In the subsequent chapters, we investigate the geometric, dynamical, and spectral information contained in the complex dimensions. The main theme of these chapters is that the oscillations in the geometry or in the spectrum of a fractal string are reflected in the presence of oscillatory terms in the explicit formulas associated with the fractal string.

In Chapter 6, we work out the necessary computations to find the oscillatory terms in the explicit formulas of a fractal string. We define the spectral operator, which relates the spectrum of a fractal string with its

---

[2] We have $\mathrm{Li}(x) = \frac{x}{\log x}(1 + o(1))$ as $x \to \infty$, but the approximation $\mathrm{Li}(x)$ for $f(x)$ is better than $\frac{x}{\log x}$.

geometry and obtain an Euler product representation for it, which provides a counterpart in this context to the usual Euler product expansion for the Riemann zeta function, but is convergent in the critical strip $0 < \operatorname{Re} s < 1$. We also illustrate our results by considering a variety of examples of geometric and direct spectral problems. We study the geometric and spectral counting functions as well as the geometric and spectral partition functions of fractal strings. In particular, we analyze in detail the geometry and the spectrum of self-similar strings, both in the lattice and the nonlattice case.

We show in Chapter 7 that the geometric zeta function of a self-similar string coincides with a suitably defined dynamical (or Ruelle) zeta function, and hence that it admits an appropriate Euler product; see Section 7.2. For example, the zeta function of the self-similar flow (or dynamical system) with weights $\mathfrak{w} = w_1, w_2, \dots, w_N$ is given by

$$\zeta_{\mathfrak{w}}(s) = \frac{1}{1 - \sum_{j=1}^{N} e^{-w_j s}},$$

which is the geometric zeta function of a self-similar fractal string with scaling ratios $r_j = e^{-w_j}$ $(j = 1, \dots, N)$ and a single gap (see Chapter 2). The connection in this dynamical context with Marc Frantz's more general self-similar strings with multiple gaps still remains to be clarified.[3] We apply the theory of Chapters 5 and 6 to obtain a suitable explicit formula and an associated Prime Orbit Theorem for self-similar flows: the function which counts primitive periodic orbits with their weight has the asymptotic expansion

$$\psi_{\mathfrak{w}}(x) = G(x)\frac{x^D}{D} + \sum_{\operatorname{Re}\omega < D} \frac{x^\omega}{\omega} + \mathcal{R}(x),$$

where the sum is over all the (dynamical) complex dimensions of the flow, repeated according to their multiplicity, and $\mathcal{R}(x) = O(1)$ as $x \to \infty$ (see Section 7.4). Here, $G(x) = 1$ in the nonlattice case, when D is the only complex dimension with real part D, and $G(x)$ is multiplicatively periodic in the lattice case. As was alluded to earlier, using the results of Section 3.6 on dimension-free regions, we deduce from this formula a prime orbit theorem *with error term*. We may analyze this error term (beyond the leading term) in terms of the Diophantine properties of the weights w_j. For example, self-similar flows with weights that are badly approximable by rationals have a larger dimension-free region and hence a better (i.e., smaller) error term in the above asymptotic formula. (See Section 7.5.)

[3] Such self-similar strings, the boundary of which corresponds to self-similar sets in $\mathbb{R}$, were introduced in [Fra1, 2] after the publication of [Lap-vF5], where the prototypical case of a single gap was studied. They are discussed in Chapter 2. With the exception of Chapter 7, our theory can be applied immediately to this more general setting.

Analogous comments hold with regard to all our explicit formulas when they are applied in the self-similar case.

In Chapter 8, we derive an explicit formula for the volume $V(\varepsilon)$ of the inner[4] ε-neighborhood of the boundary of a fractal string. For example, when the complex dimensions of $\mathcal{L}$ are simple, we obtain the following key formula:

$$V(\varepsilon) = \sum_{\omega} c_\omega \frac{(2\varepsilon)^{1-\omega}}{\omega(1-\omega)} + R(\varepsilon), \tag{**}$$

where ω runs over the complex dimensions of the fractal string $\mathcal{L}$, c_ω denotes the residue of $\zeta_{\mathcal{L}}(s)$ at $s = \omega$, and $R(\varepsilon)$ is an error term of lower order. Formula $(**)$ yields a new criterion for the Minkowski measurability of a fractal string in terms of the absence of nonreal complex dimensions with real part D, the dimension of the string. (See Section 8.3.) This extends the joint work of the first author with C. Pomerance [LapPo1; LapPo2, Theorem 2.2], in which a characterization of Minkowski measurability was obtained in terms of the absence of geometric oscillations in the string. A comparison, in Section 8.2, of our formula with Hermann Weyl's formula for tubes in Riemannian geometry [BergGo, p. 235] suggests what kind of geometric information may be contained in the complex dimensions of a fractal string.

The last part of Chapter 8 (Section 8.4) is devoted to a detailed discussion of the tube formulas for the class of self-similar strings, both in the lattice and nonlattice case. The tube formula $(**)$ takes a particularly concrete form for lattice strings, while the error term $R(\varepsilon)$ in $(**)$ can be estimated by using some of the results of Chapter 3. In particular, it follows that a self-similar string is Minkowski measurable if and only if it is nonlattice.

In Chapters 9, 10, and 11, we shift the emphasis from the geometry of fractal strings to the relationship between the geometry and the spectrum of fractal strings.

In Chapter 9, we study the inverse spectral problem, the problem of deducing geometric information from the spectrum of a fractal string: Does the absence of oscillations in the spectrum of a fractal string imply the absence of oscillations in its geometry? In other words, we consider the question (à la Mark Kac [Kac]) *"Can one hear the shape of a fractal string?"* This inverse spectral problem has been considered before by the first author jointly with H. Maier in [LapMa1–2], where it was shown that the

[4]We use the inner tubular neighborhood (see Equation (1.3) in Chapter 1) so that $V(\varepsilon)$ does not depend on the placement of the lengths l_j in space (the geometric realization of the fractal string). Likewise, the Minkowski dimension and Minkowski content are independent of the geometric realization of the fractal string, as opposed to the Hausdorff dimension and measure. This is the main reason why in [Lap1, LapPo2], the Minkowski dimension is used as well.

audibility of oscillations in the geometry of a fractal string of (Minkowski) dimension $D \in (0,1)$ is equivalent to the absence of zeros of the Riemann zeta function $\zeta(s)$ on the line $\operatorname{Re} s = D$. In our framework, this becomes the question of inverting the spectral operator. We deduce, in particular, that the spectral operator is invertible for all fractal strings of dimension $D \neq \frac{1}{2}$ if and only if the Riemann hypothesis holds, i.e., if and only if the Riemann zeta function $\zeta(s)$ does not vanish if $\operatorname{Re} s \neq \frac{1}{2}$, $\operatorname{Re} s > 0$.

By considering (generalized) fractal sprays, instead of fractal strings, we extend the above criterion for zeros of $\zeta(s)$ in the critical strip to a large class of zeta functions, including all those for which the analogue of the generalized Riemann hypothesis is expected to hold. We thus characterize the generalized Riemann hypothesis as a natural inverse spectral problem for fractal sprays. In addition to the Epstein zeta functions, this class includes all Dedekind zeta functions and Dirichlet L-series, and more generally, all Hecke L-series associated with an algebraic number field. It also includes all zeta functions associated with algebraic varieties over a finite field. We refer the interested reader to Appendix A for a brief review of such number-theoretic zeta functions.

In Chapter 10, we make an extensive study of the geometry and the spectrum of generalized Cantor strings. The complex dimensions of such strings form an infinite sequence in vertical arithmetic progression, with real part the Minkowski dimension D of the string. We show that these strings always have oscillations of order D in both their geometry and their spectrum. In Chapter 11, we deduce from this result that the explicit formulas for the geometry and the spectrum of Cantor strings always contain oscillatory terms of order D. On the other hand, if $\zeta(s)$ had a vertical arithmetic progression of zeros coinciding with the arithmetic progression of complex dimensions of $\mathcal{L}$, then, by formula $(*)$, the explicit formula for the frequencies would only contain the term corresponding to D, and not any oscillatory term. Thus we prove that $\zeta(s)$ does not have such a sequence of zeros. This theorem was first obtained by Putnam [Pu1, 2] in 1954. However, his methods do not apply to prove the extension to more general zeta functions. We also apply this idea in Section 11.1.1 to the geometry and the spectrum of the truncated generalized Cantor strings, to deduce an explicit upper bound on the maximal possible length of a vertical arithmetic sequence of zeros.

By considering (generalized) Cantor sprays, we extend this result to a large subclass of the aforementioned class of zeta functions. This class includes all the Dedekind and Epstein zeta functions, as well as many Dirichlet series not satisfying a functional equation. It does not, however, include the zeta functions associated with varieties over a finite field, for which this result does not hold (this is explained in Section 11.5). Indeed, we show that every Dirichlet series with positive coefficients and with only finitely many poles has no infinite sequence of zeros forming a vertical arithmetic progression. In Section 11.4.1, we present Mark Watkins' extension to finite shifted arithmetic progressions.

We conclude this book with a chapter of a more speculative nature, Chapter 12, in which we make several suggestions for the direction of future research in this area. Our results suggest that important information about the fractality of a string is contained in its complex dimensions. In Section 12.2, we propose as a new definition of fractality the presence of at least one nonreal complex dimension with positive real part. In this new sense, every self-similar set in the real line is fractal. On the other hand, in agreement with geometric intuition, certain compact subsets of $\mathbb{R}$, associated with the so-called a-string, are shown here to be nonfractal, whereas they are fractal according to the definition of fractality based on the notion of Minkowski dimension. We suggest one possible way of defining the complex dimensions of higher-dimensional fractals, and we discuss the examples of the Devil's staircase and of the snowflake drum (see Figures 12.1 and 12.6). In particular, the Devil's staircase is not fractal according to the traditional definition based on the Hausdorff dimension. However, there is general agreement among fractal geometers that it should be called fractal. (See [Man1, p. 84].) We show that our new definition of fractality does indeed resolve this problem satisfactorily. In spite of this positive outcome, we stress that the theory of the complex dimensions of higher-dimensional fractals still needs to be further developed.

The first steps towards such a theory are discussed in Sections 12.3.1 and 12.3.2. In Section 12.3.1 (based on [LapPe1]), a tube formula for the Koch snowflake curve is given and the corresponding complex dimensions are inferred, by analogy with formula (∗∗). In Section 12.3.2, we briefly discuss aspects of a work in progress ([Pe], [LapPe2–4]) where a theory of complex dimensions of self-similar fractals (and tilings) is developed.

In Section 12.4, we discuss two types of extensions of the main framework of this book, that of fractal strings. Namely, in Section 12.4.1—based on the paper [HamLap]—we give an overview of some of the main results on random fractal strings, such as random self-similar strings and stable random strings (which, in a special case, have for boundary the zero set of Brownian motion), and their associated complex dimensions. Moreover, in Section 12.4.2, we discuss the notion of fractal membrane (or quantized fractal string) introduced in [Lap10] and further developed in [LapNes1, 2].

In Sections 12.2 through 12.7, we discuss several conjectures and open problems regarding possible extensions and geometric, spectral, or dynamical interpretations of the present theory of complex dimensions, both for fractal strings and their higher-dimensional analogue, fractal drums, for which much research remains to be carried out in this context. We explain the connection with geometries over finite fields in Section 12.7.1. We also propose in Section 12.7.2 to develop a suitable fractal cohomology theory in the context of the theory of self-similar strings and their associated complex dimensions.

1

Complex Dimensions of Ordinary Fractal Strings

In this chapter, we recall some basic definitions pertaining to the notion of (ordinary) fractal string and introduce several new ones, the most important of which is the notion of complex dimension. We also give a brief overview of some of our results in this context by discussing the simple but illustrative example of the Cantor string. In the last section, we discuss fractal sprays, which are a higher-dimensional analogue of fractal strings.

1.1 The Geometry of a Fractal String

LapMa1–2, HeLap1–2]. A (*standard* or *ordinary*) *fractal string* $\mathcal{L}$ is a bounded open subset Ω of $\mathbb{R}$. It is well known that such a set consists of countably many open intervals, the lengths of which will be denoted by $l_1, l_2, l_3, \ldots$, called the *lengths* of the string. Note that $\sum_{j=1}^{\infty} l_j$ is finite and equal to the Lebesgue measure of Ω. From the point of view of this work, we can and will assume without loss of generality that

$$l_1 \geq l_2 \geq \cdots > 0, \tag{1.1}$$

where each length is counted according to its multiplicity. We allow for Ω to be a finite union of open intervals, in which case the sequence of lengths is finite.

An ordinary fractal string can be thought of as a one-dimensional drum with fractal boundary. Actually, we have given here the usual terminology that is found in the literature. A different terminology may be more sug-

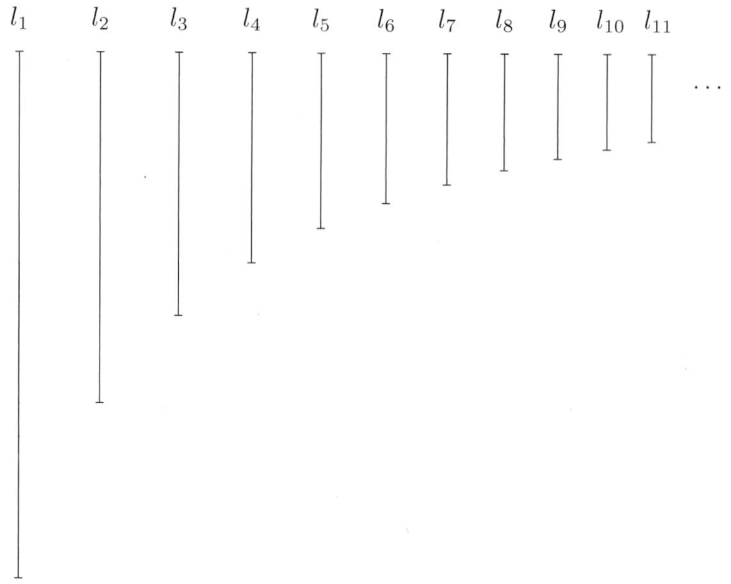

Figure 1.1: A fractal harp.

gestive: the open set Ω could be called a *fractal harp,* and each connected interval of Ω could be called a string of the harp; see Figure 1.1.

The *counting function of the reciprocal lengths,* also called the *geometric counting function* of $\mathcal{L}$, is the function

$$N_{\mathcal{L}}(x) = \#\{j \geq 1\colon l_j^{-1} \leq x\} = \sum_{j\geq 1,\, l_j^{-1}\leq x} 1, \tag{1.2}$$

for $x > 0$.[1]

Estimates for this function are closely related to estimates for the sequence of lengths, as the following proposition shows.

Proposition 1.1. *Let $\mathcal{L}$ be an ordinary fractal string with sequence of lengths $l_1, l_2, l_3, \ldots$. Then*

$$N_{\mathcal{L}}(x) = O\big(x^D\big), \text{ as } x \to \infty \quad \text{if and only if} \quad l_j = O\big(j^{-1/D}\big), \text{ as } j \to \infty.$$

Proof. Suppose we have the estimate

$$N_{\mathcal{L}}(x) \leq C \cdot x^D.$$

[1]Beginning in Chapter 4, Definition 4.1, we adopt the convention that the integers j such that $l_j^{-1} = x$ must be counted with the weight 1/2. A similar convention will be assumed for the spectral counting functions (such as, e.g., in Equation (1.34) below). For the moment, the reader may ignore this convention. With this convention, the explicit formula without error term also holds at jumps of the counting function (see, for example, Equation (1.31) below, for $x = 3^n$).

Taking $x = l_j^{-1}$, we find $j \leq C \cdot l_j^{-D}$. Therefore $l_j = O\big(j^{-1/D}\big)$.

On the other hand, if

$$l_j \leq C \cdot j^{-1/D}$$

for all $j = 1, 2, \ldots$, then given $x > 0$, we have $l_j^{-1} \geq x$ for $j \geq (Cx)^D$. We conclude that $N_{\mathcal{L}}(x) \leq (Cx)^D$. □

The *boundary* of $\mathcal{L}$, denoted $\partial\mathcal{L}$, is defined as the boundary $\partial\Omega$ of Ω. Important geometric information about $\mathcal{L}$ is contained in its *Minkowski dimension* $D = D_{\mathcal{L}}$ and its *Minkowski content* $\mathcal{M} = \mathcal{M}(D; \mathcal{L})$, defined respectively as the inner Minkowski dimension[2] and the inner Minkowski content of $\partial\Omega$. To define these quantities, let $d(x, A)$ denote the distance of $x \in \mathbb{R}$ to a subset $A \subset \mathbb{R}$ and let vol_1 denote the one-dimensional Lebesgue measure on $\mathbb{R}$. For $\varepsilon > 0$, let $V(\varepsilon)$ be the volume of the inner tubular neighborhood of $\partial\Omega$ with radius ε:

$$V(\varepsilon) = \mathrm{vol}_1\{x \in \Omega \colon d(x, \partial\Omega) < \varepsilon\}. \tag{1.3}$$

Definition 1.2. The *dimension* of a fractal string $\mathcal{L}$ is defined as the inner Minkowski dimension of $\partial\mathcal{L}$,

$$D = D_{\mathcal{L}} = \inf\{\alpha \geq 0 \colon V(\varepsilon) = O(\varepsilon^{1-\alpha}) \text{ as } \varepsilon \to 0^+\}. \tag{1.4}$$

The fractal string $\mathcal{L}$ is said to be *Minkowski measurable,* with *Minkowski content*

$$\mathcal{M} = \mathcal{M}(D; \mathcal{L}) = \lim_{\varepsilon \to 0^+} V(\varepsilon)\varepsilon^{-(1-D)}, \tag{1.5}$$

if this limit exists in $(0, \infty)$. The *upper* and *lower Minkowski content* are respectively defined by

$$\mathcal{M}^* = \mathcal{M}^*(D; \mathcal{L}) = \limsup_{\varepsilon \to 0^+} V(\varepsilon)\varepsilon^{-(1-D)}, \tag{1.6a}$$

and

$$\mathcal{M}_* = \mathcal{M}_*(D; \mathcal{L}) = \liminf_{\varepsilon \to 0^+} V(\varepsilon)\varepsilon^{-(1-D)}. \tag{1.6b}$$

Thus $0 \leq \mathcal{M}_* \leq \mathcal{M}^* \leq \infty$, and $\mathcal{L}$ is Minkowski measurable if and only if $\mathcal{M}^* = \mathcal{M}_* = \mathcal{M}$ is a nonzero real number.

Remark 1.3. The definitions of Minkowski dimension and content of $\partial\Omega$ extend naturally to the higher-dimensional case when Ω is an open bounded subset of $\mathbb{R}^d$, with $d \geq 1$, provided that we substitute the exponent $d - \alpha$ for $1 - \alpha$ in (1.4) and $d - D$ for $1 - D$ in (1.5) and (1.6), and that in (1.3),

---

[2]The Minkowski dimension is also called the capacity dimension or the (upper) box dimension in the literature on fractal geometry.

vol_1 is replaced by vol_d, the d-dimensional Lebesgue measure on $\mathbb{R}^d$; see, e.g., [Lap1, Definition 2.1 and §3]. The meaning of the exponent $d-D$ is the codimension of the boundary $\partial\Omega$ in the ambient space $\mathbb{R}^d$. Thus $V(\varepsilon)$ decreases as $\varepsilon \downarrow 0$ like ε to the power the codimension.

Remark 1.4 (Independence of the Geometric Realization). Observe that the Minkowski dimension of a self-similar set coincides with its Hausdorff dimension. The basic reason for using the Minkowski dimension is that it is invariant under displacements of the intervals of which a fractal string is composed. This is not the case of the Hausdorff dimension (see [BroCa], [Lap1, Example 5.1, pp. 512–514], [LapPo2]). See also Remark 2.22 below for further comparison between the various notions of fractal dimensions and for additional justification of the choice of the notion of Minkowski dimension in the context of fractal strings. *Throughout this work, an ordinary fractal string $\mathcal{L}$ is completely determined by the sequence of its lengths.* Hence, we will often denote such a string by $\mathcal{L} = \{l_j\}_{j=1}^{\infty}$.

Remark 1.5. The more irregular the boundary $\partial\Omega$, the larger D. Moreover, we always have $d-1 \leq H \leq D \leq d$, where H denotes the Hausdorff dimension of $\partial\Omega$, and d is the dimension of the ambient space, as in Remark 1.3. Intuitively, D corresponds to coverings of $\partial\Omega$ by d-dimensional cubes of size exactly equal to ε, whereas H corresponds to coverings by sets of size at most ε. As is discussed in [Lap1], this key difference explains why—from the point of view of harmonic analysis and spectral theory—the Minkowski dimension should be used instead of the more familiar Hausdorff dimension in this setting. In the case of fractal strings, both the Minkowski dimension and content depend only on the lengths l_j, and hence are invariant under arbitrary rearrangements of the intervals I_j (i.e., of the connected components of Ω). See, in particular, Equation (1.9).

1.1.1 The Multiplicity of the Lengths

Another way of representing a fractal string $\mathcal{L}$ is by listing its different lengths l, together with their multiplicity w_l:

$$w_l = \#\{j \geq 1 \colon l_j = l\}. \tag{1.7}$$

Thus, for example,

$$N_{\mathcal{L}}(x) = \sum_{l^{-1} \leq x} w_l. \tag{1.8}$$

In Chapter 4, we will introduce a third way to represent a fractal string, similar to this one, namely, by a measure.

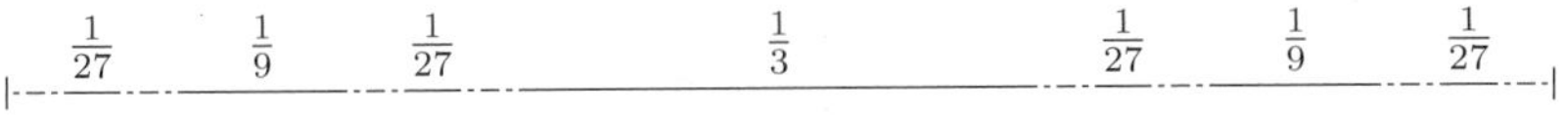

$\frac{1}{27}$ $\frac{1}{9}$ $\frac{1}{27}$ $\frac{1}{3}$ $\frac{1}{27}$ $\frac{1}{9}$ $\frac{1}{27}$

Figure 1.2: The Cantor string.

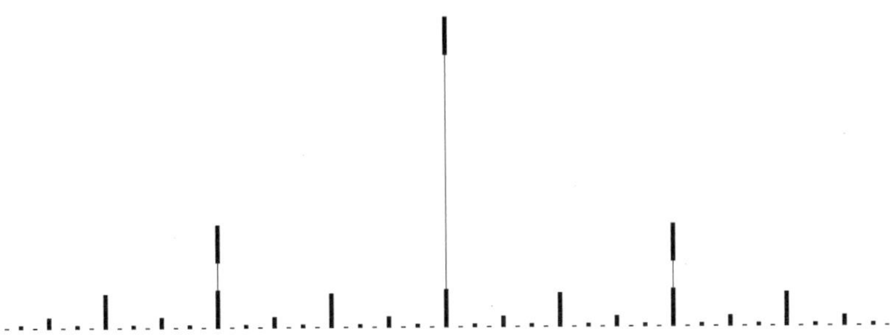

Figure 1.3: The .037-tubular neighborhood of the Cantor string.

1.1.2 *Example: The Cantor String*

We consider the ordinary fractal string $\Omega = \mathrm{CS}$, the complement in $[0,1]$ of the usual ternary Cantor set (Figure 1.2). Thus

$$\mathrm{CS} = \left(\tfrac{1}{3}, \tfrac{2}{3}\right) \cup \left(\tfrac{1}{9}, \tfrac{2}{9}\right) \cup \left(\tfrac{7}{9}, \tfrac{8}{9}\right) \cup \left(\tfrac{1}{27}, \tfrac{2}{27}\right) \cup \left(\tfrac{7}{27}, \tfrac{8}{27}\right) \cup \left(\tfrac{19}{27}, \tfrac{20}{27}\right) \cup \left(\tfrac{25}{27}, \tfrac{26}{27}\right) \cup \cdots,$$

so that $l_1 = 1/3$, $l_2 = l_3 = 1/9$, $l_4 = l_5 = l_6 = l_7 = 1/27, \dots$, or alternatively, the lengths are the numbers 3^{-n-1} with multiplicity $w_{3^{-n-1}} = 2^n$, for $n = 0, 1, 2, \dots$. We note that by construction, the boundary $\partial\Omega$ of the Cantor string is equal to the ternary Cantor set.

In general, the volume of the tubular neighborhood of the boundary of $\mathcal{L}$ is given by (see [LapPo2, Eq. (3.2), p. 48])

$$V(\varepsilon) = \sum_{j:\, l_j \geq 2\varepsilon} 2\varepsilon + \sum_{j:\, l_j < 2\varepsilon} l_j = 2\varepsilon \cdot N_{\mathcal{L}}\left(\frac{1}{2\varepsilon}\right) + \sum_{j:\, l_j < 2\varepsilon} l_j. \tag{1.9}$$

We explain the first equality in formula (1.9).[3] When the two endpoints of an interval of length l_j are covered by intervals of radius ε, then these discs overlap if $l_j < 2\varepsilon$, covering a length l_j, or they do not overlap if $l_j \geq 2\varepsilon$, in which case they cover a length of 2ε.

Applying (1.9) to the Cantor string, we find, for $0 < \varepsilon \leq 1/2$,

$$V_{\mathrm{CS}}(\varepsilon) = 2\varepsilon \cdot (2^n - 1) + \sum_{k=n}^{\infty} 2^k \cdot 3^{-k-1} = 2\varepsilon \cdot 2^n + \left(\frac{2}{3}\right)^n - 2\varepsilon,$$

where n is such that $3^{-n} \geq 2\varepsilon > 3^{-n-1}$; i.e., $n = [-\log_3(2\varepsilon)]$.[4] See Figure 1.3 for a picture of the tubular neighborhood of inner radius .037. In this picture, the lengths of the Cantor string have been put vertically.

---

[3]In view of (1.2), the second equality is obvious.

[4]For $x \in \mathbb{R}$, we write $x = [x] + \{x\}$, where $[x]$ is the integer part and the fractional part of x; i.e., $[x] \in \mathbb{Z}$ and $0 \leq \{x\} < 1$.

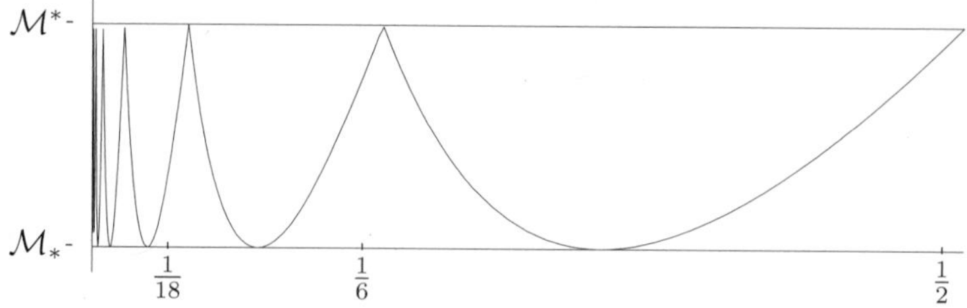

Figure 1.4: The function $\varepsilon^{D-1}(V_{\mathrm{CS}}(\varepsilon)+2\varepsilon)$, additively.

The above formula is valid, provided $n \geq 0$; i.e., $\varepsilon \leq 1/2$. We also note that $V_{\mathrm{CS}}(\varepsilon)$ is independent of the geometric realization. For example, we could have taken $(0,\frac{1}{3})\cup(\frac{1}{3},\frac{4}{9})\cup(\frac{4}{9},\frac{5}{9})\cup(\frac{5}{9},\frac{16}{27})\cup\ldots$ for the Cantor string.

To determine the minimal α such that $V_{\mathrm{CS}}(\varepsilon)=O(\varepsilon^{1-\alpha})$ as $\varepsilon \to 0^+$, we write

$$b^n = b^{[-\log_3(2\varepsilon)]} = b^{-\log_3(2\varepsilon)-\{-\log_3(2\varepsilon)\}} = (2\varepsilon)^{-\log_3 b} b^{-\{-\log_3(2\varepsilon)\}}$$

for $b=2$ and for $b=2/3$. Putting

$$D=\log_3 2 := \frac{\log 2}{\log 3}, \tag{1.10}$$

we find, for all positive $\varepsilon \leq 1/2$,

$$V_{\mathrm{CS}}(\varepsilon) = (2\varepsilon)^{1-D}\left(\left(\frac{1}{2}\right)^{\{-\log_3(2\varepsilon)\}} + \left(\frac{3}{2}\right)^{\{-\log_3(2\varepsilon)\}}\right) - 2\varepsilon. \tag{1.11}$$

The function between parentheses is bounded, nonconstant, and multiplicatively periodic: it takes the same value at ε and $\varepsilon/3$ (see Figures 1.4 and 1.5). It does not have a limit for $\varepsilon \to 0^+$. This is a simple example of what we call *geometric oscillations.* It follows that the Cantor string has Minkowski dimension $D=\log_3 2$, and that it is not Minkowski measurable. The upper and lower Minkowski content are computed in [LapPo2, Theorem 4.6, p. 65]:

$$\begin{aligned} \mathcal{M}^* &= 2^{2-D} = 2.5830, \\ \mathcal{M}_* &= 2^{1-D}D^{-D}(1-D)^{-(1-D)} = 2.4950. \end{aligned} \tag{1.12}$$

Figure 1.4 shows a graph of the periodic factor between the parentheses of formula (1.11), with a linear scale on the horizontal (i.e., ε-) axis. To make clearer the multiplicative periodicity of this function, in Figure 1.5, we present a graph of the same function with a logarithmic scale on the horizontal axis.

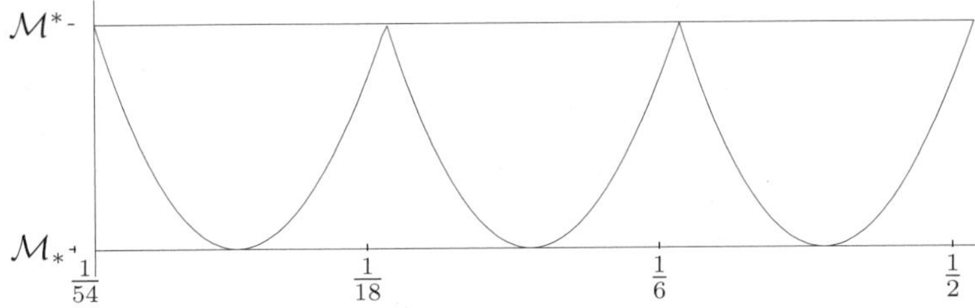

Figure 1.5: The function $\varepsilon^{D-1}(V_{\mathrm{CS}}(\varepsilon)+2\varepsilon)$, multiplicatively.

We continue our analysis of the inner tubular neighborhood $V_{\mathrm{CS}}(\varepsilon)$ of the Cantor string by using the Fourier series of the periodic function $u \mapsto b^{-\{u\}}$, for $b>0$, $b \neq 1$:[5]

$$b^{-\{u\}} = \frac{b-1}{b} \sum_{n \in \mathbb{Z}} \frac{e^{2\pi i n u}}{\log b + 2\pi i n}. \tag{1.13}$$

Writing $\mathbf{p} = 2\pi/\log 3$ and substituting (1.13) into formula (1.11), we find, for all positive $\varepsilon \leq 1/2$,

$$V_{\mathrm{CS}}(\varepsilon) = \frac{1}{2\log 3} \sum_{n=-\infty}^{\infty} \frac{(2\varepsilon)^{1-D-in\mathbf{p}}}{(D+in\mathbf{p})(1-D-in\mathbf{p})} - 2\varepsilon. \tag{1.14}$$

The number $\mathbf{p} = 2\pi/\log 3$ is called the oscillatory period of the Cantor string; see Definition 2.14. A counterpart of formula (1.14) will be derived for a general fractal string in Section 8.1.

Remark 1.6 (Reality Principle). Note that (1.14) expresses $V(\varepsilon)$ as an infinite sum of complex numbers. Likewise, for real values of b, (1.13) expresses the real-valued function $b^{-\{u\}}$ as an infinite sum of complex values. These sums are in fact real-valued, as can be seen by combining the terms for n and $-n$ into one, for $n \geq 1$. Indeed, these two terms are the complex conjugate of one another, and hence their sum is real-valued. Thus we find the following alternative expression for $V(\varepsilon)$:

$$V_{\mathrm{CS}}(\varepsilon) = \frac{2^{-D}\varepsilon^{1-D}}{D(1-D)\log 3} + \frac{1}{\log 3} \sum_{n=1}^{\infty} \mathrm{Re}\left(\frac{(2\varepsilon)^{1-D-in\mathbf{p}}}{(D+in\mathbf{p})(1-D-in\mathbf{p})}\right) - 2\varepsilon. \tag{1.15}$$

Using $(2\varepsilon)^{-in\mathbf{p}} = \cos(n\mathbf{p}\log(2\varepsilon)) - i\sin(n\mathbf{p}\log(2\varepsilon))$, one could continue to derive a formula involving the real-valued functions sine and cosine, thus exposing more clearly the oscillatory behavior of the terms in this sum. In this way, we find $V_{\mathrm{CS}}(\varepsilon)$ as a sum of the term at D and an infinite sum of real-valued terms. In the next section, we define the complex dimensions of a fractal string, and in Section 1.2.2, we will find that the complex dimensions of the Cantor string are the numbers $D+in\mathbf{p}$ ($n \in \mathbb{Z}$).

---

[5] By Dirichlet's Theorem [Fol, Theorem 8.43, p. 266], this series converges pointwise.

Thus the real-valuedness of (1.14) ultimately follows from the fact that the complex dimensions of the Cantor string come in complex conjugate pairs $\{D + in\mathbf{p}, D - in\mathbf{p}\}$ $(n \in \mathbb{N}^*)$. This holds in general: the nonreal complex dimensions of a fractal string come in complex conjugate pairs (see Remark 1.15). We call this the *reality principle* for ordinary fractal strings.

A similar comment applies to the expressions that we will obtain for $V(\varepsilon)$ in more general situations, as in Chapter 8.

Remark 1.7. Note that in view of (1.14), we have

$$\varepsilon^{-(1-D)} V_{\mathrm{CS}}(\varepsilon) = G(\varepsilon) - 2\,\varepsilon^D, \tag{1.16}$$

as $\varepsilon \to 0^+$, where G is a nonconstant multiplicatively periodic function of multiplicative period $3 = e^{2\pi/\mathbf{p}}$ which is bounded away from zero and infinity:

$$0 < \mathcal{M}_* \le G(\varepsilon) \le \mathcal{M}^* < \infty,$$

where $\mathcal{M}_*$ and $\mathcal{M}^*$ are given by (1.12) above. Alternatively, G can be viewed as an additively periodic function of $\log \varepsilon^{-1}$ with additive period $\log 3 = 2\pi/\mathbf{p}$. Namely,

$$G(\varepsilon) = F\left(\log \varepsilon^{-1}\right)$$

with[6]

$$F(u) = \frac{1}{2\log 3} \sum_{n=-\infty}^{\infty} \frac{2^{1-D-in\mathbf{p}} e^{in\mathbf{p}u}}{(D + in\mathbf{p})(1 - D - in\mathbf{p})}. \tag{1.17}$$

1.2 The Geometric Zeta Function of a Fractal String

Let $\mathcal{L}$ be a fractal string with sequence of lengths $\{l_j\}_{j=1}^\infty$. The sum $\sum_{j=1}^\infty l_j^\sigma$ converges for $\sigma = 1$. It follows that the (generalized) Dirichlet series

$$\zeta_{\mathcal{L}}(s) = \sum_{j=1}^\infty l_j^s$$

defines a holomorphic function for $\operatorname{Re} s > 1$. We show in Theorem 1.10 below that this series converges in the open right half-plane $\operatorname{Re} s > D$, defined by the Minkowski dimension D, but that it diverges at $s = D$.

We refer to Equation (1.7) above for the definition of the multiplicity w_l in the following definition.

[6]Note that since $\mathbf{p} = 2\pi/\log 3$ and $u = -\log \varepsilon$, we have $e^{i\mathbf{p}u} = \varepsilon^{-i\mathbf{p}}$. Further, the Fourier series in (1.17) is absolutely convergent.

Definition 1.8. Let $\mathcal{L}$ be a fractal string. The *geometric zeta function* of $\mathcal{L}$ is defined as

$$\zeta_{\mathcal{L}}(s) = \sum_{j=1}^{\infty} l_j^s = \sum_{l} w_l \cdot l^s, \tag{1.18}$$

for $\operatorname{Re} s > \sigma = D_{\mathcal{L}}$ (see Theorem 1.10 below).

Some values of the geometric zeta function of a string $\mathcal{L}$ have a special interpretation. If there are only finitely many lengths l, then $\zeta_{\mathcal{L}}(0)$ equals the number of lengths of the string. Similarly, the *total length* of the fractal string $\mathcal{L}$ is

$$L := \zeta_{\mathcal{L}}(1) = \operatorname{vol}_1(\Omega) = \sum_{j=1}^{\infty} l_j. \tag{1.19}$$

In Definition 1.2, we defined the (Minkowski) dimension $D = D_{\mathcal{L}}$ of $\mathcal{L}$ as $1 - c$, where c is the exponent of ε in the asymptotic volume of the tubular neighborhood of $\mathcal{L}$ with radius ε. We now show directly that this dimension coincides with the abscissa of convergence of $\zeta_{\mathcal{L}}$ in case the number of lengths of $\mathcal{L}$ is infinite. On the other hand, if $\mathcal{L}$ has finitely many lengths, then $D = 0$ but $\sigma = -\infty$ since the series for $\zeta_{\mathcal{L}}(s)$ converges for all $s \in \mathbb{C}$.

Definition 1.9. The *abscissa of convergence* of the series $\sum_{j=1}^{\infty} l_j^s$ is defined by

$$\sigma = \inf\left\{\alpha \in \mathbb{R} \colon \sum_{j=1}^{\infty} l_j^{\alpha} < \infty\right\}. \tag{1.20}$$

Thus $\{s \in \mathbb{C} \colon \operatorname{Re} s > \sigma\}$ is the largest open half-plane on which this series converges. The function $\zeta_{\mathcal{L}}(s) = \sum_{j=1}^{\infty} l_j^s$ is holomorphic in this half-plane; see, e.g., [Ser, §VI.2].

Theorem 1.10. *Suppose $\mathcal{L}$ has infinitely many lengths. Then the abscissa of convergence of the geometric zeta function of $\mathcal{L}$ coincides with D, the Minkowski dimension of $\partial\mathcal{L}$.*

Proof. We write σ for the abscissa of convergence of $\zeta_{\mathcal{L}}$. Let $d > D$. In view of definition (1.4) of the Minkowski dimension D, there exists a constant C_1 such that $V(\varepsilon) \leq C_1 \varepsilon^{1-d}$. For $n \geq 1$ we choose $\varepsilon = l_n/2$ to obtain

$$n l_n \leq n l_n + \sum_{j=n+1}^{\infty} l_j = V(l_n/2) \leq C_1 (l_n/2)^{1-d}. \tag{1.21}$$

It follows that for every positive number s, $l_n^s \leq C_2 n^{-s/d}$, for some positive constant C_2. Hence the series (1.18) converges for $s > d$, so that $\sigma \leq d$. Since this holds for every $d > D$, we obtain $\sigma \leq D$.

If $\sigma = 1$, we conclude that $D = \sigma$, since $V(\varepsilon) \leq \mathrm{vol}_1(\Omega)$ is bounded. Otherwise, let s be such that $\sigma < s < 1$. Then, by the definition of σ the series (1.18) converges. Since the sequence of lengths is nonincreasing, we find that

$$nl_n^s \leq \sum_{j=1}^{n} l_j^s \leq \zeta_{\mathcal{L}}(s).$$

Since s is fixed, we deduce that $l_n \leq (C_3/n)^{1/s}$ for all $n \geq 1$, for some positive constant C_3. Given $\varepsilon > 0$, it follows that $l_n < 2\varepsilon$ for $n > C_3(2\varepsilon)^{-s}$. For $j \leq C_3(2\varepsilon)^{-s}$, we estimate the j-th term in the first or the second sum in formula (1.9) for $V(\varepsilon)$ by 2ε, and for $j > C_3(2\varepsilon)^{-s}$, we estimate this term in the second sum by $(C_3/j)^{1/s}$. Thus we have

$$V(\varepsilon) \leq C_4(2\varepsilon)^{1-s},$$

for some positive constant C_4. It follows that $D \leq s$. Since this holds for every real number $s > \sigma$, we conclude that $D \leq \sigma$. Therefore, these two quantities coincide. □

Using this connection with the volume $V(\varepsilon)$, and noting (as in the foregoing proof) that $V(\varepsilon) \leq \mathrm{vol}_1(\Omega)$ is bounded, we derive the following corollary:

Corollary 1.11. *The dimension of an ordinary fractal string satisfies*

$$0 \leq D \leq 1.$$

1.2.1 The Screen and the Window

In general, $\zeta_{\mathcal{L}}$ may not have an analytic continuation to all of $\mathbb{C}$. We therefore introduce the *screen* S as the contour

$$S\colon S(t) + it \qquad (t \in \mathbb{R}), \tag{1.22}$$

where $S(t)$ is a continuous function $S\colon \mathbb{R} \to [-\infty, D_{\mathcal{L}}]$. (See Figure 1.6.)

The set

$$W = \{s \in \mathbb{C}\colon \operatorname{Re} s \geq S(\operatorname{Im} s)\} \tag{1.23}$$

is called the *window*, and we assume that $\zeta_{\mathcal{L}}$ has a meromorphic extension[7] to a neighborhood of W, with the set of poles $\mathcal{D} = \mathcal{D}_{\mathcal{L}}(W) \subset W$, called the (visible) complex dimensions of $\mathcal{L}$. We also require that $\zeta_{\mathcal{L}}$ does not have any pole on the screen S.

In Chapter 4, we will define generalized fractal strings, which also have a screen and a window associated to them. See Section 5.1.1 and the beginning of Section 5.2 for an explanation of the role of the screen in our explicit formulas.

[7] Following traditional usage, we will continue to denote by $\zeta_{\mathcal{L}}$ the meromorphic extension of the Dirichlet series given by (1.18).

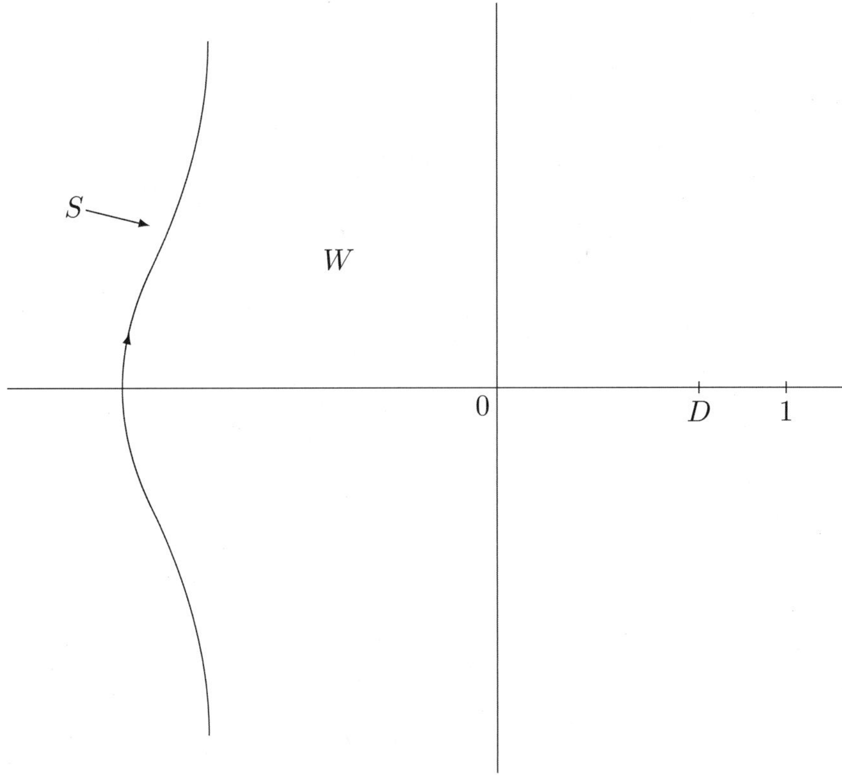

Figure 1.6: The screen S and the window W.

Definition 1.12. (i) The set of *visible complex dimensions* of the fractal string $\mathcal{L}$ is defined as

$$\mathcal{D}_{\mathcal{L}} = \mathcal{D}_{\mathcal{L}}(W) = \{\omega \in W : \zeta_{\mathcal{L}} \text{ has a pole at } \omega\}. \tag{1.24}$$

(ii) If $W = \mathbb{C}$ (that is, if $\zeta_{\mathcal{L}}$ has a meromorphic extension to all of $\mathbb{C}$), we call

$$\mathcal{D}_{\mathcal{L}} = \mathcal{D}_{\mathcal{L}}(\mathbb{C}) = \{\omega \in \mathbb{C} : \zeta_{\mathcal{L}} \text{ has a pole at } \omega\} \tag{1.25}$$

the set of *complex dimensions* of $\mathcal{L}$.

Complex dimensions may also be called *complex fractal dimensions*. This is partly justified by our proposal in Chapter 12 to define "fractality" in terms of this notion. (See especially Section 12.2.) For conciseness, however, we usually refer to them in the briefer way.

Remark 1.13. According to Theorem 1.10, $\zeta_{\mathcal{L}}(s)$ is holomorphic in the half-plane $\operatorname{Re} s > D$ and hence

$$\mathcal{D}_{\mathcal{L}} = \mathcal{D}_{\mathcal{L}}(W) \subset \{s \in W \colon \operatorname{Re} s \leq D\}. \tag{1.26}$$

Moreover, since it is the set of poles of a meromorphic function, $\mathcal{D}_{\mathcal{L}}(W)$ is a discrete subset of $\mathbb{C}$. Hence its intersection with any compact subset of $\mathbb{C}$ is finite. When $\mathcal{L}$ consists of finitely many lengths, we have $\mathcal{D}_{\mathcal{L}} = \emptyset$, since then, $\zeta_{\mathcal{L}}(s)$ is an entire function.

Remark 1.14. In general, the limit of $\zeta_{\mathcal{L}}(s)$ as $s \to D_{\mathcal{L}}$ from the right is ∞ (this follows, e.g., from [Ser, Proposition 7, p. 67] and is due to the fact that the Dirichlet series (1.18) has positive coefficients). Thus $s = D_{\mathcal{L}}$ is always a singularity of $\zeta_{\mathcal{L}}(s)$, but not necessarily a pole. If $D = D_{\mathcal{L}}$ is a pole and $D_{\mathcal{L}} \in W$, then

$$D_{\mathcal{L}} = \max\{\operatorname{Re}\omega \colon \omega \in \mathcal{D}_{\mathcal{L}}\}, \tag{1.27}$$

since $\zeta_{\mathcal{L}}$ is holomorphic for $\operatorname{Re} s > D$.

Remark 1.15. Assume that W is symmetric with respect to the real axis. Since $\zeta_{\mathcal{L}}(\overline{s}) = \overline{\zeta_{\mathcal{L}}(s)}$, it then follows that the nonreal complex dimensions of a (standard) fractal string always come in complex conjugate pairs $\omega, \overline{\omega}$; that is, $\omega \in \mathcal{D}_{\mathcal{L}}$ if and only if $\overline{\omega} \in \mathcal{D}_{\mathcal{L}}$.

The general theme of this monograph is that *the complex dimensions describe oscillations in the geometry and the spectrum of a fractal string.* The following theorem is a simple illustration of this philosophy.

Theorem 1.16. *Let $\mathcal{L}$ be a fractal string of dimension D, and assume that $\zeta_{\mathcal{L}}$ has a meromorphic extension to a neighborhood of D. If*

$$N_{\mathcal{L}}(x) = O\big(x^D\big), \quad \text{as } x \to \infty,$$

or if the volume of the tubular neighborhoods satisfies

$$V(\varepsilon) = O\big(\varepsilon^{1-D}\big), \quad \text{as } \varepsilon \to 0^+,$$

then $\zeta_{\mathcal{L}}$ has a simple pole at D.

Proof. First, recall from Remark 1.14 or from [Pos] or [Wid] that D is a singularity of $\zeta_{\mathcal{L}}$. Suppose $N_{\mathcal{L}}(x) \leq C \cdot x^D$ and $N_{\mathcal{L}}(x) = 0$ for $x \leq x_0$. Then for $s > D$,

$$\zeta_{\mathcal{L}}(s) = s \int_0^\infty N_{\mathcal{L}}(x) x^{-s-1}\, dx \leq \frac{Cs}{s-D} x_0^{D-s}.$$

It follows that the singularity at D is at most a simple pole. Since by assumption, $\zeta_{\mathcal{L}}$ has a meromorphic extension to a neighborhood of D, it follows that D is a simple pole of $\zeta_{\mathcal{L}}$.

The second part follows from the first part and Proposition 1.1 since

$$V(\varepsilon) = 2\varepsilon N_{\mathcal{L}}\left(\frac{1}{2\varepsilon}\right) + \sum_{j:\, l_j < 2\varepsilon} l_j = O\big(\varepsilon^{1-D}\big), \quad \text{as } \varepsilon \to 0^+,$$

implies that $N_{\mathcal{L}}(x) = O\big(x^D\big)$ as $x \to \infty$. □

The following example shows that the condition that the geometric zeta function of a fractal string has a meromorphic continuation is not always satisfied, and indeed the vertical line $\operatorname{Re} s = D$ can be a natural boundary for the analytic continuation of $\zeta_{\mathcal{L}}$. It also shows that the function $(s-D)\zeta_{\mathcal{L}}(s)$ can have a finite and positive limit as $s \to D^+$, even if $N_{\mathcal{L}}(x)$ is not of order x^D as $x \to \infty$ and $V(\varepsilon)$ is not of order ε^{1-D} as $\varepsilon \to 0^+$. This does not contradict the results of Chapters 6 and 8, which hold for strings having a geometric zeta function that does have a meromorphic continuation beyond $\operatorname{Re} s = D$. The example is formulated in the language of generalized fractal strings of Chapter 4, for which the multiplicities of the lengths may be nonintegral.

Example 1.17. Let

$$\zeta_{\mathcal{L}}(s) = \sum_{j=1}^{\infty} j e^{Dj^2} e^{-j^2 s}. \tag{1.28}$$

That is, we consider the generalized fractal string with lengths e^{-j^2}, each repeated with multiplicity je^{Dj^2}, for $0 < D < 1$ (we obtain an ordinary fractal string with the same properties if we choose for the multiplicity the integer nearest to je^{Dj^2}). Thus

$$N_{\mathcal{L}}(x) = \sum_{j=1}^{n} j e^{Dj^2} \geq n e^{Dn^2}$$

for $e^{n^2} \leq x < e^{(n+1)^2}$. For $x = e^{n^2}$, $N_{\mathcal{L}}(x) \geq x^D \sqrt{\log x}$, and hence $N_{\mathcal{L}}(x)$ is not of order x^D as $x \to \infty$.

On the other hand, $(s-D)\zeta_{\mathcal{L}}(s)$ has limit $1/2$ as $s \to D^+$, as we now show. For $t > 0$, we have

$$t\zeta_{\mathcal{L}}(D+t) = t\sum_{j=1}^{\infty} j e^{-j^2 t} = \sum_{j=1}^{\infty} j\sqrt{t} e^{-(j\sqrt{t})^2} \sqrt{t}.$$

This is a Riemann sum for the integral $\int_0^\infty x e^{-x^2}\, dx$, with mesh width $\sqrt{t}$. The value of this integral is $1/2$.

We conclude, in particular, that the vertical line $\operatorname{Re} s = D$ is a natural boundary for the analytic continuation of $\zeta_{\mathcal{L}}$.

1.2.2 The Cantor String (continued)

The geometric zeta function of the Cantor string is

$$\zeta_{\mathrm{CS}}(s) = \sum_{n=0}^{\infty} 2^n \cdot 3^{-(n+1)s} = \frac{3^{-s}}{1 - 2 \cdot 3^{-s}}. \tag{1.29}$$

We choose $W = \mathbb{C}$. The complex dimensions of the Cantor string are found by solving the equation $1 - 2 \cdot 3^{-\omega} = 0$, with $\omega \in \mathbb{C}$. (Note that they are all simple poles of ζ_{CS}.) Thus

$$\mathcal{D}_{\mathrm{CS}} = \{D + in\mathbf{p} \colon n \in \mathbb{Z}\}, \tag{1.30}$$

where $D = \log_3 2$ is the dimension of CS and $\mathbf{p} = 2\pi/\log 3$ is its oscillatory period.[8] (See Figure 2.4 in Section 2.3.1.) Formula (1.14) above expresses $V(\varepsilon)$ as a sum of terms proportional to $\varepsilon^{1-\omega}$, where $\omega = D + in\mathbf{p}$ runs over the complex dimensions of the Cantor string. (See also Sections 8.1.1, 8.4.1 and 8.4.2.)

In Section 8.3, we will extend and reinterpret in terms of complex dimensions the criterion for Minkowski measurability obtained in [LapPo1; LapPo2, Theorem 2.2, p. 46]. In particular, under mild growth conditions on $\zeta_{\mathcal{L}}$, we will show in Theorem 8.15 that an ordinary fractal string is Minkowski measurable if and only if D is a simple pole of $\zeta_{\mathcal{L}}$ and the only complex dimension of $\mathcal{L}$ on the vertical line $\operatorname{Re} s = D$ is D itself. Heuristically, nonreal complex dimensions above D would create oscillations in the geometry of $\mathcal{L}$, and therefore in the volume of the tubular neighborhood, $V(\varepsilon)$; see (1.31) below and (1.14)–(1.17) above, and compare with the intuition expressed in [LapMa1] or [LapMa2, esp. §3.3].

In view of (1.29) and (1.30), this makes particularly transparent the non-Minkowski measurability of the Cantor string (deduced above and established earlier in [LapPo2, Theorem 4.6]). As we will see in Chapter 8, a similar argument can be used in many other situations. (See, in particular, Theorem 8.23 in Section 8.4.2.)

As another example of an explicit formula involving the complex dimensions of the Cantor string, we compute the counting function of the reciprocal lengths. There are $1 + 2 + 4 + \cdots + 2^{n-1}$ lengths greater than x^{-1}, where $n = [\log_3 x]$. Thus $N_{\mathrm{CS}}(x) = 2^n - 1$. Using the Fourier series (1.13), we obtain

$$N_{\mathrm{CS}}(x) = \frac{1}{2\log 3} \sum_{n \in \mathbb{Z}} \frac{x^{D+in\mathbf{p}}}{D + in\mathbf{p}} - 1 = \frac{1}{2\log 3} \sum_{\omega \in \mathcal{D}_{\mathrm{CS}}} \frac{x^{\omega}}{\omega} - 1. \tag{1.31}$$

[8] Compare [Lap2, Example 5.2(i), p. 170] and [Lap3, p. 150], where our present terminology was not used.

Again, we find a sum extended over the complex dimensions of the Cantor string. In Chapter 5, we will formulate and prove the explicit formulas that allow us to derive such formulas in much greater generality.

1.3 The Frequencies of a Fractal String and the Spectral Zeta Function

Given a fractal string $\mathcal{L}$, we can listen to its sound. In mathematical terms, we consider the bounded open set $\Omega \subset \mathbb{R}$, together with the (positive) Dirichlet Laplacian $\Delta = -d^2/dx^2$ on Ω. An eigenvalue λ of Δ corresponds to the (normalized) frequency $f = \sqrt{\lambda}/\pi$ of the fractal string.

The frequencies of the unit interval are $1, 2, 3, \ldots$ (each counted with multiplicity one), and the frequencies of an interval of length l are l^{-1}, $2l^{-1}, 3l^{-1}, \ldots$ (also counted with multiplicity one). Thus the frequencies of $\mathcal{L}$ are the numbers

$$f = k \cdot l_j^{-1}, \tag{1.32}$$

where $k, j = 1, 2, 3, \ldots$; that is, they are the integer multiples of the reciprocal lengths of $\mathcal{L}$. The total multiplicity of the frequency f is equal to

$$w_f^{(\nu)} = \sum_{j\colon f \cdot l_j \in \mathbb{N}^*} 1 = \sum_{l\colon f \cdot l \in \mathbb{N}^*} w_l = w_{1/f} + w_{2/f} + w_{3/f} + \cdots . \tag{1.33}$$

To study the frequencies, we introduce the spectral counting function and the spectral zeta function. Note that N_ν and ζ_ν depend on $\mathcal{L}$. However, for simplicity, we do not indicate this explicitly in our notation.

Definition 1.18. The *counting function of the frequencies*, also called the *spectral counting function* of $\mathcal{L}$, is

$$\begin{aligned} N_\nu(x) &= \#\{f \leq x\colon \text{frequency of } \mathcal{L}, \text{ counted with multiplicity}\} \\ &= \sum_{f \leq x} w_f^{(\nu)}, \end{aligned} \tag{1.34}$$

for $x > 0$.[9]

The *spectral zeta function* of $\mathcal{L}$ is

$$\zeta_\nu(s) = \sum_{k,j=1}^{\infty} \left(k \cdot l_j^{-1}\right)^{-s} = \sum_f w_f^{(\nu)} f^{-s}, \tag{1.35}$$

which converges for $\operatorname{Re} s$ sufficiently large. In the second equality of (1.34) and (1.35), the sum is extended over all distinct frequencies of $\mathcal{L}$.

[9] According to footnote 1, the frequency $f = x$ should be counted with the weight $1/2$.

Let $\zeta(s)$ be the Riemann zeta function defined by $\zeta(s) = \sum_{n=1}^{\infty} n^{-s}$ for $\operatorname{Re} s > 1$. It is well known that $\zeta(s)$ has an extension to the whole complex plane as a meromorphic function, with one simple pole at $s = 1$, with residue 1.

The following theorem relates the spectrum of an ordinary fractal string with its geometry. It will become very helpful when we study direct or inverse spectral problems.

Theorem 1.19. *The spectral counting function of $\mathcal{L}$ is given by*

$$N_\nu(x) = N_{\mathcal{L}}(x) + N_{\mathcal{L}}\left(\frac{x}{2}\right) + N_{\mathcal{L}}\left(\frac{x}{3}\right) + \dots \tag{1.36}$$

$$= \sum_{j=1}^{\infty} [l_j x], \tag{1.37}$$

and the spectral zeta function of $\mathcal{L}$ is given by

$$\zeta_\nu(s) = \zeta_{\mathcal{L}}(s)\zeta(s), \tag{1.38}$$

where $\zeta(s)$ is the Riemann zeta function. Thus $\zeta_\nu(s)$ is holomorphic for $\operatorname{Re} s > 1$. It has a pole at $s = 1$ with residue L, the total length of $\mathcal{L}$ (see (1.19)). Moreover, it has a meromorphic extension to a neighborhood of the window W of $\mathcal{L}$.

Proof. For the spectral counting function, this follows from the following computation:

$$N_\nu(x) = \sum_{k=1}^{\infty} \sum_{j:\, k\cdot l_j^{-1} \leq x} 1 = \sum_{k=1}^{\infty} \#\{j\colon l_j^{-1} \leq x/k\} = \sum_{k=1}^{\infty} N_{\mathcal{L}}\left(\frac{x}{k}\right).$$

Observe that this is a finite sum, since $N_{\mathcal{L}}(y) = 0$ for $y < l_1^{-1}$. The second expression is derived similarly:

$$N_\nu(x) = \sum_{j=1}^{\infty} \sum_{k \leq l_j x} 1 = \sum_{j=1}^{\infty} [l_j x].$$

For the spectral zeta function, we have successively

$$\zeta_\nu(s) = \sum_{k,j=1}^{\infty} k^{-s} l_j^s = \sum_{j=1}^{\infty} l_j^s \sum_{k=1}^{\infty} k^{-s} = \zeta_{\mathcal{L}}(s)\zeta(s).$$

This completes the proof of the theorem. □

Using formula (1.37) for $N_\nu(x)$, we can derive Weyl's asymptotic law for fractal strings. More precise results will be obtained in Chapter 6, Section 6.3, and in Chapters 9 and 10. (See Appendix B, especially formula (B.2), for a formulation of Weyl's asymptotic law. See also [Lap1] and [LapPo2].)

Theorem 1.20 (Weyl's Asymptotic Law). *Let $\mathcal{L}$ be a fractal string of dimension D and of total length*

$$\mathrm{vol}_1(\mathcal{L}) = \sum_{j=1}^{\infty} l_j = \zeta_{\mathcal{L}}(1)$$

(as defined in (1.19)). Then, for every $\delta > 0$,

$$N_\nu(x) = \mathrm{vol}_1(\mathcal{L})x + O\left(x^{D+\delta}\right), \quad \text{as } x \to \infty. \tag{1.39}$$

The leading term,

$$W_{\mathcal{L}}(x) = \mathrm{vol}_1(\mathcal{L})\, x, \tag{1.40}$$

is called the Weyl term.

Proof. We write $\{x\}$ for the fractional part of x. By formula (1.37),

$$N_\nu(x) = \sum_{j=1}^{\infty} l_j x - \sum_{j=1}^{\infty} \{l_j x\}.$$

Both sums are convergent, and the first sum equals $W_{\mathcal{L}}(x)$. Let $\delta > 0$. By formula (1.21), we have that $l_j \leq C_2 j^{-1/(D+\delta)}$, for some positive constant C_2. It follows that $l_j x < 1$ for $j > (C_2 x)^{D+\delta}$. In the second sum, we estimate $\{l_j x\} \leq 1$ for $j \leq (C_2 x)^{D+\delta}$, and $\{l_j x\} = l_j x$ for $j > (C_2 x)^{D+\delta}$. Thus

$$\sum_{j=1}^{\infty} \{l_j x\} \leq (C_2 x)^{D+\delta} + x \int_{(C_2 x)^{D+\delta}}^{\infty} C_2 j^{-1/(D+\delta)}\, dj = O\left(x^{D+\delta}\right),$$

as claimed. □

One of the problems we are interested in is the inverse spectral problem for fractal strings (as considered in [LapMa1–2] in connection with the Riemann hypothesis); i.e., to derive information about the geometry of an ordinary fractal string from certain information (for example, asymptotics) about its spectrum. We study this problem in Chapters 9 and 11, with the help of explicit formulas for the frequency counting function, which will be established in Chapters 6 and 10, using the results of Chapter 5. These explicit formulas express the various functions associated with the geometry or the spectrum of a fractal string as a sum of oscillatory terms of the form a constant times x^ω, where ω runs over the complex dimensions of $\mathcal{L}$. In particular, in the case of the direct spectral problem where $N_\nu(x)$ is expressed in terms of the geometry of $\mathcal{L}$, the spectral counting function $N_\nu(x)$ is given by (6.23). We stress that if ω is simple, then the coefficient of x^ω in the explicit formula for $N_\nu(x)$ is a multiple of $\zeta(\omega)$, the value of the Riemann zeta function at the complex dimension ω; see formula (6.24b).

We note here that if the Riemann zeta function had a zero at one of the complex dimensions of a fractal string, then the expansion of the counting function of the frequencies (given in (6.23) and (6.24)) would no longer have the corresponding term. Thus, for example, the Cantor string would sound similar to a string without the complex dimensions $D \pm 37i\mathbf{p}$, if $\zeta(D \pm 37i\mathbf{p}) = \zeta(\log_3 2 \pm 74\pi i/\log 3)$ happened to vanish. This line of reasoning—based on the now rigorous notion of complex dimensions and the use of the explicit formulas—enables us to reformulate (and extend to many other zeta functions) the characterization of the Riemann hypothesis obtained in [LapMa1–2]. (See Chapter 9.)

Observe that in particular if $\zeta(s)$ were to vanish at all the points $D + in\mathbf{p}$ for $n \in \mathbb{Z}\backslash\{0\}$, the Cantor string would sound the same as a Minkowski measurable fractal string of the same dimension $D = \log_3 2$. (See, in particular, Theorem 8.15.) In Chapter 11, we will show, however, that this is not the case for a rather general class of zeta functions, and for any arithmetic sequence of points $D + in\mathbf{p}$ $(n \in \mathbb{Z})$. These zeta functions can be thought of as being associated with fractal sprays, as discussed in the next section. In Chapter 10, we will study generalized Cantor strings, which can have any sequence $\{D + in\mathbf{p}\}_{n\in\mathbb{Z}}$ (for arbitrary $D \in (0,1)$ and $\mathbf{p} > 0$) as their complex dimensions. We note that such strings can no longer be realized geometrically as subsets of Euclidean space.

1.4 Higher-Dimensional Analogue: Fractal Sprays

Fractal sprays were introduced in [LapPo3] (see also [Lap2, §4] announcing some of the results in [LapPo3]) as a natural higher-dimensional analogue of fractal strings and as a tool to explore various conjectures about the spectrum (and the geometry) of drums with fractal boundary[10] in $\mathbb{R}^d$. In the present book, fractal sprays and their generalizations (to be introduced later on) will continue to be a useful exploratory tool and will enable us to extend several of our results to zeta functions other than the Riemann zeta function. (See especially Chapters 9 and 11.)

For example, we will consider the spray of Figure 1.7, obtained by scaling an open square B of size 1 by the lengths of the Cantor string $\mathcal{L} = \mathrm{CS}$. Thus Ω is a bounded open subset of $\mathbb{R}^2$ consisting of one open square of size 1/3, two open squares of size 1/9, four open squares of size 1/27, and so on. The spectral zeta function for the Dirichlet Laplacian on the square is the function

$$\zeta_B(s) = \sum_{n_1,n_2=1}^{\infty} \left(n_1^2 + n_2^2\right)^{-s/2}, \tag{1.41}$$

[10]Of course, when $d \geq 2$, a drum with fractal boundary in $\mathbb{R}^d$ can be much more complicated than a fractal spray.

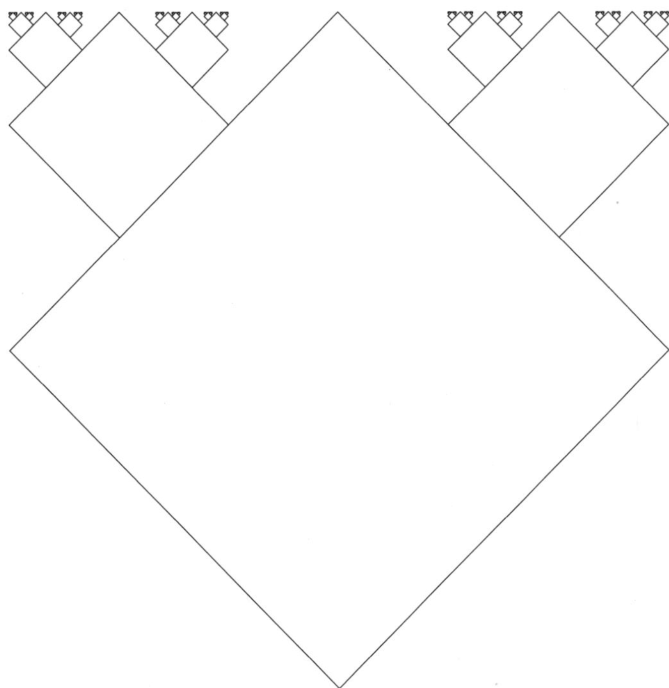

Figure 1.7: A Cantor spray: a fractal spray Ω in the plane, with basic shape B, the unit square, scaled by the Cantor string.

and hence the spectral zeta function of this spray is given by

$$\zeta_\nu(s) = \zeta_{\mathrm{CS}}(s) \cdot \zeta_B(s). \tag{1.42}$$

A fractal spray Ω in $\mathbb{R}^d$ ($d \geq 1$) is given by a nonempty bounded open set $B \subset \mathbb{R}^d$ (called the *basic shape*), scaled by a fractal string $\mathcal{L}$. More precisely, we call a *fractal spray of $\mathcal{L}$ on B* (or with basic shape B) any bounded open set Ω in $\mathbb{R}^d$ which is the disjoint union of open sets Ω_j for $j = 1, 2, \ldots$, where Ω_j is congruent to $l_j B$ (the homothetic of B by the ratio l_j) for each j. (See [LapPo3, §2].) Note that a fractal string $\mathcal{L} = \{l_j\}_{j=1}^\infty$ can be viewed as a fractal spray of $\mathcal{L}$ with basic shape $B = (0, 1)$, the unit interval.

Let $\{\lambda_k(B)\}_{k=1}^\infty$ be the sequence of nonzero eigenvalues (counted with multiplicity and written in nondecreasing order) of the positive Laplacian (with Dirichlet or other suitable boundary conditions) $\Delta = -\ sum_{q=1}^d \partial^2/\partial x_q^2$ on B. That is,

$$0 < \lambda_1(B) \leq \lambda_2(B) \leq \cdots \leq \lambda_k(B) \leq \cdots \to \infty, \quad \text{as } k \to \infty.$$

Then the (normalized) frequencies of the Laplacian Δ on B are the numbers $f_k = f_k(B) = \pi^{-1}\sqrt{\lambda_k(B)}$ (for $k = 1, 2, 3, \dots$),[11] and the spectral zeta function of B is

$$\zeta_B(s) := \sum_{k=1}^{\infty} (f_k(B))^{-s} = \pi^s \sum_{k=1}^{\infty} (\lambda_k(B))^{-s/2}. \tag{1.43}$$

Traditionally, as in [Se1, Gi], one writes $(\lambda_k(B))^{-s}$ instead of $(\lambda_k(B))^{-s/2}$. We adopt here the convention of [Lap2–3] and of Appendix B. As in Equation (1.32), a simple calculation (see, e.g., [LapPo3, §3])—relying on the invariance of the spectrum of the Laplacian under isometries—shows that the frequencies of the spray (that is, of the Laplacian Δ on Ω) are the numbers

$$f = f_k(B) \cdot l_j^{-1}, \tag{1.44}$$

where $k, j = 1, 2, 3, \dots$. The total multiplicity of the frequency f is given by the analogue of (1.33) in this more general context:

$$\begin{aligned} w_f^{(\nu)} &= \#\left\{(k,j)\colon f = f_k(B) \cdot l_j^{-1}\right\} \\ &= w_{f_1(B)/f} + w_{f_2(B)/f} + w_{f_3(B)/f} + \cdots . \end{aligned} \tag{1.45}$$

It follows, exactly as in the proof of formula (1.38), that the spectral zeta function $\zeta_\nu(s)$ of the fractal spray—defined by (1.35), with $w_f^{(\nu)}$ as in (1.45)—is given by

$$\zeta_\nu(s) = \zeta_{\mathcal{L}}(s) \cdot \zeta_B(s). \tag{1.46}$$

For example, if $B \subset \mathbb{R}^d$ is the d-dimensional unit cube $(0,1)^d$, with its opposite sides identified (so that Δ is the Laplacian with periodic boundary conditions on B), then[12]

$$\zeta_B(s) = \zeta_d(s) = \sum_{(n_1,\dots,n_d)\in\mathbb{Z}^d\setminus\{0\}} \left(n_1^2 + \cdots + n_d^2\right)^{-s/2}. \tag{1.47}$$

Thus $\zeta_d(s)$ is the classical Epstein zeta function, and hence admits a meromorphic continuation to all of $\mathbb{C}$, with a single simple pole at $s = d$ (see [Ter, §1.4, p. 58] or Section A.4 in Appendix A). Note that $\zeta_1(s) = 2\zeta(s)$.

As an example, in light of (1.29), (1.46) and (1.47), for the fractal spray of the Cantor string CS on the unit square B (as represented in Figure 1.7),

---

[11] The spectrum of the Dirichlet Laplacian Δ is always discrete if B is bounded (or more generally, if B has finite volume); further, 0 is never an eigenvalue. For the Neumann Laplacian, we assume that B has a locally Lipschitz boundary or more generally, that B satisfies the extension property, to ensure that the spectrum is discrete.

[12] Recall that we exclude the eigenvalue 0.

the corresponding spectral zeta function for the Laplacian with periodic boundary conditions on each square is given by

$$\zeta_\nu(s) = \frac{3^{-s}}{1 - 2 \cdot 3^{-s}} \zeta_2(s). \tag{1.48}$$

It is meromorphic in all of $\mathbb{C}$. Further, it has a simple pole at $s = 2$ and at each point $s = \log_3 2 + 2\pi in/\log 3$ $(n \in \mathbb{Z})$ where $\zeta_2(s)$ does not vanish.

Later on, we will use generalized fractal sprays to extend several of our results to number-theoretic zeta functions and many other Dirichlet series. (See especially Sections 9.3 and 11.2.)

Remark 1.21. The next step towards a theory of complex dimensions of higher-dimensional fractals would be to include more general drums with fractal boundary or drums with fractal membrane. For example, the complex dimensions of the Koch snowflake curve have been calculated in [LapPe1], based on a detailed analysis of $V(\varepsilon)$, the volume of the inner tubular neighborhoods of this curve (see Section 12.3.1). Furthermore, the beginning of a higher-dimensional theory of the geometric complex dimensions of self-similar fractal boundaries (and systems) is developed in the work [Pe] and [LapPe2–4] (see Section 12.3.2). See also, e.g., [BroCa, Lap1–4, HeLap2] as well as Sections 12.3 and 12.5 of Chapter 12 for a discussion of drums with fractal boundaries. The papers [Lap1, 3, FlLeVa, LapPan, LapNeuRnGri, GriLap, LapPe1–2] study from a variety of viewpoints the Koch snowflake drum (see Figure 12.6), the higher-dimensional drum with fractal boundary that has been studied most extensively so far. See [KiLap1–2, Ki2, Lap5–6; Lap-vF10, Part 1] and the relevant references therein for information on the geometry and the spectrum of drums with fractal membrane.

1.5 Notes

The notion of fractal string was formally introduced in [LapPo1–2], building in particular on the study of the a-string in [Lap1, Example 5.1 and Appendix C], and further extensively studied from different points of view in [Lap2–3, LapMa1–2, LapPo3, HeLap1–2, Lap-vF4–7, 9, Lap10], among other references.

Since fractal strings—viewed as vibrating objects—are one-dimensional drums with fractal boundary, their investigation involves a special (but surprisingly fruitful) case of the notion of fractal drum. We refer to [Berr1–2, BroCa, Lap1–6, LapPo1–3, LapMa1–2, HeLap1–2, FlVa, KiLap1–2, Lap-vF5, Tep1–2, Lap10] and the many relevant references therein for further mathematical and physical motivations for the investigation of fractal drums. See also, in particular, Sections 9.1, 12.3, 12.5 and 12.8, along with Appendix B in the present book for further discussion and references.[13]

Some of the earlier mathematical works in which the idea of complex dimension plays a role, if not defined explicitly, are [LapPo1], [LapPo2, §4.4b], [LapMa1], [LapMa2, §3.3], [Lap2, Figure 3.1 and §5], as well as [Lap3, §2.1, §2.2 and p. 150]. See also Remark 9.1, Figure 9.1, and the notes to Chapter 9. The primary motivations of the authors of those papers came from the investigation of the oscillatory phenomena in the geometry and the spectrum of fractal drums [Lap3] (including self-similar drums) and, in particular, of fractal strings, where the connections between direct or inverse spectral problems and the Riemann zeta function or the Riemann hypothesis were first discovered in [LapPo1, 2] or [LapMa1, 2], respectively.

See Remark 12.17 for a sample of references in the physics literature—of which we have become aware recently and with rather different motivations, coming from the study of turbulence, lacunarity, biophysics, and other applications.

The mathematical theory of complex dimensions of fractal strings, as discussed in this chapter, was first developed in [Lap-vF5] (partly announced in [Lap-vF1–4]) and was then pursued in [Lap-vF6, 7, 9]. It is, of course, broadly expanded in this monograph.

Section 1.1: for further information about the notions of Minkowski dimension and content in a related context, we refer to [BroCa, Lap1–3, LapPo1–3, LapMa1–2, LapFl, FlVa, Ger, GerSc1–2, Ca1–2, vB, vB-Le, HuaSl, FlLeVa, LeVa, MolVa, Fa4, vB-Gi, HeLap1–2]. For the notion of Minkowski(–Bouligand) dimension—which was extended by Bouligand from integer to real values of D—see also [Bou, Fed2, KahSa, Man1, MartVu, Tr1–3; Fa3, Chapter 3]. See [Fa3, Chapters 2 and 3], [Lap1, §2.1 and §3], [Mat, Chapter 5], [Rog] and [Tr1–2] or [Tr3, Chapters 2 and 3] for a detailed discussion of the Hausdorff and Minkowski dimensions.

Further arguments for the use of the Minkowski instead of the Hausdorff dimension are given in [Lap1, Example 5.1, pp. 512–514]; see also [LapPo2]. They exclude similarly another notion of fractal dimension, the packing dimension P [Su, Tr2]. This dimension satisfies the inequality $H \leq P \leq D$. Somewhat paradoxically, as is pointed out in [Lap1, Remark 5.1, p. 514], the Hausdorff and packing dimensions are ruled out in the context of fractal strings (or drums) precisely because they are good mathematical notions. Indeed, they are both associated with countably additive measures, which implies that countable sets have zero dimension. By contrast, the Minkowski content is only finitely subadditive, so that countable sets can have positive Minkowski dimension, as in the example of the a-string in Section 6.5.1 (see especially Equation (6.65)).

[13]Note that a fractal string is a drum with fractal boundary, and its boundary is a drum with fractal membrane; see, e.g., [Lap3, 6, 10] and the recent papers [Tep1, 2].

The Cantor string was studied, in particular, in [LapPo1] and [LapPo2, Example 4.5, pp. 65–67], where it was first shown, among other things, that it is not Minkowski measurable. However, the techniques developed in this book enable us to obtain more precise results for this and related Cantor-type strings; see, e.g., Sections 6.4.1 and 8.4.1, along with Chapter 10.

Theorem 1.10: the fact that the Minkowski dimension coincides with the abscissa of convergence was first observed by the first author in [Lap2, Eq. (5.4), p. 169] using a key result of Besicovich and Taylor [BesTa]; see also [LapPo2, LapMa2].

Section 1.3: we refer to [Da, Edw, In, Ivi, Pat, Ti] for the classical theory of the Riemann zeta function. The factorization formula (1.38) of Theorem 1.19 was established in [Lap2, Eqs. (5.2) and (5.3), p. 169] and used, in particular, in [Lap2–3, LapPo1–3, LapMa1–2, HeLap1–2].

See [BiSo, Met, EdmEv], [Maz, §1.5.1], or [Lap1, esp. Chapter 2 and pp. 510–511], as well as the relevant references therein for more information on spectral geometry.

Section 1.4: the notion of fractal spray was introduced in [LapPo3]. The main examples studied, e.g., in [BroCa], [Lap2, §4 and §5.2], [Lap3, §4.4.1b], [FlVa, Ger, GerSc1–2, LapPo3, LeVa], are special cases of ordinary fractal sprays.

2

Complex Dimensions of Self-Similar Fractal Strings

Throughout this book, we use an important class of ordinary fractal strings, the self-similar fractal strings, to illustrate our theory. These strings are constructed in the usual way via contraction mappings. In this and the next chapter, we give a detailed analysis of the structure of the complex dimensions of such fractal strings.

2.1 Construction of a Self-Similar Fractal String

Given a closed interval I of length L (called the *initial interval*), we construct a *self-similar string* $\mathcal{L}$ as follows. Let $N \geq 2$ and let $\Phi_1, \Phi_2, \ldots, \Phi_N$ be N contraction similitudes mapping I to I, with respective *scaling factors* $r_1, r_2, \ldots, r_N$ satisfying

$$1 > r_1 \geq r_2 \geq \ldots \geq r_N > 0. \tag{2.1}$$

Assume that

$$\sum_{j=1}^{N} r_j < 1, \tag{2.2}$$

and that the images $\Phi_j(I)$ of I, for $j = 1, \ldots, N$, do not overlap, except possibly at the endpoints. (This is the open set condition, see Section 2.1.1.)

In a procedure reminiscent of the construction of the Cantor set, subdivide the interval I into the pieces $\Phi_j(I)$. The remaining pieces in between

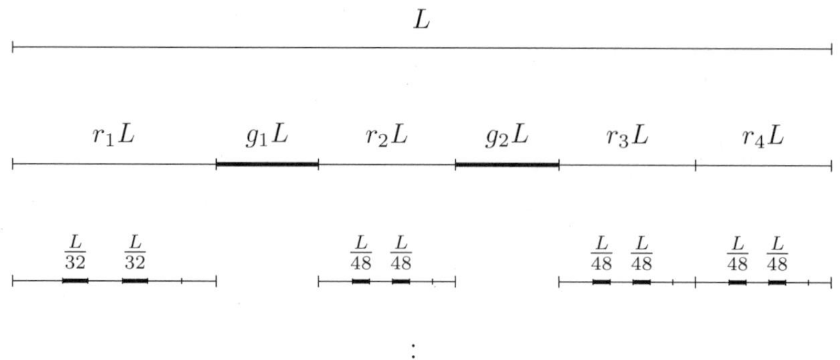

Figure 2.1: The construction of a self-similar string with four scaling ratios $r_1 = \frac{1}{4}$, $r_2 = r_3 = r_4 = \frac{1}{6}$, and two gaps $g_1 = g_2 = \frac{1}{8}$. The lengths of the string are indicated by the bold lines.

are the first intervals[1] of the string, of length $l_k = g_k L$, for $k = 1, \ldots, K$, where the scaling factors $g_1, \ldots, g_K$ of the *gaps* satisfy

$$1 > g_1 \geq \ldots \geq g_K > 0 \tag{2.3}$$

and

$$\sum_{j=1}^{N} r_j + \sum_{k=1}^{K} g_k = 1. \tag{2.4}$$

Repeat this process with the remaining intervals $\Phi_j(I)$, for $j = 1, \ldots, N$. As a result, we obtain a self-similar fractal string $\mathcal{L} = l_1, l_2, l_3, \ldots$, consisting of intervals of length l_n given by

$$r_{\nu_1} r_{\nu_2} \ldots r_{\nu_q} g_k L, \tag{2.5}$$

for $k = 1, \ldots, K$ and all choices of $q \in \mathbb{N}$ and $\nu_1, \ldots, \nu_q \in \{1, \ldots, N\}$; see Figure 2.1. The lengths are of the form (with $e_1, \ldots, e_N \in \mathbb{N}$)

$$r_1^{e_1} \cdots r_N^{e_N} g_k L.$$

The number of ways to write $r_1^{e_1} \ldots r_N^{e_N}$ as a product $r_{\nu_1} \ldots r_{\nu_q}$ of the scaling ratios (where $q = e_1 + \cdots + e_N$ and for each $\mu = 1, \ldots, N$, the exponent e_μ equals the number of j such that $\nu_j = \mu$), is given by the multinomial coefficient

$$\binom{q}{e_1 \ldots e_N} = \frac{q!}{e_1! \cdots e_N!} \qquad \left(\text{with } q = \textstyle\sum_{j=1}^N e_j\right). \tag{2.6}$$

[1]It would be better to call them the first strings of the fractal harp, see Section 1.1 and Figure 1.1.

Therefore the total multiplicity of a length l is the sum of all such multinomial coefficients for every choice of $k = 1, \ldots, K$ and $e_1, \ldots, e_N$ which satisfies the equation $r_1^{e_1} \ldots r_N^{e_N} g_k L = l$. In other words, we have

$$w_l = \sum_{k=1}^{K} \sum_{(e_1,\ldots,e_N)\in E_k} \binom{\sum_{j=1}^{N} e_j}{e_1 \ldots e_N}, \tag{2.7}$$

where $E_k = \{(e_1, \ldots, e_N) \colon r_1^{e_1} \cdots r_N^{e_N} g_k L = l\}$.

Remark 2.1. Note that we allow that the gaps and the images of I do not alternate. In particular, there may be more gaps than scaling factors.

Remark 2.2. Throughout this book, *we will always assume that a self-similar string is nontrivial; that is, we exclude the trivial case when $\mathcal{L}$ is composed of a single interval.* This will permit us to avoid having to consider separately this obvious exception to some of our theorems.

2.1.1 Relation with Self-Similar Sets

It may be helpful for the reader to recall that a self-similar set in $\mathbb{R}^d$ is the union of N scaled copies of itself. More precisely, a compact subset F of $\mathbb{R}^d$ is said to be a *self-similar set* if there exist N similarity transformations Φ_j ($j = 1, \ldots, N$, $N \geq 2$) of $\mathbb{R}^d$ with scaling ratios $r_j \in (0, 1)$ such that

$$F = \bigcup_{j=1}^{N} \Phi_j(F).$$

(See, for example, [Mor, Hut] or [Fa3, Sections 9.1 and 9.2].) As explained below, the boundary of a self-similar string is a self-similar set in $\mathbb{R}$. Conversely, every self-similar set in $\mathbb{R}$ satisfying the open set condition can essentially be obtained in this way.

Let N be an integer ≥ 2. For $j = 1, \ldots, N$, let

$$\Phi_j \colon [0, 1] \to [0, 1]$$

be a contraction similitude of $\mathbb{R}$ with scaling ratio r_j ($0 < r_j < 1$); i.e.,

$$|\Phi_j(x) - \Phi_j(y)| = r_j |x - y|, \qquad \text{for all } x, y.$$

We assume that the system of maps

$$\{\Phi_j \colon j = 1, \ldots, N\}$$

satisfies the *open set condition*; i.e., there exists a nonempty open subset U of $[0, 1]$, such that

$$\Phi_j(U) \cap \Phi_{j'}(U) = \emptyset \text{ for all } j \neq j', \; j, j' \in \{1, \ldots, N\},$$

and $\Phi_j(U) \subset U$ for all $j \in \{1, \dots, N\}$. (See, for example, [Hut] or [Fa3, p. 118].) Then the associated self-similar set F is the unique nonempty compact subset of $[0, 1]$ satisfying the fixed point equation

$$F = \bigcup_{j=1}^{N} \Phi_j(F). \tag{2.8}$$

The existence and uniqueness of the fixed point F in (2.8) follows from the Contraction Mapping Principle, applied to the complete metric space of nonempty compact subsets of $[0, 1]$, equipped with the Hausdorff metric. It is well known that F can also be obtained as the limit of repeated applications of the maps Φ_j. (We assume that F is not a single interval.) More precisely,

$$F = \bigcap_{n=0}^{\infty} \bigcup_{J \in \mathcal{J}_n} \Phi_J([0, 1]), \tag{2.9}$$

where for each integer[2] $n \geq 0$,

$$\mathcal{J}_n = \{1, \dots, N\}^n$$

denotes the set of all finite sequences of length n in the symbols $1, \dots, N$, and for $J = (j_1, \dots, j_n) \in \mathcal{J}_n$,

$$\Phi_J := \Phi_{j_n} \circ \cdots \circ \Phi_{j_2} \circ \Phi_{j_1}.$$

Let the open intervals G_k ($k = 1, \dots, K$) be the connected components of $(0, 1) \backslash \bigcup_{j=1}^{N} \Phi_j([0, 1])$ (we exclude the case $K = 0$ or $N = 1$ in order to avoid the trivial situation when F is an interval). Then

$$[0, 1] = \bigcup_{j=1}^{N} \Phi_j([0, 1]) \cup \bigcup_{k=1}^{K} \overline{G_k}.$$

(See Figure 2.2.) Without loss of generality, we may assume that the compact set F spans the interval $[0, 1]$, so that $\min F = 0$ and $\max F = 1$. Also, we may assume that the closed intervals $\Phi_j([0, 1])$ do not intersect, except possibly at their endpoints (which means that the open set condition is satisfied with $U = (0, 1)$). Thus 0 is the left endpoint of one of the intervals $\Phi_j([0, 1])$ and 1 is the right endpoint of one of them.

The length of the open interval G_k ($k = 1, \dots, K$) is denoted by

$$g_k = \mathrm{vol}_1(G_k)$$

and called a *gap*.[3] Clearly, the intervals $\Phi_j([0, 1])$ ($j = 1, \dots, N$) have length r_j. Note that by construction, the identity (2.4) is satisfied.

---

[2] By convention, $\mathcal{J}_0 = \{\emptyset\}$ consists of the empty sequence and $\Phi_\emptyset$ is the identity map.

[3] In [Fra2], these intervals are called the *initial gaps*.

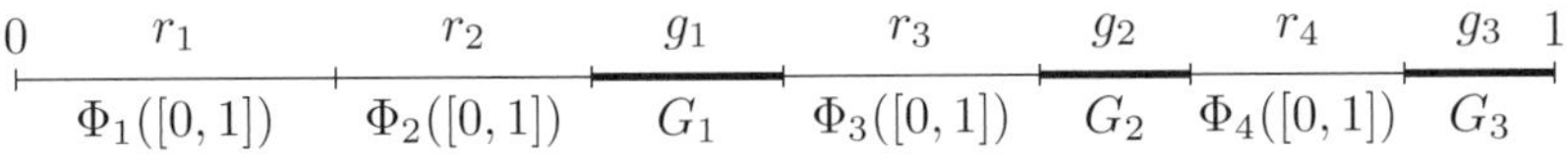

$0 \quad r_1 \quad r_2 \quad g_1 \quad r_3 \quad g_2 \quad r_4 \quad g_3 \quad 1$

$\Phi_1([0,1]) \quad \Phi_2([0,1]) \quad G_1 \quad \Phi_3([0,1]) \quad G_2 \quad \Phi_4([0,1]) \quad G_3$

Figure 2.2: The first iteration in the recursive construction of a self-similar set F with scaling ratios $r_1, \ldots, r_4$ and initial gaps G_k of length g_k $(k = 1, \ldots, 3)$. Here, $N = 4$, $K = 3$, and $\min F = 0$, $\max F = 1$.

Much as in the recursive construction of the standard (middle third) Cantor set,[4] we see that each iteration of the system of maps gives rise to a new collection of open intervals (the deleted intervals)

$$G_{Jk} = \Phi_J(G_k),$$

with $n \geq 0$, $J = (j_1, \ldots, j_n) \in \{1, \ldots, N\}^n$ and $k \in \{1, \ldots, K\}$, of length

$$l_{Jk} = g_k \prod_{\nu=1}^{n} r_{j_\nu}.$$

(See Equation (2.9) above.) Then the self-similar string $\mathcal{L}$ is defined as the sequence $\{l_{Jk}\}_{J,k}$, written in nonincreasing order.

In our terminology, the fractal string $\mathcal{L}$ is a self-similar string with scaling ratios $r_1, \ldots, r_N$ and gaps $g_1, \ldots, g_K$. It is called the self-similar string associated with the self-similar set F. Note that F is the boundary of the open set Ω defining $\mathcal{L}$ in the above construction: $\partial\Omega = F$, where

$$\Omega = [0,1] \backslash F$$

is the disjoint union of the deleted intervals. Conversely, it is clear from the above discussion that every self-similar string $\mathcal{L}$ of total length $L = 1$ determines a self-similar set $F \subset [0,1]$ (with $\min F = 0$ and $\max F = 1$) satisfying the open set condition (with the open set $U = (0,1)$) and with boundary equal to $\mathcal{L}$.

In summary, the class of self-similar strings $\mathcal{L}$ with (possibly) multiple gaps essentially corresponds (via the correspondence $\mathcal{L} = \Omega \leftrightarrow F = \partial\Omega$) to the class of self-similar sets F of $\mathbb{R}$ satisfying the open set condition.

In closing, we note that the special (but prototypical) case of a self-similar string with a single gap (i.e., $K = 1$), discussed in Section 2.2.1, is the one studied in [Lap-vF5, Chapter 2].

---

[4] For the Cantor set, $N = 2$, $K = 1$, and $r_1 = r_2 = g_1 = 1/3$; see Section 2.3.1 below.

2.2 The Geometric Zeta Function of a Self-Similar String

In Section 2.1, we have explained the construction of a self-similar string with scaling ratios $r_1, \ldots, r_N$ and gaps scaled by $g_1, \ldots, g_K$, satisfying the conditions (2.1)–(2.4).

Remark 2.3. The gaps of $\mathcal{L}$ have lengths $g_1 L, \ldots, g_K L$. By abuse of language, we will usually refer to the quantities $g_1, \ldots, g_K$ as the *gaps* of the self-similar string $\mathcal{L}$.

Theorem 2.4. *Let $\mathcal{L}$ be a self-similar string as in Section 2.1. Then the geometric zeta function of this string has a meromorphic continuation to the whole complex plane, given by*

$$\zeta_{\mathcal{L}}(s) = \frac{L^s \sum_{k=1}^{K} g_k^s}{1 - \sum_{j=1}^{N} r_j^s}, \quad \textit{for } s \in \mathbb{C}. \tag{2.10}$$

Here, $L = \zeta_{\mathcal{L}}(1)$ is the total length of $\mathcal{L}$, which is also the length of I, the initial interval from which $\mathcal{L}$ is constructed.

Proof. Indeed, we have

$$\sum_{\nu_1=1}^{N} \cdots \sum_{\nu_q=1}^{N} \left(r_{\nu_1} \cdots r_{\nu_q}\right)^s = \sum_{\nu_1=1}^{N} \cdots \sum_{\nu_q=1}^{N} r_{\nu_1}^s \cdots r_{\nu_q}^s = \left(\sum_{j=1}^{N} r_j^s\right)^q.$$

Hence, in view of (1.18) and the discussion surrounding (2.5), we deduce that

$$\begin{aligned}\zeta_{\mathcal{L}}(s) &= \sum_{k=1}^{K} \sum_{q=0}^{\infty} \left(\sum_{\nu_1=1}^{N} \cdots \sum_{\nu_q=1}^{N} \left(r_{\nu_1} \cdots r_{\nu_q} g_k L\right)^s \right) \\ &= \sum_{k=1}^{K} (g_k L)^s \sum_{q=0}^{\infty} \left(\sum_{j=1}^{N} r_j^s \right)^q.\end{aligned}$$

Let D be the unique real solution of $\sum_{j=1}^{N} r_j^s = 1$ (see also the first part of the proof of Theorem 2.17, in particular Equations (2.46) and (2.47) below). For $\operatorname{Re} s > D$, we have $\left|\sum_{j=1}^{N} r_j^s\right| < 1$, so that the above sum converges (hence D is the dimension of $\mathcal{L}$). We obtain that

$$\zeta_{\mathcal{L}}(s) = \frac{L^s \sum_{k=1}^{K} g_k^s}{1 - \sum_{j=1}^{N} r_j^s}.$$

This computation is valid for $\operatorname{Re} s > D$, but now, by the Principle of Analytic Continuation, the meromorphic continuation of $\zeta_{\mathcal{L}}$ to all of $\mathbb{C}$ exists and is given by the last formula, as desired. □

Throughout this chapter, we choose W, the window of $\mathcal{L}$ (as defined in Section 1.2.1), to be the entire complex plane, so that $W = \mathbb{C}$ and $\mathcal{D}_\mathcal{L} = \mathcal{D}_\mathcal{L}(\mathbb{C})$ can be called (without any ambiguity) the set of complex dimensions of $\mathcal{L}$ (as in Definition 1.12(ii)). This choice of W is justified by Theorem 2.4.

Note that in view of Remark 1.15, the nonreal complex dimensions of a self-similar string come in complex conjugate pairs ω, $\overline{\omega}$. This is also clear in view of the following corollary.

Corollary 2.5. *Let $\mathcal{L}$ be a self-similar string constructed with a single gap size $g_1 = g_2 = \cdots = g_K$. Then the set of complex dimensions $\mathcal{D}_\mathcal{L}$ of $\mathcal{L}$ is the set of solutions of the equation*

$$\sum_{j=1}^{N} r_j^\omega = 1, \qquad \omega \in \mathbb{C}, \tag{2.11}$$

with the same multiplicity.

In general, when the gaps have different sizes, $\mathcal{D}_\mathcal{L}$ is contained in the set of solutions of Equation (2.11), *and each complex dimension has a multiplicity of at most that of the corresponding solution.*

Equation (2.11) could be called the complexified Moran equation. As is discussed in Remark 2.22 below, the usual Moran equation is given by (2.11) restricted to real values of s. It has a unique real solution, necessarily equal to the Minkowski dimension of $\mathcal{L}$.

Remark 2.6. The length L of the initial interval I of a self-similar string may be normalized so that the first length of $\mathcal{L}$ equals 1 by choosing

$$L = g_1^{-1}, \tag{2.12}$$

where g_1 is the largest gap (see (2.3)). This assures that $\lim_{s\to+\infty} \zeta_\mathcal{L}(s)$ equals the multiplicity of the first length, which is also the multiplicity of the largest gap. It does not affect the complex dimensions of the string.

Remark 2.7. For a self-similar string, the total length of $\mathcal{L}$ is also the length L of the initial interval I in the construction given in Section 2.1. Further, the complement of $\mathcal{L}$ in I is the boundary $\partial\mathcal{L}$, and $\partial\mathcal{L}$ has length 0, since the Hausdorff dimension of $\partial\mathcal{L}$ is less than one. Note that in the above construction, the Hausdorff dimension and the Minkowski dimension of $\partial\mathcal{L}$ coincide. This is no longer the case if we were to represent $\mathcal{L}$ as a sequence of intervals of lengths $l_1, l_2, \ldots$, so that $\partial\mathcal{L}$ is a sequence of points with a single limit point. In that case, only the Minkowski dimension of this set of points gives the dimension of the fractal string. See also Remark 1.4.

Remark 2.8. Note that the zeros of the geometric zeta function $\zeta_\mathcal{L}(s)$ given by Equation (2.10) correspond to the solutions of the *Dirichlet poly-*

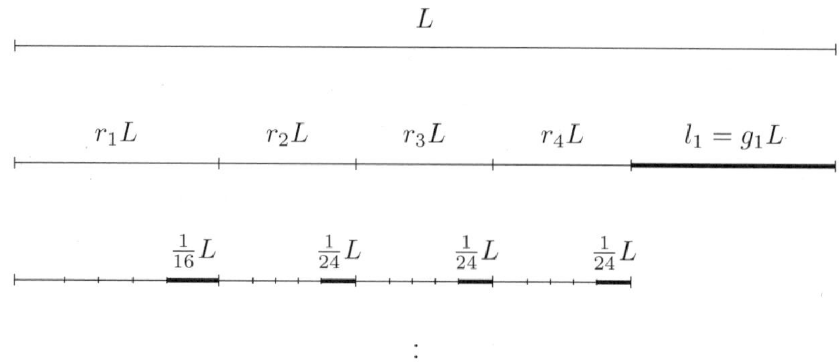

Figure 2.3: The construction of a self-similar string with $N = 4$ and similitudes with scaling ratios $r_1 = \frac{1}{4}$, $r_2 = r_3 = r_4 = \frac{1}{6}$ and a single gap $g_1 = \frac{1}{4}$. The first group of scaled intervals each have size $r_jL = \text{vol}_1(\Phi_j(I))$, and the first five lengths of the string are $l_1 = \frac{1}{4}L$, $l_2 = \frac{1}{16}L$, $l_3 = \frac{1}{24}L$, $l_4 = \frac{1}{24}L$, and $l_5 = \frac{1}{24}L$.

nomial equation

$$\sum_{k=1}^{K} g_k^s = 0, \qquad s \in \mathbb{C}, \tag{2.13}$$

to be studied in Chapter 3. In Chapter 7, we apply our explicit formulas of Chapter 5 in a dynamical context. The resulting expressions will involve both the zeros and poles of $\zeta_{\mathcal{L}}$.

2.2.1 Self-Similar Strings with a Single Gap

In the case where there is only one gap (see Figure 2.3) the situation is somewhat simpler.

Theorem 2.9. *Let $\mathcal{L}$ be a self-similar string, constructed as above with scaling ratios $r_1, \dots, r_N$, and a single gap g_1. Then the geometric zeta function of this string has a meromorphic continuation to the whole complex plane, given by*

$$\zeta_{\mathcal{L}}(s) = \frac{(g_1L)^s}{1 - \sum_{j=1}^{N} r_j^s}, \quad \textit{for } s \in \mathbb{C}. \tag{2.14}$$

Here, L is the total length of $\mathcal{L}$ (as defined by (1.19)), which is also the length of I, the initial interval. The first length of $\mathcal{L}$ is $l_1 = g_1L$, of multiplicity one.

In particular, Corollary 2.5 applies to this case. Thus, the complex dimensions of a self-similar string with a single gap are given by Equation (2.11), with the same multiplicity.

Note that if $\mathcal{L}$ is normalized as in Remark 2.6, then its first length equals 1 and therefore

$$\zeta_{\mathcal{L}}(s) = \frac{1}{1 - \sum_{j=1}^{N} r_j^s}, \qquad \text{for } s \in \mathbb{C}. \tag{2.15}$$

For notational simplicity, we will often normalize our examples in this way (see, for instance, Section 2.3).

Remark 2.10. In Section 7.2, we will find an Euler product for $\zeta_{\mathcal{L}}$, coming from a dynamical system associated with the string, also called a self-similar flow. We only define this dynamical system in the case of a single gap, but it would be very interesting to have a general definition. The lengths correspond to the periodic orbits of this dynamical system. See also [Lap-vF6, Remark 2.15].

Remark 2.11. An interesting new feature of self-similar strings with multiple gaps is that in general (i.e., when $K \geq 2$), the geometric zeta function may have zeros as well as poles, and that in some cases, certain cancellations may occur. (See Remark 2.8 above and Section 2.3.3 below.) Furthermore, in [HamLap], Ben Hambly and the first author have considered an even broader extension to random recursive constructions (in $\mathbb{R}$), including random self-similar strings. See Section 12.4.1.

2.3 Examples of Complex Dimensions of Self-Similar Strings

In each of these examples, the given self-similar string has a single gap, as in Section 2.2.1, and we normalize it so that its first length is 1, as was explained in Remark 2.6.

2.3.1 The Cantor String

We take two equal scaling factors $r_1 = r_2 = 1/3$ and one gap $g_1 = 1/3$. The self-similar string CS with total length 3 and these scaling factors is the *Cantor string*.[5] It consists of lengths 3^{-n} with multiplicity 2^n, $n \geq 0$. The geometric zeta function of this string is

$$\zeta_{\mathrm{CS}}(s) = \frac{1}{1 - 2 \cdot 3^{-s}}. \tag{2.16}$$

The complex dimensions are found by solving the equation

$$2 \cdot 3^{-\omega} = 1 \qquad (\omega \in \mathbb{C}). \tag{2.17}$$

[5]It differs from the Cantor string discussed in Chapter 1 in the normalization of the first length. Alternatively, one could say that we have used another unit to measure the lengths. The initial interval of the Cantor string introduced in Section 1.1.2 has length 1.

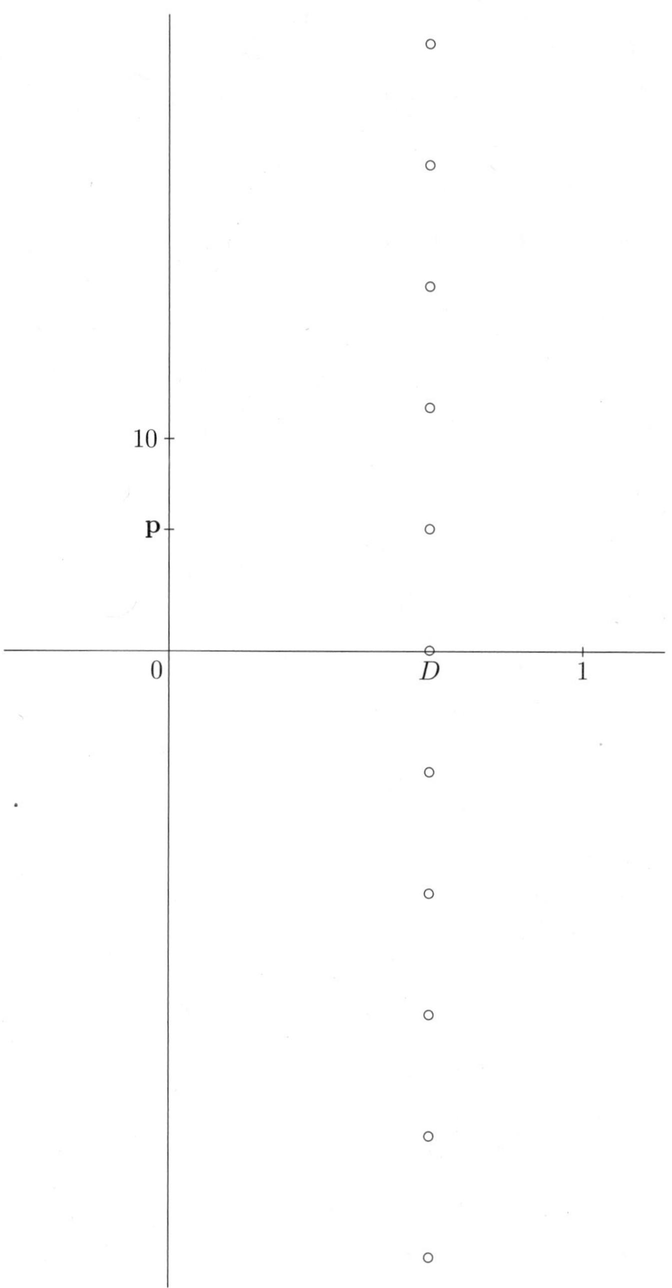

Figure 2.4: The complex dimensions of the Cantor string. $D = \log_3 2$ and $\mathbf{p} = 2\pi/\log 3$.

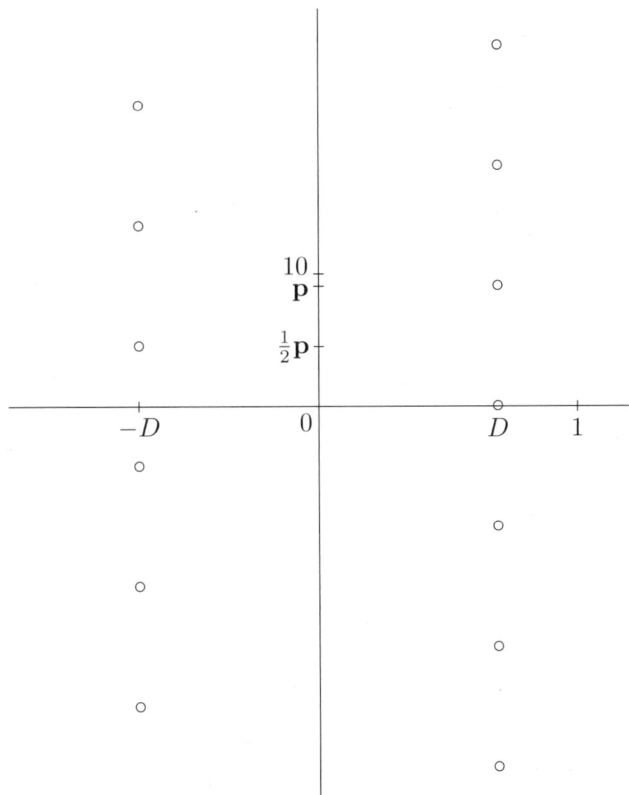

Figure 2.5: The complex dimensions of the Fibonacci string. $D = \log_2 \phi$ and $\mathbf{p} = 2\pi/\log 2$.

We find

$$\mathcal{D}_{\mathrm{CS}} = \{D + in\mathbf{p} \colon n \in \mathbb{Z}\}, \tag{2.18}$$

with $D = \log_3 2$ and $\mathbf{p} = 2\pi/\log 3$. (See Figure 2.4.) All poles are simple. Further, the residue at each pole is $1/\log 3$.

Remark 2.12. This example will be studied in much more detail and extended to generalized Cantor strings in Chapter 10. See also Chapter 1, Sections 6.4.1 and 8.4.1, along with Example 8.22 below.

2.3.2 The Fibonacci String

Next we consider a self-similar string with two lines of complex dimensions. The *Fibonacci string* is the string Fib with total length 4 and scaling factors $r_1 = 1/2$, $r_2 = 1/4$ and a gap $g_1 = 1/4$. Its lengths are

$$1, \frac{1}{2}, \frac{1}{4}, \frac{1}{8}, \dots, \frac{1}{2^n}, \dots,$$

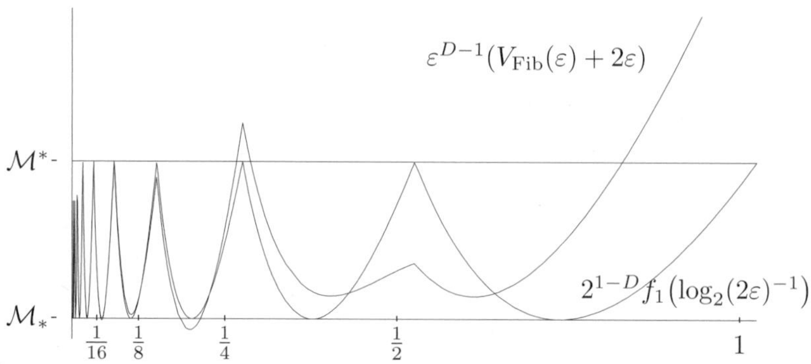

Figure 2.6: The functions $2^{1-D}f_1\left(\log_2(2\varepsilon)^{-1}\right)$ and $\varepsilon^{D-1}(V_{\mathrm{Fib}}(\varepsilon)+2\varepsilon)$, additively.

with multiplicity respectively

$$1,\ 1,\ 2,\ 3,\dots,\ F_{n+1},\dots,$$

the Fibonacci numbers. Recall that these numbers are defined by the following recursive formula:

$$F_{n+1}=F_n+F_{n-1},\ \text{ and } F_0=0,\ F_1=1. \tag{2.19}$$

The geometric zeta function of the Fibonacci string is given by

$$\zeta_{\mathrm{Fib}}(s)=\frac{1}{1-2^{-s}-4^{-s}}. \tag{2.20}$$

The complex dimensions are found by solving the quadratic equation

$$(2^{-\omega})^2+2^{-\omega}=1 \qquad (\omega\in\mathbb{C}). \tag{2.21}$$

We find $2^{-\omega}=\left(-1+\sqrt{5}\right)/2=\phi^{-1}$ and $2^{-\omega}=-\phi$, where

$$\phi=\frac{1+\sqrt{5}}{2} \tag{2.22}$$

is the golden ratio. Hence

$$\mathcal{D}_{\mathrm{Fib}}=\{D+in\mathbf{p}\colon n\in\mathbb{Z}\}\cup\{-D+i(n+1/2)\mathbf{p}\colon n\in\mathbb{Z}\}, \tag{2.23}$$

with $D=\log_2\phi$ and $\mathbf{p}=2\pi/\log 2$. (See Figure 2.5.)[6] Again, all the poles are simple. Further, for all $n\in\mathbb{Z}$, the residue at the poles $D+in\mathbf{p}$ equals $\frac{\phi+2}{5\log 2}$ and the residue at $-D+i(n+1/2)\mathbf{p}$ equals $\frac{3-\phi}{5\log 2}$.

The volume $V_{\mathrm{Fib}}(\varepsilon)$ of the tubular neighborhood of the Fibonacci string can be computed directly, as we did in Section 1.1.2 for the Cantor string. It

[6]By abuse of language, we may say that the Fibonacci string has two lines of complex dimensions (by which we mean discrete lines).

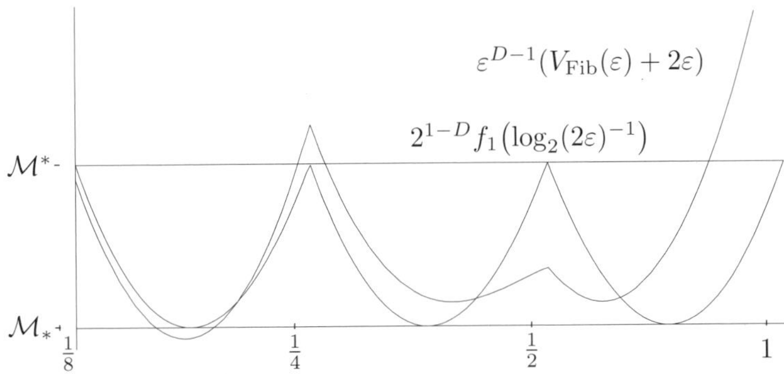

Figure 2.7: The functions $2^{1-D}f_1\left(\log_2(2\varepsilon)^{-1}\right)$ and $\varepsilon^{D-1}(V_{\rm Fib}(\varepsilon)+2\varepsilon)$, multiplicatively.

is well known, and it follows by solving the linear recurrence relation (2.19), that

$$F_n=\frac{\phi^n-(1-\phi)^n}{\sqrt{5}}. \tag{2.24}$$

By (1.9), we have

$$V_{\rm Fib}(\varepsilon)=2\varepsilon\sum_{2^{-n}\geq 2\varepsilon}F_{n+1}+\sum_{2^{-n}<2\varepsilon}F_{n+1}2^{-n},$$

provided $\varepsilon<1$ (for $1/2<\varepsilon<1$, the first sum is empty). By formula (2.24), both sums are geometric. We write $x=\log_2(2\varepsilon)^{-1}$, so that x increases by one unit if ε is halved in value. Evaluating the above sums, we find that, for $\varepsilon<1$,

$$V_{\rm Fib}(\varepsilon)=(2\varepsilon)^{1-D}f_1(x)-2\varepsilon+(2\varepsilon)^{1+D}f_2(x),$$

where $x=\log_2(2\varepsilon)^{-1}$ and

$$f_1(x):=\frac{1}{\sqrt{5}}\left(\phi^3\phi^{-\{x\}}+\phi^4(\phi/2)^{-\{x\}}\right),$$

$$f_2(x):=\frac{(-1)^{[x]}}{\sqrt{5}}\left(\phi^{-3}\phi^{\{x\}}-\phi^{-4}(2\phi)^{\{x\}}\right).$$

The functions f_1 and f_2 are periodic and continuous. See Figures 2.6 and 2.7.

Computing the Fourier series of the functions f_1 and f_2 by using (1.13), we find the explicit formula

$$\begin{aligned}V_{\rm Fib}(\varepsilon)&=\frac{\phi}{\sqrt{5}\log 2}\sum_{k=-\infty}^{\infty}\frac{(2\varepsilon)^{1-D-ik\mathbf{p}}}{(D+ik\mathbf{p})(1-D-ik\mathbf{p})}-2\varepsilon\\&\quad+\frac{\phi-1}{\sqrt{5}\log 2}\sum_{k=-\infty}^{\infty}\frac{(2\varepsilon)^{1+D-i\mathbf{p}/2-ik\mathbf{p}}}{(-D+i\mathbf{p}/2+ik\mathbf{p})(1+D-i\mathbf{p}/2-ik\mathbf{p})},\end{aligned} \tag{2.25}$$

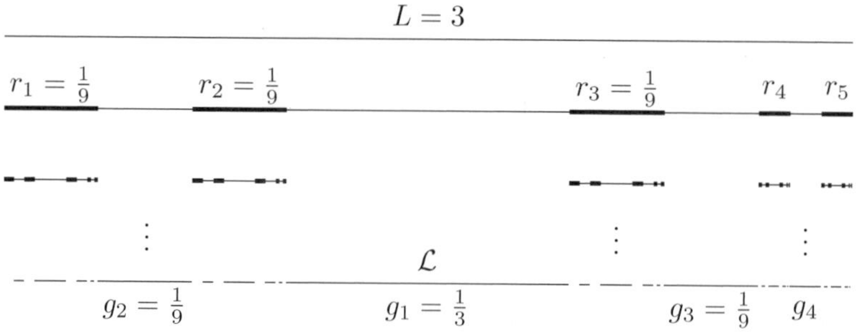

Figure 2.8: The construction of the modified Cantor string, with five scaling ratios $r_1 = r_2 = r_3 = \frac{1}{9}$, $r_4 = r_5 = \frac{1}{27}$, and four gaps $g_1 = g_3 = \frac{1}{9}$, $g_2 = \frac{1}{3}$, $g_4 = \frac{1}{27}$.

for $\varepsilon < 1$, where, as earlier, $D = \log_2 \phi$ and $\mathbf{p} = 2\pi/\log 2$.

In Section 8.4.2, Theorem 8.25 and Corollary 8.27, we will derive the counterpart of Equation (2.25) for general lattice strings, by using the pointwise tube formula without error term of Section 8.1.1, given in the second part of Theorem 8.7. We note that the result (2.25) of the above direct computation of $V(\varepsilon)$ is in agreement with this result. See especially Example 8.32.

2.3.3 *The Modified Cantor and Fibonacci Strings*

The string with five scaling ratios $r_1 = r_2 = r_3 = 1/9$, $r_4 = r_5 = 1/27$ and four gaps $g_1 = 1/3$, $g_2 = g_3 = 1/9$, $g_4 = 1/27$ in an interval of length 3 (see Figure 2.8) has for geometric zeta function

$$\zeta_{\mathcal{L}}(s) = 3^s \frac{3^{-s} + 2 \cdot 3^{-2s} + 3^{-3s}}{1 - 3 \cdot 3^{-2s} - 2 \cdot 3^{-3s}}. \tag{2.26}$$

The denominator factors as $1 - 3x^2 - 2x^3 = (1-2x)(1+x)^2$, with $x = 3^{-s}$, and the numerator as $x(1+x)^2$. Hence the zeta function simplifies to the geometric zeta function (2.16) of the Cantor string, with two scaling ratios $r_1 = r_2 = 1/3$ and one gap $g_1 = 1/3$, in an interval of length 3. Thus the sequence of its lengths coincides with the sequence of lengths of the Cantor string, as defined in Section 2.3.1. The poles (of (2.26) or (2.16)) are simple, located at $D + ik\mathbf{p}$ ($k \in \mathbb{Z}$), with $D = \log_3 2$ and $\mathbf{p} = 2\pi/\log 3$. The residue at each pole is equal to $1/\log 3$.

Remark 2.13. The initiator of a self-similar fractal string is not unique since one can always find a Dirichlet polynomial $L^s(g_1^s + \cdots + g_K^s)$ with positive coefficients such that the product of this Dirichlet polynomial by the denominator of $\zeta_{\mathcal{L}}$ is again of the form $1 - r_1^s - \cdots - r_N^s$. Indeed, repeated application of the identity

$$\frac{1}{1 - r_1^s - \cdots - r_N^s} = \frac{1 + r_1^s + \cdots + r_N^s}{1 - (r_1^s + \cdots + r_N^s)^2}$$

shows that every fractal string has infinitely many initiators.

The following example gives an alternative initiator for the Fibonacci string. Let $r_1 = r_2 = g_1 = 1/4$, $r_3 = g_2 = 1/8$ in an interval of length 4. This initiator generates a self-similar string with geometric zeta function

$$\zeta_{\mathcal{L}}(s) = 2^{2s} \frac{2^{-2s} + 2^{-3s}}{1 - 2 \cdot 2^{-2s} - 2^{-3s}} = \frac{1}{1 - 2^{-s} - 2^{-2s}}.$$

Hence the sequence of lengths of this string coincides with that of the Fibonacci string of Section 2.3.2.

2.3.4 A String with Multiple Poles

Let $\mathcal{L}$ be the self-similar string with scaling factors $r_1 = r_2 = r_3 = 1/9$, $r_4 = r_5 = 1/27$, one gap $g_1 = 16/27$ and with total length $L = \frac{27}{16}$ (so that its first length is 1). Then the geometric zeta function of $\mathcal{L}$ is given by

$$\zeta_{\mathcal{L}}(s) = \frac{1}{1 - 3 \cdot 9^{-s} - 2 \cdot 27^{-s}}. \tag{2.27}$$

The complex dimensions are found by solving the cubic equation

$$2z^3 + 3z^2 = 1, \quad z = 3^{-\omega} \qquad (\text{with } z, \omega \in \mathbb{C}). \tag{2.28}$$

This factors as $(2z-1)(z+1)^2 = 0$. Thus we see that there is one sequence of simple poles $\omega = D + in\mathbf{p}$ ($D = \log_3 2$ and $\mathbf{p} = 2\pi/\log 3$, $n \in \mathbb{Z}$), each with residue $\frac{4}{9\log 3}$, corresponding to the solution $z = 1/2$, and another sequence of double poles $\omega = \frac{1}{2}i\mathbf{p} + in\mathbf{p}$ ($n \in \mathbb{Z}$) corresponding to the double solution $z = -1$. (See Figure 2.9.) The Laurent series at a double pole is

$$\frac{1}{3(\log 3)^2}\left(s - \tfrac{1}{2}i\mathbf{p} - in\mathbf{p}\right)^{-2} + \frac{5}{9\log 3}\left(s - \tfrac{1}{2}i\mathbf{p} - in\mathbf{p}\right)^{-1} + O(1),$$

as $s \to \frac{1}{2}i\mathbf{p} + in\mathbf{p}$.

2.3.5 Two Nonlattice Examples: the Two-Three String and the Golden String

The above examples are all self-similar lattice strings, as will be defined in Section 2.4. The reader may get the mistaken impression that in general it is easy to find the complex dimensions of a self-similar string. However, in the nonlattice case, in the sense of Definition 2.14 below, it is practically impossible to obtain complete information about the complex dimensions. Nevertheless, in Sections 2.5–2.6 and Chapter 3 (see Theorem 2.17 below), we obtain a large amount of information about the location and the density of the complex dimensions of a nonlattice string.

We now give our first explicit example of a nonlattice string, which may be called the *Two-Three string*. (See Remark 2.16 and Section 3.1.1 for some further information.) We take two scaling factors $r_1 = 1/2$, $r_2 = 1/3$.

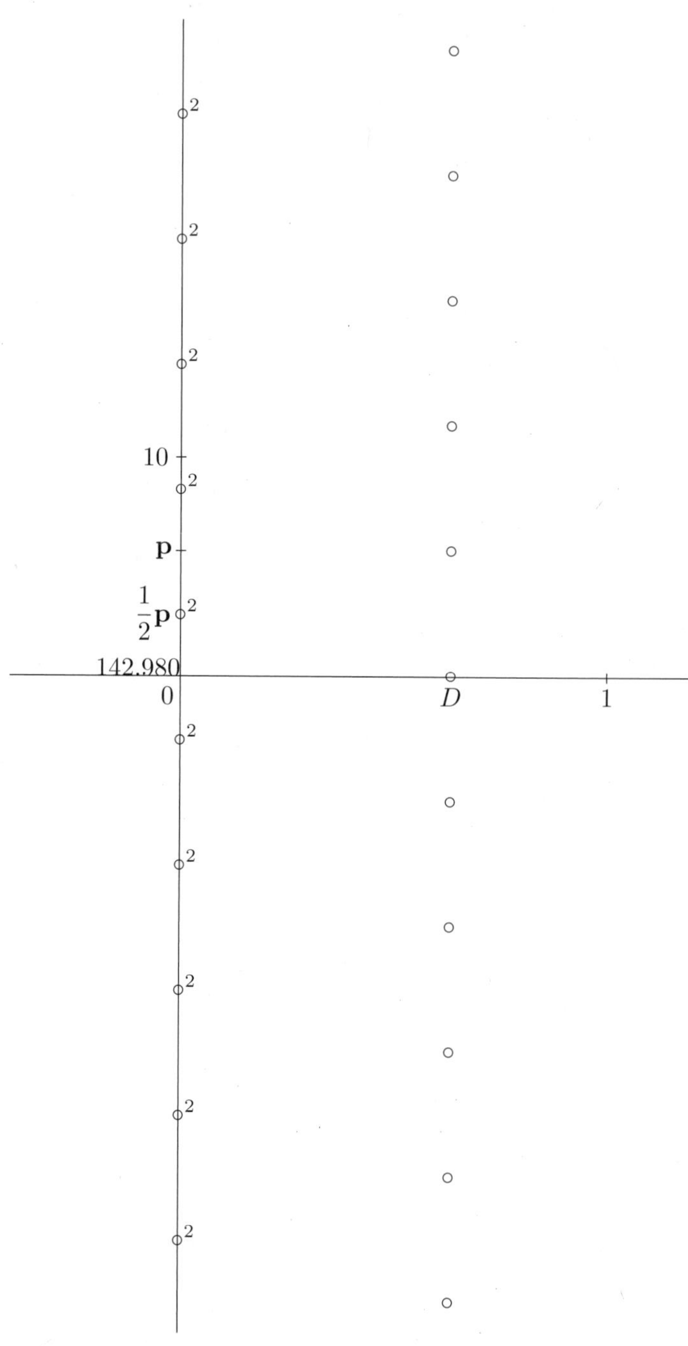

Figure 2.9: The complex dimensions of a string with multiple poles. $D = \log_3 2$ and $\mathbf{p} = 2\pi/\log 3$. Here, the symbol $\circ^2$ denotes a multiple pole of order two.

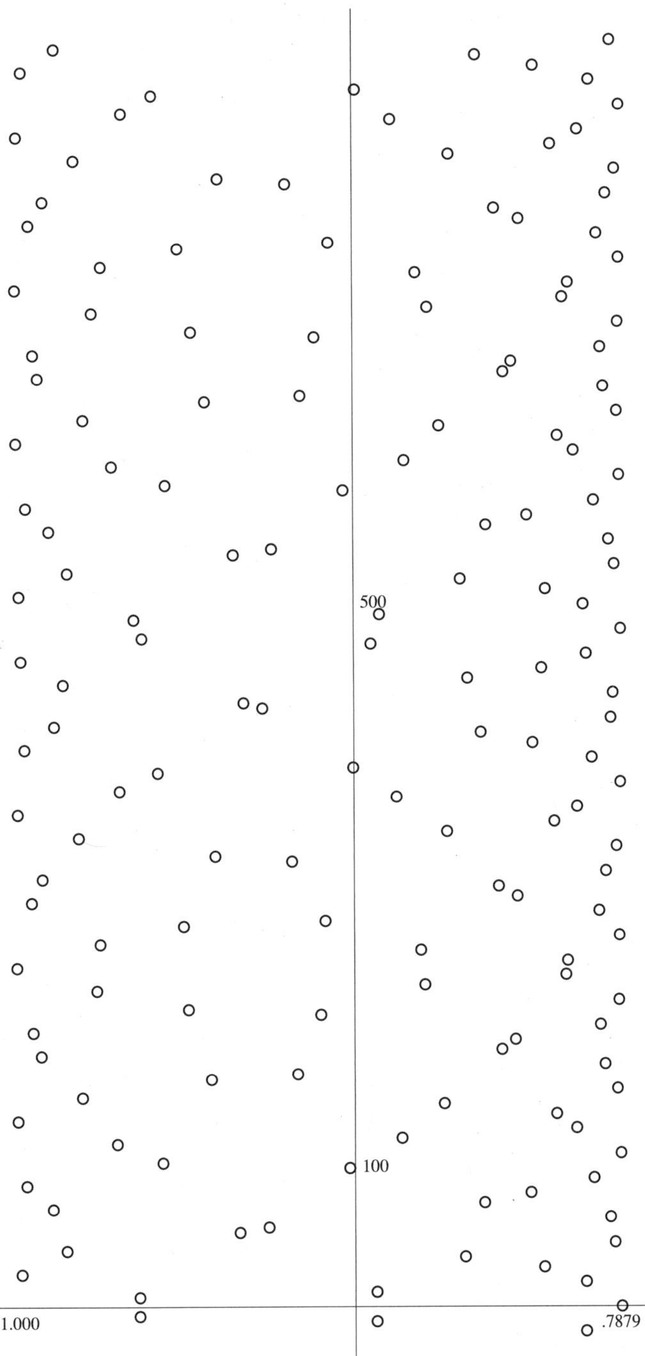

Figure 2.10: The complex dimensions of the nonlattice string with scaling ratios $r_1 = 1/2$, $r_2 = 1/3$ and one gap $g_1 = 1/6$.

The self-similar string $\mathcal{L}$ with total length 6, a single gap $g_1 = 1/6$ (so that $\mathcal{L}$ is normalized as in Remark 2.6) and these scaling factors is nonlattice. It consists of lengths $2^{-m}3^{-n}$ with multiplicity the binomial coefficient $\binom{m+n}{m}$ for $m, n = 0, 1, \ldots$ (see Section 2.1). The geometric zeta function of this string is

$$\zeta_{\mathcal{L}}(s) = \frac{1}{1 - 2^{-s} - 3^{-s}}. \tag{2.29}$$

The complex dimensions are found by solving the *transcendental* equation

$$2^{-\omega} + 3^{-\omega} = 1 \qquad (\omega \in \mathbb{C}). \tag{2.30}$$

However, we cannot solve this equation exactly. We cannot even find the exact value of the dimension D of this string; i.e., the precise value of the unique real solution of Equation (2.30). But it can be verified that all complex dimensions have real part $\geq -1 = D_l$ (see Theorem 2.17), and that $D \approx .78788\ldots$. With the help of a computer (and Theorems 3.14 and 3.18 of the next chapter), we have sketched in Figure 2.10 an approximate plot of the set of complex dimensions $\mathcal{D} = \mathcal{D}_{\mathcal{L}}(\mathbb{C})$. See Section 3.8 for some information about these computations.

The Golden String

Next, we consider the nonlattice string GS with a single gap and scaling factors $r_1 = 2^{-1}$ and $r_2 = 2^{-\phi}$, where ϕ is the golden ratio given by formula (2.22). We call this string the *golden string*. As usual, we normalize it so that the first length equals 1, by choosing the total length to be

$$L = \frac{1}{1 - 2^{-1} - 2^{-\phi}}.$$

Its geometric zeta function is

$$\zeta_{\mathrm{GS}}(s) = \frac{1}{1 - 2^{-s} - 2^{-\phi s}}, \tag{2.31}$$

and its complex dimensions are the solutions of the transcendental equation

$$2^{-\omega} + 2^{-\phi\omega} = 1 \qquad (\omega \in \mathbb{C}). \tag{2.32}$$

A diagram of the complex dimensions of GS is given in Figure 2.11. We have not obtained it by directly solving (2.32) numerically. Instead, we have obtained it by applying Theorems 3.14 and 3.18 of Chapter 3, in which the complex dimensions of a nonlattice string are approximated by those of a lattice string with a large oscillatory period: we chose the approximation $\phi \approx 987/610$ to approximate $\mathcal{L}$ by the lattice string with

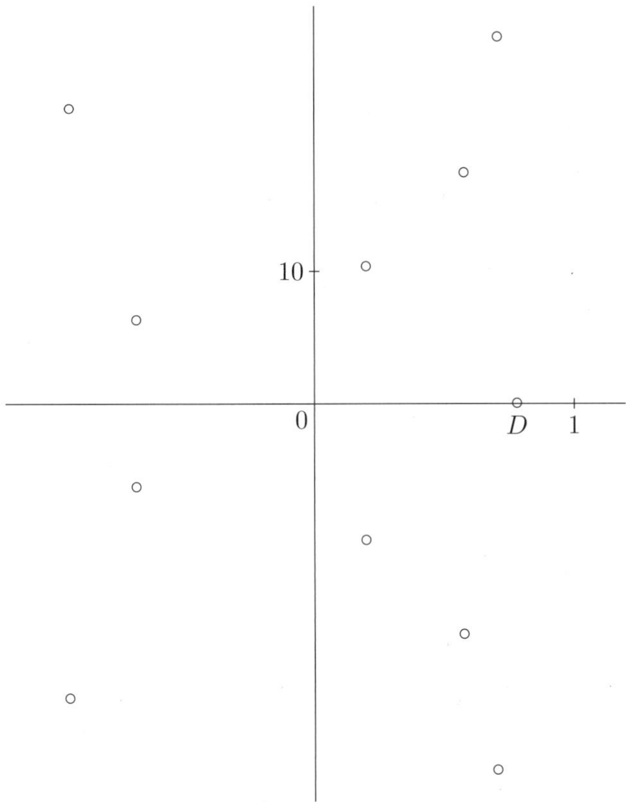

Figure 2.11: The complex dimensions of the golden string (the nonlattice string with scaling ratios $r_1 = 2^{-1}$ and $r_2 = 2^{-\phi}$).

scaling factors $r_1 = r^{610}$, $r_2 = r^{987}$, where $r = 2^{-1/610}$, and hence with multiplicative generator $2^{-1/610}$.

We note that the theory of Chapter 3, in particular Lemma 3.29, provides a concrete method for obtaining more and more accurate approximations of $\mathcal{D} = \mathcal{D}_{\mathrm{GS}}(\mathbb{C})$. In particular, the dimension D of the golden string is approximately equal to $D = .77921\ldots$. See also Section 3.8.

Our numerical investigations—and the theoretical information contained in Lemma 3.29 and Theorem 3.18—indicate that the complex dimensions of the golden string have a very interesting and beautiful structure; see Figure 2.12 and also Figure 3.6 to see the development of the quasiperiodic pattern. In Chapter 3, we make a detailed study of the complex dimensions of nonlattice strings.

2.4 The Lattice and Nonlattice Case

There is an important dichotomy regarding the scaling ratios $r_1, \ldots, r_N$ with which a self-similar string is constructed. Indeed, recall that an ad-

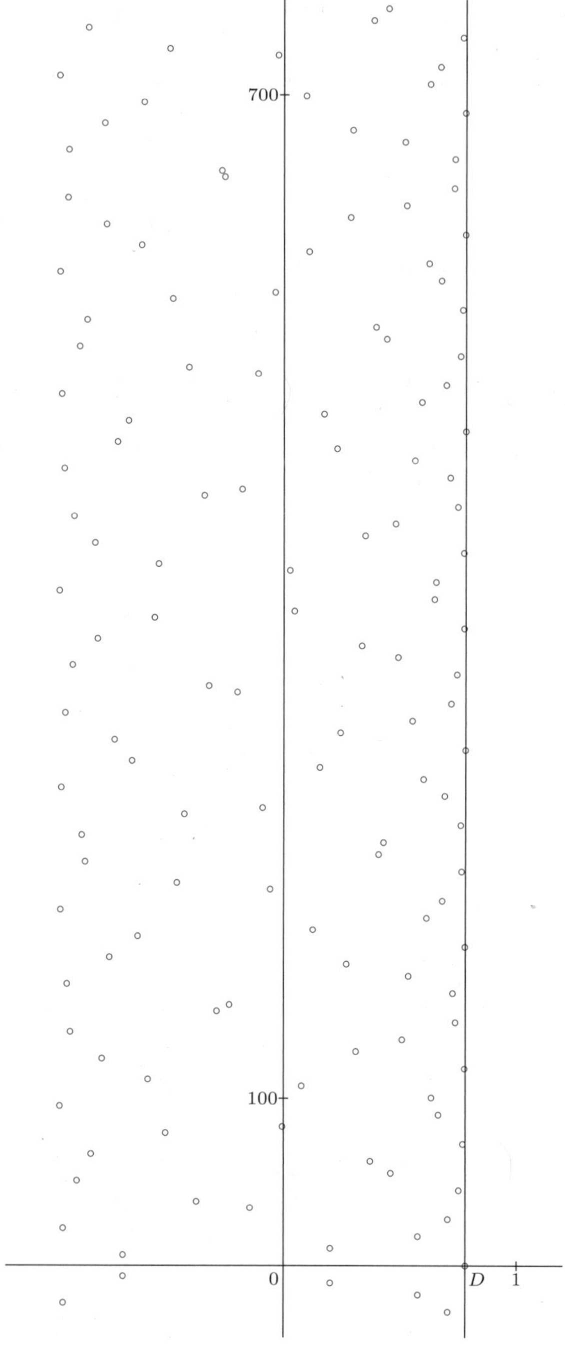

Figure 2.12: The quasiperiodic behavior of the complex dimensions of the golden string.

ditive subgroup of the real numbers is either dense in $\mathbb{R}$ or else discrete. In the latter case, if the group is not trivial, there exists a number $w > 0$, called the *additive generator*, such that the subgroup is equal to $w\mathbb{Z}$.

We now apply this basic fact to the additive group

$$A = \sum_{j=1}^{N} (\log r_j)\mathbb{Z},$$

generated by the logarithms $\log r_j$ ($j = 1, \ldots, N$) of the scaling factors. Alternatively, in light of the isomorphism $x \mapsto e^x$ between the additive group $(\mathbb{R}, +)$ and the multiplicative group $(\mathbb{R}_+^*, \cdot)$, we formulate the next definition in terms of the multiplicative subgroup

$$G = \prod_{j=1}^{N} r_j^{\mathbb{Z}} \tag{2.33}$$

of $\mathbb{R}_+^*$, the positive real line.

Definition 2.14. The case when G is dense in $\mathbb{R}_+^*$ is called the *nonlattice case.* We then say that $\mathcal{L}$ is a *nonlattice string.*

The case when G is not dense (and hence discrete) in $\mathbb{R}_+^*$ is called the *lattice case.* We then say that $\mathcal{L}$ is a *lattice string.* In this situation there exist a unique real number r, $0 < r < 1$, called the *multiplicative generator* of the string, and positive integers $k_1, \ldots, k_N$ without common divisor, such that $1 \leq k_1 \leq \cdots \leq k_N$ and

$$r_j = r^{k_j}, \tag{2.34}$$

for $j = 1, \ldots, N$. The positive number

$$\mathbf{p} = \frac{2\pi}{\log r^{-1}} \tag{2.35}$$

is called the *oscillatory period* of the lattice string $\mathcal{L}$.

We note that r is the generator of the multiplicative group $\prod_{j=1}^{N} r_j^{\mathbb{Z}}$ such that $r < 1$. Similarly, by taking logarithms, we see that $\log r^{-1}$ is the positive generator of the additive group $\sum_{j=1}^{N} (\log r_j)\mathbb{Z}$. Thus $w = \log r^{-1}$, in the above notation.

Remark 2.15. The nonlattice case is the generic case, in the sense that if one randomly picks at least two numbers $r_1, \ldots, r_N \in (0, 1)$, then, with probability 1, they will generate a nonlattice string. A key objective of the rest of this chapter and the next chapter is to demonstrate the relationship between lattice and nonlattice equations as well as to understand the qualitative and quantitative differences between various nonlattice equations in terms of the Diophantine properties of their scaling ratios (or weights). In Section 3.6, we apply our results to the case of the complex dimensions of a nonlattice string.

Remark 2.16. For Dirichlet polynomials, to be introduced in the next chapter, we refine the definition of the nonlattice case by distinguishing the generic nonlattice case. See Definition 3.1 of Section 3.1. The two nonlattice strings discussed in Section 2.3.5 are generic nonlattice.

2.5 The Structure of the Complex Dimensions

The following key theorem summarizes many of the results that we obtain about the complex dimensions of self-similar strings. It provides, in particular, a useful criterion for distinguishing between a lattice and a nonlattice string, by looking at the right-most elements in its set of complex dimensions $\mathcal{D}$. It also gives the basic structure of $\mathcal{D}$ in the lattice case. We establish most of the facts stated in Theorem 2.17 in Chapter 3, where we formulate the counterpart, Theorem 3.6, for the more general case of Dirichlet polynomial equations. The statement regarding Minkowski measurability is proved in Chapter 8. More precise and additional information can be found in the latter sections as well as in Theorems 8.23 and 8.36 (along with Remark 8.38, Corollary 8.27 and Theorem 8.25) of Sections 8.4.2 and 8.4.4.

Theorem 2.17. *Let $\mathcal{L}$ be a self-similar string of dimension D, with scaling ratios $r_1, \dots, r_N$ and gaps $g_1, \dots, g_K$, as defined in Section 2.1. Then all the complex dimensions of $\mathcal{L}$ lie to the left of or on the line* $\operatorname{Re} s = D$:

$$\mathcal{D} = \mathcal{D}_{\mathcal{L}}(\mathbb{C}) \subset \{s \in \mathbb{C}\colon \operatorname{Re} s \leq D\}. \tag{2.36}$$

The value $s = D$ is the only pole of $\zeta_{\mathcal{L}}$ on the real line. (In particular, Equation (1.27) *holds, with $D = D_{\mathcal{L}}$.) Moreover, $0 < D < 1$ and D is equal to the Minkowski dimension of the boundary of the string. The set of complex dimensions of $\mathcal{L}$ is symmetric with respect to the real axis and is contained in a horizontally bounded strip* $D_l \leq \operatorname{Re} s \leq D$, *for some real number D_l. It is infinite, with density at most*

$$\#\left(\mathcal{D}_{\mathcal{L}} \cap \{\omega \in \mathbb{C}\colon |\operatorname{Im} \omega| \leq T\}\right) \leq \frac{\log r_N^{-1}}{\pi} T + O(1), \tag{2.37}$$

as $T \to \infty$. Here, the elements of $\mathcal{D}_{\mathcal{L}}$ are to be counted according to their multiplicity as poles of $\zeta_{\mathcal{L}}$. The dimension D is never canceled. If there are no other cancellations, for example if the gaps are all equal to a single value, then the density of the complex dimensions is exactly given by the right-hand side of (2.37).

In the lattice case, *the complex dimensions ω are obtained by finding the complex solutions z of the polynomial equation (of degree k_N)*

$$\sum_{j=1}^{N} z^{k_j} = 1, \quad \text{with } r^{\omega} = z. \tag{2.38}$$

In that case, the poles lie periodically on finitely many vertical lines, and on each line they are separated by the oscillatory period $\mathbf{p} = 2\pi/\log r^{-1}$ *of the string (see Definition 2.14, Equation* (2.35)*). In other words, there exist finitely many poles* $\omega_1(= D), \omega_2, \dots, \omega_q$ *(with* $\operatorname{Re}\omega_q \leq \dots \leq \operatorname{Re}\omega_2 < D$*) such that*

$$\mathcal{D}_{\mathcal{L}} = \{\omega_u + in\mathbf{p} \colon n \in \mathbb{Z}, u = 1, \dots, q\}. \tag{2.39}$$

The multiplicity of the complex dimensions corresponding to one value of $z = r^\omega$ *is the same as that of* z*. In particular, the poles on the line above* D *are all simple. It follows that a lattice string is not Minkowski measurable: its geometry has oscillations of order* D.

Finally, if the gap sizes g_jL *are also integral powers of the multiplicative generator* r *of* $\mathcal{L}$*, i.e.,*[7]

$$g_jL = r^{k_j'}, \quad \text{where } k_j' \in \mathbb{Z} \text{ for } j = 1, \dots, K, \tag{2.40}$$

then $\zeta_{\mathcal{L}}(s)$ *is a rational function of* r^s*, and hence is a periodic function of* s *with period* $i\mathbf{p}$*. Furthermore, the residue of* $\zeta_{\mathcal{L}}(s)$ *at* $s = D + in\mathbf{p}$ *is independent of* $n \in \mathbb{Z}$ *and equal to*

$$\operatorname{res}(\zeta_{\mathcal{L}}(s); D + in\mathbf{p}) = \frac{\sum_{j=1}^{K} r^{k_j' D}}{\log r^{-1} \sum_{j=1}^{N} k_j r^{k_j D}}. \tag{2.41}$$

Also, for $u = 1, \dots, q$*, the principal part of the Laurent series of* $\zeta_{\mathcal{L}}(s)$ *at* $s = \omega_u + in\mathbf{p}$ *does not depend on* $n \in \mathbb{Z}$*. In general, when* (2.40) *is not necessarily assumed, we have*

$$\operatorname{res}(\zeta_{\mathcal{L}}(s); D) = \frac{\sum_{j=1}^{K} (g_jL)^D}{\log r^{-1} \sum_{j=1}^{N} k_j r^{k_j D}}, \tag{2.42}$$

and moreover, for each $u = 1, \dots, q$*, the residue of* $\zeta_{\mathcal{L}}(s)$ *at* $s = \omega_u + in\mathbf{p}$ *(with* $n \in \mathbb{Z}$*) is given in Remark 2.19 below, provided* ω_u *is simple.*

In the nonlattice case, D *is simple and is the unique pole of* $\zeta_{\mathcal{L}}$ *on the line* $\operatorname{Re} s = D$*. The residue of* $\zeta_{\mathcal{L}}(s)$ *at* $s = D$ *is equal to*

$$\operatorname{res}(\zeta_{\mathcal{L}}(s); D) = \frac{\sum_{j=1}^{K} (g_jL)^D}{\sum_{j=1}^{N} r_j^D \log r_j^{-1}}. \tag{2.43}$$

Moreover, there is an infinite sequence of simple complex dimensions of $\mathcal{L}$ *coming arbitrarily close (from the left) to the line* $\operatorname{Re} s = D$*. These complex dimensions are not canceled by zeros of the numerator of* $\zeta_{\mathcal{L}}$.

---

[7] This is true, in particular, if $\mathcal{L}$ has a single gap and is normalized (i.e., $K = 1$ and $g_1L = 1$; see Remark 2.6).

Finally, the complex dimensions of $\mathcal{L}$ can be approximated (via an explicit procedure) by the complex dimensions of a sequence of lattice strings, with larger and larger oscillatory period. Hence the complex dimensions of a nonlattice string have a quasiperiodic structure.

Remark 2.18. In Section 8.4.3, Theorem 8.30, we also compute the average Minkowski content of a lattice string, as given in Definition 8.29. Moreover, it follows from the fact that a nonlattice string has no nonreal complex dimensions with real part D that a nonlattice string is Minkowski measurable: its geometry does not have oscillations of order D. Further, the Minkowski content and average Minkowski content of a nonlattice string are given by

$$\mathcal{M} = \mathcal{M}_{\text{av}} = \frac{2^{1-D} \sum_{j=1}^{K} (g_j L)^D}{D(1-D) \sum_{j=1}^{N} r_j^D \log r_j^{-1}}. \tag{2.44}$$

Proof of part of Theorem 2.17. The statement about the existence of complex dimensions approaching $\operatorname{Re} s = D$ is proved in Theorem 3.23 below, with a refinement in Theorem 3.25. Also, the statement about Minkowski measurability is most naturally stated in the general context of fractal strings, and is proved in Chapter 8. (See Theorems 8.36 and 8.23, along with Remark 8.38, Corollary 8.27 and Theorem 8.25, in Sections 8.4.2 and 8.4.4.) The other statements—concerning the asymptotic density estimate (2.37) and the fact that the complex dimensions lie in a horizontally bounded strip—are most naturally formulated and proved in the more general context of Dirichlet polynomials of Chapter 3 (see Theorem 3.6). A weaker form of this result is formulated and proven in Section 2.6, using the Nevanlinna Theory of Appendix C.

We prove the existence and uniqueness of D (on the real line and on the line $\operatorname{Re} s = D$ in the nonlattice case). Recall from Corollary 2.5 that the complex dimensions of $\mathcal{L}$ are given by the complex solutions of Equation (2.11):

$$1 - \sum_{j=1}^{N} r_j^s = 0. \tag{2.45}$$

Consider the Dirichlet polynomial

$$f(s) = 1 - \sum_{j=1}^{N} r_j^s \tag{2.46}$$

for real values of s. Because $0 < r_j < 1$, f is strictly increasing. Since $f(0) = 1 - N < 0$ and $f(1) = g_1 + \cdots + g_K > 0$, there exists a unique value D strictly between 0 and 1 such that $f(D) = 0$. This is the only real value of s where $\zeta_{\mathcal{L}}$ has a pole. Since $\sum_{j=1}^{K} g_j^D > 0$, this pole is not canceled by a zero of the numerator of $\zeta_{\mathcal{L}}$.

Consider now a complex number s with real part $\sigma > D$. For this number,

$$|1 - f(s)| \leq \sum_{j=1}^{N} r_j^{\sigma} < \sum_{j=1}^{N} r_j^{D} = 1. \tag{2.47}$$

It follows that $f(s)$ cannot vanish, hence all the poles of $\zeta_{\mathcal{L}}$ lie to the left of or on the line $\operatorname{Re} s = D$.

Suppose there is another pole on the line vertical $\operatorname{Re} s = D$, say at $s = D + it$. Without loss of generality, we may assume that $t > 0$. The first inequality in (2.47) is an equality if and only if all numbers r_j^s have the same argument (i.e., point in the same direction), and then $f(s)$ itself vanishes if and only if all the numbers r_j^s are real and positive. This means that $r_j^{it} > 0$, and hence is equal to 1, for all $j = 1, \ldots, N$. Thus there exist a positive integer l and positive integers k_j without common divisor, such that

$$t \log r_j^{-1} = 2\pi k_j l \qquad (j = 1, \ldots, N).$$

This means that we are in the lattice case. Note that $1 \leq k_1 \leq \ldots \leq k_N$ and that $s = D + it/l$ is also a pole of $\zeta_{\mathcal{L}}$.

In this case, we proceed as follows. Write $z = r^s$, so that $r_j^s = r^{k_j s} = z^{k_j}$, for $j = 1, \ldots, N$. Then, by (2.46), $f(s) = 0$ is equivalent to

$$\sum_{j=1}^{N} z^{k_j} = 1. \tag{2.48}$$

This is a polynomial equation of degree k_N. Therefore, it has k_N complex solutions, counted with multiplicity. To every solution $z = |z|e^{i\theta}$ $(-\pi < \theta \leq \pi)$ of this equation there corresponds a unique solution

$$s_0 = -\frac{\log|z|}{\log r^{-1}} - \frac{i\theta}{\log r^{-1}}$$

with imaginary part $-\pi/\log r^{-1} \leq \operatorname{Im} s_0 < \pi/\log r^{-1}$, and a sequence of solutions

$$s = -\frac{\log|z|}{\log r^{-1}} - \frac{i(\theta - 2n\pi)}{\log r^{-1}} = s_0 + in\mathbf{p},$$

with $n \in \mathbb{Z}$. If all gap sizes are equal, there is no cancellation from the numerator of $\zeta_{\mathcal{L}}$, and these values are all poles of $\zeta_{\mathcal{L}}$; i.e., complex dimensions of the string. Note that the orders of these poles are all equal, and coincide with the multiplicity of the corresponding solution z of the algebraic equation (2.48). In general, some poles may be canceled by zeros of the numerator of $\zeta_{\mathcal{L}}$. (See the examples given in Section 2.3.3 above.) However, in the lattice case, D itself is never canceled, as was shown above. By the methods of Chapter 3 (Diophantine approximation, see Section 3.4), the sum $\sum_{j=1}^{K} g_j^{D+in_\mu \mathbf{p}}$ is close to the nonvanishing value $\sum_{j=1}^{K} g_j^D$ for a subsequence $\{n_\mu\}_{\mu=-\infty}^{\infty}$ of the integers, and hence does not vanish itself. Therefore, only a strict subsequence[8] of the poles $D+in\mathbf{p}$ $(n \in \mathbb{Z})$ may be canceled by a zero of the numerator of $\zeta_{\mathcal{L}}(s)$. Hence, $\mathcal{L}$ always has infinitely many complex dimensions on the line $\operatorname{Re} s = D$ (of the form $D + in_\mu \mathbf{p}$, with $n_\mu \in \mathbb{Z}$, $n_\mu \to \pm\infty$ as $\mu \to \pm\infty$).[9] Note that a similar argument does not work in general for the other lines of complex dimensions (through the complex dimension ω_u for $u \neq 1$), since the examples of Section 2.3.3 above show that an entire line of complex dimensions may be canceled.

Since D is simple, we find by a direct computation, analogous to that carried out for the Cantor and the Fibonacci strings in Sections 2.3.1 and 2.3.2,

$$\operatorname{res}\left(\zeta_{\mathcal{L}}(s); D\right) = \lim_{s \to D} \frac{(s-D)\sum_{j=1}^{K}(g_j L)^s}{1-\sum_{j=1}^{N} r_j^s} = \frac{\sum_{j=1}^{K}(g_j L)^D}{\sum_{j=1}^{N}(\log r_j^{-1}) r_j^D}. \tag{2.49}$$

If also each $g_j L$ is an integral power of r, then $\zeta_{\mathcal{L}}(s)$ is periodic, with period $i\mathbf{p}$. It follows that for $u = 1, \dots, q$, the principal part of the Laurent series of this function at $s = \omega_u + in\mathbf{p}$ does not depend on $n \in \mathbb{Z}$. In particular, the residue at $s = D + in\mathbf{p}$ is equal to the residue at $s = D$. □

The rest of the proof of Theorem 2.17 can be found in several places in the book: The statement about the density of the poles is proved in Chapter 3 (see Theorem 3.6), where we will also show that the poles of $\zeta_{\mathcal{L}}$ lie to the right of some vertical line; see Equations (3.8a) and (3.8b), along with Remark 3.10. In the nonlattice case, the statement about poles of $\zeta_{\mathcal{L}}$ close to the line $\operatorname{Re} s = D$ is proved in Theorem 3.23. Further, that about the Minkowski measurability is proved in Sections 8.4.2 and 8.4.4, Theorems 8.23 and 8.36. (In the nonlattice case, it was proved by different means in [Lap3, §4.4] and later on, independently, in [Fa4]; see Remark 8.40.)

The last statement of the nonlattice case of Theorem 2.17—concerning the approximation of the complex dimensions of nonlattice strings by those of a sequence of lattice strings—follows from Theorems 3.14 and 3.18 in Section 3.4. The explicit procedure is provided by the method of proof of Theorem 3.18 and by Lemma 3.29 or 3.39.

Remark 2.19. In the lattice case, if ω_u is a simple pole of $\zeta_{\mathcal{L}}(s)$, then a computation similar to that leading to formula (2.49) shows that for

---

[8]Of course, if $\mathcal{L}$ has a single gap (i.e., $K = 1$), then none of the poles above D (or more generally, above ω_u, for $u = 1, \dots, q$) can be canceled, since the numerator of $\zeta_{\mathcal{L}}(s)$ does not have any zeros in that case.

[9]It follows that a lattice string always has (multiplicatively periodic) oscillations of order D. In Chapter 8, Theorem 8.23 (along with Corollary 8.27), we deduce from this fact that a lattice string cannot be Minkowski measurable.

$u = 1, \ldots, q$,

$$\operatorname{res}\left(\zeta_{\mathcal{L}}(s); \omega_u + in\mathbf{p}\right) = \frac{\sum_{j=1}^{K} (g_j L)^{in\mathbf{p}+\omega_u}}{\log r^{-1} \sum_{j=1}^{N} k_j r^{k_j \omega_u}}, \quad \text{for } n \in \mathbb{Z}. \tag{2.50}$$

If, in addition, $g_j L$ is an integral power of r for each $j = 1, \ldots, K$, as in Equation (2.40), then $(g_j L)^{in\mathbf{p}} = 1$ and so this residue only depends on ω_u, but not on n. More precisely, in that case, (2.50) becomes, for $u = 1, \ldots, q$,

$$\operatorname{res}\left(\zeta_{\mathcal{L}}(s); \omega_u + in\mathbf{p}\right) = \frac{\sum_{j=1}^{K} r^{k'_j \omega_u}}{\log r^{-1} \sum_{j=1}^{N} k_j r^{k_j \omega_u}}, \quad \text{for } n \in \mathbb{Z}. \tag{2.50$'$}$$

In particular, if we set $u = 1$ (so that $\omega_1 = D$), we recover formula (2.41) of Theorem 2.17.

Remark 2.20. For a general lattice string, the degree k_N of the polynomial equation (2.38) can be any positive integer. Indeed, given any integer $k \geq 1$, the lattice string with $N = 2$ and scaling ratios $r_1 = r$ and $r_2 = r^k$, with $0 < r < 1$ (and only one gap), gives rise to the algebraic equation of degree k

$$z^k + z = 1 \tag{2.51}$$

for the complex dimensions.

The following corollary of Theorem 2.17 will show that every (nontrivial) self-similar string is fractal in the new sense that we will introduce in Section 12.2. Namely, it has at least one (and hence at least two complex conjugate) nonreal complex dimensions with positive real part. Recall from Remark 2.2 that we exclude the trivial case when $\mathcal{L}$ consists of a single interval.

Corollary 2.21. *Every self-similar string has infinitely many complex dimensions with positive real part.*

Proof. In the lattice case, this follows from the fact that $\mathcal{D}$ always contains the vertical line of complex dimensions $D + in\mathbf{p}$, for $n \in \mathbb{Z}$ (see formula (2.39) in Theorem 2.17).

In the nonlattice case, this follows from the fact that $D > 0$, combined with the statement about the complex dimensions approaching $\operatorname{Re} s = D$ from the left in the nonlattice part of Theorem 2.17 (which will be established and made more precise in Theorem 3.23). □

We first comment on some aspects of Corollary 2.21.

Remark 2.22. It is well known that for a self-similar set satisfying the open set condition—as is the case of the boundary of a self-similar string $\mathcal{L}$ considered here, see Section 2.1.1 above—the Minkowski dimension is the unique real solution of Equation (2.11). (See, for example, [Fa3, Theorems 9.1 and 9.3, pp. 114 and 118].) In other words, D is equal to the similarity dimension [Man1], defined by the Moran equation [Mor]

$$\sum_{j=1}^{N} r_j^s = 1 \qquad (s > 0). \tag{2.52}$$

In our terminology, this is the unique complex dimension of $\mathcal{L}$ located on the real axis. In view of Theorem 2.4 and Corollary 2.5, what is new and much less easy to establish in the statement of Corollary 2.21 is that Equation (2.11) (or equivalently, (2.52)) has infinitely many complex solutions with positive real part.

We briefly comment on some aspects of the lattice vs. nonlattice dichotomy exhibited by Theorem 2.17.

Remark 2.23. As was noted earlier, the Cantor string, the Fibonacci string and the example of a self-similar string with multiple poles given in Sections 2.3.1, 2.3.2 and 2.3.4, respectively, are all lattice strings. Their multiplicative generator r is equal to $1/3$, $1/2$ and $1/3$, respectively. Further, the algebraic equation (2.38) giving rise to their complex dimensions is provided, respectively, by (2.17), (2.21) and (2.28), which are, respectively, of degree 1, 2 and 3.

Remark 2.24. As was mentioned in Section 2.3.5, in the nonlattice case, the poles of the geometric zeta function $\zeta_{\mathcal{L}} = \zeta_{\mathcal{L}}(s)$ are usually intractable. For example, we expect that in general all, or almost all, poles are simple and that there are no, or hardly any, cancellations. This issue is studied in Chapter 3, but the question is still very open. (As a first step, we show in Section 3.5.2, Theorem 3.30, that when $N = 2$—or, more generally, when $\mathcal{L}$ is defined by two distinct scaling ratios with positive multiplicities—all the complex dimensions are indeed simple.)

As for the real parts of the poles, they could be dense in the critical interval (i.e., the intersection of the real line with the narrowest strip that contains all the poles). Or there could be dense pieces (or a discrete set of points) with pole-free regions in between; i.e., substrips of the critical strip that contain no poles. Regarding these questions, we formulate two conjectures in Section 3.7.1; see also Problem 3.4 in Section 3.2.1 and Problem 3.22 in Section 3.4.2. However, thanks to the algebraic equation (2.38), we can say a lot more in the lattice case.

2.6 The Asymptotic Density of the Poles in the Nonlattice Case

In the lattice case, the algebraic equation (2.38) allows one to gain a complete understanding of the structure of the poles. To study the poles in the nonlattice case, we may use Nevanlinna theory to obtain information about the solutions of the equation $\sum_{j=1}^N r_j^s = 1$. We will prove that the solutions of this equation lie in a bounded strip, and that they have the same asymptotic density as in the lattice case; that is, a linear density given by (2.37). A stronger version of Theorem 2.25 will be formulated in Theorem 3.6 of the next chapter.

We apply Theorem C.1 of Appendix C, choosing for $a_1, \dots, a_M$ the sequence of distinct numbers among $r_1, \dots, r_N$, and letting $m_0 = -1$ and m_j be the number of integers k in $\{1, \dots, N\}$ such that $r_k = a_j$. In Theorem C.1, $r_j = e^{-w_j}$ for $j = 1, \dots, N$ and $r_0 = 1$. This theorem applies since $N \geq 2$ ensures that $\sum_{j=1}^N r_j^0 = N \neq 1$. We then obtain the following result:

Theorem 2.25. *Let $\mathcal{L}$ be a self-similar string with a single gap size. Then the number of complex dimensions (counted with multiplicity) of the nonlattice string with scaling ratios $r_1 \geq r_2 \geq \cdots \geq r_N > 0$ in the disc $|s| \leq T$ is asymptotically given by*

$$\frac{\log r_N^{-1}}{\pi} T + O\big(\sqrt{T}\big), \qquad \text{as } T \to +\infty. \tag{2.53}$$

In general, when the gaps have different sizes, complex dimensions may be cancelled by zeros of the numerator of $\zeta_{\mathcal{L}}$, in which case (2.53) provides an upper bound for the asymptotic density.

In the next chapter, Theorem 3.6, we show by a different method that the error is actually a bounded function of T. Clearly, that last result is best possible in the sense that the error is at least $1/2$ when one approximates an integer-valued function by a continuous one.

Remark 2.26. Let $\mathcal{L}$ be a self-similar string with scaling ratios $r_1, \dots, r_N$. Then, in view of Corollary 2.5, Theorem 2.25 yields the asymptotic density (2.37) of the set of complex dimensions $\mathcal{D}_{\mathcal{L}}$, as stated in Theorem 2.17. Moreover, the fact (also stated in Theorem 2.17) that the complex zeros of the equation $1 - \sum_{j=1}^N r_j^s$ (and hence the poles of $\zeta_{\mathcal{L}}(s)$) lie in a bounded strip $D_l \leq \operatorname{Re} s \leq D$, also follows from Theorem C.1, with the aforementioned choice of the numbers m_j and a_j. (Recall that we already know that $\mathcal{D}_{\mathcal{L}}$ is contained in the left half-plane $\{s\colon \operatorname{Re} s \leq D\}$.) Observe that Theorem C.1 implies that $D_l \geq \sigma_l$, where σ_l is the real number given by (C.7),

$$\sigma_l = -\frac{1}{w_M - w_{M-1}} \log\left(2\sum_{j=0}^M \frac{|m_j|}{|m_M|}\right)$$

(recall that w_j is defined by $r_j = e^{-w_j}$ for $j = 0, \ldots, N$, with $r_0 := 1$); i.e., σ_l is uniquely determined by Equation (C.6a). Even though, in general, σ_l does not give the optimal choice for D_l, it provides an effective estimate for D_l. See also Theorem 3.6 of Section 3.3 and Remark 3.10, where we obtain a better left and right estimate for the critical strip.

2.7 Notes

Section 2.1.1: for the construction of self-similar sets (via possibly nonlinear contractions), see the original articles by Moran [Mor] and Hutchinson [Hut]. See also the exposition in [Fa1] and [Fa3, Sections 9.1 and 9.2].

In [Lap-vF4] and [Lap-vF5], only self-similar strings with a single gap were considered. The present theory of self-similar strings with multiple gaps was initiated by Mark Frantz in [Fra1, 2]. Following Frantz, we have extended the results of [Lap-vF2, Chapter 2] to the multiple gap setting in the papers [Lap-vF7, Section 5] and [Lap-vF9].

The dichotomy lattice vs. nonlattice comes from (probabilistic) renewal theory [Fel, Chapter XI] and was used in a related context by Lalley in [Lal1–3]. It was then introduced by the first author in [Lap3, Section 4] in the present setting of self-similar strings—or, more generally, of self-similar drums. (In [Fel], the lattice and nonlattice case are called the arithmetic and nonarithmetic case, respectively.) See also, for example, [Lap3, Section 5] and [KiLap1, Lap5–6] for the case of self-similar drums with fractal membrane (rather than with fractal boundary), [Fa4, LeVa, Gat, HamLap] and the relevant references therein, as well as [Str1–2] for a different but related context.

That the Minkowski dimension is the unique real solution to (2.11) is essentially due to Moran [Mor] in the one-dimensional situation, and in higher dimensions, it follows from the results of Hutchinson in [Hut], along with the fact that for a self-similar set satisfying the open set condition, the Hausdorff and Minkowski dimensions coincide (see, e.g., [Fa3, Theorem 9.3, p. 118]).

Section 2.3.2: the Fibonacci string (or harp), introduced in [Lap-vF4] and [Lap-vF5, Section 2.2.2], was recently given a geometric realization and shown in [CranMH] to be connected with a well-known dynamical system defined by the baker's map $x \mapsto 2x \pmod 1$ of the unit interval. We expect analogous results to hold for other lattice strings, using some of the results and methods of Section 4.4.

3

Complex Dimensions of Nonlattice Self-Similar Strings: Quasiperiodic Patterns and Diophantine Approximation

The study of the complex dimensions of nonlattice self-similar strings is most naturally carried out in the more general setting of Dirichlet polynomials. In this chapter, we study the solutions in s of a *Dirichlet polynomial equation*

$$m_1 r_1^s + \cdots + m_M r_M^s = 1.$$

In view of Corollary 2.5 of Chapter 2, this includes as a special case the equation satisfied by the complex dimensions of a self-similar string. We distinguish two cases. In the lattice case, when the numbers r_j are integral powers of a common base r, $r_j = r^{k_j}$ for all $j \geq 1$, the equation corresponds to a polynomial equation, which is readily solved numerically by using a computer. In the nonlattice case, when some ratio $\log r_j / \log r_1$, $j \geq 2$, is irrational, we obtain valuable information by approximating the (nonlattice) equation by lattice equations of higher and higher degree. This is accomplished by means of suitable Diophantine approximation techniques (Section 3.4). In that section, we show that the set of lattice equations is dense in the set of all equations, and deduce that the roots of a nonlattice Dirichlet polynomial equation have a quasiperiodic structure, which we study in detail both theoretically and numerically (see, especially, Theorem 3.18 and the comments following it).

An important consequence is that nonlattice strings are approximated by lattice strings with larger and larger oscillatory periods, which explains how the quasiperiodic patterns of the complex dimensions of a nonlattice string emerges progressively out of the periodic patterns of the complex dimensions of the approximating lattice strings.

We also establish a number of specific theorems that enable us to obtain a very good quantitative and qualitative understanding of the quasiperiodic patterns of complex dimensions in terms of the arithmetic properties of the scaling ratios r_j (Section 3.5). Further, we analyze the asymptotic density of the real parts of the complex dimensions and discuss (in Section 3.7) several open problems motivated by our theoretical and computational work.

In Section 3.6, we obtain concrete dimension-free regions for nonlattice self-similar strings, and estimate their size in terms of the Diophantine approximation properties of the scaling ratios. The significance of such estimates is that—once combined with the explicit formulas of Chapter 5—they will enable us to obtain corresponding error terms in the asymptotic expansions of various geometric, spectral or dynamical quantities associated with self-similar strings (see Chapters 6–8 for many applications). For instance, it will follow that if the logarithms of the scaling ratios r_j are badly approximable by rational numbers, then the resulting dimension-free regions will be better (i.e., larger) and hence the associated asymptotic error estimate will be better as well.

The results in this chapter suggest, in particular, that a nonlattice string possesses a set of complex dimensions with countably many real parts (fractal dimensions) which are dense in a connected interval. We give several examples and, as was alluded to above, formulate conjectures in Section 3.7.

Apart from being natural generalizations of the equations satisfied by the complex dimensions of self-similar strings (cf. Corollary 2.5), Dirichlet polynomial equations arise in several other parts of mathematics (see, for example, [JorLan1–3]). It is therefore worthwhile to extend the scope of our investigations to this broader setting.

The main results of this chapter go well beyond those obtained in the corresponding part of [Lap-vF5, Chapter 2, especially Section 2.6]. They were first published in [Lap-vF7], and a preliminary version of that paper was provided, in lesser generality, as an MSRI preprint. We have, moreover, included some additional material and several new examples and consequences of our approximation techniques.

3.1 Dirichlet Polynomial Equations

Let $1 > r_1 \geq \cdots \geq r_N > 0$ be N positive real numbers. The equation

$$r_1^s + \cdots + r_N^s = 1 \tag{3.1}$$

has one real root, called D, and many complex roots. More generally, for $M+1$ *scaling ratios* $r_0 > r_1 > \cdots > r_M > 0$ (which we now assume to be unequal), and *multiplicities* $m_j \in \mathbb{C}$ $(j = 0, \ldots, M)$, an expression of the form

$$m_0 r_0^s + \cdots + m_M r_M^s \tag{3.2}$$

is called a *Dirichlet polynomial.* In this chapter, we study the complex solutions of the *Dirichlet polynomial equation*

$$m_0 r_0^s + \cdots + m_M r_M^s = 0. \tag{3.3}$$

Without loss of generality, we assume the normalization

$$r_0 = 1 \quad \text{and} \quad m_0 = -1.$$

We also define the *weights*

$$w_j = -\log r_j \tag{3.4}$$

for $j = 0, \ldots, M$. Thus $w_0 = 0$ and $w_j > 0$ for $j = 1, \ldots, M$. (In Chapter 7, these numbers are interpreted as the weights of a self-similar flow associated with a self-similar string $\mathcal{L}$.) Let

$$f(s) = 1 - \sum_{j=1}^{M} m_j r_j^s = 1 - \sum_{j=1}^{M} m_j e^{-w_j s}. \tag{3.5}$$

We refer to f as an *integral* (respectively, *positive*) Dirichlet polynomial if all multiplicities m_j ($j = 1, \ldots, M$) are integers (respectively, positive).

As was mentioned above, in this chapter we will study in detail the complex roots of the Dirichlet polynomial equation

$$f(s) = 0, \tag{3.6}$$

and later deduce from it suitable dimension-free regions for nonlattice self-similar strings (Section 3.6), as well as corresponding error estimates for tube formulas (Chapter 8) or various counting functions (Chapters 6 and 7).

3.1.1 The Generic Nonlattice Case

Recall Definition 2.14 of the lattice and nonlattice case. Thus the lattice case occurs when $f(s)$ is a polynomial of r^s for some $r > 0$, and the nonlattice case occurs when f cannot be so written.

We now refine the definition of the nonlattice case. Let

$$A = \sum_{j=1}^{M} w_j \mathbb{Z} \tag{3.7}$$

be the group introduced in Section 2.4. Since A is a free abelian group, another way of phrasing Definition 2.14 is as follows: the lattice case is when the rank of A equals 1, and the nonlattice case is when this rank is ≥ 2.

Definition 3.1. The *generic nonlattice* case is when the above-defined group A has rank M, the number of different scaling ratios, and $M \geq 2$. In other words, $M \geq 2$ and $w_1, \ldots, w_M$ are rationally independent.

Remark 3.2. This definition is different from the one used in [Lap-vF5]. There, a nonlattice string was called generic nonlattice if the rank of the group A equals N, where $N = \sum_{j=1}^{M} m_j$ is the sum of the multiplicities of r_j, in case $m_j \in \mathbb{N}^*$ for $j = 1, \ldots, M$. The present definition is better suited to distinguish the different behaviors of the complex dimensions of nonlattice strings.

3.2 Examples of Dirichlet Polynomial Equations

The examples of self-similar strings discussed in Section 2.3 immediately provide examples of Dirichlet polynomials, since the denominator of the geometric zeta function of the corresponding fractal string is a Dirichlet polynomial. In the examples below, we assume for simplicity that the numerator of the geometric zeta function is trivial; i.e., the corresponding self-similar string has a single gap. See Section 2.2.1.

Figure 2.4 in Section 2.3 gives a diagram of the complex roots of the linear lattice equation

$$1 - 2 \cdot 3^{-s} = 0,$$

and Figure 2.5 is a diagram of the complex roots of the quadratic lattice equation

$$1 - 2^{-s} - 4^{-s} = 0.$$

These correspond, respectively, to the complex dimensions of the Cantor string of Section 2.3.1, and of the Fibonacci string of Section 2.3.2. Moreover, the example of Section 2.3.4 shows that lattice equations can have multiple roots.

3.2.1 Generic and Nongeneric Nonlattice Equations

The following example is closely related to the 2-3 nonlattice string, discussed at the beginning of Section 2.3.5. Indeed, it can be regarded as the nongeneric nonlattice counterpart of this string. Interestingly, there is a correspondence between the complex roots of this 2-3-4 equation and the complex roots (or dimensions) of the 2-3 equation, *but only for the roots with positive real part.* This can be checked by laying Figure 2.10 over Figure 3.1.

Let us consider three scaling factors $r_1 = 1/2$, $r_2 = 1/3$ and $r_3 = 1/4$, each with multiplicity one. Figure 3.1 gives the complex roots of the assoc-

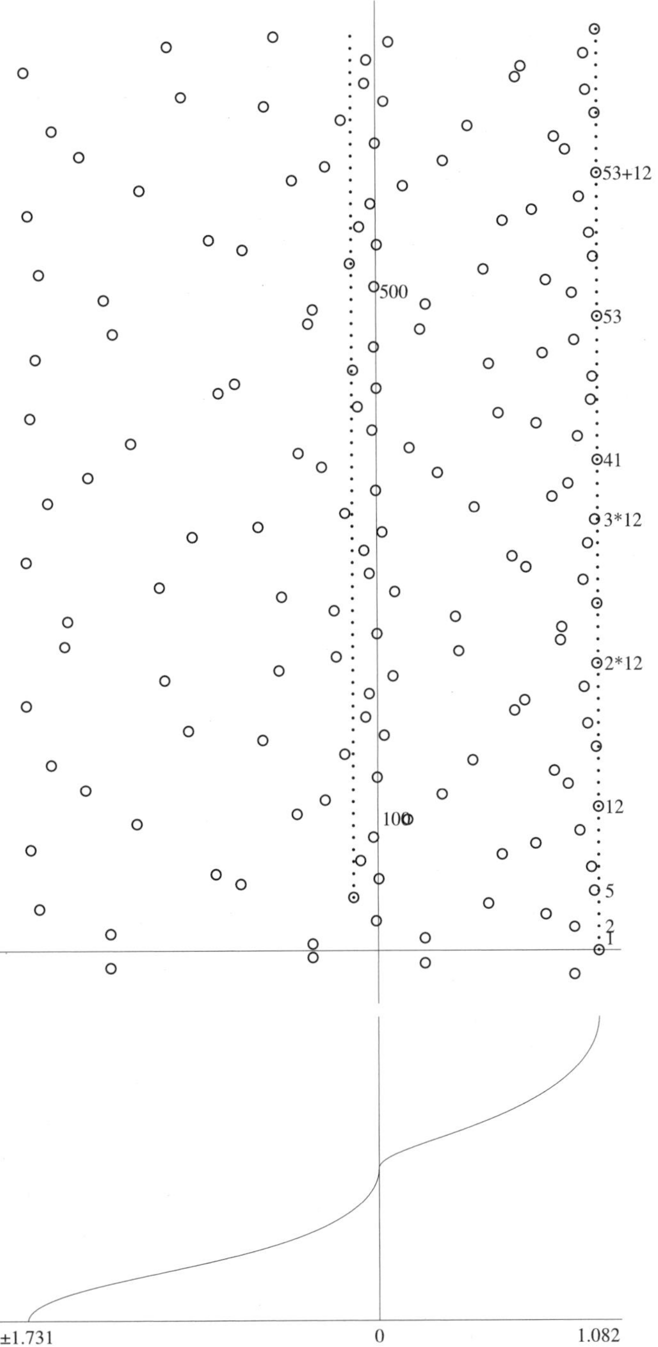

Figure 3.1: The complex roots of the nongeneric nonlattice equation $2^{-s} + 3^{-s} + 4^{-s} = 1$. The accumulative density of the real parts of the complex roots. The dotted lines and the associated markers are explained in Example 3.46.

iated nongeneric,[1] nonlattice Dirichlet polynomial

$$f(s) = 1 - 2^{-s} - 3^{-s} - 4^{-s}.$$

The dotted lines and the associated markers are explained in Example 3.46 on page 102 in the context of the golden string. (See also Remarks 3.33 and 3.41.) The markers 1, 2, 5, 12, 41, 53, ... come from the denominators of the convergents of the continued fraction of $\log 3/\log 2$.

The graph below the diagram of the complex roots in Figure 3.1 gives a plot of the density of the real parts of these roots. It will be explained in more detail in Section 3.5, Theorem 3.36 and Remark 3.42.

We observe the interesting phenomenon that the complex roots of the nongeneric nonlattice equation $f(s) = 0$ tend to be denser at the boundaries $\operatorname{Re} s = 1.082$ and $\operatorname{Re} s = -1.731$ of the critical strip, *and around* $\operatorname{Re} s = 0$. As mentioned above, comparing the complex roots of Figure 2.10 and Figure 3.1 more closely, one does indeed observe that each complex root of Figure 2.10 has its counterpart in Figure 3.1, in the half strip $\operatorname{Re} s > 0$, and extra complex dimensions are found to the left of $\operatorname{Re} s = 0$ to bring the average density of the roots to $\log 4 / 2\pi$ instead of $\log 3 / 2\pi$. This phenomenon is further illustrated by six stages of approximation to the nongeneric nonlattice string with scaling ratios $r_1 = 1/2$, $r_2 = 1/4$, $r_3 = 2^{-1-\sqrt{2}}$ of Figure 3.2 and Example 3.3, which should be contrasted with the two generic nonlattice strings of Example 3.5 below, illustrated in Figures 3.3 and 3.7. See Theorem 3.6, Equation (3.10).

Example 3.3 (A Nongeneric Nonlattice String). Consider the self-similar string with scaling ratios $r_1 = 1/2$, $r_2 = 1/4$, $r_3 = 2^{-1-\sqrt{2}}$ and one gap $g = 1/4 - r_3$. Figure 3.2 gives six approximations to its complex dimensions (along with an associated graph of the density of their real parts in the lower diagrams, to be explained below) corresponding to the successive approximations to $\sqrt{2}$ (obtained by the continued fraction expansion [HardW], see also Sections 3.4.1 and 3.5.1 below):

$$\frac{p_n}{q_n} = \frac{3}{2}, \frac{7}{5}, \frac{17}{12}, \frac{41}{29}, \frac{99}{70}, \frac{239}{169}, \dots \to \sqrt{2}.$$

For example, the first approximation gives the lattice string with scaling ratios $\tilde{r}_1 = 2^{-1}$, $\tilde{r}_2 = 2^{-2}$, $\tilde{r}_3 = 2^{-5/2}$, the complex dimensions of which are the solutions to the equation $z^2 + z^4 + z^5 = 1$, $2^{-\omega/2} = z$. The oscillatory period of this lattice approximation is $\mathbf{p} = 4\pi/\log 2$. One sees the development of a quasiperiodic pattern: the complex dimensions of the nonlattice string are well approximated by those of a lattice string for a certain *finite* number of periods of the lattice approximation. Then that periodic pattern gradually disappears, and a new periodic pattern, approximated by the next lattice approximation, emerges. See Figure 3.7 on page 86 (the left diagram) for an impression on a larger scale of the quasiperiodic behavior of the complex dimensions of this nongeneric nonlattice string.

[1]Note that the rank of the group A is equal to 2 in this case, whereas $M = N = 3$.

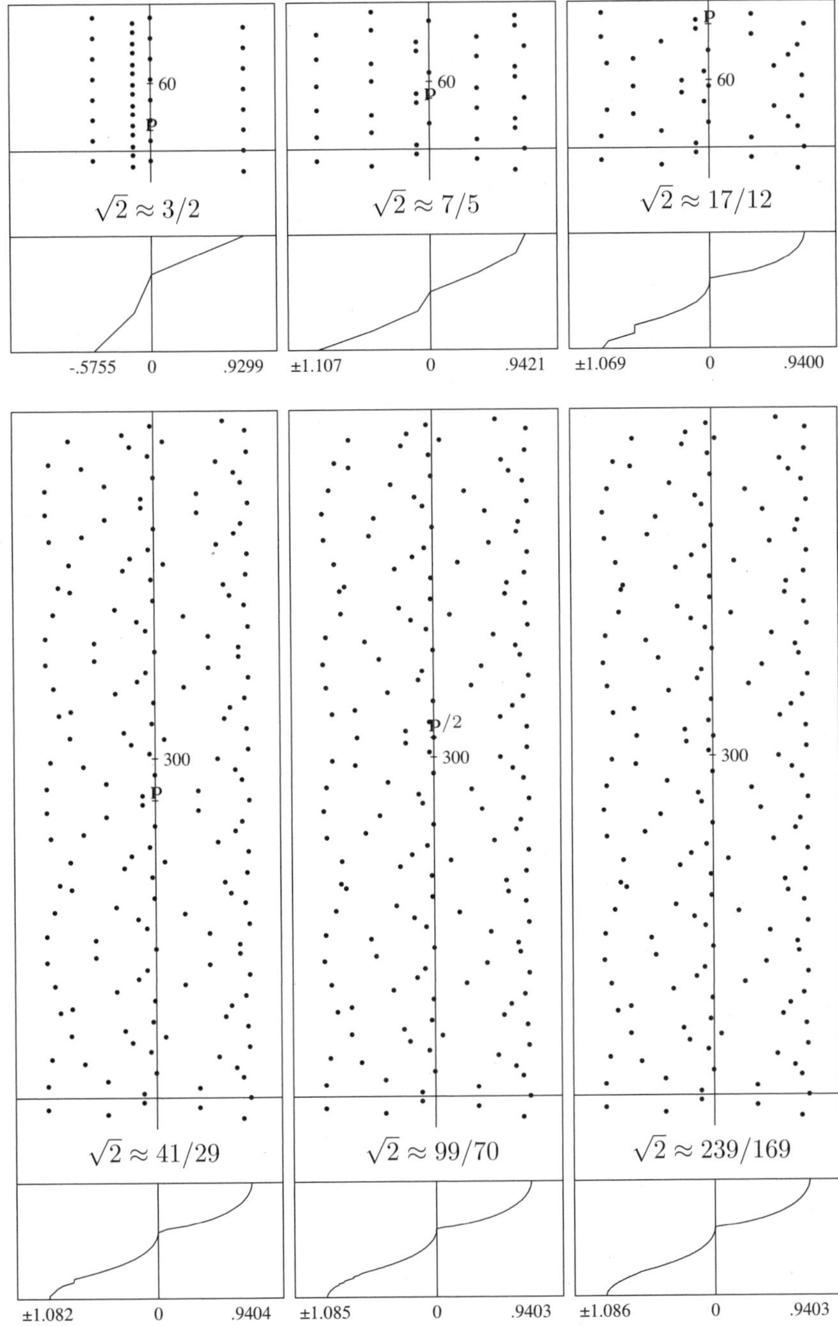

Figure 3.2: Six stages of approximation of the complex dimensions of the nongeneric nonlattice string of Example 3.3, with $r_1 = 2^{-1}$, $r_2 = 2^{-2}$ and $r_3 = 2^{-1-\sqrt{2}}$. *Lower parts:* The density graph of the real parts of the complex dimensions.

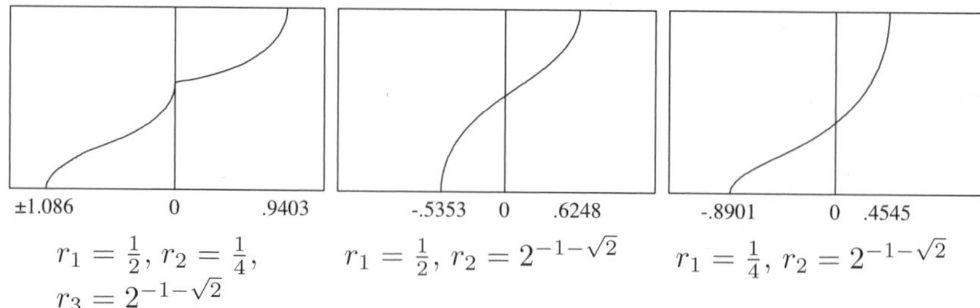

Figure 3.3: Comparison of the densities of the real parts for the nongeneric nonlattice string of Example 3.3 (left) and the two generic nonlattice strings of Example 3.5 (center and right).

The middle and right diagram in Figure 3.7 give the complex dimensions of the two *generic* nonlattice strings of Example 3.5 below. One sees that the complex dimensions in the left diagram of the nongeneric nonlattice string of the above example are much denser to the left of vanishing real part. One sees this even more clearly in Figure 3.3, where the cumulative density of the real parts is graphed for these three self-similar strings. We have no explanation for this apparent phase transition, and we formulate this question as a problem (see Theorem 3.6 and Equation (3.8b) for the definition of D_l):

Problem 3.4 (Transition in the Nongeneric Nonlattice Case). A nongeneric nonlattice string has a vertical line of transition inside the vertical strip $D_l \leq \operatorname{Re} s \leq D$, to the left of which the density of the real parts is infinitely higher than to the right. Such a transition does not occur for generic nonlattice strings.

Thus, for the nongeneric nonlattice string of Example 3.3, this transition occurs at $\operatorname{Re} s = 0$, as indicated by the corner of the density graph at this point, and the vertical part of the graph to the left of $\operatorname{Re} s = 0$. Moreover, from other numerical evidence, it seems that this line of transition often occurs at $\operatorname{Re} s = 0$. However, it is clear that by replacing the multiplicity m_j of r_j by $m_j r_j^{-a}$ ($j = 1, \dots, N$), we can shift this line to any position $\operatorname{Re} s = a$.

Another surprising phenomenon is that in the right diagram in Figure 3.7, and also to a lesser extent in the middle diagram, the complex roots appear to lie on sinusoidal curves. The quasiperiodicity only partially explains this pattern. This does not seem to be a general phenomenon, as the left diagram in Figure 3.7 shows, but we have no explanation for it when it occurs.

Example 3.5 (Two Generic Nonlattice Strings). We also include in Figures 3.3 and 3.7 the analogous diagrams for two generic nonlattice strings.

These are respectively the nonlattice fractal strings with two scaling ratios $r_1 = 1/2$ and $r_2 = 2^{-1-\sqrt{2}}$, in the middle diagrams in both figures, and with $r_1 = 1/4$ and $r_2 = 2^{-1-\sqrt{2}}$, in the right diagrams in both figures. Each of these strings has a single gap of length respectively $g_1 = 1/2 - 2^{-1-\sqrt{2}}$ and $g_1 = 3/4 - 2^{-1-\sqrt{2}}$. Note that for the second string, the one with scaling ratios $r_1 = 1/4$ and $r_2 = 2^{-1-\sqrt{2}}$, the approximation $1+\sqrt{2} \approx \frac{408}{169}$ leads to the equation

$$z^{2\cdot 169} + z^{408} = 1, \qquad 2^{-\omega/169} = z.$$

Since this is an equation in z^2, the oscillatory period of the lattice equation, namely, $169 \cdot \pi/\log 2$, is only slightly larger than that of the lattice equation corresponding to the previous approximation, $1+\sqrt{2} \approx \frac{169}{70}$, which is equal to $70 \cdot 2\pi/\log 2$.

3.2.2 *The Complex Roots of the Golden Plus Equation*

In Figure 3.4, we give a diagram of the complex dimensions of the *golden+ string*, defined as the nongeneric nonlattice string with $M = N = 3$ and scaling ratios 2^{-1}, $2^{-\phi}$ and 2^{-2}, so that

$$\zeta_{\mathcal{L}}(s) = \frac{1}{1 - 2^{-s} - 2^{-\phi s} - 2^{-2s}},$$

where $\phi = (1+\sqrt{5})/2$ is the golden ratio of Equation (2.22). The complex dimensions of $\mathcal{L}$ are found by solving the Dirichlet polynomial equation

$$2^{-s} + 2^{-\phi s} + 2^{-2s} = 1.$$

For the real parts, we observe the same phenomenon of phase transition in the complex dimensions as discussed in Example 3.3 and Problem 3.4. The complex dimensions with *positive* real part again correspond to those of the golden string of Section 2.3.5, like in the 2-3-4 equation discussed at the beginning of Section 3.2.1.

To produce this diagram, we approximated $1 - 2^{-s} - 2^{-\phi s} - 2^{-2s}$ by $1 - z^{2584} - z^{4181} - z^{5168}$, for $z = r^s$, $r = 2^{-1/2584}$. See Section 3.8 for more details.

3.3 The Structure of the Complex Roots

The simplest example of a Dirichlet polynomial equation is

$$1 - m_1 r_1^s = 0,$$

with $M = 1$ and one scaling ratio with multiplicity m_1. In that case, the complex roots are

$$\omega = \frac{\log m_1}{w_1} + \frac{2\pi i k}{w_1} \qquad (k \in \mathbb{Z}).$$

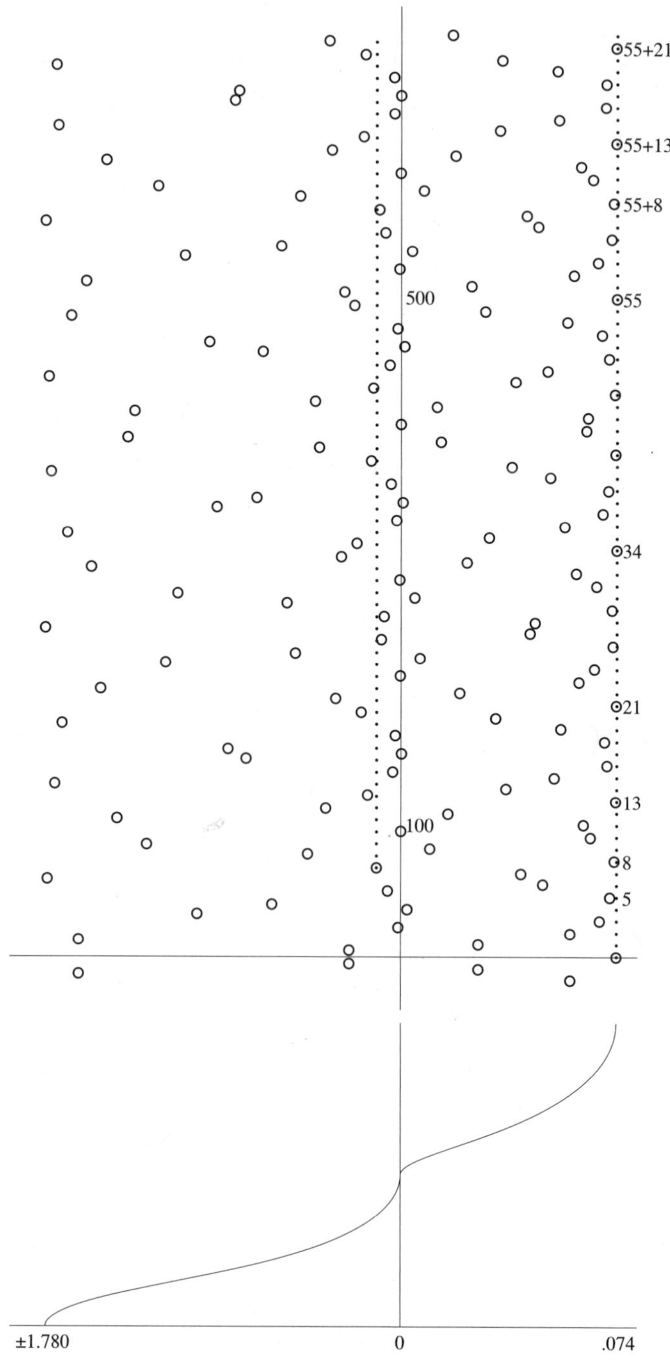

Figure 3.4: The complex dimensions of the golden+ string; the accumulative density of the real parts. The dotted lines and the associated markers are explained in Example 3.46.

Hence the complex roots lie on the vertical line $\operatorname{Re} s = (\log|m_1|)/w_1$, and they are separated by $2\pi i/w_1$.

For $M \geq 1$, the complex roots of a Dirichlet polynomial equation always lie in a horizontally bounded strip $D_l \leq \operatorname{Re} s \leq D_r$, determined as follows: D_r is the unique real number such that

$$\sum_{j=1}^{M} |m_j| r_j^{D_r} = 1, \tag{3.8a}$$

and D_l is the unique real number such that

$$1 + \sum_{j=1}^{M-1} |m_j| r_j^{D_l} = |m_M| r_M^{D_l}. \tag{3.8b}$$

The following theorem describes the structure of the complex roots of Dirichlet polynomials and is the counterpart of Theorem 2.17 in the present more general context.

Theorem 3.6. *Let f be a Dirichlet polynomial with M different scaling ratios $1 > r_1 > \cdots > r_M > 0$ and complex multiplicities m_j as in Equation (3.5). Then, both in the lattice and the nonlattice case, the set $\mathcal{D}_f$ of complex roots of f is contained in the horizontally bounded strip $D_l \leq \operatorname{Re} s \leq D_r$ defined by (3.8):*

$$\mathcal{D}_f = \mathcal{D}_f(\mathbb{C}) \subseteq \{s \in \mathbb{C} \colon D_l \leq \operatorname{Re} s \leq D_r\}. \tag{3.9}$$

It has density $\frac{w_M}{2\pi}$ (with $w_M = \log r_M^{-1}$):

$$\#\left(\mathcal{D}_f \cap \{\omega \in \mathbb{C} \colon 0 \leq \operatorname{Im} \omega \leq T\}\right) = \frac{w_M}{2\pi} T + O(1), \tag{3.10}$$

as $T \to \infty$. Here, the elements of $\mathcal{D}_f$ are counted according to multiplicity.

If all the numbers m_j are real, for $j = 1, \ldots, M$, then the set of complex dimensions is symmetric with respect to the real axis. Furthermore, if all the multiplicities m_j in (3.5) are positive, for $j = 1, \ldots, M$, then the value $s = D = D_r$ is the only complex root of f on the real line, and it is simple. If, moreover, the multiplicities are integral (i.e., $m_j \in \mathbb{N}^*$ for $j = 1, \ldots, M$), and $M \geq 2$ or $m_1 > 1$, then $D > 0$.

In the lattice case, *$f(s)$ is a polynomial function of $r^s = e^{-ws}$, where r is the multiplicative generator of f. Hence, as a function of s, it is periodic with period $2\pi i/w$. The positive number*

$$\mathbf{p} := \frac{2\pi}{w} = \frac{2\pi}{\log r^{-1}} \tag{3.11}$$

is called the oscillatory period *of the lattice equation $f(s) = 0$. The complex roots ω of this equation are obtained by finding the complex solutions z of* *the polynomial equation (of degree k_M)*

$$\sum_{j=1}^{M} m_j z^{k_j} = 1, \quad \textit{with } e^{-w\omega} = z. \tag{3.12}$$

Hence there exist finitely many roots $\omega_1, \omega_2, \dots, \omega_q$ such that

$$\mathcal{D}_f = \{\omega_u + in\mathbf{p} \colon n \in \mathbb{Z}, u = 1, \dots, q\}. \tag{3.13}$$

In other words, the complex roots of f lie periodically on finitely many vertical lines, and on each line they are separated by $\mathbf{p} = 2\pi/w$. The multiplicity of the complex roots corresponding to $z = e^{-w\omega}$ is equal to the multiplicity of z as a solution of (3.12).

In the nonlattice case, *all roots of f, except D, have real part less than D. The complex roots of f can be approximated (via an explicit procedure specified in Theorem 3.18 below) by the complex roots of a sequence of lattice equations with larger and larger oscillatory period. Hence, the complex roots of a nonlattice equation have a quasiperiodic structure. Furthermore, there exists a screen[2] S to the left of the line* $\operatorname{Re} s = D$, *such that $1/f$ satisfies* **L1** *and* **L2** *with $\kappa = 0$ (see Chapter 5, Equations (5.19) and (5.20)), and the complex roots of f in the corresponding window W are simple.*

Finally, in the generic nonlattice case *(i.e., if $M \geq 2$ and the weights $w_1, \dots, w_M$ are independent over the rationals), we have*

$$D_l = \inf\{\operatorname{Re}\omega \colon \omega \text{ is a complex root of } f\} \tag{3.14}$$

and

$$D_r = \sup\{\operatorname{Re}\omega \colon \omega \text{ is a complex root of } f\}. \tag{3.15}$$

Otherwise, the infimum of the real parts of the complex roots may be larger than D_l and the supremum may be smaller than D_r.

Corollary 3.7. *Every integral positive[3] Dirichlet polynomial has infinitely many complex roots with positive real part.*

Remark 3.8. Since by Theorem 2.4, the denominator of $\zeta_{\mathcal{L}}(s)$ is an integral positive Dirichlet polynomial, Theorem 3.6 (along with Corollary 3.7) can be applied to deduce a corresponding statement regarding the complex dimensions of a self-similar string $\mathcal{L}$. In that case, we have $D_r = D = D_{\mathcal{L}}$, the Minkowski dimension of $\mathcal{L}$.

---

[2]See Section 1.2.1.

[3]i.e., such that $m_j \in \mathbb{N}^*$ for $j = 1, \dots, M$.

Remark 3.9. Note that in the case of the geometric zeta function of a self-similar fractal with multiple gaps of unequal size, as we have seen in Section 2.3.3, some poles may be canceled by zeros of the numerator, so that the density (3.10) provides only an upper bound for the density of the complex dimensions. A weaker density estimate was proved using Nevanlinna Theory in [Lap-vF5] (see also Section 2.6 and Appendix C). The present argument used to estimate the density of the complex roots was first published in [Lap-vF7, Theorem 2.5]. We provide here a lot more details of that argument.

Proof of Theorem 3.6. The case of positive and integral weights was proved in Section 2.5. There, the real numbers D_l and D_r were not defined, but their main property (3.9) can be deduced by an argument similar to that used for D. Indeed, let s be a complex number with real part $\sigma > D_r$. Then

$$|1 - f(s)| \leq \sum_{j=1}^{M} |m_j| r_j^{\sigma} < 1,$$

by the defining property (3.8a) of D_r. Hence $f(s) \neq 0$ for $\operatorname{Re} s > D_r$. Similarly, let s be a complex number with $\sigma = \operatorname{Re} s < D_l$. Then

$$\begin{aligned} \left|r_M^{-s} f(s) + m_M\right| &= \left|r_M^{-s} - \sum_{j=1}^{M-1} m_j (r_j/r_M)^s\right| \\ &\leq r_M^{-\sigma} + \sum_{j=1}^{M-1} |m_j| (r_j/r_M)^{\sigma} < |m_M|, \end{aligned}$$

by the defining property (3.8b) of D_l. Hence again $f(s)$ cannot vanish for $\operatorname{Re} s < D_l$.

To establish (3.15), assume that $M \geq 2$ and the weights $w_1, \dots, w_M$ are rationally independent. Using Diophantine approximation (see Lemma 3.16 below), we can find real values t such that $m_j r_j^{it}$ is very close to $|m_j|$ for every $j = 1, \dots, M$. We obtain that

$$f(D_r + it) = 1 - \sum_{j=1}^{M} m_j r_j^{D_r + it}$$

almost vanishes. By Rouché's Theorem [Ahl, Corollary to Theorem 18, p. 153], applied to $f(s+it)$ and the approximation $1 - \sum_{j=1}^{M} |m_j| r_j^s$, which has a root at D_r, there exists a root of f near $D_r + it$. The same argument applies to D_l, since the numbers $w_M, w_1 - w_M, \dots, w_{M-1} - w_M$ are also rationally independent, in order to establish (3.14).

In order to establish the density estimate (3.10), we will estimate the winding number of the function $f(s) = 1 - \sum_{j=1}^{M} m_j r_j^s$ when s runs around the contour $C_1 + C_2 + C_3 + C_4$, where C_1 and C_3 are the vertical line segments $c_1 - iT \to c_1 + iT$ and $c_3 + iT \to c_3 - iT$, with $c_1 > D_r$ and $c_3 < D_l$, respectively, and C_2 and C_4 are the horizontal line segments $c_1 + iT \to c_3 + iT$ and $c_3 - iT \to c_1 - iT$, with $T > 0$. (See Figure 3.5.)

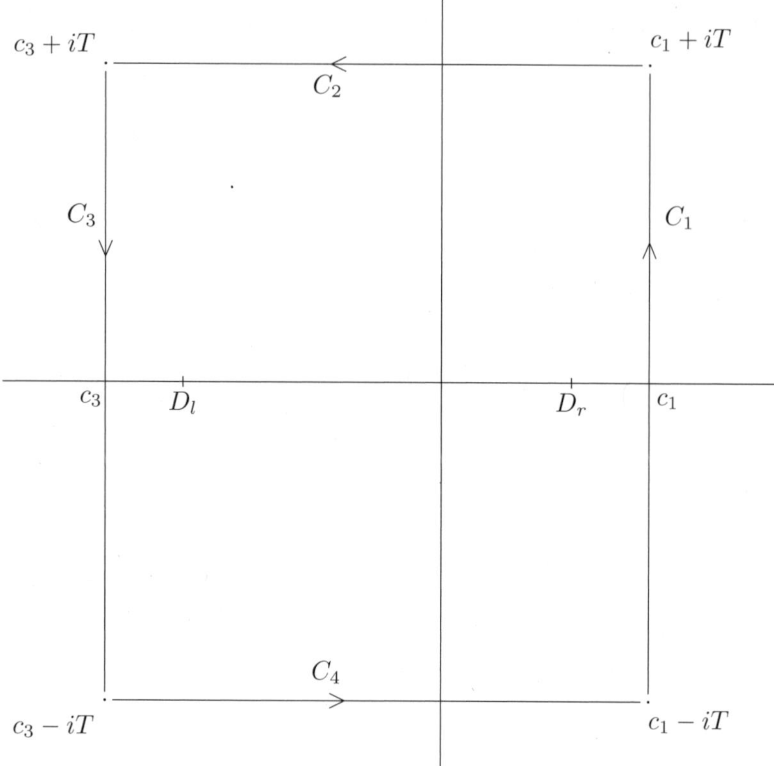

Figure 3.5: The contour $C_1 + C_2 + C_3 + C_4$.

For $\operatorname{Re} s = c_1$, we have

$$\left|\sum\nolimits_{j=1}^{M} m_j r_j^s\right| < 1;$$

hence the winding number along C_1 is at most $1/2$. Likewise, for $\operatorname{Re} s = c_3$, we have

$$\left|1 - \sum\nolimits_{j=1}^{M-1} m_j r_j^s\right| \leq 1 + \sum_{j=1}^{M-1} |m_j| r_j^{c_3} < |m_M| r_M^{c_3};$$

so the winding number along C_3 is that of the last term $m_M r_M^s$ of $f(s)$, up to at most $1/2$. Hence the winding number along the contour $C_1 + C_3$ equals $(T/\pi) \log r_M^{-1} = w_M T/\pi$, up to at most 1.

We will now show that the winding number along $C_2 + C_4$ is bounded, using a classical argument, originally applied to the Riemann zeta function (see [In, p. 69]). Let n be the number of distinct points s on C_2 at which $\operatorname{Re} f(s) = 0$. For real values of z,

$$2 \operatorname{Re} f(z + iT) = 2 - \left(\sum_{j=1}^{M} m_j r_j^{z+iT} + \sum_{j=1}^{M} \overline{m}_j r_j^{z-iT} \right).$$

Hence, putting

$$g(z) = 2 - \left(\sum_{j=1}^{M} m_j r_j^{z+iT} + \sum_{j=1}^{M} \overline{m}_j r_j^{z-iT} \right),$$

we see that n is bounded by the number of zeros of g in a disc containing the interval $[D_l, D_r]$. We take the disc centered at $D_r + 1$, with radius $D_r - D_l + 2$. We have

$$|g(D_r + 1)| \geq 2 - 2 \sum_{j=1}^{M} |m_j| r_j^{D_r+1} \geq 2(1 - r_1) > 0,$$

since $\sum_{j=1}^{M} m_j r_j^{D_r} = 1$. Furthermore, let G be the maximum of g on the disc with the same center and radius $e \cdot (D_r - D_l + 2)$. Then

$$G \leq 2 + 2 \sum_{j=1}^{M} r_j^{D_r+1-e(D_r-D_l+2)}.$$

By [In, Theorem D, p. 49], it follows that $n \leq \log |G/g(D_r + 1)|$. This gives a uniform bound on the winding number over C_2. The winding number over C_4 is estimated in the same manner.

We conclude from the above discussion that the winding number of f over the closed contour $C_1 + C_2 + C_3 + C_4$ equals $w_M T/\pi$, up to a constant (depending on f), from which the asymptotic density estimate (3.10) follows.

The approximation of a nonlattice equation by lattice equations—along with the quasiperiodic structure of the complex roots of a nonlattice equation mentioned at the end of the statement of the theorem—is discussed in Section 3.4 below. See especially Theorem 3.18. □

Remark 3.10. As in Appendix C (see σ_l and σ_r of formula (C.7)), we determine a left and a right bound for the critical strip. Recall that the weights are ordered in increasing order: $w_0 = 0 < w_1 < \cdots < w_M$ and that $m_0 = -1$. Define the real numbers d_l and d_r by the equations

$$e^{(w_M - w_{M-1})d_l} \sum_{j=0}^{M} |m_j| = |m_M| \quad \text{and} \quad e^{-w_1 d_r} \sum_{j=0}^{M} |m_j| = |m_0|, \tag{3.16}$$

respectively. In other words, we have

$$d_l = -\frac{1}{w_M - w_{M-1}} \log \sum_{j=0}^{M} \frac{|m_j|}{|m_M|} \quad \text{and} \quad d_r = \frac{1}{w_1} \log \sum_{j=0}^{M} |m_j|. \tag{3.17}$$

Then the vertical strip of complex numbers s whose real part satisfies

$$d_l \leq \operatorname{Re} s \leq d_r \tag{3.18}$$

contains all the complex roots of f, with strict inequality if $M \geq 2$. This can be seen as in Appendix C, Section C.2, noting that the values of d_l and d_r ensure that $|f(s)| > 0$ for $\operatorname{Re} s < d_l$ or $\operatorname{Re} s > d_r$, and for $\operatorname{Re} s \leq d_l$ or $\operatorname{Re} s \geq d_r$ if $M \geq 2$. The relationship between these values, the numbers D_l and D_r defined in (3.8), and an arbitrary root ω of f is given by

$$\sigma_l < d_l < D_l \leq \operatorname{Re} \omega \leq D_r < d_r < \sigma_r.$$

See also Remark C.3 in Appendix C.

3.4 Approximating a Nonlattice Equation by Lattice Equations

The generic self-similar string is nonlattice. Such a string has a slightly more evenly distributed behavior in its geometry because it is Minkowski measurable. This fact is explained in Sections 8.4.4 and 8.3, Theorem 8.15, by the absence of nonreal complex dimensions with real part D. On the other hand, a lattice string is never Minkowski measurable and always has periodic oscillations of order D in its geometry because its complex dimensions with real part D form an infinite, vertical arithmetic progression. (See Theorems 8.23 and 8.36 for more details.)

In the present section, we show, in particular, that the set of lattice strings is dense (in a suitable sense) in the set of all self-similar strings: every nonlattice string can be approximated by lattice strings. This approximation is such that it results in an approximation of the complex dimensions: given any fixed $T > 0$, there exists a lattice string such that the complex dimensions of the nonlattice string with imaginary part less than T (in absolute value) are approximated by those of the lattice string. Moreover, the oscillatory period of this lattice string is much smaller than T. This means that the complex dimensions of all self-similar strings exhibit a quasiperiodic behavior.

We use the language of Dirichlet polynomial equations in order to formulate our results. We begin by stating several definitions and a result regarding the convergence of a sequence of meromorphic functions and of the associated (zero minus pole) divisors. In Section 3.4.1 below, we will study in more detail the particular situation of Dirichlet polynomials.

We measure the convergence on the Riemann sphere $\mathbb{P}^1(\mathbb{C})$, using the metric

$$\|a, b\| = \frac{|a - b|}{\sqrt{1 + |a|^2}\sqrt{1 + |b|^2}},$$

for $a, b \in \mathbb{P}^1(\mathbb{C})$ (see also Appendix C, Equation (C.1)). More precisely, the meromorphic functions (e.g., zeta functions) involved are viewed as taking their values in $\mathbb{P}^1(\mathbb{C})$, and their value at a pole is the point at infinity (e.g., the north pole) of the Riemann sphere.

Definition 3.11. Let f be a meromorphic function on the closed[4] subset W of $\mathbb{C}$, and let $f_1, f_2, \dots$ be a sequence of meromorphic functions on $W_1, W_2, \dots$. The *sequence* f_n *converges to* f (notation: $f_n \to f$) if for every compact set $C \subseteq W$, we have that $C \subseteq W_n$ for all sufficiently large n, and $\|f_n(s), f(s)\| \to 0$ uniformly on C.

The (*visible*) *divisor* of a meromorphic function f defined on W, is the set of zeros and poles of f in W, counted with multiplicity. A zero of f is counted with a positive multiplicity, and a pole with a negative multiplicity. Thus, the divisor of f is the formal sum

$$\mathfrak{D} = \mathfrak{D}(W) = \sum_{s \in W} \operatorname{ord}(f; s)(s), \tag{3.19}$$

where the order of f at s is defined as the integer m such that the function $f(z)(z-s)^{-m}$ is bounded away from 0 in a neighborhood of s. Hence, if s is a zero of (positive) order m, then $\operatorname{ord}(f; s) = m$, whereas if s is a pole of (positive) order n, then $\operatorname{ord}(f; s) = -n = m$ is negative. In particular, $\operatorname{ord}(f; s) = 0$ if s is neither a pole nor a zero of f. The set of zeros and poles of f is called the *support*, $\operatorname{supp} \mathfrak{D}(W)$, of $\mathfrak{D}(W)$.

Definition 3.12. Let $\mathfrak{D} = \mathfrak{D}(W)$ be the divisor of a meromorphic function f and let $\mathfrak{D}_n = \mathfrak{D}(W_n)$ be the sequence of divisors of the meromorphic functions $\{f_n\}_{n=1}^{\infty}$. We say that the *sequence of divisors of* f_n *converges locally to the divisor of* f *(and write* $\mathfrak{D}_n \to \mathfrak{D}$), if for every compact set $C \subseteq W$ and every $\varepsilon > 0$, there exists an integer n_0 such that for all integers $n \geq n_0$ there exists a finite cover of C by open sets U of diameter at most ε such that for each set U in the cover, the multiplicities of the points in $\operatorname{supp} \mathfrak{D}_n \cap U$ add up to the sum of those in $\operatorname{supp} \mathfrak{D} \cap U$, and for any two such covering sets U and V, there are no points of the support of $\mathfrak{D}$ or $\mathfrak{D}_n$ in $U \cap V$.

This applies to the zeta function of a self-similar fractal string with a nonvanishing numerator; i.e., there is a single gap ($K = 1$) or more generally, all the gaps have the same size ($g_1 = g_2 = \cdots = g_K$). In that case, the zero divisor is trivial, and $\mathfrak{D}(W)$ is simply the set of (visible) complex dimensions $\mathcal{D}(W)$, relative to the window W, where each complex dimension is counted with a (negative) multiplicity. In general, $\zeta_{\mathcal{L}_n}$ could have poles that are canceled in the limit by a root of the numerator of $\zeta_{\mathcal{L}}$.

---

[4]This means that f is meromorphic on an open neighborhood of W; similarly, f_n is meromorphic on an open neighborhood of W_n.

Definition 3.13. Let $\mathcal{L}$ be a fractal string with window $W \subseteq \mathbb{C}$, and let $\{\mathcal{L}_n\}_{n=1}^{\infty}$ be a sequence of fractal strings with windows W_n. We say that the *sequence* $\mathcal{L}_n$ *converges to* $\mathcal{L}$ (and write $\mathcal{L}_n \to \mathcal{L}$) if for every compact set $C \subseteq W$, we have that $C \subseteq W_n$ for all sufficiently large n, and $\|\zeta_{\mathcal{L}_n}(s), \zeta_{\mathcal{L}}(s)\| \to 0$ uniformly on C.

In the next theorem, we use the notations $f_n \to f$ and $\mathfrak{D}_n \to \mathfrak{D}$ of Definitions 3.11 and 3.12. A similar theorem holds for a convergent sequence of fractal strings $\mathcal{L}_n \to \mathcal{L}$. (See Corollary 3.15 below.)

Theorem 3.14. *Let f be a meromorphic function and let $\{f_n\}_{n=1}^{\infty}$ be a sequence of meromorphic functions such that $f_n \to f$. Then $\mathfrak{D}_n \to \mathfrak{D}$.*

Proof. Let $C \subset W$ be compact and choose circles T_0 and T_∞ around 0 and ∞, respectively, so small that the pre-image $f^{-1}(T_0 \cup T_\infty) \cap C$ is the union of disjoint small circles (really, closed Jordan curves) around each point ω in $C \cap \operatorname{supp} \mathfrak{D}$. Let $\varepsilon > 0$ and assume that ε is smaller than the radius of T_0 and T_∞. Let n_0 be such that

$$\|f_n(s), f(s)\| \leq \frac{\varepsilon}{2},$$

for all $n \geq n_0$ and all $s \in C$. On the circle around ω, $f_n(s)$ is away from 0 (respectively, ∞) by at most $\varepsilon/2$. By the Maximum Modulus Principle [Ahl, Theorem 12′, p. 134], f_n (respectively, $1/f_n$) has a zero (respectively, pole) inside the circle in $f^{-1}(T_0)$ (respectively, $f^{-1}(T_\infty)$) around ω. Refining this argument, by comparing the complex arguments of f_n and f, we see that f_n has the same number of zeros (respectively, poles) inside the circle around ω as the multiplicity of ω. Thus we can use the preimages $f^{-1}(T_0)$ and $f^{-1}(T_\infty)$ to define the cover as in Definition 3.12. We conclude that $\mathfrak{D}_n \to \mathfrak{D}$. □

If a zero of f_n cancels a pole in the limit f, then $\mathfrak{D}_n \to \mathfrak{D}$ in the sense of Definition 3.12, but it is not true that $f_n \to f$ in the sense of Definition 3.11. Thus, if $f_n \to f$, then $\mathfrak{D}_n \to \mathfrak{D}$ in the stronger and more straightforward sense that to every zero (or pole) of f there corresponds a nearby zero (or pole) of f_n. In particular, the converse of Theorem 3.14 is false.

Note that by the above proof, the divisors of zeros and of poles converge separately. In the following corollary of Theorem 3.14, we use the notation $\mathfrak{D}_n(\infty) \to \mathfrak{D}(\infty)$ to denote that the divisors of poles converge in the sense of Definition 3.12. We also use the notation $\mathcal{L}_n \to \mathcal{L}$ of Definitions 3.13.

Corollary 3.15. *Let $\mathcal{L}$ be a fractal string and let $\mathcal{L}_n$ be a sequence of fractal strings such that $\mathcal{L}_n \to \mathcal{L}$. Then $\mathfrak{D}_n(\infty) \to \mathfrak{D}(\infty)$.*

3.4.1 Diophantine Approximation

We now focus our attention on the case of self-similar strings or, what amounts to the same thing, Dirichlet polynomials. Our main objective in the present section is to show that in the sense of Definition 3.11, every nonlattice string can be approximated by a sequence of lattice strings (Theorem 3.18). In view of Theorem 3.14 and Corollary 3.15 above, this will show that the complex dimensions of a nonlattice string can be approximated by those of a sequence of lattice strings. It will imply, in particular, that the complex dimensions of a nonlattice string have a quasiperiodic structure, and that they come arbitrarily close to the line $\operatorname{Re} s = D$ (Theorem 3.23), as was stated in Theorem 3.6.

The results of this section will be used in Chapter 6 to establish our explicit formulas for nonlattice strings, and in Chapters 7 and 8 to derive a good error term in the explicit formula for respectively the counting function of the periodic orbits of a nonlattice self-similar flow and for the volume of the tubular neighborhoods of a nonlattice string.

We continue to use the language of Dirichlet polynomials. In other words, we focus on the denominator of $\zeta_{\mathcal{L}}$. Let scaling ratios $r_1 > r_2 > \cdots > r_M$ be given that generate a nonlattice Dirichlet polynomial f; i.e., by Definition 2.14, the dimension of the $\mathbb{Q}$-vector space generated by the numbers $w_j = -\log r_j$ (for $j = 1, \dots, M$) is at least 2.

The following lemma on simultaneous Diophantine approximation can be found in [Schm, Theorem 1A and the remark following Theorem 1E]. It says that if at least one of the real numbers $\alpha_1, \dots, \alpha_M$ is irrational, then one can approximate these numbers by rational numbers *with a common denominator*. Thus one can find integers q such that for each $j = 1, \dots, M$, the multiple $q\alpha_j$ has a small distance to the nearest integer.

Lemma 3.16. *Let $w_1, w_2, \dots, w_M$ be positive weights (see Equation (3.4)) such that at least one ratio w_j/w_1 is irrational, for some $j = 2, \dots, M$. Then for every $Q > 1$, there exist integers q and $k_1, \dots, k_M$ such that $1 \le q < Q^{M-1}$ and*

$$|qw_j - k_j w_1| \le w_1 Q^{-1}$$

for $j = 1, \dots, M$. In particular,

$$|qw_j - k_j w_1| < w_1 q^{-1/(M-1)}$$

for $j = 1, \dots, M$.

Remark 3.17. Note that $|qw_j - k_j w_1| \neq 0$ when w_j/w_1 is irrational, so that $q \to \infty$ as $Q \to \infty$.

Theorem 3.18. *Let weights $w_0 = 0 < w_1 < \cdots < w_M$ and multiplicities $m_0 = -1$ and $m_1, \dots, m_M \in \mathbb{C}$ be given as in (3.5), such that at least* *one ratio* w_j/w_1 *is irrational, for some* $j = 2, \dots, M$. *Let*

$$f(s) = 1 - \sum_{j=1}^{M} m_j r_j^s,$$

with $r_j = e^{-w_j}$ *for* $j = 1, \dots, M$, *be the corresponding nonlattice Dirichlet polynomial. Let* $Q > 1$, *and let* q *and* k_j $(j = 1, \dots, M)$ *be as in Lemma 3.16. Further, let* $\tilde{w} = w_1/q$. *Then the lattice Dirichlet polynomial*

$$\tilde{f}(s) = 1 - \sum_{j=1}^{M} m_j \tilde{r}_j^s, \qquad \log \tilde{r}_j = -\tilde{w}_j = -k_j \tilde{w}$$

approximates f *in the sense of Definition 3.11. This approximation is such that for every given* $\varepsilon > 0$, *the complex roots of* f *are approximated (in the sense of Definition 3.12) up to order* ε *for* $2\varepsilon CQ$ *(positive and negative) periods of* $\tilde{f}$, *where*

$$C = \frac{\sum_{j=1}^{M} |m_j|}{2\pi} \left(\frac{\sum_{j=0}^{M} |m_j|}{\min\{1, |m_M|\}} \right)^{-2w_M / \min\{w_1, w_M - w_{M-1}\}}.$$

Proof. Let $r_j = e^{-w_j}$ for $j = 1, \dots, M$, and $\tilde{r} = e^{-\tilde{w}}$. To show that f is well approximated by $\tilde{f}$, we consider the expression

$$r_j^s - \tilde{r}^{k_j s} = -s \int_{k_j \tilde{w}}^{w_j} e^{-sx}\, dx.$$

Using $k_j \tilde{w} = \tilde{w}_j$ and $|w_j - \tilde{w}_j| \leq w_1/(qQ)$, we obtain

$$\begin{aligned} \left| r_j^s - \tilde{r}^{k_j s} \right| &\leq |s| |w_j - \tilde{w}_j| e^{-\sigma w_j} \max\{1, e^{-\sigma(\tilde{w}_j - w_j)}\} \\ &\leq |s| \frac{w_1}{qQ} e^{(w_M + w_1/(qQ))|\sigma|} \end{aligned}$$

for $j = 1, \dots, M$ and with $\sigma := \operatorname{Re} s$. We simplify this bound further, using $w_1/(qQ) < w_M$ in the exponent, to find

$$\left| f(s) - \tilde{f}(s) \right| \leq \sum_{j=1}^{M} \left| m_j r_j^s - m_j \tilde{r}^{k_j s} \right| \leq |s| \frac{w_1}{qQ} e^{2w_M |\sigma|} \sum_{j=1}^{M} |m_j|.$$

By Equation (3.18) in Remark 3.10, we may restrict $s = \sigma + it$ to

$$-\frac{\log \sum_{j=0}^{M} |m_j/m_M|}{w_M - w_{M-1}} \leq \sigma \leq \frac{\log \sum_{j=0}^{M} |m_j|}{w_1}.$$

For such s,

$$
\begin{aligned}
|f(s) - \tilde{f}(s)| &\leq |s| \frac{w_1}{qQ} \sum_{j=1}^{M} |m_j| \left(\frac{\sum_{j=0}^{M} |m_j|}{\min\{1, |m_M|\}} \right)^{2w_M / \min\{w_1, w_M - w_{M-1}\}} \\
&= |s| \frac{w_1}{2\pi q Q} C^{-1}.
\end{aligned}
$$

Thus, if $|s| < \varepsilon C Q \frac{2\pi q}{w_1}$ then $|f(s) - \tilde{f}(s)| < \varepsilon$. Since $\frac{2\pi i q}{w_1}$ is the period of $\tilde{f}$, the theorem follows. □

Note that the sequence f_n is obtained by constructing approximations of the scaling ratios of f. The following theorem shows that this is the only way to approximate a given Dirichlet polynomial by a sequence of Dirichlet polynomials. It is of independent interest even though it will not be used in the sequel.

Theorem 3.19. *Let f be a Dirichlet polynomial, with scaling ratios*

$$r_0 = 1 > r_1 > \cdots > r_M > 0$$

and multiplicities $m_j \in \mathbb{C}$. Let $\{f_n\}_{n=1}^{\infty}$ be a sequence of Dirichlet polynomials, with scaling ratios

$$1 > r_1^{(n)} > \cdots > r_{M_n}^{(n)} > 0$$

and multiplicities $m_j^{(n)} \in \mathbb{C}$. Let $W = W_n = \mathbb{C}$ for all n. If $f_n \to f$, then the scaling ratios converge with the correct multiplicity; i.e., for every $\varepsilon > 0$, there exists n_0 such that for all $n \geq n_0$ there exists a j' for each j such that $|r_j^{(n)} - r_{j'}| < \varepsilon$, and for each j',

$$\left| \sum\nolimits_j m_j^{(n)} - m_{j'} \right| < \varepsilon,$$

where the sum is over those j between 1 and M_n for which $|r_j^{(n)} - r_{j'}| < \varepsilon$.

Remark 3.20. Note that the statement of Theorem 3.19 only starts to be interesting for

$$\varepsilon \leq \min_{1 \leq j < M} (r_j - r_{j+1})/2;$$

that is, when ε is so small that $|r_j^{(n)} - r_{j'}| < \varepsilon$ uniquely determines j'. Then it says that the numbers $r_j^{(n)}$ start to cluster around the $r_{j'}$ in the sense that for each j there is a unique j', and the corresponding multiplicities add up to approximate $m_{j'}$.

Proof of Theorem 3.19. Suppose first that f_n and f are lattice polynomials with the same multiplicative generator r. Since, then, both functions are polynomials in r^s, the theorem follows from [Ahl, §5.5]. In general, we choose lattice strings approximating everything on the level of the scaling ratios. Then the lattice strings are close, from which we deduce the convergence of the scaling ratios. It follows that the original scaling ratios are close, with multiplicities that are close. □

3.4.2 The Quasiperiodic Pattern of the Complex Dimensions

The number of periods for which the approximation in Theorem 3.18 remains good tends to infinity as $Q \to \infty$. Thus the complex dimensions of a nonlattice string are almost periodically distributed like that of a lattice string for a certain (large) number of periods of that lattice string. Then the complex dimensions of the nonlattice string start to deviate from this periodic pattern, and a new periodic pattern, associated with the next lattice approximation, gradually emerges (see Figure 3.6). Each lattice approximation corresponds to a value of the denominator q in Theorem 3.18. See also Examples 3.3 and 3.5 in Section 3.2 above (and the accompanying Figures 3.2 and 3.7) for more examples of the emergence of the quasiperiodic pattern of complex dimensions.

It takes fairly large approximations to see this quasiperiodic pattern emerge. For example, the complex dimensions of the three nonlattice strings of Examples 3.3 and 3.5 are respectively approximated to within a distance 0.1 by the points in the three diagrams of Figure 3.7 for only about two periods $\mathbf{p}$ of the respective lattice strings (four periods for the last one, since $\mathbf{p}$ is half the size, as explained in Example 3.5). This is found by adapting the proof of Theorem 3.18 to these strings, since a direct application of this theorem would give only half a period of good approximation. The number of periods (of the lattice string) for which the approximation is good does grow linearly in the denominator of the approximation. Note that the period itself also grows like this denominator, so that the vertical range for which the approximation is good (i.e., better than some prescribed bound) grows like the square of the denominator of the chosen lattice approximation.

Remark 3.21. In [Lap-vF5] and in [Lap-vF4, 6], we have used the term "almost periodic". Then, in later papers [Lap-vF7, 9] we have used the term "quasiperiodic" instead, which we continue to use in this book. However, neither of these expressions refers to its counterpart encountered in the standard literature in function theory (almost periodicity as in [Bohr]) or in the theory of dynamical systems (quasiperiodicity). In fact, we are not able at this point to formalize the notion of quasiperiodic pattern encountered in the present situation. However, we now have a substantial amount of mathematical and numerical information (much more than in [Lap-vF5]), provided here and throughout this chapter, about the nature of such patterns. It would be desirable in the future to give an appropriate formal definition of quasiperiodicity, as encountered in this context.

The reader may wonder if there is any connection between our notion of quasiperiodicity in this chapter and the beautiful theory of mathematical

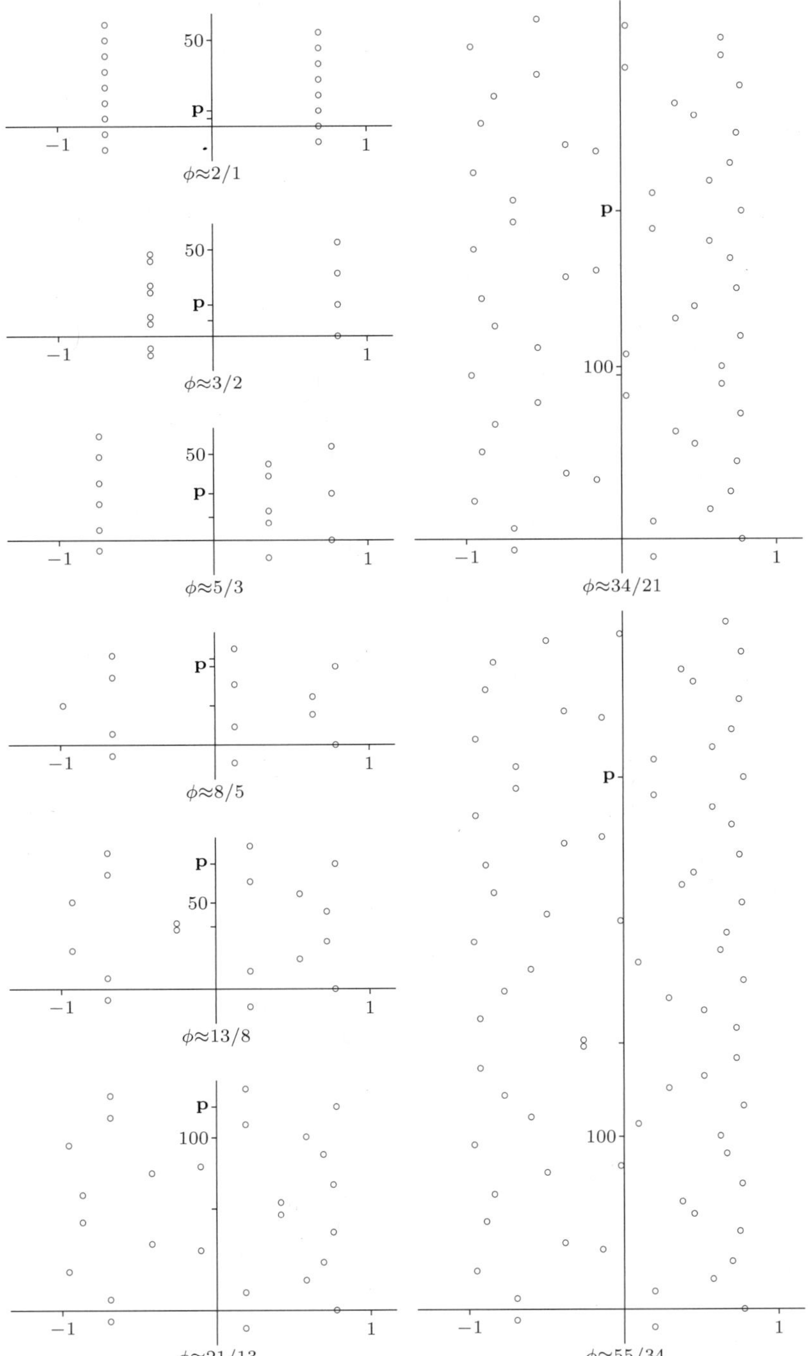

Figure 3.6: Consecutive approximations of the complex dimensions of the golden string. Emergence of the quasiperiodic pattern.

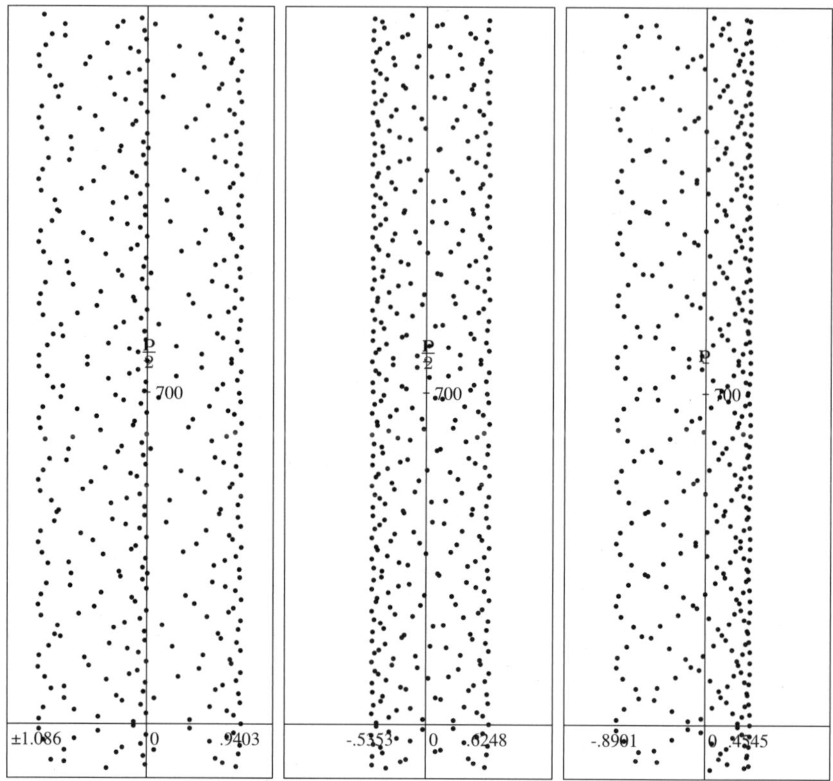

Figure 3.7: An impression of the quasiperiodic pattern of complex dimensions on a larger scale, for the self-similar string of Example 3.3 (left) and those of Example 3.5 (center and right).

and physical quasicrystals and of quasiperiodic tilings (see, e.g., [Sen, Moo, LagP] and [Lap10, Appendix F]). We suspect that this might be the case but have not yet been able to find a precise relationship. Therefore, we formulate the following open problem.

Problem 3.22. Is there a natural way in which the quasiperiodic pattern of the set of complex dimensions of a nonlattice self-similar string can be understood in terms of a suitable (generalized) quasicrystal or of an associated quasiperiodic tiling?

Recall that the simplest and most common examples of quasicrystals in $\mathbb{R}^m$—often referred to as model sets or cut-and-project sets—are obtained by suitably projecting a higher-dimensional periodic lattice onto $\mathbb{R}^m$ (see [Sen, Chapter 3] and [Moo]). The most important examples of quasicrystals are usually both model sets and substitution tilings. Preliminary investigation by the authors seems to indicate that these notions of quasicrystals are too restrictive to understand the type of quasiperiodicity encountered in this and the previous chapter. More general notions of mathematical quasicrystals include Delone sets and Meyer sets (see [Sen, Moo, LagP] and [Lap10, Appendix F]), but further extensions of the classical notion of quasicrystal may need to be introduced in this context in order to address Problem 3.22.

3.4.3 Application to Nonlattice Strings

We give three applications of the foregoing theory of approximation of nonlattice strings that will be used in Chapters 5, 6 and 8. For the application to a fractal string with scaling ratios $1 > r_1 > \cdots > r_M > 0$ and gaps scaled by $g_1, \dots, g_K$ (see Remark 2.3), we fix in this section $m_j \in \mathbb{N}^*$ to be the multiplicity of r_j (for $j = 1, \dots, M$).

In Theorems 3.23 and 3.26 below, we apply the construction of the proof of Theorem 3.18 in the following way. We choose $\varepsilon > 0$ and a large value for Q. Let f be the function defined by (3.5) and determine an approximation $\tilde{f}$ as in Theorem 3.18. Then f is approximated by $\tilde{f}$ to within an error of ε for $2\varepsilon CQ$ (positive and negative) periods of $\tilde{f}$. The period of $\tilde{f}$ is $i\mathbf{p} = 2\pi i q/\log r_1$.

The next theorem completes the proof of the nonlattice part of Theorem 2.17 (with the exception of the statement regarding the Minkowski measurability, which will be established in Theorem 8.36). Recall that we write

$$\zeta_{\mathcal{L}}(s) = \frac{\sum_{k=1}^{K} (g_k L)^s}{1 - \sum_{j=1}^{M} m_j r_j^s} \tag{3.20}$$

and $f(s) = 1 - \sum_{j=1}^{M} m_j r_j^s$.

Theorem 3.23. *Let $\mathcal{L}$ be a nonlattice self-similar string. Then there exists a sequence of simple complex dimensions of $\mathcal{L}$ approaching the line $\operatorname{Re} s = D$ from the left.*

Proof. Note that $\lim_{\sigma \to +\infty} f(\sigma + it) = 1$ and $\lim_{\sigma \to -\infty} |f(\sigma + it)| = \infty$, for any given real value of t. Given $t \in \mathbb{R}$, let

$$l(t) = \inf_{\sigma \in \mathbb{R}} |f(\sigma + it)| \tag{3.21}$$

be the infimum of $|f(s)|$ on the horizontal line $s = \sigma + it$ ($\sigma \in \mathbb{R}$). Thus for every $t \in \mathbb{R}$, $0 \le l(t) \le 1$. Further, $l(0) = 0$, and in general, $l(t) = 0$ if and only if $f(\sigma + it) = 0$ for some $\sigma \in \mathbb{R}$. In particular, the function l does not vanish identically. Choose t_0 such that $l(t) \neq 0$ for $0 < t \le t_0$. Take an approximation $\tilde{f}$ of f to within $\varepsilon \le l(t_0)/3$. This means in particular that the real value $\tilde{D}$ for which $\tilde{f}$ vanishes is very close to D, and that D and $\widetilde{D}$ are the unique zeros of f and $\tilde{f}$, respectively, in the region $-\sigma_0 \leq \sigma \leq 1$, $|t| \leq t_0$. By Rouché's Theorem [Ahl, Corollary to Theorem 18, p. 153], D and $\widetilde{D}$ actually lie within the circle (really, Jordan curve) $\{s \in \mathbb{C} \colon |\tilde{f}(s)| = \varepsilon\}$ in this region. Again by Rouché's Theorem, the translate of this circle by a period of $\tilde{f}$ also contains a zero of f, at least for the first $[2\varepsilon CQ]$ (positive and negative) periods of $\tilde{f}$. Now we can make sure that this circle is arbitrarily small around D, by choosing ε small. This shows that f has a sequence of simple zeros approaching $\operatorname{Re} s = D$, and this proves the theorem, since (by (3.5) and Corollary 2.5) zeros of f are complex dimensions of $\mathcal{L}$ (with the same or a lower multiplicity).

To see that, at least for a subsequence, there will be no cancellations from the zeros of the numerator of $\zeta_{\mathcal{L}}$ (in Equation (3.20)), we note that $\sum_{k=1}^{K}(g_k L)^D$ is positive. Hence if we also approximate the numbers $g_k L$, along with the scaling ratios r_j, then $\sum_{k=1}^{K}(g_k L)^{D+in\mathbf{p}}$ will be close to a positive number. As such, it does not vanish, and it will not cancel a pole. □

Remark 3.24. Note that the plot of the complex dimensions of the golden string $\mathcal{L}$ given in Figure 2.12 on page 52 is in agreement with Theorem 3.23. As was mentioned earlier, further application of the method of proof of Theorems 3.14 and 3.18 provides increasingly more accurate plots of $\mathcal{D}$. It yields, in particular, two sequences of complex dimensions of $\mathcal{L}$ (symmetric with respect to the real axis) converging from the left to the vertical line $\operatorname{Re} s = D$. (See the rightmost part of Figure 2.11 on page 51 for the beginning of that process.)

The last two theorems of this section are of a technical nature and verify conditions that will be needed (in Sections 6.4 and 8.4.4) in order to apply the explicit formulas of Chapter 5 to the case of self-similar strings.

The following theorem can be applied when one cannot choose a screen passing between $\operatorname{Re} s = D$ and the complex dimensions to the left of this line. (See especially Example 5.32 and Section 8.4.4.)

Theorem 3.25. *Let $\mathcal{L}$ be a nonlattice self-similar string with scaling ratios $r_1, \dots, r_N$ and gaps $g_1, \dots, g_K$. Then there exists a screen S such that $\zeta_{\mathcal{L}}$ is bounded on S and all complex dimensions to the right of S are simple with uniformly bounded residue.*

Proof. Let $\widetilde{\mathcal{L}}$ be a lattice string with scaling ratios $\tilde{r}_1, \dots, \tilde{r}_N$, approximating $\mathcal{L}$. Let D be the dimension of $\mathcal{L}$, and assume[5] that D is also the dimension of $\widetilde{\mathcal{L}}$. Since the complex dimensions of $\widetilde{\mathcal{L}}$ lie on finitely many vertical lines, there exists $\delta > 0$ such that $\widetilde{\mathcal{L}}$ has no complex dimensions

[5]If this is not the case, then $\sum_{j=1}^{N} \tilde{r}_j^{\widetilde{D}} = 1$ for an approximation $\widetilde{D}$ of D. We then replace each $\tilde{r}_j$ by its $\widetilde{D}/D$-th power. The lattice string with these new scaling ratios approximates $\mathcal{L}$ and has the same Minkowski dimension as $\mathcal{L}$.

with $D - 2\delta < \operatorname{Re}\omega < D$. Let $\omega = \alpha + i\gamma$ be a complex dimension of $\mathcal{L}$, with $D - \delta < \alpha < D$. Then $r_1^{\alpha+i\gamma} + \cdots + r_N^{\alpha+i\gamma} = 1$. Hence, by Rouché's Theorem, $r_1^{i\gamma}\tilde{r}_1^s + \cdots + r_N^{i\gamma}\tilde{r}_N^s = 1$ for s close to α. But this means that the coefficients $r_1^{i\gamma}, \ldots, r_N^{i\gamma}$ are close to 1. Again by Rouché's Theorem, the order of ω is the same as the order of D, which is 1. Since the derivatives are also approximated, it follows that the residue of $\zeta_{\mathcal{L}}$ at ω is close to that of $\zeta_{\tilde{\mathcal{L}}}$ at D, say $|\operatorname{res}(\zeta_{\mathcal{L}}(s); \omega)| \leq 2 \operatorname{res}\left(\zeta_{\tilde{\mathcal{L}}}(s); \omega\right)$.

We now construct a screen S as follows: Initially, we choose $S(t) = D - \delta/2$. But each time that $S(t) + it$ comes within a distance of $\delta/4$ of a complex dimension of $\mathcal{L}$, we go around this complex dimension along the shortest arc of radius $\delta/4$ to the left or to the right. Since the residue of $\zeta_{\mathcal{L}}$ at ω is bounded, it follows that $\zeta_{\mathcal{L}}$ is bounded along this screen. □

Theorem 3.26. *Let $\mathcal{L}$ be a self-similar string as in Theorem 3.25. Then there exists a sequence of positive numbers $T_1, T_2, \ldots$, tending to infinity, such that $|\zeta_{\mathcal{L}}(s)|$ is uniformly bounded from above on each horizontal line $\operatorname{Im} s = T_n$ and $\operatorname{Im} s = -T_n$, for $n = 1, 2, \ldots$.*

Proof. If $\mathcal{L}$ is a lattice string, then the existence of such a sequence $\{T_n\}_{n=1}^{\infty}$ follows easily from the fact that the denominator of $\zeta_{\mathcal{L}}$ is periodic (with period $i\mathbf{p}$, where $\mathbf{p}$ is the oscillatory period).

If $\mathcal{L}$ is a nonlattice string, we proceed as follows. Let the function l be given by Equation (3.21). Choose some t_0 such that $l(t_0) \neq 0$, and let $\varepsilon < \frac{1}{3} l(t_0)$. Construct a positive integer q and a function $\tilde{f}$ approximating f, as was explained before the statement of Theorem 3.23. Then on the line $s = \sigma + it_0$, the function $\tilde{f}$ is bounded away from 0 by at least $\frac{2}{3} l(t_0)$. By periodicity, we have $|\tilde{f}(s)| \geq \frac{2}{3} l(t_0)$ on $2\varepsilon CQ$ different horizontal lines of the form

$$s = \sigma + it_0 + in\mathbf{p} \quad (\text{for } -\varepsilon CQ \leq n \leq \varepsilon CQ).$$

It follows that $|f(\sigma + it_0 + in\mathbf{p})|$ is bounded from below by $\frac{1}{3} l(t_0)$ on these lines. By choosing ε smaller, and consequently q larger, we find infinitely many lines $s = \sigma + iT_n$ on which $|f(s)|$ is uniformly bounded from below. Since $\zeta_{\mathcal{L}}(s) = L^s \sum_{k=1}^{K} g_k^s / f(s)$, by Equations (3.20) and (3.5), this implies that $\zeta_{\mathcal{L}}(s)$ is uniformly bounded from above on these lines. □

Remark 3.27. For future reference, we note that the conclusion of Theorem 3.26 clearly applies to the generalized Cantor strings studied in Chapter 10 and used in Chapter 11. In fact, these generalized fractal strings are lattice strings and hence their geometric zeta function is periodic (see Equation (10.2)). In conjunction with Section 6.4 below, this justifies the application of our explicit formulas to these generalized Cantor strings (or to the associated generalized Cantor sprays) in Chapters 10 and 11.

3.5 Complex Roots of a Nonlattice Dirichlet Polynomial

In this section, we carry out a detailed study of the complex dimensions close to the line $\operatorname{Re} s = D$ of a nonlattice string. The main results are formula (3.33) (for $M = 2$) and formula (3.41) (for $M > 2$). This is applied in Section 3.6 to obtain dimension-free regions for nonlattice strings, which in turn allows us to obtain good error terms in the explicit formulas for nonlattice strings. Analogous results can be obtained for any vertical line $\operatorname{Re} s = \operatorname{Re} \omega$ through a complex dimension ω. In Section 3.7, we formulate some of the expected results that such an analysis may yield.

A nonlattice Dirichlet polynomial has weights $w_1 < \cdots < w_M$, where at least one ratio w_j/w_1 is irrational. Let

$$f(s) = 1 - \sum_{j=1}^{M} m_j e^{-w_j s}. \tag{3.22}$$

Assume that all multiplicities m_j are positive. Recall from Theorem 3.6 that in this case $D = D_r$ is the unique real solution of the equation $f(s) = 0$. Moreover, the derivative

$$f'(s) = \sum_{j=1}^{M} m_j w_j e^{-w_j s} \tag{3.23}$$

does not vanish at D since D is simple. We first consider the case $M = 2$.

3.5.1 *Continued Fractions*

Let $\alpha > 1$ be an irrational number with continued fraction expansion

$$\alpha = [[a_0, a_1, a_2, \dots]] := a_0 + \cfrac{1}{a_1 + \cfrac{1}{a_2 + \cdots}}.$$

Here, the integers a_j (also called partial quotients in [HardW]) are determined recursively by

$$\alpha_0 := \alpha,\ a_j = [\alpha_j],\ \alpha_{j+1} = \frac{1}{\alpha_j - a_j}, \qquad \text{for } j = 0, 1, 2, \dots.$$

The convergents p_n/q_n of α are defined by

$$\frac{p_n}{q_n} = [[a_0, a_1, \dots, a_n]]. \tag{3.24}$$

Recall that these rational numbers are given by the recursion relations

$$\begin{aligned} p_{-2} = 0,\ p_{-1} = 1,\ p_{n+1} = a_{n+1}p_n + p_{n-1}, \\ q_{-2} = 1,\ q_{-1} = 0,\ q_{n+1} = a_{n+1}q_n + q_{n-1}. \end{aligned} \tag{3.25}$$

We also define $q_n' = \alpha_1 \cdot \alpha_2 \cdot \dots \cdot \alpha_n$, and note that $q_{n+1}' = \alpha_{n+1}q_n + q_{n-1}$. Then

$$q_n\alpha - p_n = \frac{(-1)^n}{q_{n+1}'}. \tag{3.26}$$

For all $n \geq 1$, we have $q_n \geq \phi^{n-1}$, where $\phi = (1+\sqrt{5})/2$ is the golden ratio. We refer the interested reader to [HardW, Chapter X] for an introduction to the theory of continued fractions.

Let $n \in \mathbb{N}$ and choose l such that $q_{l+1} > n$. We can successively apply division with remainder to compute (see [Os])

$$n = d_l q_l + n_l,\ n_l = d_{l-1}q_{l-1} + n_{l-1}, \dots, n_1 = d_0 q_0,$$

where d_ν is the quotient and $n_\nu < q_\nu$ is the remainder of the division of $n_{\nu+1}$ by q_ν. We set $d_{l+1} = d_{l+2} = \ldots = 0$. Then

$$n = \sum_{\nu=0}^{\infty} d_\nu q_\nu. \tag{3.27}$$

We call this the *α-adic expansion* of n. Note that $0 \leq d_\nu \leq a_{\nu+1}$ and that $d_{\nu-1} = 0$ if $d_\nu = a_{\nu+1}$. Also $d_0 < a_1$. It is not difficult to show that these properties uniquely determine the sequence $d_0, d_1, \dots$ of *α-adic digits* of n.

Lemma 3.28. *Let n be given by formula (3.27). Suppose that the last k digits of n vanish: $k \geq 0$ is such that $d_k \neq 0$ and $d_{k-1} = \cdots = d_0 = 0$. Put $m = \sum_{\nu=k}^{\infty} d_\nu p_\nu$. Then $n\alpha - m$ lies strictly between*

$$\frac{(-1)^k}{q_{k+1}'}\left(d_k - 1 + \alpha_{k+2}^{-1}\right) \quad \text{and} \quad \frac{(-1)^k}{q_{k+1}'}\left(d_k + \alpha_{k+2}^{-1}\right).$$

In particular, $n\alpha - m$ lies strictly between $(-1)^k/q_{k+2}'$ and $(-1)^k/q_k'$.

Proof. We have $n\alpha - m = \sum_{\nu=k}^{\infty} d_\nu(\alpha q_\nu - p_\nu)$, which is close to the first term $d_k(-1)^k/q_{k+1}'$ by Equation (3.26). Again by this equation, the terms in this sum are alternately positive and negative, and it follows that $n\alpha - m$ lies between the sum of the odd numbered terms and the sum of the even numbered terms. To bound these sums, we use the inequalities $d_\nu \leq a_{\nu+1}$ for $\nu > k$. Moreover, $d_k \geq 1$, hence $d_{k+1} \leq a_{k+2} - 1$. It follows that $n\alpha - m$ lies strictly between

$$d_k(\alpha q_k - p_k) + a_{k+3}(\alpha q_{k+2} - p_{k+2}) + a_{k+5}(\alpha q_{k+4} - p_{k+4}) + \ldots$$

and

$$d_k(\alpha q_k - p_k) + (a_{k+2} - 1)(\alpha q_{k+1} - p_{k+1}) + a_{k+4}(\alpha q_{k+3} - p_{k+3}) + \dots .$$

Now $a_{\nu+1}(\alpha q_\nu - p_\nu) = (\alpha q_{\nu+1} - p_{\nu+1}) - (\alpha q_{\nu-1} - p_{\nu-1})$, so both sums are telescopic. The first sum evaluates to

$$d_k(\alpha q_k - p_k) - (\alpha q_{k+1} - p_{k+1}) = (-1)^k(d_k + \alpha_{k+2}^{-1})/q'_{k+1},$$

and the second sum to

$$d_k(\alpha q_k - p_k) - (\alpha q_{k+1} - p_{k+1}) - (\alpha q_k - p_k) = (-1)^k(d_k - 1 + \alpha_{k+2}^{-1})/q'_{k+1}.$$

The cruder bounds follow on noting that $1 \le d_k \le a_{k+1}$, and using the relations $q'_{k+2} = \alpha_{k+2} q'_{k+1}$ and $a_{k+1} + \alpha_{k+2}^{-1} = \alpha_{k+1}$. □

3.5.2 Two Generators

Assume that $M = 2$, and let f be defined as in (3.22) with positive multiplicities m_1 and m_2 and weights w_1 and $w_2 = \alpha w_1$, for some irrational number $\alpha > 1$.[6] Since $m_1, m_2 > 0$, there exists a unique real number D such that $m_1 e^{-w_1 D} + m_2 e^{-w_2 D} = 1$. We want to study the complex solutions to the equation $f(\omega) = 0$ that lie close to the line $\operatorname{Re} s = D$. First of all, such solutions must have $e^{-w_1\omega}$ close to $e^{-w_1 D}$, so ω will be close to $D + 2\pi i q/w_1$, for an integer q. We write Δ for the difference $\omega - D - 2\pi i q/w_1$, so that

$$\omega = D + \frac{2\pi i q}{w_1} + \Delta.$$

Then we write $\alpha q = p + \frac{x}{2\pi i}$, hence

$$x = 2\pi i(q\alpha - p),$$

for an integer p, which we will specify below. With these substitutions, the equation $f(\omega) = 0$ becomes

$$1 - m_1 e^{-w_1 D} e^{-w_1 \Delta} - m_2 e^{-w_2 D} e^{-x} e^{-w_2 \Delta} = 0.$$

This equation defines Δ as a function of x.

Lemma 3.29. *Let $w_1, w_2 > 0$ and $\alpha = w_2/w_1 > 1$. Let $\Delta = \Delta(x)$ be the function of x, defined implicitly by*

$$m_1 e^{-w_1 D} e^{-w_1 \Delta} + m_2 e^{-w_2 D} e^{-x} e^{-w_2 \Delta} = 1, \tag{3.28}$$

[6] In the terminology of dynamical systems of Chapter 7, this case corresponds to Bernoulli flows.

and $\Delta(0) = 0$. Then Δ is analytic in x, in a disc of radius at least π around $x = 0$, with power series

$$\Delta(x) = -\frac{m_2 e^{-w_2 D}}{f'(D)} x + \frac{m_1 m_2 w_1^2 e^{-w_1 D} e^{-w_2 D}}{2f'(D)^3} x^2 + O(x^3), \quad \text{as } x \to 0. \tag{3.29}$$

All the coefficients in this power series are real. Further, the coefficient of x^2 is positive.

Proof. Write $e^{-w_1 \Delta} = y(x)$, so that y is defined by

$$m_1 e^{-w_1 D} y + m_2 e^{-w_2 D} e^{-x} y^\alpha = 1 \quad \text{and} \quad y(0) = 1.$$

Since $y = 0$ is not a solution of this equation, it follows that if $y(x)$ is analytic in a disc centered at $x = 0$, then Δ will be analytic in that same disc. Moreover, y is real-valued and positive when x is real. Thus Δ is real-valued as well when x is real. Further, $y(x)$ is locally analytic in x, with derivative

$$y'(x) = \frac{m_2 e^{-w_2 D} y^\alpha e^{-x}}{m_1 e^{-w_1 D} + \alpha m_2 e^{-w_2 D} y^{\alpha-1} e^{-x}}.$$

Hence there is a singularity at those values of x at which the denominator vanishes, which is at

$$y = \frac{\alpha}{m_1 e^{-w_1 D}(\alpha - 1)}$$

and

$$e^{-x} = -\alpha^{-\alpha}(\alpha-1)^{\alpha-1}\frac{m_1^\alpha}{m_2}. \tag{3.30}$$

Since this latter value is negative, the disc of convergence of the power series for $y(x)$ is

$$|x| < |-\alpha \log \alpha + (\alpha - 1)\log(\alpha - 1) + \alpha \log m_1 - \log m_2 + \pi i|.$$

This is a disc of radius at least π.

To compute the first two terms in the power series for $\Delta(x)$, we use the formula for $y'(x)$ above, to obtain $y'(0) = w_1 m_2 e^{-w_2 D}/f'(D)$, and also

$$y''(0) = -\frac{w_1^2 m_2 e^{-w_2 D}}{f'(D)^3}\left(m_1^2 w_1 e^{-2w_1 D} - m_2^2 w_2 e^{-2w_2 D}\right).$$

Since $-w_1\Delta(x) = \log y(x)$, we obtain

$$-w_1 \Delta(x) = y'(0)x + \left(y''(0) - y'(0)^2\right)\frac{x^2}{2} + O(x^3).$$

Simplifying, we now readily obtain the first two terms of the power series for $\Delta(x)$ as given above. □

The proof of the following theorem is in the same spirit as the above proof. Note that this theorem implies that *the complex dimensions of a nonlattice self-similar string with two scaling ratios are all simple.* Recall the assumption that the multiplicities m_1 and m_2 are positive.

Theorem 3.30. *Let $w_2 = \alpha w_1$, with $\alpha > 1$ and irrational. Then the complex roots of the Dirichlet polynomial equation*

$$m_1 e^{-w_1 s} + m_2 e^{-w_2 s} = 1$$

are simple.

Proof. Writing $A = m_1 e^{-w_1 s}$ and $B = m_2 e^{-w_2 s}$, the assumption that s is a double root of this Dirichlet polynomial leads to the equations $A + B = 1$ and $A + \alpha B = 0$. Thus $A = \frac{\alpha}{\alpha - 1}$ and $B = -\frac{1}{\alpha - 1}$. It follows that $e^{-w_1 s}$ is positive. Hence, the imaginary part of $w_1 s$ is an integer multiple of 2π. On the other hand, since B is negative, the imaginary part of $w_2 s$ is an odd integer multiple of π. But then α is rational, contrary to our assumption. □

Remark 3.31. Lemma 3.29 and its proof remain valid if we replace D by any root $s = \omega_0$ of the equation $f(s) = 0$, provided that $f'(\omega_0) \neq 0$. In particular, the singularity of $\Delta(x)$ occurs at the same value of x satisfying (3.30), hence the power series for $\Delta(x)$ has radius of convergence at least π. Replacing D by ω_0 in (3.29) above, we obtain the first few terms of this power series,

$$\Delta(x) = -\frac{m_2 e^{-w_2 \omega_0}}{f'(\omega_0)} x + \frac{m_1 m_2 w_1^2 e^{-w_1 \omega_0} e^{-w_2 \omega_0}}{2 f'(\omega_0)^3} x^2 + O(x^3), \quad \text{as } x \to 0. \tag{3.31}$$

Since ω_0 is not real if $\omega_0 \neq D$, the coefficients of this power series are in general not real-valued.

Finally, note that $f'(\omega_0) = 0$ implies that $m_2 = -m_1^\alpha \alpha^{-\alpha} (\alpha - 1)^{\alpha - 1}$. Thus if $m_1, m_2 > 0$, then the roots of $f(s) = 0$ are all simple.

Substituting (3.29) in $\omega = D + 2\pi i q / w_1 + \Delta$, we find

$$\omega = D + 2\pi i \frac{q}{w_1} - \frac{m_2 e^{-w_2 D}}{f'(D)} x + \frac{m_1 m_2 w_1^2 e^{-w_1 D} e^{-w_2 D}}{2 f'(D)^3} x^2 + O(x^3), \tag{3.32}$$

as $x = 2\pi i (q\alpha - p) \to 0$. We view this formula as expressing ω as an initial approximation $D + 2\pi i q / w_1$, which is corrected by each additional term in the power series. The first corrective term is in the imaginary direction, as are all the odd ones, and the second corrective term, along with all the even ones, are in the real direction. The second term decreases the real part of ω.

Theorem 3.32. *Let α be an irrational number with convergents p_ν/q_ν defined by (3.24) and (3.25). Let $q = \sum_{\nu=k}^{\infty} d_\nu q_\nu$ be the α-adic expansion of the positive integer q, as in Lemma 3.28. Assume $k \geq 2$, or $k = 1$ and $a_1 \geq 2$, and put $p = \sum_{\nu=k}^{\infty} d_\nu p_\nu$. Then there exists a complex root of f at*

$$\omega = D + 2\pi i \frac{q}{w_1} - 2\pi i \frac{m_2 e^{-w_2 D}}{f'(D)} (q\alpha - p) - 2\pi^2 \frac{m_1 m_2 w_1^2 e^{-w_1 D} e^{-w_2 D}}{f'(D)^3} (q\alpha - p)^2 + O\big((q\alpha - p)^3\big). \quad (3.33)$$

The imaginary part of this complex root is approximately $2\pi i q/w_1$, and its distance to the line $\operatorname{Re} s = D$ is at least $C/q_{k+2}'^2$, where

$$C = 2\pi^2 m_1 m_2 w_1^2 e^{-(w_1+w_2)D} / f'(D)^3.$$

The number C depends only on w_1 and w_2 and the multiplicities m_1 and m_2.

Moreover, $|f(s)| \gg q_{k+2}'^{-2}$ around $s = D + 2\pi i q/w_1$ on the line $\operatorname{Re} s = D$, and $|f(s)|$ reaches a minimum of size $C'(q\alpha - p)^2$, where C' depends only on the weights w_1 and w_2 and on m_1 and m_2.

Proof. By Lemma 3.28, the quantity $q\alpha - p$ lies between $(-1)^k/q_{k+2}'$ and $(-1)^k/q_k'$. Under the given conditions on k, $q_k' > q_k \geq 2$. Hence $x = 2\pi i(q\alpha - p)$ is less than π in absolute value. Then (3.32) gives the value of ω. The estimate for the distance of ω to the line $\operatorname{Re} s = D$ follows from this formula.

Since the derivative of f is bounded on the line $\operatorname{Re} s = D$, and f does not vanish on this line except at $s = D$, this also implies that $f(s)$ reaches a minimum of order $(q\alpha - p)^2$ on an interval around $s = D + 2\pi i q/w_1$ on the line $\operatorname{Re} s = D$. □

Remark 3.33. By Remark 3.31 above, each root ω_0 also gives rise to a sequence of roots

$$\omega = \omega_0 + 2\pi i \frac{q}{w_1} + O(q\alpha - p)$$

close to the sequence $\omega_0 + 2\pi i q/w_1$. This is illustrated in Figures 3.9, 3.10 and 3.11 below for the repetitions of D and one other complex root.

We obtain more precise information when q comes from the continued fraction of α; i.e., $q = q_k$ and ω is close to $D + 2\pi i q_k/w_1$.

Theorem 3.34. *For every integer $k \geq 0$ (or $k \geq 1$ if $a_1 = 1$), there exists a complex root ω of f of the form*

$$\omega = D + 2\pi i \frac{q_k}{w_1} - 2\pi i(-1)^k \frac{m_2 e^{-w_2 D}}{f'(D) q'_{k+1}} - 2\pi^2 m_1 m_2 w_1^2 \frac{e^{-(w_1+w_2)D}}{f'(D)^3 q'^2_{k+1}} + O\left(q'^{-3}_{k+1}\right), \quad (3.34)$$

as $k \to \infty$.

Moreover, $|f(s)| \gg q'^{-2}_{k+1}$ around $s = D + 2\pi i q_k / w_1$ on the line $\operatorname{Re} s = D$, and $|f(s)|$ reaches a minimum of size $C' q'^{-2}_{k+1}$, where C' is as in the last part of Theorem 3.32.

Proof. In this case, $q = q_k$ is the α-adic expansion of q. Put $p = p_k$. Then $x = 2\pi i(-1)^k / q'_{k+1}$, which is less than π in absolute value. The rest of the proof is the same as the proof of Theorem 3.32. □

Remark 3.35. Theorem 3.32 implies that the density of complex roots in a small strip around $\operatorname{Re} s = D$ is $w_1/(2\pi)$. For Cantor-like lattice strings, with $M = 1$ (i.e., such that there is only one nonzero weight, so that $w_M = w_1$), there is only one line of complex roots, and this density coincides with formula (3.10) of Theorem 3.6. However, it is unclear how wide the strip around $\operatorname{Re} s = D$ should be. For example, in Figure 3.9, the strip extends to the left of $\operatorname{Re} s = 0$.

The following theorem was first published in [Lap-vF7, Theorem 4.7], with an incorrect value of the limit. Below follows the corrected statement and proof.

Theorem 3.36. *Let*

$$C = \frac{f'(D)^{3/2}}{\pi w_2 \sqrt{2 m_1 m_2}} e^{(w_1+w_2)D/2}. \quad (3.35)$$

The relative density of the real parts of the complex roots close to the line $\operatorname{Re} s = D$,

$$\frac{\#\{\omega \colon \omega \in \mathcal{D}, 0 \leq \operatorname{Im} \omega \leq T, \operatorname{Re} \omega \geq x\}}{\#\{\omega \colon \omega \in \mathcal{D}, 0 \leq \operatorname{Im} \omega \leq T\}}, \quad (3.36)$$

has a limit as $T \to \infty$. This limit equals

$$2C\sqrt{D - x}, \quad (3.37)$$

for values of $x \leq D$ close to D.

Proof. By Theorem 3.6, there are $\frac{w_2}{2\pi} T + O(1)$ complex roots with imaginary part $0 \leq \operatorname{Im} \omega \leq T$. Assume that T is of the form $2\pi q_{m+1}/w_1$ for some integer m. Given $x < D$, we will count the number of these roots with

$\operatorname{Re}\omega \geq x$. By Theorem 3.32, for every q with $0 \leq q < \frac{w_1}{2\pi}T = q_{m+1}$ (and p the nearest integer to $q\alpha$), we find a complex root with $0 \leq \operatorname{Im}\omega \leq T$ and lying slightly to the left of the line $\operatorname{Re} s = D$ with real part

$$\operatorname{Re}\omega = D - C_1(q\alpha - p)^2 + O\big((q\alpha - p)^4\big),$$

where

$$C_1 = \pi^2 w_1^2 (2m_1 m_2) e^{-(w_1+w_2)D} f'(D)^{-3} = \left(C^{-1} w_1/w_2\right)^2. \tag{3.38}$$

If $\operatorname{Re}\omega \geq x$, then we find that q is restricted by

$$|q\alpha - p| \leq \sqrt{(D-x)/C_1}\,(1 + O(D-x)).$$

Determine n such that

$$1/q'_{n+1} \leq \sqrt{(D-x)/C_1} < 1/q'_n.$$

Let $q = \sum_{\nu=k}^{\infty} d_\nu q_\nu$ be the α-adic expansion of q as in Equation (3.27). Then

$$|q\alpha - p| = d_k/q'_{k+1} - d_{k+1}/q'_{k+2} + \dots.$$

If $k \leq n-2$ then $|q\alpha - p| > 1/q'_{k+2} \geq 1/q'_n$ by Lemma 3.28. Hence $|q\alpha - p| > \sqrt{(D-x)/C_1}$. Likewise, if $k = n-1$ and $d_k \geq 2$, then, by the finer estimate in Lemma 3.28, $|q\alpha - p| > (1 + \alpha_{n+1}^{-1})/q'_n$. We find that the expansion of q must be of the form $q = d_{n-1}q_{n-1} + d_n q_n + \dots + d_m q_m$, where d_{n-1} is either 0 or 1.

If $d_{n-1} = 1$, then $|q\alpha - p| = 1/q'_n - d_n/q'_{n+1} + \dots$, which is approximately equal to $(\alpha_{n+1} - d_n)/q'_{n+1}$. We find that $d_n \geq \alpha_{n+1} - q'_{n+1}\sqrt{(D-x)/C_1}$. Since $d_n < \alpha_{n+1}$, we find

$$\left\lceil q'_{n+1}\sqrt{(D-x)/C_1}\right\rceil$$

possibilities for d_n.

Also, if $d_{n-1} = 0$, we find $d_n < q'_{n+1}\sqrt{(D-x)/C_1}$, and we find again the same number of possibilities for d_n.

We fix d_n, and $d_{n-1} = 0$ or 1, and count the number of positive integers q satisfying these requirements. Let N_{m+1} be the number of such q. Since $0 \leq d_\nu \leq a_{\nu+1}$, and $d_{\nu-1} = 0$ if $d_\nu = a_{\nu+1}$, we find that

$$N_{m+1} = a_{m+1}N_m + N_{m-1},$$

with initial conditions $N_{n+2} = a_{n+2}$ if $d_n \neq 0$, or $N_{n+2} = a_{n+2}+1$ if $d_n = 0$, and $N_{n+1} = 1$. Since N_m satisfies the same recursion as q_m, but with different initial conditions, these two sequences grow at the same rate, and we

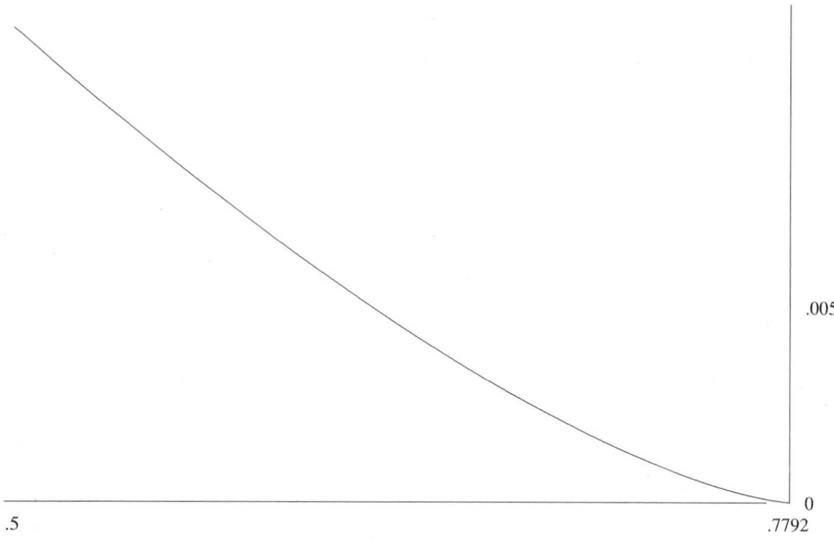

Figure 3.8: The error in the prediction of Theorem 3.36, for the golden Dirichlet polynomial equation $2^{-s} + 2^{-\phi s} = 1$.

find the approximation $N_{m+1} = q_{m+1}/q'_{n+1}$. Together with the total number of possibilities for d_n, we find approximately $2q_{m+1}\sqrt{(D-x)/C_1}$ possibilities for q. Since the total number of roots is approximately $q_{m+1}w_2/w_1$, we obtain a relative density of

$$2(w_1/w_2)\sqrt{(D-x)/C_1},$$

which by (3.38) equals $2C\sqrt{D-x}$, where C is given by (3.35). □

Remark 3.37. This theorem is illustrated in the diagrams at the bottom of Figures 3.9 and 3.10 (pages 103 and 105, repectively). These diagrams show in one figure the graph of the accumulated density function (3.36) and the graph of the function (3.37). The function (3.37) approximates the accumulative density only in a small neighborhood of D. Figure 3.8 gives a graph of the difference of the two graphs in Figure 3.9 (page 103) for the complex roots with real parts between 1/2 and D.

3.5.3 *More than Two Generators*

When there are three or more generators, i.e., $M \geq 3$, the construction of approximations p_j/q of w_j, for $j = 1, \ldots, M$, is much less explicit than for $M = 2$ since there does not exist a continued fraction algorithm for simultaneous Diophantine approximation. We use Lemma 3.16 as a substitute for this algorithm. The number Q then plays the role of q'_{k+1} in Theorem 3.34 above. In particular, if q is often much smaller than Q, then $w_1, \ldots, w_M$ is well approximable by rationals, and we find a small root-free region.

Remark 3.38. The LLL-algorithm of [LeLeLo] allows one to find good denominators. However, the problem of finding the best denominator is NP-complete [Lag1, Theorem C]. See also [RösS] and the references therein, in particular [HasJLS]. Further, it may be possible to adapt the algorithm in [El] to solve Dirichlet polynomial equations.

Again, we are looking for a solution of $f(\omega) = 0$ close to $D + 2\pi i q/w_1$, where f is defined by (3.22). We write $\omega = D + 2\pi i q/w_1 + \Delta$ and

$$w_j q = w_1 p_j + \frac{w_1}{2\pi i} x_j,$$

for $j = 1, \ldots, M$. For $j = 1$, we take $p_1 = q$ and consequently $x_1 = 0$. In general, for $j = 2, \ldots, M$, we have

$$x_j = 2\pi i (q w_j / w_1 - p_j).$$

Then $f(\omega) = 0$ is equivalent to $\sum_{j=1}^{M} m_j e^{-w_j D} e^{-x_j - w_j \Delta} = 1$.

The following lemma is the several variables analogue of Lemma 3.29. In the present case, however, we do not know the radius of convergence with respect to the variables $x_2, \ldots, x_M$.

Lemma 3.39. *Let $0 < w_1 < w_2 < \cdots < w_M$, let D be the real number such that $\sum_{j=1}^{M} m_j e^{-w_j D} = 1$, and let $\Delta = \Delta(x_2, \ldots, x_M)$ be implicitly defined by*

$$\sum_{j=1}^{M} m_j e^{-w_j D} e^{-x_j - w_j \Delta} = 1, \tag{3.39}$$

with $x_1 = 0$ and $\Delta(0, \ldots, 0) = 0$. Then Δ is analytic in $x_2, \ldots, x_M$, with power series

$$\begin{aligned} \Delta = &-\sum_{j=2}^{M} \frac{m_j e^{-w_j D}}{f'(D)} x_j + \frac{1}{2} \sum_{j=2}^{M} \frac{m_j e^{-w_j D}}{f'(D)} x_j^2 \\ &- \frac{1}{2} \sum_{j,k=2}^{M} \left(\frac{f''(D)}{f'(D)^3} + \frac{w_j + w_k}{f'(D)^2} \right) m_j m_k e^{-(w_j + w_k) D} x_j x_k \\ &+ O\left(\sum_{j=2}^{M} |x_j|^3 \right), \end{aligned} \tag{3.40}$$

where $f''(D) = -\sum_{j=1}^{M} m_j w_j^2 e^{-w_j D}$ and $f'(D)$ is given by formula (3.23) for $s = D$. This power series has real coefficients. Moreover, the terms of degree two yield a positive definite homogenous quadratic form.

Proof. The proof is analogous to that of Lemma 3.29. Taking the derivative with respect to x_k $(k \geq 2)$ gives

$$m_k e^{-w_k D} + f'(D) \frac{\partial \Delta}{\partial x_k}(0, \ldots, 0) = 0.$$

We thus find the coefficients of the linear part of Δ.

The positive definiteness of the quadratic form follows from the fact that the complex roots lie to the left of $\operatorname{Re} s = D$, see Theorem 3.6. It can also be verified directly. □

We substitute formula (3.40) into $\omega = D + 2\pi i q/w_1 + \Delta$ to find

$$\omega = D + 2\pi i \frac{q}{w_1} - \sum_{j=2}^{M} \frac{m_j e^{-w_j D}}{f'(D)} x_j + \frac{1}{2} \sum_{j=2}^{M} \frac{m_j e^{-w_j D}}{f'(D)} x_j^2$$
$$- \frac{1}{2} \sum_{j,k=2}^{M} \left(\frac{f''(D)}{f'(D)^3} + \frac{w_j + w_k}{f'(D)^2} \right) m_j m_k e^{-(w_j+w_k)D} x_j x_k$$
$$+ O\left(\sum_{j=2}^{M} |x_j|^3 \right), \tag{3.41}$$

where $x_j = 2\pi i (q w_j / w_1 - p_j)$, for $j = 2, \dots, M$. Again, this formula expresses ω as an initial approximation $D + 2\pi i q/w_1$, which is corrected by each term in the power series. The corrective terms of degree one are again in the imaginary direction, as are all the odd degree ones, and the corrective terms of degree two, along with all the even ones, are in the real direction. The degree two terms decrease the real part of ω.

Remark 3.40. As in Remark 3.31, we have a formula analogous to (3.41) corresponding to any complex root ω_0. Thus every complex root ω_0 gives rise to a sequence of complex roots close to the points $\omega_0 + 2\pi i q/w_1$. In this case it may be possible that f has multiple roots, even if the multiplicities m_j are positive for $m = 1, \dots, M$.

Theorem 3.41. *Let $M \geq 2$ and let $w_1, \dots, w_M$ be weights of a nonlattice equation. Let Q and q be as in Lemma 3.16. Then f has a complex root close to $D + 2\pi i q/w_1$ at a distance of at most $O(Q^{-2})$ from the line $\operatorname{Re} s = D$, as $Q \to \infty$. The function $|f|$ reaches a minimum of order Q^{-2} on the line $\operatorname{Re} s = D$ around the point $s = D + 2\pi i q/w_1$.*

Proof. Again, for $j = 2, \dots, M$, the numbers x_j are purely imaginary, so the terms of degree 1 (and of every odd degree) give a correction in the imaginary direction, and only the terms of even degree will give a correction in the real direction. Since $|x_j| < 2\pi/Q$, the theorem follows. □

Remark 3.42. By analogy with Theorem 3.36, we find $C(D - x)^{(M-1)/2}$ as an approximation of the density function of the complex roots close to $x = D$, for some positive constant C. However, in this case we do not know the value of C. It may depend on the properties of Diophantine approximation of the weights $w_1, \dots, w_M$.

Figure 3.11 suggests the approximation $C(D - x)$ for the accumulated density function of the Two-Three-Five equation (for which $M = 3$, so that $(M - 1)/2 = 1$), for some positive constant C.

3.6 Dimension-Free Regions

We discuss an application of the above results to the theory of self-similar fractal strings, as in Chapter 2. Thus, a Dirichlet polynomial corresponds to the geometric zeta function of a self-similar string, while the complex roots of such a polynomial correspond to the complex dimensions of this self-similar string.

Definition 3.43. An open neighborhood of the line $\operatorname{Re} s = D$ in the complex plane is called a *dimension-free region* for the string $\mathcal{L}$ if the only pole of $\zeta_{\mathcal{L}}$ in that region is $s = D$.

We assume that we are in the setting of Section 3.5.2; that is, we consider fractal strings with scaling ratios $r_1 = r$ and $r_2 = r^\alpha$. Recall the function $f(s) = 1 - m_1 r^s - m_2 r^{\alpha s}$, so that

$$f'(D) = m_1 w_1 r^D + m_2 w_2 r^{\alpha D} > 0. \tag{3.42}$$

From Theorem 3.34, and with the notation of Sections 3.5.1 and 3.5.2, we deduce the following result, in case $M = 2$.

Corollary 3.44. *Assume that the partial quotients $a_0, a_1, \dots$ of the continued fraction of α are bounded by a constant b. Put*

$$B = \frac{1}{2}\pi^4 e^{-(w_1+w_2)D} f'(D)^{-3}, \tag{3.43}$$

so that $B > 0$ by (3.42). Then $\mathcal{L}$ has a dimension-free region of the form

$$\left\{\sigma + it \in \mathbb{C} \colon \sigma > D - \frac{B}{b^2 t^2}\right\}. \tag{3.44}$$

Further, the function $\zeta_{\mathcal{L}}$ satisfies hypotheses **L1** *and* **L2** *(i.e., $\mathcal{L}$ is languid) with $\kappa = 2$ on the screen $\operatorname{Re} s = D - B/(b \operatorname{Im} s)^2$; see (5.19) and (5.20).*

More generally, let $b \colon \mathbb{R}^+ \to [1, \infty)$ be a function such that the partial quotients $\{a_k\}_{k=0}^\infty$ of the continued fraction of α satisfy

$$a_{k+1} \leq b(q_k) \quad \text{for every } k \geq 0. \tag{3.45}$$

Then $\mathcal{L}$ has a dimension-free region of the form

$$\left\{\sigma + it \in \mathbb{C} \colon \sigma > D - \frac{B}{t^2 b^2(tw_1 / 2\pi)}\right\}. \tag{3.46}$$

If $b(q)$ grows at most polynomially, then $\zeta_{\mathcal{L}}$ satisfies hypotheses **L1** *and* **L2** *on the screen $S \colon S(t) + it$, $S(t) = D - Bt^{-2} b^{-2}(tw_1 / 2\pi)$, with κ such that $t^\kappa \geq t^2 b^2(tw_1 / 2\pi)$ for all $t \in \mathbb{R}$.*

Proof. We have $q'_{k+1} = \alpha_{k+1} q'_k \le 2b(q_k) q'_k \le 4b(q_k) q_k$. By Theorem 3.34, the complex dimension close to $D + it$ for $t = 2\pi q_k / w_1$, is located at

$$\omega = D + i\left(t + O\left(q'^{-1}_{k+1}\right)\right) - (w_1^2/\pi^2) B q'^{-2}_{k+1} + O\left(q'^{-4}_{k+1}\right),$$

where the big-O's denote real-valued functions. The real part of this complex dimension is less than $D - Bt^{-2} b^{-2}(tw_1 / 2\pi)$. □

Remark 3.45. By Khintchine's theory of continued fractions (see [HardW, Theorem 197, p. 168]), if $\sum_{k=1}^{\infty} 1/b_k$ is convergent, then almost all numbers have a continued fraction for which $a_k \le b_k$ eventually. Since q_k grows at least exponentially ($q_k \ge \phi^{k-1}$; see Section 3.5.1), condition (3.45) is satisfied already for a function that grows very slowly, such as, for example, $b(x) = (\log\log x)^2$. The set of numbers for which this condition is violated for infinitely many integers k and every function b of polynomial growth has measure zero.

Example 3.46. One of the simplest nonlattice strings is the *golden string,* introduced in Section 2.3.5. It is the nonlattice string with $M = 2$ and $\alpha = \phi$, $w_1 = \log 2$. The continued fraction of the golden ratio is

$$\phi = \frac{1+\sqrt{5}}{2} = [[1, 1, 1, \dots]].$$

Hence $q_0, q_1, \dots$ is the sequence of Fibonacci numbers

$$1,\ 1,\ 2,\ 3,\ 5,\ 8,\ 13,\ 34,\ 55, \dots, \tag{3.47}$$

and $q'_k = \phi^k$ for all $k \ge 0$.

Numerically, we find $D \approx .7792119034$ and the following approximation to the power series $\Delta(x)$:

$$-\,.47862\,x + .08812\,x^2 + .00450\,x^3 - .00205\,x^4 - .00039\,x^5 + \dots.$$

For every $k \ge 0$, we find a complex dimension close to $D + 2\pi i q_k / \log 2$. For example, $q_9 = 55$, and we find a complex dimension at $D - .00023 + 498.58i$. More generally, for numbers that can be written as a sum of Fibonacci numbers of relatively high index, like $q = 55 + 5$ or $q = 55 - 5$, we find a complex dimension close to $D + 2\pi i q / \log 2$. The distance to the line $\operatorname{Re} s = D$ is then determined by the smallest Fibonacci number in the sum (or difference), the number 5 in these two examples. For these two values for q, we find respectively a complex dimension at $D - .023561 + 543.63i$ and at $D - .033919 + 453.53i$. In both cases, the distance of these complex dimensions to the line $\operatorname{Re} s = D$ is comparable to the distance of the complex dimension close to $D + 2\pi i \cdot (5/\log 2)$ (for $q = 5$) to this line, which is located at $D - .028499 + 45.05i$. See Figure 3.9, where the markers indicate the Fibonacci numbers (3.47).

The pattern persists for other complex dimensions as well. Indeed, by Remark 3.33, every complex dimension repeats itself according to the pattern of the Fibonacci numbers. This is illustrated for one other complex dimension.

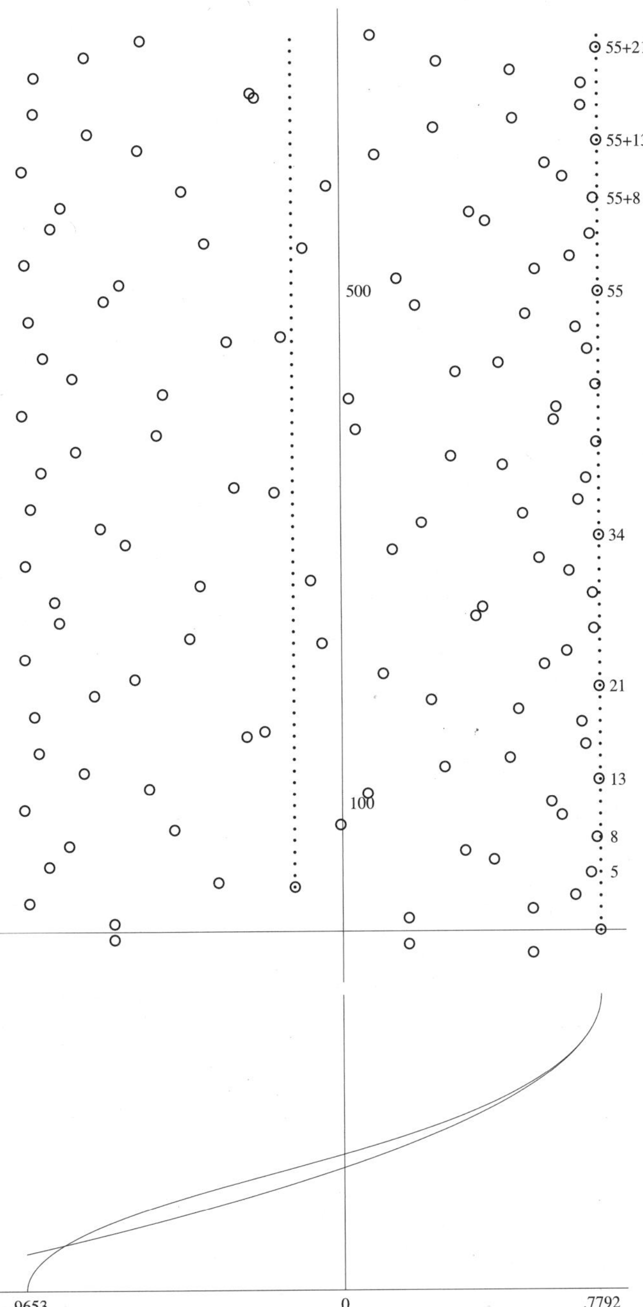

Figure 3.9: The complex dimensions of the golden string; the accumulative density of the real parts, compared with the theoretical prediction of Theorem 3.36. The dotted lines and the associated markers are explained in Example 3.46.

The second graph, at the bottom of Figure 3.9, shows the accumulative density of the real parts of the complex dimensions, compared with the predicted density (graphed for $D_l \leq x \leq D$) of Theorem 3.36.

Figure 3.9 was obtained by approximating ϕ by 4181/2584. Thus, the figure shows the complex dimensions of the lattice string with generator $w = (\log 2)/2584$, $k_1 = 2584$, $k_2 = 4181$. The oscillatory period of this string is $2\pi/w \approx 23423.23721$. Figure 3.6 gives intermediate stages of this approximation.

Example 3.47. Figure 3.10 gives the complex dimensions and the density of their real parts of the generic nonlattice string with weights $w_1 = \log 2$ and $w_2 = \alpha \log 2$, where α is the positive real number with continued fraction $[[1, 2, 3, 4, \ldots]]$. One can compute that

$$\alpha = \frac{\sum_{n=0}^{\infty} \frac{1}{(n!)^2}}{\sum_{n=0}^{\infty} \frac{1}{n!(n+1)!}} = \frac{I_0(2)}{I_1(2)}, \tag{3.48}$$

where $I_k(z) = \sum_{n=0}^{\infty} \frac{(z/2)^{2n+k}}{n!(n+k)!}$ is the modified Bessel function. Indeed, by [WhW, pp. 359, 373], we have the recursion relation

$$\frac{I_{n-1}(z)}{I_n(z)} = \frac{2n}{z} + \frac{1}{\frac{I_n(z)}{I_{n+1}(z)}},$$

for $n = 1, 2, \ldots$. Thus $I_0(z)/I_1(z) = [[2/z, 4/z, 6/z, \ldots]]$ and so α is given by Equation (3.48), as claimed.[7] (See also [BorCP, p. 499] and [Wats].) For this reason, we propose to call this string the *Bessel string*. Even though α is better approximable by rationals than numbers with bounded partial quotients, qualitatively there does not seem to be a significant difference with the golden string.

Figure 3.10 is obtained by approximating α by 1393/972. The markers illustrate again the different periodic patterns. The repetitions occur at denominators of convergents of α, which are the numbers $1, 2, 7, 30, 157, \ldots$, and combinations of these (in the sense of the α-adic expansion of Equation (3.27)).

Example 3.48. The Two-Three-Five string is an example of a generic nonlattice string with $M > 2$. It has one gap and $M = N = 3$ scaling ratios $r_1 = 1/2$, $r_2 = 1/3$, and $r_3 = 1/5$. See Figure 3.11 for a diagram of the complex dimensions and of the density of their real parts. Figure 3.11 is obtained using the approximation $\log_2 3 \approx 2826/1783, \log_2 5 \approx 4140/1783$. Note that now the markers are not related to the continued fractions of these numbers. Instead, they are the denominators of the rational numbers that *simultaneously* approximate these numbers. The same is true for the markers in Figures 3.12 and 3.13.

---

[7]The second author learned this argument from Ronald Kortram (personal communication, 1985).

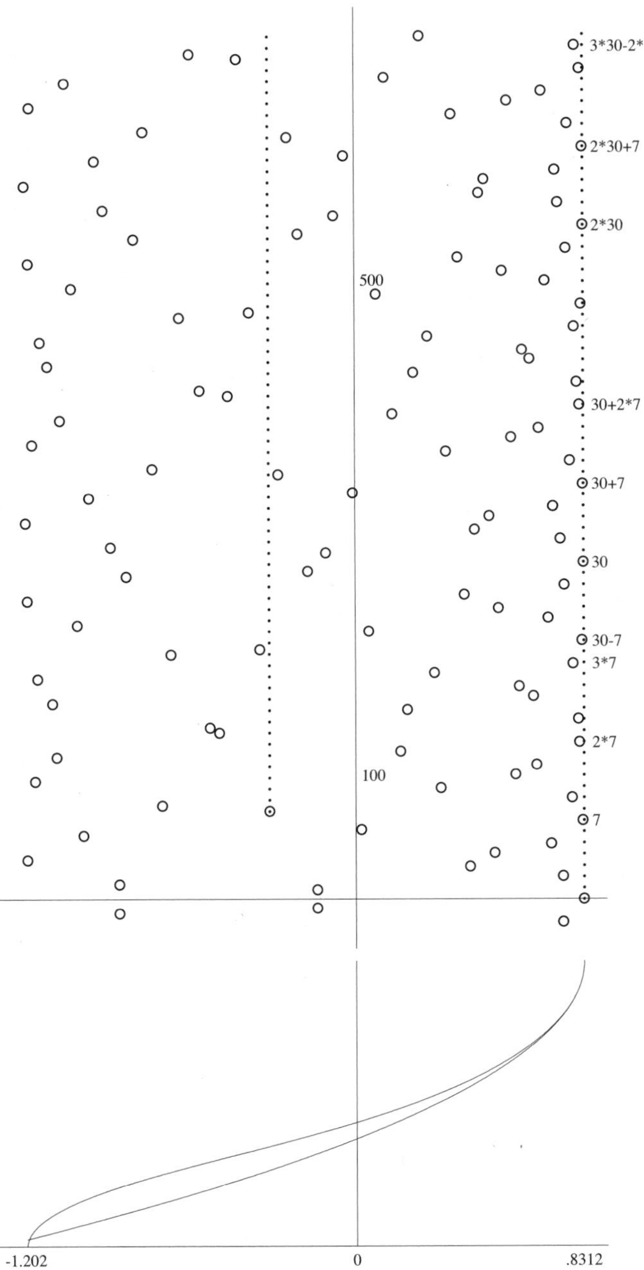

Figure 3.10: The complex dimensions of the Bessel string; the accumulative density of the real parts, compared with the theoretical prediction of Theorem 3.36. The dotted lines and the associated markers are explained in Example 3.47.

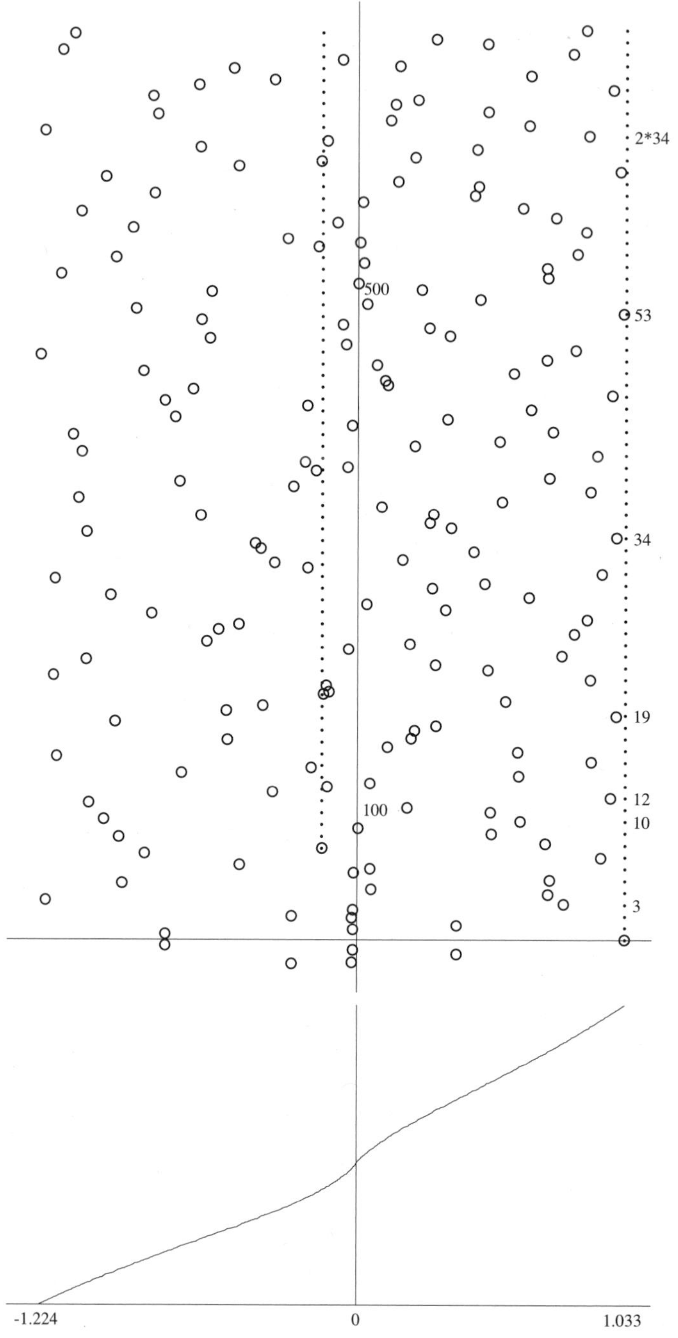

Figure 3.11: The quasiperiodic behavior of the complex dimensions of the Two-Three-Five string of Example 3.48; the accumulative density of the real parts of the complex dimensions. The dotted lines and the associated markers are explained in Examples 3.46 and 3.48.

In the following corollary, given $M \geq 2$, we say that $w_1, \ldots, w_M$ is *b-approximable* if $b\colon [1, \infty) \to \mathbb{R}^+$ is an increasing function such that for every $q \geq 1$, $j \in \{1, \ldots, M\}$ and integers p_j,

$$|qw_j - p_j w_1| \geq \frac{w_1}{b(q)} q^{-1/(M-1)}.$$

Since this means that $Q \ll q^{-1/(M-1)}$ in Lemma 3.16, we deduce the following result from Theorem 3.41:

Corollary 3.49. *Let $M \geq 2$. The best dimension-free region that $\mathcal{L}$ can have is of size*

$$\left\{\sigma + it \in \mathbb{C}\colon \sigma \geq D - O\left(t^{-2/(M-1)}\right)\right\}, \tag{3.49}$$

where the implied constant is positive and depends only on $w_1, \ldots, w_M$.

If $w_1, \ldots, w_M$ is b-approximable, then the dimension-free region has the form

$$\left\{\sigma + it \in \mathbb{C}\colon \sigma \geq D - O\left(b^{-2}(w_1 t/2\pi) t^{-2/(M-1)}\right)\right\}, \tag{3.50}$$

where the O-term is a positive function, bounded as indicated.

Remark 3.50. The best dimension-free region is obtained when the system $w_1, \ldots, w_M$ is badly approximable by rationals; i.e., when the weights are only b-approximable for a constant function $b = b(q)$. This corresponds to the case of two generators when α has bounded partial quotients in its continued fraction (Corollary 3.44). When $w_1, \ldots, w_M$ is better approximable by rationals, the dimension-free region is smaller and there are complex dimensions closer to the line $\operatorname{Re} s = D$. Such nonlattice strings behave more like lattice strings (compare Theorem 3.6).

Remark 3.51. Regarding the dependence on M, we see from comparing Corollaries 3.44 and 3.49 that in general, depending of course on the properties of simultaneous Diophantine approximation of the weights, nonlattice strings with a larger number of scaling ratios have a wider (better) dimension-free region.

Remark 3.52. By formulas (8.70) and (8.71), we deduce that for a nonlattice self-similar string $\mathcal{L}$, the volume $V(\varepsilon)$ of the tubular neighborhood of $\mathcal{L}$ is approximated by its leading term up to an error of order $|\log \varepsilon|^{-(M-1)/2}$, or worse if the weights are well approximable. See the end of Section 8.4.4 for further discussion of this point. More generally, the usefulness of obtaining a dimension-free region is that when combined with the explicit formulas of Chapter 5 (or Chapters 7 and 8), it enables us to obtain suitable error estimates in the corresponding asymptotic formulas. The better (i.e., the wider) the dimension-free regions, the better the error estimate. This principle will be illustrated in several places in Chapters 6–8.

3.7 The Dimensions of Fractality of a Nonlattice String

In this section, we obtain a rigorous result and make several conjectures regarding the real parts of the complex dimensions of nonlattice strings. It is noteworthy that both the theorem and the conjectures were themselves suggested by computer experimentation guided by the theoretical investigations in this chapter.

Recall from Remark 3.40 that the analogue of formula (3.41) holds for any complex dimension ω, besides D.

Theorem 3.53. *The set of real parts of the complex dimensions of a nonlattice string has no isolated points.*

Proof. By Remark 3.40, every complex dimension gives rise to a sequence of complex dimensions close to the points $\omega + 2\pi i q/w_1$, for integers q. Since the corrective terms are not all purely imaginary, we find complex dimensions with real parts close to $\operatorname{Re}\omega$. When q increases through a sequence of integers such that for each $j = 1, \ldots, M$, $x_j = 2\pi i(qw_j/w_1 - p_j) \to 0$ (see Lemma 3.16), we find a sequence of complex dimensions whose real parts approach $\operatorname{Re}\omega$. □

Remark 3.54. We call a string *fractal in dimension* α if it has a complex dimension with real part α. Each complex dimension gives rise to oscillations in the geometry of the fractal string. The frequencies of these oscillations are determined by the imaginary part of the complex dimension, and the real part of the complex dimension determines the amplitude of the oscillations. We define the set of *dimensions of fractality* of a fractal string as the closure of the set of real parts of its complex dimensions. Thus, Theorem 3.53 can be interpreted as saying that a nonlattice string is fractal in a perfect set of fractal dimensions. (See also Chapter 12, Section 12.2.)

3.7.1 The Density of the Real Parts

The density of the real parts of the complex dimensions of six different nonlattice strings are plotted in Figures 3.1–3.4 and 3.9–3.13. These figures show the graph of the function given by formula (3.36) for some large value of T. Since there are no horizontal pieces in these graphs, we conjecture that the real parts are dense in $[D_l, D]$. More generally, we expect the same to be true for the real parts of the complex dimensions of any nonlattice string.

We summarize this discussion by stating the following conjecture:

Conjecture 3.55. *The set of dimensions of fractality of a nonlattice string, as defined in Remark 3.54 above, is a bounded connected interval $[\sigma_l, D]$, where D is the Minkowski dimension of the string; in other words, the set* *of real parts of the complex dimensions is dense in* $[\sigma_l, D]$, *for some real number* σ_l. *In the generic nonlattice case,* $\sigma_l = D_l$ *is defined by Equation* (3.8b).

Sketch of a possible proof. By Remark 3.31 (for $M = 2$) and Remark 3.40 (for $M > 2$), there exist complex dimensions with real part arbitrarily close to any complex dimension ω_0. The main correction term, of order x, will in general not be purely imaginary (as it is for $\omega_0 = D$), and for suitable values of q, the real part of x will be negative. For $\omega_0 = D$, the second corrective term is negative. Thus starting at $\omega_0 = D$, we find a sequence of complex dimensions with decreasing real parts, which can be chosen to be spaced arbitrarily densely. This would prove the density of real parts in an interval. It remains to show that the real part of x can be chosen negative until we reach the left boundary of the critical strip (which is the vertical line $\operatorname{Re} s = D_l$ in the generic nonlattice case, see the end of Theorem 3.6). □

The graphs of the densities of the complex dimensions are qualitatively different for $M = 2$ and $M = 3$ (i.e., for either two or three different scaling factors). Indeed, compare Figures 3.4 and 3.11, where $M = 3$, with Figures 3.9 and 3.10, where $M = 2$. In the case of the Two-Three-Five string of Example 3.48 (Figure 3.11), the density for negative real parts is approximately .77 (i.e., below average density), while for positive real parts it is 1.08, slightly above average density. The density around vanishing real part becomes as large as 2.1. Thus the complex dimensions of the Two-Three-Five string show a phase transition between negative and positive real part. It seems that for larger M, this phenomenon persists. Therefore, we make the following conjecture:

Conjecture 3.56. *As* $M \to \infty$, *there exists a vertical line such that the density of the complex dimensions off this line vanishes in the limit.*

We close this chapter with two examples of nonlattice strings with four scaling ratios. The first is a diagram of the complex dimensions and the density of the real parts of the generic nonlattice string with $r_1 = 1/2$, $r_2 = 1/3$, $r_3 = 1/5$ and $r_4 = 1/7$. See Figure 3.12. As was suggested in Remark 3.51, the complex dimensions tend to be more concentrated in the middle, away from $\operatorname{Re} s = D$ and $\operatorname{Re} s = D_l$. We have used the approximation $\log_2 3 \approx 699/441$, $\log_2 5 \approx 1024/441$ and $\log_2 7 \approx 1238/441$.

Figure 3.13 gives a diagram of the complex dimensions of its nongeneric counterpart, the self-similar string with scaling ratios $r_1 = 1/2$, $r_2 = 1/3$, $r_3 = 1/4$, and $r_4 = 1/6$. Note that the group generated by these scaling ratios, $2^{\mathbb{Z}}3^{\mathbb{Z}}$, has rank 2.

Clearly, more mathematical experimentation—guided by our theoretical investigations—is needed to determine the generality of this phenomenon

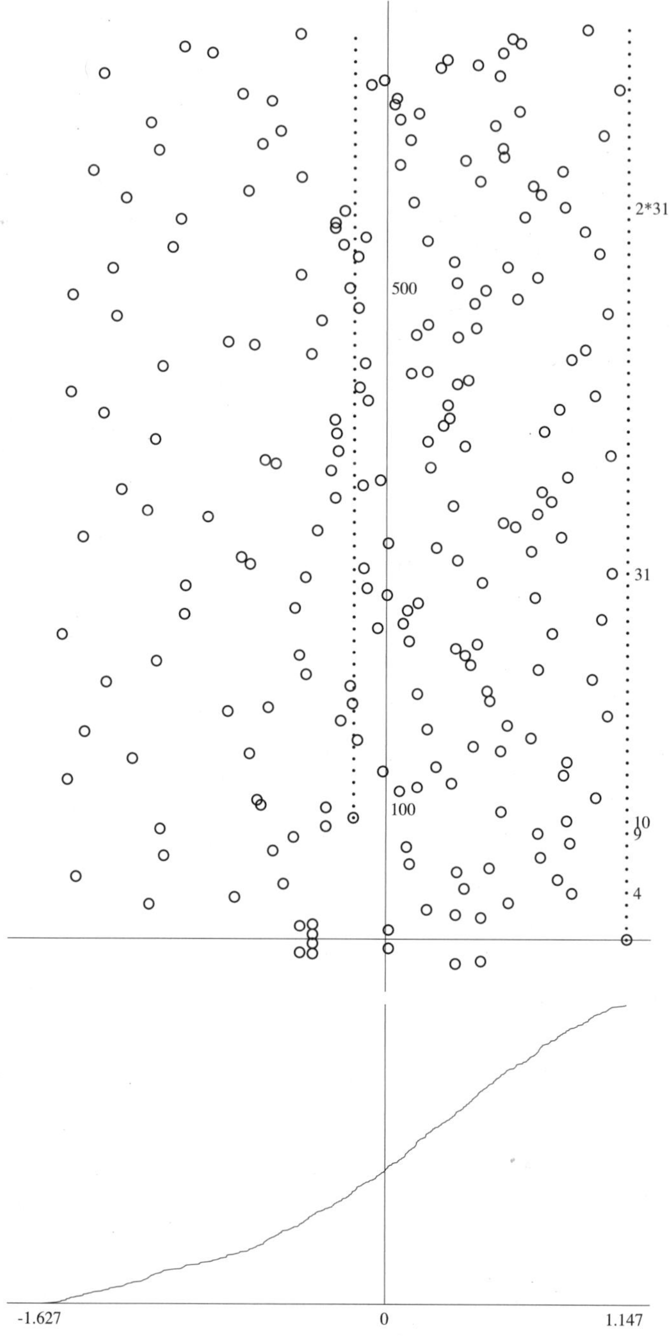

Figure 3.12: The complex dimensions of the nonlattice string with $M = 4$ and $r_1 = 1/2$, $r_2 = 1/3$, $r_3 = 1/5$, $r_4 = 1/7$; the accumulative density of the real parts. The dotted lines and the associated markers are explained in Examples 3.46 and 3.48.

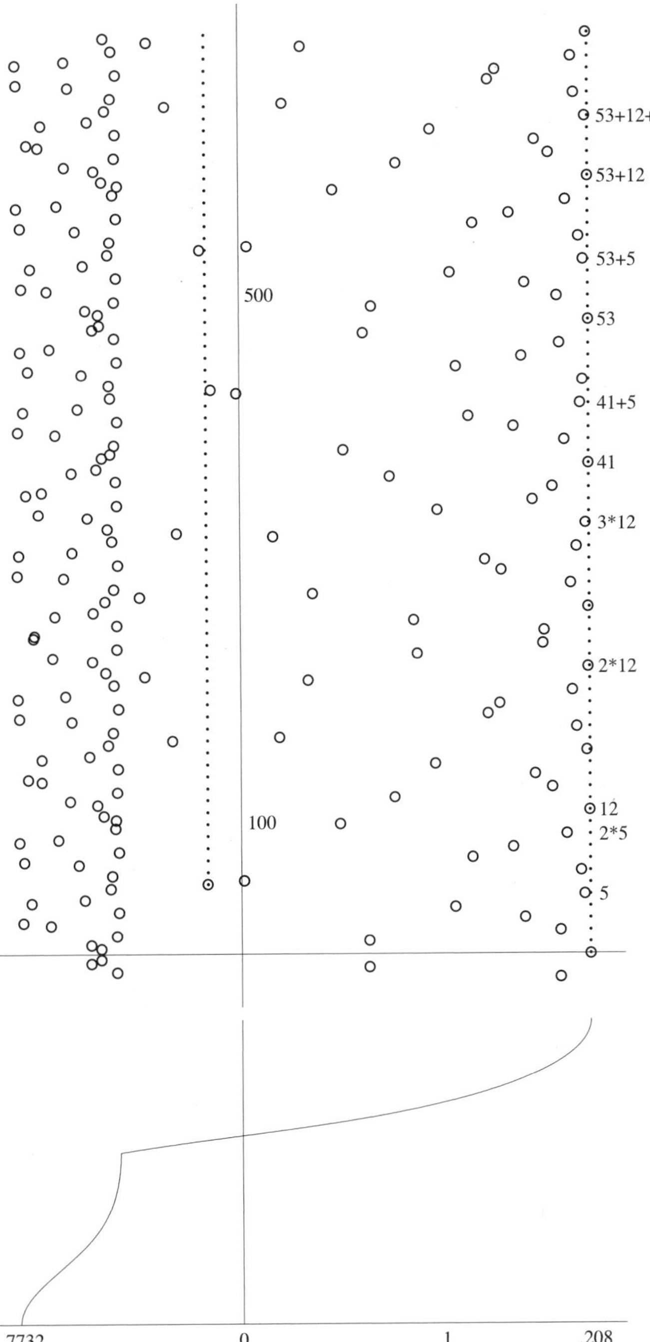

Figure 3.13: The complex dimensions of the nongeneric nonlattice string with $M = 4$ and $r_1 = 1/2$, $r_2 = 1/3$, $r_3 = 1/4$, $r_4 = 1/6$; the accumulative density of the real parts. The dotted lines and the associated markers are explained in Examples 3.46 and 3.48.

and to formulate suitable additional conjectures regarding the fine structure of the complex dimensions of nonlattice strings.

3.8 A Note on the Computations

The computations for the nonlattice examples were done using Maple.[8] In each case, we approximated the nonlattice equation by a lattice one, resulting in a polynomial equation of degree d between 400 and 5000. Solving the corresponding polynomial equation yields d complex numbers z in an annulus, and the roots ω are given by $\omega = \log z/\log r + 2k\pi i/\log r$, for $k \in \mathbb{Z}$.

Remark 3.57. Note that Maple can also directly solve the nonlattice equation numerically. However, Maple will just return one root, usually D or a root in a specified rectangular region. To obtain all complex roots in a given range, one would have to subdivide this range into many small rectangles. This would have to be supplemented with an evaluation of f'/f around each rectangle to assure that all roots in that rectangle have been found. On the other hand, when solving a polynomial equation numerically, Maple returns a list of all roots, with their multiplicity. Also, on a more theoretical level, there are advantages to solving a lattice equation corresponding to a lattice approximation, as is done in this chapter. Indeed, this technique allowed us to discover the quasiperiodic pattern of the roots.

In order to obtain Figure 2.10, we have approximated $f(s) = 1-2^{-s}-3^{-s}$ by

$$p(z) = 1 - z^{306} - z^{485},$$

with $z = 2^{-s/306}$ and $r = 2^{-1/306}$. The corresponding lattice equation with ratios r^{306} and r^{485} has a period of $\mathbf{p} = 2\pi \cdot 306/\log 2 \approx 2773.8$. For example, the real root

$$D \approx .7878849110$$

of f is approximated by

$$\widetilde{D} \approx D - .1287 \cdot 10^{-5}.$$

We have $|f(\widetilde{D})| \approx .11 \cdot 10^{-5}$ and $|f(\widetilde{D} + i\mathbf{p})| \approx .39 \cdot 10^{-2}$. As another example, the root

$$\tilde{\omega} = .7675115443 + 45.55415979i$$

approximates a root ω of f, with an error

$$\omega - \tilde{\omega} \approx (-.12 + .75i) \cdot 10^{-4}.$$

Both $|p(r^{\omega})|$ and $|f(\tilde{\omega})|$ are approximately equal to $.64 \cdot 10^{-4}$, and $|f(\tilde{\omega} + i\mathbf{p})|$ approximately equals $.40 \cdot 10^{-2}$. Lemmas 3.29 and 3.39, along with Theorems 3.32, 3.34 and 3.41 give theoretical information about the error of approximation.

---

[8]The programs, and some documentation, are available on the webpage of the second author at `http://www.research.uvsc.edu/~machiel/programs.html`.

Since in the applications we consider only equations with real values for the multiplicities m_j, the roots come in complex conjugate pairs. Maple normalizes $\log z$ so that the imaginary part lies between $-\pi i$ and πi. For the density graphs, we have taken the real parts of the roots $\log z / \log r$ for those z with $-\pi i \leq \operatorname{Im} z \leq 0$, and ordered these values. This way, we have obtained a sequence of $(d+1)/2$ or $d/2+1$ real parts $v_1 \leq v_2 \leq \ldots$ in nondecreasing order. The density graph is a plot of the points (v_j, j) for $1 \leq j \leq d/2+1$.

Interestingly, if one takes the roots not in one full period, but up to some bound for the imaginary part, the density graph is not smooth but seems to exhibit a fractal pattern. For $M = 2$, this pattern can be predicted from the α-adic expansion of T, see Section 3.5.1. This reflects the fact that roots come in quasiperiodic arrays, each one slightly shifted from the previous one, reaching completion only at a period.

The maximal degree 5000 is the limit of computation: It took several hours with our software on a Sun workstation to compute the golden diagram, Figure 3.9, which involved solving a polynomial equation of degree 4181. However, finding the roots of the polynomial is the most time-consuming part of the computation. Since these polynomials contain only a few monomials, there may exist ways to speed up this part of the computation.

4

Generalized Fractal Strings Viewed as Measures

In this chapter, we develop the notion of generalized fractal string, viewed as a measure on the half-line. This is more general than the notion of fractal string considered in Chapter 1 and in the earlier work on this subject (see the notes to Chapter 1). We will use this notion in Chapter 5 to formulate the explicit formulas which will be applied throughout the remaining chapters. Besides ordinary fractal strings, generalized fractal strings enable us to deal with strings whose lengths vary continuously or whose multiplicities are nonintegral or even infinitesimal. In Section 4.2, we discuss the spectrum of a generalized fractal string, and in Section 4.3, we briefly discuss the notion of generalized fractal spray, which will be used in Chapters 9 and 11.

The conceptual difficulties associated with the notion of frequency with noninteger multiplicity led us to introduce the formalism of generalized fractal strings presented in this chapter. The flexibility of the language of measures allows us to deal in a natural way with nonintegral multiplicities, in the case of discrete measures, as in Example 4.8 and Chapter 10, and even to formalize the intuitive notion of infinitesimal multiplicity in the case of continuous measures, as in Sections 9.2 and 10.3.

In Section 4.4, we study the properties of the measure associated with a self-similar string (defined as in Chapter 2). Although Section 4.4 is of interest in its own right, it may be omitted on a first reading since it will not be used in the rest of this book.

4.1 Generalized Fractal Strings

For a measure η, we denote by $|\eta|$ the total variation measure associated with η (see, e.g., [Coh, p. 126] or [Ru2, p. 116]),

$$|\eta|(A) = \sup \left\{ \sum_{k=1}^{m} |\eta(A_k)| \right\}, \tag{4.1}$$

where $m \geq 1$ and $\{A_k\}_{k=1}^{m}$ ranges over all finite partitions of A into disjoint measurable subsets of $(0, \infty)$. Recall that $|\eta|$ is a positive measure and that $|\eta| = \eta$ if η is positive. (See Remark 4.3 below and [Coh, Chapter 4] or [Ru2, Chapter 6].)

Definition 4.1. (i) A *generalized fractal string* is either a local complex or a local positive measure η on $(0, \infty)$, such that

$$|\eta|(0, x_0) = 0$$

for some positive number x_0. (See Remark 4.3 below.)

(ii) The *counting function of the reciprocal lengths*, or *geometric counting function* of η, is defined as $N_\eta(x) = \int_0^x d\eta$. If the measure η has atoms, it is necessary to specify how the endpoint is counted. Throughout the book, we adopt the convention that x is counted half; i.e.,

$$N_\eta(x) = \int_0^x d\eta := \eta(0, x) + \frac{1}{2}\eta(\{x\}). \tag{4.2}$$

(iii) The *dimension* of η, denoted $D = D_\eta$, is the abscissa of convergence of the Dirichlet integral $\zeta_{|\eta|}(\sigma) = \int_0^\infty x^{-\sigma}\, |\eta|(dx)$. In other words, it is the infimum of the real numbers σ such that the improper Riemann–Lebesgue integral $\int_0^\infty x^{-\sigma}\, |\eta|(dx)$ converges and is finite:

$$D = D_\eta = \inf\left\{\sigma \in \mathbb{R}\colon\ \int_0^\infty x^{-\sigma}\, |\eta|(dx) < \infty\right\}. \tag{4.3}$$

(iv) The *geometric zeta function* is defined as the Mellin transform of η,

$$\zeta_\eta(s) = \int_0^\infty x^{-s}\, \eta(dx), \tag{4.4}$$

for $\operatorname{Re} s > D_\eta$.

By convention, $D_\eta = \infty$ means that $x^{-\sigma}$ is not $|\eta|$-integrable for any σ, and $D_\eta = -\infty$ means that $x^{-\sigma}$ is $|\eta|$-integrable for all σ in $\mathbb{R}$. In this last case, ζ_η is a holomorphic function, defined by its Dirichlet integral (4.4) on the whole complex plane. (See, for example, [Pos, Sections 2–4] or [Wid].)

Note that if η is a continuous measure (i.e., $\eta(\{x\}) = 0$ for all $x > 0$), then $N_\eta(x) = \eta(0, x) = \eta(0, x]$. On the other hand, if η is discrete, say

$$\eta = \sum_l w_l \delta_{\{l^{-1}\}}$$

(see Section 1.1.1 for a discussion of the multiplicities w_l), then

$$N_\eta(x) = \sum_{l^{-1}<x} w_l + \frac{1}{2} w_{1/x},$$

with the convention that $w_{1/x}$ vanishes if x is not one of the reciprocal lengths l^{-1}.

Remark 4.2. In general, for example in a fractal spray, the lengths play the role of scales.

Remark 4.3. There is a simple technical point that needs to be clarified here. Indeed, since we have to use integrals of the type $\int_0^\infty f(x)\,\eta(dx)$, for suitable functions f (see, for example, Equations (4.2) and (4.4) above), some caution is necessary in dealing with measures on $(0, \infty)$.

A *local positive measure* (or a locally bounded positive measure) is just a standard positive Borel measure on $(0, \infty)$ which satisfies the following local boundedness condition:

$$\eta(J) < \infty, \text{ for all bounded subintervals } J \text{ of } (0, \infty). \tag{4.5}$$

In this case, we have $\eta = |\eta|$ on every bounded Borel subset of $(0, \infty)$.

More generally, we will say, much as in [DolFr, Definition 6.1, p. 179], that a (complex-valued or else $[0, +\infty]$-valued) set function η on the half-line $(0, \infty)$ is a *local complex measure* on $(0, \infty)$ if, for every compact subinterval $[a, b]$ of $(0, \infty)$, the following conditions are satisfied: (i) $\eta(A)$ is well defined for any Borel subset A of $[a, b]$, and (ii) the restriction of η to the Borel subsets of $[a, b]$ is a complex measure on $[a, b]$ in the traditional sense. (See, e.g., [Coh, Chapter 4] or [Ru2, Chapter 6]. It follows, by [Ru2, Theorems 6.2 and 6.4, pp. 117 and 118], that η is of bounded total variation on $[a, b]$.) According to the aforementioned results, the measure $|\eta|$ is finite and positive on each bounded subinterval of $(0, \infty)$. Note that a local positive measure is simply a local complex measure which takes its values in $[0, +\infty]$ (rather than in $\mathbb{C}$).

In addition to [DolFr, §5.5 and §1.6], we refer to the later work [JohLap, Chapters 15–19, esp. Sections 17.6, 15.2.F and 19.1] and the relevant references therein, where similar issues had to be dealt with.

A reader unconcerned with such technicalities or else unfamiliar with the notion of complex measure may assume throughout that a generalized fractal string is a positive Borel measure on $(0, \infty)$ satisfying condition (4.5) above (and which does not carry mass near 0). In Sections 9.3 and 11.4, however, we will use generalized fractal strings associated with local complex measures.

For simplicity, we will drop from now on the adjective "local" when referring to the measure associated with a general fractal string.

As in Chapter 1, we are interested in the meromorphic continuation of the function ζ_η. We define the screen S and the window W as in Section 1.2.1. Let $S\colon \mathbb{R} \to [-\infty, D]$ be a bounded continuous function. The *screen,* also denoted by S, is the curve

$$S\colon S(t) + it \quad (t \in \mathbb{R}) \tag{4.6}$$

(see Figure 1.6 of Section 1.2.1), and the *window* is the closed part of the plane to the right of the screen S,

$$W = \{s \in \mathbb{C}\colon \operatorname{Re} s \geq S(\operatorname{Im} s)\}. \tag{4.7}$$

We assume that ζ_η has a meromorphic continuation to an open neighborhood of W. We also require that ζ_η does not have any pole on S. The poles of ζ_η inside W will be called the visible complex dimensions and the associated set is denoted by $\mathcal{D}_\eta(W)$; namely, *the set of visible complex dimensions* of η is

$$\mathcal{D}_\eta = \mathcal{D}_\eta(W) = \{\omega \in W\colon \zeta_\eta \text{ has a pole at } \omega\}. \tag{4.8}$$

See Section 5.1.1 for an explanation of the role of the screen in the explicit formulas.

As in the case of ordinary fractal strings, $\mathcal{D}_\eta(W)$ is a discrete subset of $\mathbb{C}$. Hence, its intersection with any compact subset of $\mathbb{C}$ is finite. Moreover, since, by definition, ζ_η is holomorphic for $\operatorname{Re} s > D$, where D is the dimension of η, it follows that $\mathcal{D}_\eta$ is contained in the closed half-plane $\operatorname{Re} s \leq D$. Also, if η is a positive measure, then the exact counterpart of Remark 1.14 holds. That is, D is a singularity of ζ_η.

In applying our explicit formulas, obtained in the next chapter, it will sometimes be useful to change the location of the screen S, even for a fixed choice of η. See Section 5.5 for examples. However, when no ambiguity arises regarding the choice of the string η and the location of the screen (and hence of the associated window W), we will simply write $\mathcal{D}_\eta$ instead of $\mathcal{D}_\eta(W)$.

If $W = \mathbb{C}$, then $\mathcal{D}_\eta(\mathbb{C})$ is called *the set of complex dimensions* of η. In this case, we set formally $S(t) \equiv -\infty$ and do not define the screen.

Remark 4.4. Assume that W is symmetric with respect to the real axis (i.e., the function $S(t)$ defining the screen is even, so that W is equal to $\overline{W}$, the complex conjugate of W). If η is a real-valued measure, then $\zeta_\eta(s)$ is real-valued for real s and hence the set of complex dimensions $\mathcal{D}_\eta(W)$ is symmetric with respect to the real axis; that is, $\omega \in \mathcal{D}_\eta(W)$ if and only if $\overline{\omega} \in \mathcal{D}_\eta(W)$. This was the case, of course, for all standard fractal strings considered in Chapters 1, 2 and 3, since the measure associated with the fractal string $\{l_j\}_{j=1}^\infty$ is $\eta = \sum_{j=1}^\infty \delta_{\{l_j^{-1}\}}$. Consequently, the explicit formulas of the next chapter are real-valued when applied to a fractal string. We call this the *reality principle*; see also Remark 1.6.

4.1.1 Examples of Generalized Fractal Strings

Our definition of a generalized fractal string includes our previous definition of Section 1.1. With an ordinary fractal string $\mathcal{L}$, composed of the sequence of lengths $\{l_j\}_{j=1}^\infty$ (counted with multiplicity), we associate the positive measure

$$\mu_{\mathcal{L}} = \sum_{j=1}^{\infty} \delta_{\{l_j^{-1}\}}. \tag{4.9}$$

Here and in the following, we use the notation $\delta_{\{x\}}$ for the point mass (or Dirac measure) at x; i.e., for a set $A \subset (0, \infty)$, $\delta_{\{x\}}(A) = 1$ if $x \in A$ and $\delta_{\{x\}}(A) = 0$ otherwise.

In the geometric situation, as considered in the first three chapters, η is a discrete measure with integer multiplicities because each length is repeated an integer number of times. In the applications, however, we will often need to consider discrete strings

$$\eta = \sum_{l} w_l \delta_{\{l^{-1}\}}$$

with noninteger multiplicities w_l. In fact, this was one of our initial motivations for introducing the notion of generalized fractal strings and for viewing them as measures. Note also that the Dirichlet polynomials (see Chapter 3) fit into this framework.

An example of such a discrete nongeometric string is provided by the *generalized Cantor string*. For $1 < b < a$, we define a string consisting of lengths a^{-n} with multiplicity b^n. The associated measure is

$$\sum_{n=0}^{\infty} b^n \delta_{\{a^n\}}. \tag{4.10}$$

If b is integral, this is an ordinary fractal string. For arbitrary b, we study the generalized Cantor string in Section 8.4.1 and Chapter 10, and use it in Chapter 11.

In Chapter 10, we also study the so-called *truncated Cantor strings,* defined for a positive integer Λ by the measure

$$\frac{1}{\Lambda \log a}\, x^{D-1} \left(\frac{\sin \pi \Lambda t}{\sin \pi t}\right)^2 dx, \qquad \text{where } x = a^t.$$

Thus each length has a nonnegative but infinitesimal multiplicity, and close to the length a^{-n}, the multiplicities add up to almost b^n, for $n = 0, 1, 2, \ldots$. See Section 10.3.

We note two additional examples. First, the *harmonic string* (introduced in [Lap2, Example 5.4(ii), pp. 171–172] and further discussed in [Lap3, pp. 144–145]) is given by the positive measure[1]

$$h = \sum_{j=1}^{\infty} \delta_{\{j\}}. \tag{4.11}$$

This string does not have finite total length. In fact, its lengths are $1, 1/2, 1/3, \ldots, 1/n, \ldots$, each counted with multiplicity one, and hence the total length of h is $\sum_{n=1}^{\infty} 1/n = \infty$. Its dimension is 1. Since, by definition, $\zeta_h(s) = \sum_{j=1}^{\infty} j^{-s}$ (for $\operatorname{Re} s > 1$), the associated geometric zeta function is equal to the Riemann zeta function (as was noted in [Lap2, p. 171]),

$$\zeta_h(s) = \zeta(s). \tag{4.12}$$

Secondly, the *prime string* is defined by the positive measure

$$\mathfrak{P} = \sum_{m \geq 1,\, p} (\log p)\, \delta_{\{p^m\}}, \tag{4.13}$$

where p runs over all prime numbers. Note that $\mathfrak{P}$ is not an ordinary fractal string because the reciprocal lengths p^m (that is, the prime powers) have noninteger multiplicity $\log p$. Next, we use (and reinterpret) a well-known identity (see, e.g., [In, Eq. (14), p. 17] or [Pat, p. 9]). By logarithmic differentiation of the Euler product representation of $\zeta(s)$,

$$\zeta(s) = \prod_p \frac{1}{1 - p^{-s}}, \tag{4.14}$$

valid for $\operatorname{Re} s > 1$, we obtain

$$-\frac{\zeta'(s)}{\zeta(s)} = \sum_{m \geq 1,\, p} (\log p)\, p^{-ms}.$$

Therefore we see that the geometric zeta function of $\mathfrak{P}$ is given by

$$\zeta_{\mathfrak{P}}(s) = -\frac{\zeta'(s)}{\zeta(s)} \qquad (s \in \mathbb{C}). \tag{4.15}$$

Thus this string is one-dimensional and its complex dimensions are the zeros of the Riemann zeta function, counted without multiplicity, and the simple pole at $s = 1$.

[1] The harmonic string could also be called the *Riemann string*, as was suggested to the first author by Victor Kac.

Recall that the zeros of ζ consist of the critical (or nontrivial) zeros, located in the critical strip $0 \leq \operatorname{Re} s \leq 1$, and the trivial zeros, which are simple and located at the even negative integers, $-2, -4, -6, \ldots$. It is well known that ζ does not have any zero on the vertical line $\operatorname{Re} s = 1$ (see, e.g., [In, Theorem 10, p. 28]), and hence it has no zeros on $\operatorname{Re} s = 0$ either. The Riemann hypothesis states that the critical zeros of ζ are all located on the critical line $\operatorname{Re} s = 1/2$. See, for example, Chapter 9, Section 11.1, and [Ti], [Edw] or [Da].

4.2 The Frequencies of a Generalized Fractal String

In general, there is no clear interpretation for the frequencies of a generalized fractal string. (See, however, Section 11.1.1.) Therefore, we simply adopt the following definition, motivated by the case of an ordinary fractal string (see Section 1.3 and Remark 4.6 below).

Definition 4.5. For a generalized fractal string ℓ, the *spectral measure*[2] ν of ℓ is defined by

$$\nu(A) = \ell(A) + \ell\left(\frac{A}{2}\right) + \ell\left(\frac{A}{3}\right) + \ldots, \tag{4.16}$$

for each bounded (Borel) set $A \subset (0, \infty)$.

The *spectral zeta function* of ℓ is defined as the geometric zeta function of the measure ν.

Note that the sum defining $\nu(A)$ is finite because A is bounded and $|\ell|$ is assumed to have no mass near 0. Indeed, choose k large enough so that $k^{-1}A \subset (0, x_0)$, where $|\ell|(0, x_0) = 0$. Then $\ell(k^{-1}A) = 0$.

For notational simplicity (see the comment following Definition 1.18), we do not explicitly indicate the dependence of ν on ℓ. In particular, ζ_ν and N_ν, the associated spectral counting function, depend on ℓ.

We will give an alternative expression for ν in Equation (4.20) below, where we should set $\eta = \ell$. By abuse of language, we will also consider ν as a generalized fractal string.[3] This way, we can conveniently formulate the explicit formulas for the geometric ($\eta = \ell$) as well as the spectral ($\eta = \nu$) situation.

---

[2] *Caution*: This should not be mistaken with the notion of spectral measure encountered in the spectral theory of self-adjoint operators (see, e.g., [ReSi1]), which will not be used in this book.

[3] Indeed, one checks that $|\nu|(0, x_0) = 0$ if $|\ell|(0, x_0) = 0$. Thus $|\nu|$ does not have mass near 0 and ν is a local measure on $(0, \infty)$.

Remark 4.6. When $\ell = \sum_{j=1}^{\infty} \delta_{\{l_j^{-1}\}}$ is the measure associated with an ordinary fractal string $\mathcal{L} = l_1, l_2, \ldots$ as in Section 1.3, then by Equation (1.38) of Theorem 1.19, ζ_ν coincides with the spectral zeta function of $\mathcal{L}$, as defined by Equation (1.35). Also, in view of (4.2), (4.16) and footnote 1 of Chapter 1, N_ν coincides with the spectral counting function of $\mathcal{L}$, as defined by Equation (1.34).

Much as in the proof of Theorem 1.19, one shows that $\zeta_\nu(s)$, the spectral zeta function, is obtained by multiplying $\zeta_{\mathcal{L}}(s)$ by the Riemann zeta function. In other words,

$$\zeta_\nu(s) = \int_0^\infty x^{-s}\, \nu(dx) = \zeta_\ell(s) \cdot \zeta(s), \tag{4.17}$$

where $\zeta(s) = \sum_{n=1}^{\infty} n^{-s}$ is the Riemann zeta function.

Definition 4.7. The *convolution*[4] of two strings η and η' is the measure $\eta * \eta'$ defined by

$$\int f\, d(\eta * \eta') = \iint f(xy)\, \eta(dx)\eta'(dy). \tag{4.18}$$

One easily checks that

$$\zeta_{\eta * \eta'}(s) = \zeta_\eta(s)\zeta_{\eta'}(s). \tag{4.19}$$

Thus the spectral zeta function of a string η is simply the zeta function associated with the string $\eta * h$. Indeed, in view of (4.16) and (4.18), one can easily check that the spectral measure of η is given by

$$\nu = \eta * h, \tag{4.20}$$

where h is the harmonic string defined by formula (4.11) above.

In particular, the spectral measure ν of an ordinary fractal string $\mathcal{L} = \{l_j\}_{j=1}^{\infty}$, viewed as the measure $\eta = \sum_{j=1}^{\infty} \delta_{\{l_j^{-1}\}}$, is given by

$$\begin{aligned} \nu &= \sum_f w_f^{(\nu)} \delta_{\{f\}} \\ &= \eta * h \\ &= \sum_{n,j=1}^{\infty} \delta_{\{n \cdot l_j^{-1}\}}, \end{aligned} \tag{4.21}$$

in agreement with (1.32). Here, f runs over the (distinct) frequencies of $\mathcal{L}$ and $w_f^{(\nu)}$ denotes the multiplicity of f, as in (1.33). Note that in view of (4.19), we recover Equation (1.35) (respectively, (1.38)) from the first (respectively, second) equality of (4.21).

---

[4] This is a multiplicative (rather than an additive) convolution of measures on $\mathbb{R}_+^*$. It is called the tensor product in [JorLan2].

Example 4.8 (The frequencies of the prime string). We compute in two different ways the frequencies of the prime string $\mathfrak{P}$, defined by (4.13). We will first do this by evaluating the spectral zeta function of $\mathfrak{P}$, the function $\zeta_\nu(s) = \zeta_{\nu,\mathfrak{P}}(s)$.

According to Equations (4.15) and (4.17), we have

$$\zeta_\nu(s) = \zeta_{\mathfrak{P}}(s) \cdot \zeta(s) = -\frac{\zeta'(s)}{\zeta(s)} \cdot \zeta(s) = -\zeta'(s),$$

where $\zeta'(s)$ is the derivative of the Riemann zeta function.

Next, since $\zeta(s) = \sum_{n=1}^{\infty} n^{-s}$ for $\operatorname{Re} s > 1$, we obtain (for $\operatorname{Re} s > 1$),

$$\zeta_\nu(s) = -\zeta'(s) = \sum_{n=1}^{\infty} (\log n)\, n^{-s}. \tag{4.22}$$

In view of Equation (4.17), we deduce from (4.22) that

$$\nu = \sum_{n=1}^{\infty} (\log n)\, \delta_{\{n\}}. \tag{4.23}$$

Hence, the prime string $\mathfrak{P}$ has for frequencies all the positive integers $1, 2, \dots, n, \dots$, and the frequency n has a noninteger multiplicity $\log n$. In contrast, both the lengths and the frequencies of an ordinary fractal string have integer multiplicities. By (4.13) and (4.23), this is not the case for $\mathfrak{P}$.

It is also instructive to recover this result by determining directly the spectral measure ν associated with $\mathfrak{P}$. Namely, in view of formula (4.18), we have

$$\delta_{\{x\}} * \delta_{\{y\}} = \delta_{\{xy\}}. \tag{4.24}$$

Then, in view of formulas (4.11), (4.13) and (4.20), we have, successively

$$\begin{aligned}\nu &= \sum_{m\geq 1,\, p} (\log p)\, \delta_{\{p^m\}} * \sum_{k\geq 1} \delta_{\{k\}} \\ &= \sum_{m,k\geq 1,\, p} (\log p)\, \delta_{\{p^m \cdot k\}} \\ &= \sum_{n\geq 1} \delta_{\{n\}} \sum_{m\geq 1,\, p\colon\, p^m \mid n} \log p.\end{aligned}$$

as was found in (4.23). Here, as before, p runs over all prime numbers. Now, for a fixed prime number p, we have that $\sum_{m\geq 1,\, p^m \mid n} \log p = \log p^m$, so that $\sum_{m\geq 1\, p\colon\, p^m \mid n} \log p = \log n$ by unique factorization of integers into prime powers. Thus we conclude that

$$\nu = \sum_{n\geq 1} (\log n)\, \delta_{\{n\}}.$$

4.2.1 Completion of the Harmonic String: Euler Product

It is noteworthy that the harmonic string itself is the infinite convolution over all prime numbers of what one could call the elementary prime strings. Define

$$h_p = \sum_{j=0}^{\infty} \delta_{\{p^j\}}, \tag{4.25}$$

for every prime number p. Then

$$h = \underset{p}{*}\, h_p, \tag{4.26}$$

where p ranges over all prime numbers. (Here, as in (4.18), $*$ denotes the multiplicative convolution of measures.) This corresponds to the Euler product of the Riemann zeta function, recalled in formula (4.14) above. Indeed,

$$\zeta_{h_p}(s) = \frac{1}{1-p^{-s}}, \tag{4.27}$$

the p-th Euler factor of $\zeta(s)$, and for $\operatorname{Re} s > 1$,

$$\zeta_{*_p h_p}(s) = \prod_p \frac{1}{1-p^{-s}} = \zeta(s). \tag{4.28}$$

To obtain $\zeta_{\mathbb{R}}(s) = \Gamma(s/2)\pi^{-s/2}$, the Euler factor at infinity of the completed Riemann zeta function, where Γ denotes the classical gamma function (see Appendix A, Section A.3), one has to convolve one more time with the continuous measure

$$h_\infty(dx) = 2e^{-\pi x^{-2}} \frac{dx}{x}, \tag{4.29}$$

which is not a string in our sense, because it has positive (albeit extremely small) mass near 0.

In summary, we have the following adelic decomposition (in the sense of Tate's

thesis [Ta]) of the *completed harmonic string* $h_c := h_\infty * h$:

$$h_c = h_\infty * \left(\underset{p}{*}\, h_p\right), \tag{4.30}$$

where h_∞ is the continuous measure defined by (4.29) and where the infinite convolution product runs over all prime numbers p. One computes that this is the measure

$$h_c(dx) = \left(\theta(x^{-1}) - 1\right) \frac{dx}{x},$$

where the *theta function*[5] is defined by

$$\theta(x) = \sum_{n=-\infty}^{\infty} e^{-\pi n^2 x^2}.$$

It is well known that $\theta(1/x) = x\theta(x)$ (see, e.g., [Ta]). Hence

$$\theta(x) = 1 + 2e^{-\pi x^2} + O\left(e^{-4\pi x^2}\right) \text{ as } x \to \infty$$

and

$$\theta(x) = \frac{1}{x}\left(1 + O\left(e^{-\pi/x^2}\right)\right) \text{ as } x \to 0.$$

Thus the measure h_c has density about $2e^{-\pi/x^2}$ near 0, and density 1 near infinity, as does h. Note, however, that h_c, like h_∞, is not a measure that fits in our framework since it does not vanish in a neighborhood of $x = 0$, even though it is very small near 0.

Correspondingly, the completed Riemann zeta function (as in [Ta], but defined slightly differently from either [Da, Chapter 12] or [Edw, §1.8]) is given by

$$\xi(s) = \zeta_{h_c}(s) = \pi^{-s/2}\Gamma(s/2)\zeta(s). \tag{4.31}$$

The function ξ is meromorphic in all of $\mathbb{C}$, with simple poles at $s = 0$ and at $s = 1$. Since the measure h_c has nonzero mass near 0, this function grows faster than exponentially as $s \to \infty$.

We leave it to the interested reader to investigate how the functional equation for ξ, namely,

$$\xi(s) = \xi(1-s) \quad (s \in \mathbb{C}), \tag{4.32}$$

translates in terms of the measure h_c.

It is noteworthy that the partition function (or theta function) of h, defined as the Laplace transform of h,

$$\theta_h(t) := \int_0^\infty e^{-tx}\, h(dx), \tag{4.33}$$

for $t > 0$, is given by the function

$$\theta_h(t) = \frac{1}{e^t - 1}. \tag{4.34}$$

We refer to Section 6.2.3 below for further discussion of the partition function of a fractal string.

---

[5]The classical theta function is defined by making the change of variable $x^2 = -i\tau$, where τ has positive imaginary part.

4.3 Generalized Fractal Sprays

As in Section 1.4, we can consider basic shapes B other than the unit interval, scaled (in a formal sense) by the generalized fractal string η. Thus we define the spectral zeta function of the generalized fractal spray of η on the basic shape B as

$$\zeta_\nu(s) = \zeta_\eta(s) \cdot \zeta_B(s). \tag{4.35}$$

When η is an ordinary fractal string $\mathcal{L}$, then we recover the notion of ordinary fractal spray (of $\mathcal{L}$ on B) from [LapPo3] and Section 1.4. Further, in view of (1.46), ζ_ν coincides with the spectral zeta function of an ordinary fractal spray, as defined by (1.43).

For later applications, we need to further extend the notion of generalized fractal spray by going beyond the geometric situation when the basic shape B is an actual region in some space. That is, we sometimes define B only virtually by its associated spectral zeta function, $\zeta_B(s)$, which can be any given generalized Dirichlet series or Dirichlet integral. Then the spectral zeta function of such a spray (of η on B) is still given by (4.35). More precisely, if

$$\zeta_B(s) = \int_0^\infty x^{-s} \rho(dx) \tag{4.36}$$

for some measure ρ, then, by definition, the spectral measure of such a virtual generalized fractal spray is given by

$$\nu = \eta * \rho, \tag{4.37}$$

from which (by (4.19)) relation (4.35) follows.

This extension will allow us to investigate the properties of any zeta function (or generalized Dirichlet series) ζ_B, as will become particularly apparent in Sections 9.3 and 11.2.

4.4 The Measure of a Self-Similar String

In this section, we investigate some of the properties of the measure associated (as in Section 4.1) with a self-similar string (with a single gap), introduced in Chapter 2. As was mentioned in the introduction to this chapter, the present section—which is of independent interest—can be omitted on a first reading.

Let $\mathcal{L}$ be a self-similar string with a single gap, as in Section 2.2.1, constructed with the scaling ratios $r_1, \dots, r_N$ and normalized such that the first length is 1; i.e., the total length is $L = g_1^{-1}$, by Remark 2.6. By Equations (4.9) and (2.7), the measure $\mu_{\mathcal{L}}$ associated with $\mathcal{L}$ is

$$\mu_{\mathcal{L}} = \sum_{e_1 \geq 0, \dots, e_N \geq 0} \binom{\Sigma \mathbf{e}}{\mathbf{e}} \delta_{\{\mathbf{r}^{-\mathbf{e}}\}}, \tag{4.38}$$

where the sum is over all N-tuples of nonnegative integers $\mathbf{e} = e_1, \dots, e_N$. Here and in the following, we use the multi-index notation

$$\Sigma \mathbf{e} = \sum_{j=1}^{N} e_j, \tag{4.39a}$$

$$\binom{\Sigma \mathbf{e}}{\mathbf{e}} = \binom{\Sigma \mathbf{e}}{e_1\ ldotse_N}, \tag{4.39b}$$

$$\mathbf{r}^{-\mathbf{e}} = \prod_{j=1}^{N} r_j^{-e_j}, \tag{4.39c}$$

and we will use $\mathbf{e} \geq 0$ for $e_1 \geq 0, \dots, e_N \geq 0$. (See formula (2.6) for the definition of the multinomial coefficient $\binom{\Sigma \mathbf{e}}{\mathbf{e}}$.)

For a scaling factor t, we let tA be the set $\{tx \colon x \in A\}$. Further, for a predicate P, we let δ_P be 1 if P is true, and 0 otherwise.

Theorem 4.9. *The measure $\mu_{\mathcal{L}}$ defined by (4.38) satisfies the following scaling property, which we call its* self-similarity property:

$$\mu_{\mathcal{L}}(A) = \delta_{1 \in A} + \sum_{j=1}^{N} \mu_{\mathcal{L}}(r_j A), \tag{4.40}$$

for every subset A of $(0, \infty)$.

Moreover, $\mu_{\mathcal{L}}$ is completely characterized by this property. In other words, every generalized fractal string satisfying this property necessarily coincides with $\mu_{\mathcal{L}}$.

Proof of Theorem 4.9. The measure of A is

$$\mu_{\mathcal{L}}(A) = \sum_{\mathbf{e} \geq 0} \binom{\Sigma \mathbf{e}}{\mathbf{e}} \delta_{\mathbf{r}^{-\mathbf{e}} \in A}, \tag{4.41}$$

and that of $r_j A$ is

$$\begin{aligned} \mu_{\mathcal{L}}(r_j A) &= \sum_{\mathbf{e} \geq 0} \binom{\Sigma \mathbf{e}}{\mathbf{e}} \delta_{\mathbf{r}^{-\mathbf{e}} \in r_j A} \\ &= \sum_{\mathbf{e} \geq 0} \binom{\Sigma \mathbf{e}}{\mathbf{e}} \delta_{\mathbf{r}^{-\mathbf{e}} r_j^{-1} \in A} \\ &= \sum_{\mathbf{e} \geq 0,\, e_j \geq 1} \binom{\Sigma \mathbf{e} - 1}{e_1, \dots, e_j - 1, \dots, e_N} \delta_{\mathbf{r}^{-\mathbf{e}} \in A}. \end{aligned}$$

We sum these expressions over j and use the following generalization for multinomial coefficients of the usual property of binomial coefficients:

$$\sum_{j=1}^{N} \delta_{e_j \geq 1} \binom{\Sigma \mathbf{e} - 1}{e_1, \dots, e_{j-1}, e_j - 1, e_{j+1}, \dots, e_N} = \binom{\Sigma \mathbf{e}}{\mathbf{e}},$$

where only e_j has been decreased by one. We find

$$\begin{aligned}
\sum_{j=1}^{N} \mu_{\mathcal{L}}(r_j A) &= \sum_{j=1}^{N} \sum_{\mathbf{e} \geq 0} \binom{\Sigma \mathbf{e} - 1}{e_1, \dots, e_j - 1, \dots, e_N} \delta_{e_j \geq 1} \delta_{\mathbf{r}^{-\mathbf{e}} \in A} \\
&= \sum_{\mathbf{e} \geq 0} \delta_{\mathbf{r}^{-\mathbf{e}} \in A} \sum_{j=1}^{N} \delta_{e_j \geq 1} \binom{\Sigma \mathbf{e} - 1}{e_1, \dots, e_j - 1, \dots, e_N} \\
&= \sum_{\mathbf{e} > 0} \binom{\Sigma \mathbf{e}}{\mathbf{e}} \delta_{\mathbf{r}^{-\mathbf{e}} \in A},
\end{aligned}$$

where the notation $\mathbf{e} > 0$ indicates that $\mathbf{e} = (0, 0, \dots, 0)$ is excluded. We recover formula (4.41) for $\mu_{\mathcal{L}}(A)$, except for the first term, corresponding to $\mathbf{e} = (0, \dots, 0)$. This term is $\delta_{1 \in A}$. Thus $\mu_{\mathcal{L}}$ satisfies relation (4.40), as claimed.

Next, let η be a string that satisfies this same relation. Then $\eta - \mu_{\mathcal{L}}$ is a measure η' that satisfies

$$\eta'(A) = \sum_{j=1}^{N} \eta'(r_j A). \tag{4.42}$$

We will show that this implies that $\eta' = 0$. Since every measure is determined by its values on bounded sets, we can assume that A is bounded. Since η' is a generalized fractal string, there exists x_0 such that $\eta' = 0$ on $(0, x_0)$. So we are done if A is contained in this interval. Suppose now that we know that $\eta' = 0$ on $(0, x_0 r_1^{-k})$. Then if A is contained in $\left(0, x_0 r_1^{-(k+1)}\right)$, the sets $r_j A$ are contained in $(0, x_0 r_1^{-k})$, because $r_j \leq r_1$, for $j = 1, \dots, N$. Hence $\eta'(A) = 0$ by relation (4.42). It follows that $\eta'(A) = 0$ for all bounded subsets A of $(0, \infty)$. Hence $\eta' = 0$ and so $\eta = \mu_{\mathcal{L}}$. This completes the proof of the theorem. □

4.4.1 Measures with a Self-Similarity Property

It follows from the above proof that a measure η on $(0, \infty)$ vanishes if it satisfies

$$\eta(A) = \sum_{j=1}^{N} \eta(r_j A) \tag{4.43}$$

and is supported away from 0. It is interesting to study how η is determined when it has mass near 0. Assume that η is absolutely continuous with respect to the measure dx/x, the Haar measure on the multiplicative group $\mathbb{R}_+^*$. Thus, there is a Borel measurable function f on $(0, \infty)$ such that $\eta(dx) = f(x)\,dx/x$. Then f satisfies the relation

$$f(x) = \sum_{j=1}^{N} f(r_j x), \tag{4.44}$$

for all $x > 0$.

Choose some $r \in (0,1)$ and write $r_j = r^{k_j}$ for positive real numbers k_j with $0 < k_1 \leq \ldots \leq k_N$. Let $g(t) = f\left(r^{-t}\right)$, for $t \in \mathbb{R}$. The function g has the following periodicity property:

$$g(t) = \sum_{j=1}^{N} g\left(t - k_j\right), \tag{4.45}$$

for all $t \in \mathbb{R}$. We cannot deal with such functions in general (see Problem 4.11 below), but we can handle them when η is a lattice string in the sense of Definition 2.14.

In the lattice case, we choose the multiplicative generator, r, of η, and hence positive integers $k_N \geq \cdots \geq k_1$, such that $r_j = r^{k_j}$ for $j = 1, \ldots, N$. (See Definition 2.14.) Then g is determined on $\mathbb{Z}$, for example, if $g(0), g(1), \ldots, g(k_N - 1)$ are chosen, and in general, g is determined on $\mathbb{R}$ when g is given on the interval $[0, k_N)$. We can solve the associated recursion by using the next proposition.

Proposition 4.10. *The solution space of the recursion relation*

$$a_n = \sum_{j=1}^{N} a_{n-k_j} \qquad (n \in \mathbb{Z})$$

has dimension k_N. For each complex solution z of the polynomial equation

$$z^{k_N} = \sum_{j=1}^{N} z^{k_N - k_j}, \tag{4.46}$$

of multiplicity $m(z)$, we obtain $m(z)$ solutions

$$n \mapsto n^q z^n \qquad (n \in \mathbb{Z}),$$

of the recursion relation, for each integer q between 0 and $m(z) - 1$.

Alternatively, for every $t \in \mathbb{R}$, we obtain $m(z)$ solutions

$$n \mapsto (n+t)^q z^{n+t} \qquad (n \in \mathbb{Z}).$$

These solutions (for fixed t, and all z and q, $0 \leq q \leq m(z) - 1$) form a basis of the solution space.

Thus, if the values $g(0), g(1), \dots, g(k_N - 1)$ are known, then there exist coefficients $c_{z,q}$ such that $g(n) = \sum_z \sum_{q=0}^{m(z)-1} c_{z,q} n^q z^n$ for all natural numbers $n \geq 0$, where z runs through the solutions of (4.46). More generally, for each $t \in [0,1)$, if $g(t), g(t+1), \dots, g(t+k_N-1)$ are known, then there exist coefficients $c_{z,q}(t)$ such that

$$g(n+t) = \sum_z \sum_{q=0}^{m(z)-1} c_{z,q}(t)(n+t)^q z^{n+t}.$$

We extend the definition of $c_{z,q}(t)$ by periodicity: Thus $c_{z,q}(t)$ is defined by $c_{z,q}(\{t\})$, where $\{t\} = t - [t]$ is the fractional part of t. The functions $c_{z,q}(t)$ have a Fourier series expansion: $c_{z,q}(t) = \sum_{n\in\mathbb{Z}} c_{z,q,n} e^{2\pi int}$. We thus find the following expansion for the general function with periodicity property (4.45):

$$g(t) = \sum_z \sum_{q=0}^{m(z)-1} \sum_{n\in\mathbb{Z}} c_{z,q,n} e^{2\pi int} t^q z^t. \tag{4.47}$$

This argument is justified if we impose some integrability condition on g. A convenient condition is that g is locally L^2; i.e., $|g|^2$ has a finite integral on every compact subset of the real line. Then the coefficients $c_{z,q}(t)$ are locally L^2, which is equivalent to the condition that for every z and q the sequence of Fourier coefficients $c_{z,q,n}$ is square-summable: for every z and q, $\sum_{n\in\mathbb{Z}} |c_{z,q,n}|^2 < \infty$. Since there are only finitely many z and q, this is equivalent to $\sum_z \sum_{q=0}^{m(z)-1} \sum_{n\in\mathbb{Z}} |c_{z,q,n}|^2 < \infty$.

In multiplicative terms, we obtain the following expansion for f:

$$f(x) = \sum_z \sum_{q=0}^{m(z)-1} \sum_{n\in\mathbb{Z}} c_{z,q,n} x^{in\mathbf{p}} (-\log_r x)^q z^{-\log_r x}, \qquad \text{for } x > 0,$$

with $\mathbf{p} = 2\pi/\log r^{-1}$, the oscillatory period of the given lattice string. Recall from Theorem 2.17 that the complex dimensions ω of the lattice string $\mathcal{L}$ are the solutions of the equation $1 = \sum_{j=1}^{N} r^{k_j\omega}$ and that they lie periodically with period $\mathbf{p}$ on finitely many lines. Thus, to every z there corresponds an ω such that $z = r^{-\omega}$. Choose complex dimensions ω_z, one for each solution z. This means that we choose one ω_z on every line of poles of $\zeta_{\mathcal{L}}$. Observe that $z^{-\log_r x} = x^{\omega_z}$ and that $\omega_z + in\mathbf{p}$ runs over all complex dimensions. Thus we find that

$$f(x) = \sum_\omega \sum_{q=0}^{m(\omega)-1} c_{\omega,q} (-\log_r x)^q x^\omega. \tag{4.48a}$$

Here, ω runs over all complex dimensions of $\mathcal{L}$, and q runs from 0 to the multiplicity of ω minus one. The coefficients are determined by $c_{\omega,q} = c_{z,q,n}$

for $\omega = \omega_z + in\mathbf{p}$, and they are square-summable. This expression is very similar to the explicit formulas that we shall derive in the next chapter. In case all complex dimensions of $\mathcal{L}$ are simple, we obtain

$$f(x) = \sum_{\omega} c_\omega x^\omega. \tag{4.48b}$$

For a nonlattice string, there is no natural choice for r. We set $r = 1/e$ and we formulate the following open problem, which may be solvable using the techniques of the next chapter.

Problem 4.11. Let $\mathcal{L}$ be a self-similar string, constructed with scaling ratios $r_1, \dots, r_N$ and a single gap g_1, normalized as in Remark 2.6. Show that every function f that is locally L^2 on $(0, \infty)$ and satisfies the self-similarity relation (4.44) has an expansion of the following form:

$$f(x) = \sum_{\omega \in \mathcal{D}_{\mathcal{L}}(\mathbb{C})} \sum_{q=0}^{m(\omega)-1} c_{\omega,q} (\log x)^q x^\omega, \qquad \text{for } x > 0, \tag{4.49}$$

where ω runs over the complex dimensions of the string $\mathcal{L}$ and $m(\omega)$ is the multiplicity of ω. Further, show that the coefficients $c_{\omega,q}$ in (4.49) are uniquely determined by f and that they are square-summable.

The reader may first try to solve the special case of simple complex dimensions:

Problem 4.12. If all the complex dimensions of $\mathcal{L}$ are simple in the situation of Problem 4.11, then show that

$$f(x) = \sum_{\omega \in \mathcal{D}_{\mathcal{L}}(\mathbb{C})} c_\omega x^\omega, \qquad \text{for } x > 0. \tag{4.50}$$

4.5 Notes

Our point of view in this chapter is more general than that of [JorLan2] in the sense that we allow for infinitesimal multiplicities, but less general in the sense that they allow for complex 'lengths'. Of course, the interpretation in [JorLan2] is not in terms of lengths and complex dimensions of fractal strings. Moreover, the notions of screen and window for the meromorphic continuation of the Dirichlet integral ζ_η are not used in that work.

Section 4.4: the measure $\mu_{\mathcal{L}}$ on $(0, \infty)$ is *not* a self-similar measure in the sense encountered in the literature on fractal geometry; compare, e.g., [Hut; Fa1, §8.3] and [Str1–2, Lap5–6].

5

Explicit Formulas for Generalized Fractal Strings

In this chapter, we obtain pointwise and distributional explicit formulas for the lengths and the frequencies of a fractal string. These explicit formulas express the counting function of the lengths or of the frequencies as a sum over the visible complex dimensions ω of the fractal string. To unify the exposition, and with a view toward later applications, we formulate our results in the language of generalized fractal strings, introduced in Chapter 4.

After having introduced some necessary notation and given a heuristic proof of one of our formulas in Section 5.1, we discuss some technical preliminaries in Section 5.2. Our pointwise explicit formulas are proved in Section 5.3, while our distributional explicit formulas and various useful extensions are established in Section 5.4. Finally, in Section 5.5, we close the chapter by explaining how to apply our explicit formulas to reprove the classical Prime Number Theorem (with error term). Many additional examples illustrating our theory will be discussed in Chapter 6 and throughout the remainder of the book.

5.1 Introduction

Our explicit formulas will usually contain an error term. In most applications, this error term will be given by an integral over the vertical line $\operatorname{Re} s = \sigma_0$, for some value of σ_0. In general, it will be given by an integral over the screen S, introduced in Section 1.2.1. When applied to a nonlattice self-similar fractal string, the theory of dimension-free regions developed in Chapter 3 (see especially Section 3.6) will allow us to maximally exploit a suitable choice of the screen.

We have defined $N_\eta(x)$, the counting function of the reciprocal lengths, in Definition 4.1(ii), formula (4.2). In our framework, it will be very useful to also consider the integrated versions of the counting function. We will denote by $N_\eta^{[k]}$ the *k-th primitive* (or *k-th antiderivative*) of η, vanishing at 0. Thus

$$N_\eta^{[k]}(x) = \int_0^x \frac{(x-y)^{k-1}}{(k-1)!}\,\eta(dy), \tag{5.1}$$

for $x > 0$ and $k = 1, 2, \ldots$. In particular, $N_\eta = N_\eta^{[1]}$ is the antiderivative of η, vanishing at 0. Note that $N_\eta^{[k]}$ is a continuous function as soon as $k \geq 2$. In general, $N_\eta^{[k]}$ is $(k-2)$ times continuously differentiable for $k \geq 2$.

Formally, and this will be completely justified distributionally,

$$N_\eta^{[0]} = \frac{d}{dx} N_\eta = \eta. \tag{5.2}$$

The pointwise formulas give an expression for $N_\eta^{[k]}(x)$, valid for all $x > 0$ (or all $x > A$ for some $A > 0$) and all $k \geq 1$ sufficiently large.

The distributional formulas describe η as a distribution: on a test function φ, the distribution η acts by

$$\langle \eta, \varphi \rangle = \int_0^\infty \varphi(x)\,\eta(dx). \tag{5.3}$$

The *k-th primitive* of this distribution will be denoted by $\mathcal{P}_\eta^{[k]}$. More precisely, $\mathcal{P}_\eta^{[k]}$ is the distribution given for all test functions φ by

$$\langle \mathcal{P}_\eta^{[k]}, \varphi \rangle = (-1)^k \langle \eta, \mathcal{P}^{[k]}\varphi \rangle, \tag{5.4}$$

where $\mathcal{P}^{[k]}\varphi$ is the k-th primitive of φ that vanishes at infinity together with its derivatives. Thus, for a test function φ,

$$\langle \mathcal{P}_\eta^{[k]}, \varphi \rangle = \int_0^\infty \int_y^\infty \frac{(x-y)^{k-1}}{(k-1)!}\,\varphi(x)\,dx\,\eta(dy), \tag{5.5}$$

and $\mathcal{P}_\eta^{[0]} = \eta$.

For the general theory of distributions (or generalized functions, in the sense of Laurent Schwartz), we refer, e.g., to [Schw1–2, Hö2, ReSi1–2]. We recall from that theory that any locally integrable function f on $(0, \infty)$ defines a distribution in the obvious manner. Specifically, for any measurable function on $(0, \infty)$ such that $\int_a^b |f(x)|\,dx$ is finite for every $[a, b] \subset (0, \infty)$,

$$\langle f, \varphi \rangle = \int_0^\infty f(x)\varphi(x)\,dx, \tag{5.6}$$

for all test functions φ with compact support contained in $(0, \infty)$. This applies in particular, for each fixed $k \geq 1$, to the k-th integrated counting function, $f(x) = N_\eta^{[k]}(x)$, associated with an arbitrary generalized fractal string η.

5.1.1 *Outline of the Proof*

In this section, we discuss heuristically how the pointwise explicit formula is established. We will derive a pointwise formula for η, even though, to make the argument rigorous, this formula has to be interpreted distributionally.

Our starting point is an expression for the Dirac delta function at y as a Mellin transform,

$$\frac{1}{2\pi i} \int_{c-i\infty}^{c+i\infty} x^{s-1} y^{-s} \, ds = \delta_{\{y\}}(x),$$

for $c > 0$. This is Lemma 5.1 below, applied formally for $k = 0$. For the moment, we interpret $\delta_{\{y\}}$ as a function, which is why the present argument is not rigorous. Viewing the measure η as a superposition of delta functions, we write

$$\eta(x) = \int_0^\infty \delta_{\{y\}}(x) \, \eta(dy) = \int_0^\infty \frac{1}{2\pi i} \int_{c-i\infty}^{c+i\infty} x^{s-1} y^{-s} \, ds \, \eta(dy).$$

For $c > D$, we interchange the order of integration and use the expression for the geometric zeta function $\zeta_\eta(s) = \int_0^\infty y^{-s} \, \eta(dy)$ to deduce that

$$\eta(x) = \frac{1}{2\pi i} \int_{c-i\infty}^{c+i\infty} x^{s-1} \zeta_\eta(s) \, ds.$$

This expresses $\eta(x)$ as the inverse Mellin transform of $\zeta_\eta(s)$. In order to obtain information about η, we need to push the line of integration $\operatorname{Re} s = c$ as far to the left as possible. When we push it to the screen S, we pick up a residue at each complex dimension ω of η. Thus, we obtain the *density of lengths* (or *density of geometric states*) formula:

$$\eta = \sum_{\omega \in \mathcal{D}_\eta(W)} \operatorname{res}\left(x^{s-1} \zeta_\eta(s); \omega\right) + \frac{1}{2\pi i} \int_S x^{s-1} \zeta_\eta(s) \, ds, \tag{5.7a}$$

where we denote the residue of a meromorphic function $g = g(s)$ at $s = \omega$ by $\operatorname{res}(g(s); \omega)$. If the complex dimensions are simple, this becomes the formula

$$\eta = \sum_{\omega \in \mathcal{D}_\eta(W)} \operatorname{res}\left(\zeta_\eta(s); \omega\right) x^{\omega-1} + \frac{1}{2\pi i} \int_S x^{s-1} \zeta_\eta(s) \, ds. \tag{5.7b}$$

In order to turn this argument into a rigorous proof, we need, in particular, to assume suitable growth conditions on ζ_η, which will be stated at the beginning of Section 5.3. We say that η is *languid* if ζ_η satisfies these conditions.

5.1.2 Examples

We shall give two versions of the explicit formula. The first is pointwise, in Section 5.3, and the second is distributional, in Section 5.4. We have in mind a number of examples to which we want to apply our explicit formulas:

1. The counting function of the lengths of a self-similar string. Then, we can choose $W = \mathbb{C}$ and therefore obtain an explicit formula involving all the complex dimensions of the string; see Sections 6.4.1 and 6.4.2.

2. The counting function of the frequencies of a self-similar string. Here, we shall obtain information up to a certain order (i.e., $W \neq \mathbb{C}$), due to the growth of the Riemann zeta function to the left of the critical strip $0 \leq \operatorname{Re} s \leq 1$; see Section 6.4.3.

3. More generally, the geometric and spectral counting functions (Sections 6.2.1, 6.2.2 and 6.3.1) and the geometric and spectral partition functions (Section 6.2.3) of an ordinary fractal string.

4. The fractal string of [Lap1, Example 5.1], also called the a-string in Section 6.5.1.

5. The ordinary fractal string of [LapMa2], with which M. L. Lapidus and H. Maier gave a characterization of the Riemann hypothesis. Again, we shall obtain information up to an error term since the geometric and spectral zeta functions of this string may not have an analytic continuation to all of $\mathbb{C}$. We will, however, improve significantly the error term obtained in [LapMa2]; see Chapter 9.

6. A continuous version of this string, discussed in Section 9.2.

7. The geometric and spectral counting functions of a generalized Cantor string (Chapter 10 and Section 11.1), and of a generalized Cantor spray (Section 11.2).

8. The geometry and the spectrum of generalized fractal sprays; see Sections 6.6, 9.3 and 11.2.

9. The volume of the tubular neighborhoods of fractal strings, as discussed in Chapter 8; see especially Section 8.1 and in the case of self-similar strings, Section 8.4.

10. The classical Prime Number Theorem and the Riemann–von Mangoldt explicit formula for the zeros of the Riemann zeta function (Section 5.5).

11. The Prime Number Theorem and the corresponding explicit formula for the primitive periodic orbits of the dynamical system naturally associated with a self-similar string (Section 7.4).

As was alluded to above, our explicit formulas can be applied either to the geometric zeta function of an ordinary fractal string $\mathcal{L}$, yielding explicit formulas for the counting functions of the lengths of $\mathcal{L}$, or to the spectral zeta function of $\mathcal{L}$, yielding explicit formulas for the counting functions of the frequencies of $\mathcal{L}$; see, for example, Chapters 6 and 10. The resulting explicit formulas show clearly the relationship between the counting function of the lengths and that of the frequencies. This relationship can be described as follows. The counting function of the frequencies is obtained by applying an operator, the spectral operator, to the explicit formula for the counting function of the lengths; see Sections 6.1, 6.2, and especially 6.3.1. This is already suggested by the results of [LapPo2] and [LapMa2], but it can be precisely formalized in our framework.

A variant of Riemann's explicit formula recalled in the introduction (page 3) is

$$\psi(x) = x - \sum_{\rho} \frac{x^{\rho}}{\rho} - \frac{\zeta'}{\zeta}(0) - \frac{1}{2}\log\left(1 - 1/x^2\right), \tag{5.8}$$

where $\psi(x) = \sum_{m\geq 1, p^m \leq x} \log p$ counts the prime powers with a weight $\log p$, and ρ runs over all zeros of the Riemann zeta function in order of increasing absolute value. It can be used to give a proof of the Prime Number Theorem with error term, using a zero-free region for ζ, first established by de la Vallée Poussin. If one uses only $\operatorname{Re}\rho < 1$, then one obtains, with some effort since the sum is not absolutely convergent, that $\psi(x) = x + o(x)$ as $x \to \infty$, which implies the Prime Number Theorem.

Later, in 1952, André Weil established the distributional formula

$$\sum_{p} W_p(F) = \Phi(0) + \Phi(1) - \sum_{\rho} \Phi(\rho), \tag{5.9}$$

where Φ is the Mellin transform of the test function F, and W_p (for primes p and for $p = \infty$) is the *Weil-distribution* (see [Wei4, p. 262]). This formula reveals more clearly the underlying structure: on the left-hand side, we have contributions from all valuations of the field of rational numbers, while the right-hand side involves the pole and all zeros of the Riemann zeta function.

The duality between the concrete side of the prime counting functions and the abstract side of the zeros and poles is also the main idea behind the explicit formulas in this book. Thus we will show that the counting function of the lengths of a fractal string,

$$N_{\mathcal{L}}(x) = \#\{j : l_j^{-1} \leq x\}$$

has the following explicit formula (if the complex dimensions are simple):

$$N_{\mathcal{L}}(x) = \sum_{\omega} \operatorname{res}(\zeta_{\mathcal{L}}(s); \omega) \frac{x^{\omega}}{\omega}.$$

If $\mathcal{L}$ is a lattice self-similar string, then there exist infinitely many complex dimensions with real part D, and one obtains $N_{\mathcal{L}}(x) = g_{\mathcal{L}}(x)x^D + o(x^D)$, for some nonconstant multiplicatively periodic function $g_{\mathcal{L}}(x)$. On the other hand, for nonlattice strings, D is the only complex dimension with real part D, and all other complex dimensions ω satisfy $\operatorname{Re}\omega < D$. Thus one obtains for nonlattice strings that $N_{\mathcal{L}}(x) = g_{\mathcal{L}}x^D + o(x^D)$, for some constant $g_{\mathcal{L}}$.

The pointwise and distributional explicit formulas that we will give (in Sections 5.3 and 5.4 below) deviate in two ways from the usual explicit formulas found in number theory. On the one hand, we consider the density of states formula to be more fundamental. This formula corresponds, for instance, to the derivative of the usual explicit formula of the prime number counting function, and exists only in a distributional sense (see Section 5.5). The integrated versions of this formula always exist as distributional formulas, and also sometimes as pointwise formulas. On the other hand, our explicit formulas will usually contain an error term (see Theorems 5.10 and 5.18). In the number-theoretic formulas, this error term is not present (or else is not considered as such), thanks to the use of the functional equation satisfied by the Riemann or other number-theoretic zeta functions.

Sometimes, as in the case of the counting function of the lengths of a self-similar string, this error term can be analyzed by pushing the screen arbitrarily far to the left, and the resulting formula is an explicit formula in the classical sense. (See Theorems 5.14 and 5.22; see also Section 6.4 for the case of self-similar strings.) But already for the counting function of the frequencies of a self-similar string—and also, for example, when the geometric zeta function of a string does not have a meromorphic continuation to the whole complex plane—there is no way to avoid the presence of an error term, and our formulas are, in some sense, best possible. (See Section 6.4 and, for example, Section 6.5.2.) The usefulness of our explicit formulas depends very much on the possibility of giving a satisfactory analysis of this error term. We will provide such asymptotic estimates both for the pointwise error term (see Theorem 5.10, Equations (5.36)–(5.38)) and, in a suitable sense to be specified in Definition 5.29, for the distributional error term (see Theorems 5.18 and 5.30).

5.2 Preliminaries: The Heaviside Function

We refine here the basic lemma of [In, pp. 31 and 75; Da, p. 105]. This extension will be needed in the proof of Lemma 5.9, the truncated pointwise formula, which itself will be used to establish both the pointwise and the distributional formulas (Theorems 5.10, 5.14 and 5.18, 5.22). We refer the interested reader to Figure 8.1 (page 246) for a diagram of the interdependence of the explicit formulas and other theorems in this book.

Define the *k-th Heaviside function* for $k \geq 1$ by

$$H^{[k]}(x) = \begin{cases} \dfrac{x^{k-1}}{(k-1)!}, & \text{for } x > 0, \\ 0, & \text{for } x < 0 \text{ or } x = 0,\ k \geq 2, \\ \dfrac{1}{2}, & \text{for } x = 0,\ k = 1. \end{cases} \tag{5.10}$$

For $k \geq 2$, $H^{[k]}(x)$ is the $(k-1)$-th antiderivative (vanishing at $x = 0$) of the classical Heaviside function $H^{[1]}(x)$, equal to 1 for $x > 0$, to 0 for $x < 0$, and taking the value $\frac{1}{2}$ at $x = 0$. Note that, in view of definition (5.1), we have

$$N_\eta^{[k]}(x) = \int_0^\infty H^{[k]}(x-t)\,\eta(dt). \tag{5.11}$$

For $k \geq 1$, we define the *Pochhammer symbol* $(s)_k$ by

$$(s)_k = s(s+1)\cdots(s+k-1). \tag{5.12}$$

The usefulness of the following lemma will become apparent upon consulting Definitions 5.2 and 5.3 and observing how the conditions **L1** and **L2** or **L2′** therein are used to obtain the subsequent results.

Lemma 5.1. *For $c > 0$, $T_- < 0 < T_+$, $x, y > 0$ and $k = 1, 2, \ldots$, the k-th Heaviside function is approximated as follows:*

$$H^{[k]}(x-y) = \frac{1}{2\pi i}\int_{c+iT_-}^{c+iT_+} x^{s+k-1}y^{-s}\frac{ds}{(s)_k} + E. \tag{5.13}$$

Putting $T_{\min} = \min\{T_+, |T_-|\}$ and $T_{\max} = \max\{T_+, |T_-|\}$, the error E of this approximation does not exceed in absolute value

$$x^{c+k-1}y^{-c}T_{\min}^{-k}\min\left\{T_{\max}, \frac{1}{|\log x - \log y|}\right\}, \quad \textit{if } x \neq y, \tag{5.14a}$$

$$x^{k-1}T_{\min}^{-k}T_{\max}, \quad \textit{if } x = y, \textit{ all } k, \tag{5.14b}$$

$$\left((c+k-1)2^{k-1} + T_{\max} - T_{\min}\right)x^{k-1}T_{\min}^{-k}, \quad \textit{if } x = y, \textit{ k is odd.} \tag{5.14c}$$

Proof. Let $x < y$, so that $H^{[k]}(x-y) = 0$. We consider the integral

$$\frac{1}{2\pi i}\int x^{s+k-1}y^{-s}\frac{ds}{(s)_k}$$

over the contour $c + iT_-, c + iT_+, U + iT_+, U + iT_-, c + iT_-$, for a large positive value of U. The integral over the left side equals $-E$, and we want to show that it is small. By the Theorem of Residues [Ahl, Theorem 17, p. 150], the contour integral vanishes. Hence the integral over the left side equals the value of the integral over the contour composed of the upper, lower and right-hand sides. The integral over the right-hand side is bounded as follows:

$$\begin{aligned}\left|\frac{1}{2\pi i}\int_{U+iT_-}^{U+iT_+} x^{s+k-1}y^{-s}\frac{ds}{(s)_k}\right| &\leq \frac{1}{2\pi}\int_{T_-}^{T_+} x^{U+k-1}y^{-U}\frac{dt}{|(U+it)_k|} \\ &\leq \frac{T_+-T_-}{2\pi}x^{U+k-1}y^{-U}U^{-k}.\end{aligned}$$

Further, the integral over the upper side satisfies the inequality

$$\begin{aligned}\left|\frac{1}{2\pi i}\int_{c+iT_+}^{U+iT_+} x^{s+k-1}y^{-s}\frac{ds}{(s)_k}\right| &\leq \frac{1}{2\pi}\frac{x^{k-1}}{T_+^k}\int_c^U (x/y)^\sigma\,d\sigma \\ &\leq \frac{x^{c+k-1}y^{-c}}{2\pi|\log x-\log y|}T_+^{-k},\end{aligned}$$

and, similarly, that over the lower side is bounded by

$$\frac{x^{c+k-1}y^{-c}}{2\pi|\log x-\log y|}|T_-|^{-k},$$

independent of U. On letting U go to infinity, the contribution of the right-hand side becomes arbitrarily small. The contribution of the upper and lower sides is now bounded by

$$\frac{x^{c+k-1}y^{-c}}{|\log x-\log y|}\frac{T_+^{-k}+|T_-|^{-k}}{2\pi} \leq \frac{x^{c+k-1}y^{-c}}{\pi|\log x-\log y|}T_{\min}^{-k}.$$

This proves the second inequality of (5.14a). To prove the first inequality of (5.14a), we integrate over the contour composed of the line segment from $c+iT_-$ to $c+iT_+$, a circular arc to the right with center the origin of radius $T_{\min}$, and a line segment from $c+iT_+$ to $c+iT_{\min}$, if $T_+=T_{\max}$ (respectively, from $c-iT_{\min}$ to $c+iT_-$, if $|T_-|=T_{\max}$). Again, the integral over the line segment $c+iT_-$ to $c+iT_+$ vanishes up to the value of the integral over the circular part and the little line segment of the contour. These two integrals are bounded by

$$\frac{1}{2\pi}\cdot 2\pi T_{\min}\cdot x^{c+k-1}y^{-c}T_{\min}^{-k} = x^{c+k-1}y^{-c}T_{\min}^{1-k},$$

and

$$x^{c+k-1}y^{-c}(T_{\max}-T_{\min})T_{\min}^{-k}.$$

Note that this estimate is also valid for $x=y$, proving (5.14b).

The estimate of the error in the case $x > y$ is derived similarly, with respectively a rectangular and a circular contour going to the left. The value of the contour integral requires some consideration now. The integrand has simple poles at the points $0, -1, -2, \ldots, 1-k$. The residue at $-j$ is

$$\frac{x^{-j+k-1}y^j}{(-j)\cdot(1-j)\cdot\cdots\cdot(-1)\cdot 1\cdot 2\cdot\cdots\cdot(k-1-j)} = \frac{1}{(k-1)!}x^{k-1-j}(-y)^j\binom{k-1}{j}.$$

Hence the sum of the residues is $(x-y)^{k-1}/(k-1)! = H^{[k]}(x-y)$.

It remains to derive the better order in T_+ and $|T_-|$ when k is odd and $x = y$ (see inequality (5.14c)). Without loss of generality, we assume that $T_+ = T_{\max}$. By a direct computation,

$$\begin{aligned}\frac{1}{2\pi i}\int_{c+iT_-}^{c+iT_+} x^{k-1}\frac{ds}{(s)_k} &= \frac{x^{k-1}}{2\pi}\int_{T_-}^{T_+}\frac{dt}{(c+it)_k}\\ &= \frac{x^{k-1}}{\pi}\int_0^{T_{\min}}\frac{\mathrm{Re}(c+it)_k}{|(c+it)_k|^2}\,dt + \frac{x^{k-1}}{2\pi}\int_{T_{\min}}^{T_+}\frac{dt}{(c+it)_k},\end{aligned}$$

where $(c+it)_k$ is given by (5.12). For $k = 1$, the second term on the right is bounded by $(T_{\max} - T_{\min})\,T_{\min}^{-1}$. Further, the first term on the right is

$$\begin{aligned}\frac{1}{\pi}\int_0^{T_{\min}}\frac{c}{c^2+t^2}\,dt &= \frac{1}{\pi}\int_0^{T_{\min}/c}\frac{du}{1+u^2}\\ &= \frac{1}{2} - \frac{1}{\pi}\int_{T_{\min}/c}^{\infty}\frac{du}{1+u^2},\end{aligned}$$

which differs from $1/2$ by at most $c/T_{\min}$.

For $k \geq 3$, the error is given by

$$\frac{x^{k-1}}{\pi}\int_{T_{\min}}^{\infty}\frac{\mathrm{Re}(c+it)_k}{|(c+it)_k|^2}\,dt + \frac{x^{k-1}}{2\pi}\int_{T_{\min}}^{T_+}\frac{dt}{(c+it)_k}. \tag{5.15}$$

The last integral is bounded by $(T_{\max} - T_{\min})T_{\min}^{-k}x^{k-1}$. Next, we expand $(c+it)_k$ in powers of t:

$$(c+it)_k = (it)^k + \sum_{j=0}^{k-1} a_j(it)^j,$$

where a_j is the sum of all products of $k-j$ factors from $c, c+1, \ldots, c+k-1$. Hence for odd k, $\mathrm{Re}(c+it)_k = \sum_{j=0,\, j\,\text{even}}^{k-1} a_j(it)^j$. One checks that a_j is bounded by $(c+k-1)^{k-j}\binom{k}{j} \leq (c+k-1)^{k-j}2^{k-1}$. Thus we find

$$|\,\mathrm{Re}(c+it)_k| \leq \sum_{j=0}^{(k-1)/2}(c+k-1)^{k-2j}2^{k-1}t^{2j}.$$

On the other hand, $|(c+it)_k|^2 \geq t^{2k}$. Thus the integrand in (5.15) is bounded by $\sum_{j=0}^{(k-1)/2}(c+k-1)^{k-2j}2^{k-1}t^{2j-2k}$. Integrating this function, we find the following upper bound for the error:

$$\sum_{j=0}^{\frac{k-1}{2}}(c+k-1)^{k-2j}2^{k-1}\frac{T_{\min}^{2j-2k+1}}{2k-2j-1} \leq (c+k-1)T_{\min}^{-k}2^{k-1}, \tag{5.16}$$

for $T > c+k-1$. This establishes inequality (5.14c) and completes the proof of the lemma. □

5.3 Pointwise Explicit Formulas

In this section, we establish two different versions of our pointwise explicit formulas: one with error term (Theorem 5.10), which will be the most useful to us in this book, as well as one without error term (Theorem 5.14). The latter requires more stringent assumptions.

Let η be a generalized fractal string as in Section 4.1, with associated geometric zeta function denoted ζ_η; see Definition 4.1(iv).

Recall from Section 4.1 and Figure 1.6 of Section 1.2 that the screen S is given as the graph of a bounded, real-valued continuous function $S(t)$, with the horizontal and vertical axes interchanged:

$$S = \{S(t) + it \colon t \in \mathbb{R}\}.$$

We assume in addition that $S(t)$ is a Lipschitz continuous function; i.e., there exists a nonnegative real number, denoted by $\|S\|_{\mathrm{Lip}}$, such that

$$|S(x) - S(y)| \leq \|S\|_{\mathrm{Lip}}|x-y|, \quad \text{for all } x, y \in \mathbb{R}.$$

We associate with the screen the following finite quantities:

$$\inf S := \inf_{t\in\mathbb{R}} S(t), \tag{5.17a}$$

and

$$\sup S := \sup_{t\in\mathbb{R}} S(t). \tag{5.17b}$$

We assume that $\sup S \leq D$; that is, $S(t) \leq D$ for every t. Further, recall from Section 4.1 that the window W is the part of the complex plane to the right of S; see formula (4.7).

Definition 5.2 (Languid). The generalized fractal string η is said to be *languid*[1] if its geometric zeta function ζ_η satisfies the following growth

---

[1]We wish to thank Erin Pearse for having suggested the term "languid" to refer to strings satisfying **L1** and **L2**. In [Lap-vF5], the hypotheses **L1** and **L2** were denoted by $(\mathbf{H}_1)$ and $(\mathbf{H}_2)$, and **L2′** of Definition 5.3 below was denoted $(\mathbf{H}_2')$.

conditions: There exist real constants κ and $C > 0$ and a two-sided sequence $\{T_n\}_{n\in\mathbb{Z}}$ of real numbers such that $T_{-n} < 0 < T_n$ for $n \geq 1$, and

$$\lim_{n\to\infty} T_n = \infty, \quad \lim_{n\to\infty} T_{-n} = -\infty, \quad \lim_{n\to+\infty} \frac{T_n}{|T_{-n}|} = 1, \tag{5.18}$$

such that

L1 For all $n \in \mathbb{Z}$ and all $\sigma \geq S(T_n)$,

$$|\zeta_\eta(\sigma + iT_n)| \leq C \cdot (|T_n| + 1)^\kappa, \tag{5.19}$$

L2 For all $t \in \mathbb{R}$, $|t| \geq 1$,

$$|\zeta_\eta(S(t) + it)| \leq C \cdot |t|^\kappa. \tag{5.20}$$

Hypothesis **L1** is a polynomial growth condition along horizontal lines (necessarily avoiding the poles of ζ_η), while hypothesis **L2** is a polynomial growth condition along the vertical direction of the screen. We will need to assume these hypotheses to establish our (pointwise and distributional) explicit formulas with error term, Theorems 5.10 and 5.18 below.

Sometimes we can obtain an explicit formula without error term; see Theorems 5.14 and 5.22. In that case, in addition to **L1** (for every $\sigma \in \mathbb{R}$), we need to assume a stronger form of hypothesis **L2**.

Definition 5.3 (Strongly languid). We say that η is *strongly languid* if its geometric zeta function ζ_η satisfies the following condition, in addition to **L1** (with $S(t) \equiv -\infty$ in (5.19) above; i.e., for every $\sigma \in \mathbb{R}$): There exists a sequence of screens $S_m\colon t \mapsto S_m(t) + it$ for $m \geq 1$, $t \in \mathbb{R}$, with $\sup S_m \to -\infty$ as $m \to \infty$ and with a uniform Lipschitz bound $\sup_{m\geq 1} \|S_m\|_{\mathrm{Lip}} < \infty$, such that

L2′ There exist constants $A, C > 0$ such that for all $t \in \mathbb{R}$ and $m \geq 1$,

$$|\zeta_\eta(S_m(t) + it)| \leq CA^{|S_m(t)|}(|t| + 1)^\kappa. \tag{5.21}$$

Clearly, condition **L2′** is stronger than **L2**. Hence, if η is strongly languid, it is also languid (for each screen S_m separately).

Sometimes, it will be convenient to say that ζ_η—rather than η—is languid (or strongly languid), in the sense of Definition 5.2 (or 5.3).

Remark 5.4. In view of (5.19) and (5.20), if η is languid for some κ, then it is also languid for every larger value of κ. In view of (5.21), the same statement holds for strongly languid strings.

Remark 5.5. The hypothesis that the generalized fractal string η is languid is needed to establish our pointwise or distributional explicit formulas with error term (Theorems 5.10, 5.18, and 5.26), while the stronger hypothesis that η is strongly languid is required to establish our pointwise or distributional explicit formulas without error term (Theorems 5.14, 5.22, and 5.27). (See also Theorems 8.1 and 8.7, for example.)

Remark 5.6. We could have formulated **L1** and **L2** with different values of κ, or even with a sequence of real exponents $\{\kappa_m\}_{m=1}^{\infty}$ in **L2′**. One reason not to do this is that if ζ_η has no poles, or finitely many poles, then, by Lindelöf's Theorem [Edw, p. 184], **L1** is implied by **L2** (or **L2′**), with the same value of κ.

Remark 5.7. As is explained at the beginning of Section 6.4, self-similar strings are always strongly languid. Thus, for a self-similar string we can let $W = \mathbb{C}$ and obtain explicit formulas without error term, at least at the geometric or dynamical (but not at the spectral) level. See for example, Sections 1.1.1, 1.2.2, 6.4.1, 6.4.2, 7.4.2, 7.4.3 and 8.4, for concrete illustrations of this statement. This is especially useful for lattice strings, where closed formulas can be obtained. On the other hand, for nonlattice strings, it is often convenient to use an appropriate screen along which the string is languid, in order to obtain suitable remainder estimates. See, for example, Sections 7.5 and 8.4.4.

The next definition will be used to formulate the truncated pointwise formula (Lemma 5.9 below), which is the first step towards our explicit formulas.

Definition 5.8. Given an integer $n \geq 1$, the *truncated screen* $S_{|n}$ is the part of the screen S restricted to the interval $[T_{-n}, T_n]$, and the *truncated window* $W_{|n}$ is the window W intersected with $\{s \in \mathbb{C} \colon T_{-n} \leq \operatorname{Im} s \leq T_n\}$. (See Figure 5.1 on page 146.)

The set of *truncated visible complex dimensions* is

$$\mathcal{D}(W_{|n}) = \mathcal{D}_\eta(W_{|n}) := \mathcal{D}_\eta(W) \cap \{s \in \mathbb{C} \colon T_{-n} \leq \operatorname{Im} s \leq T_n\}. \tag{5.22}$$

It is the set of visible complex dimensions of η with imaginary part between T_{-n} and T_n.

We begin by proving a technical lemma that summarizes the estimates that we will need in order to establish both the pointwise formulas in this section and the distributional formulas in the next section.

Note that given $\alpha, \beta \in \mathbb{R}$, with $\alpha \leq \beta$, we have

$$\max\{x^\alpha, x^\beta\} = \begin{cases} x^\alpha, & \text{if } 0 < x < 1, \\ x^\beta, & \text{if } x \geq 1. \end{cases} \tag{5.23}$$

In the following, it will also be useful to keep in mind that, in view of (5.17), we have

$$\inf S \leq \sup S \leq D \tag{5.24}$$

and

$$S \subset \{s \in \mathbb{C}\colon \inf S \leq \operatorname{Re} s \leq \sup S\}. \tag{5.25}$$

Lemma 5.9 (Truncated pointwise formula). *Let $k \geq 1$ be an integer and let η be a generalized fractal string. Then, for all $x > 0$ and $n \geq 1$, the function $N_\eta^{[k]}(x)$ is approximated by*

$$\sum_{\omega \in \mathcal{D}_\eta(W_{|n})} \operatorname{res}\left(\frac{x^{s+k-1}\zeta_\eta(s)}{(s)_k}; \omega\right) + \frac{1}{(k-1)!} \sum_{\substack{j=0 \\ -j \in W \setminus \mathcal{D}_\eta}}^{k-1} \binom{k-1}{j} (-1)^j x^{k-1-j} \zeta_\eta(-j) + \frac{1}{2\pi i} \int_{S_{|n}} x^{s+k-1} \zeta_\eta(s) \frac{ds}{(s)_k}, \tag{5.26}$$

where $(s)_k$ is given by (5.12) and where $S_{|n}$ and $\mathcal{D}_\eta(W_{|n}) = \mathcal{D}(W_{|n})$ are given as in Definition 5.8, while $\mathcal{D}_\eta = \mathcal{D}_\eta(W)(= \mathcal{D})$ is defined by (4.8).

More precisely, assume hypothesis **L1** *and let*[2]

$$T_{\max} = \max\{T_n, |T_{-n}|\} \quad \text{and} \quad T_{\min} = \min\{T_n, |T_{-n}|\}.$$

Let $c > D$. Then, for all $x > 0$ and all integers $n \geq 1$, the difference between $N_\eta^{[k]}(x)$ and the expression in (5.26) is bounded in absolute value by

$$d(x,n) := 2x^{k-1} T_{\min}^{-k} \cdot [\text{a sum of four terms}], \tag{5.27}$$

where the four terms are

$$x^c \zeta_{|\eta|}(c) T_{\min}^{1/2}, \tag{5.28a}$$

$$T_{\max} \cdot \left(|\eta|(x - xT_{\min}^{-1/2}, x) + |\eta|(x, x + xT_{\min}^{-1/2})\right), \tag{5.28b}$$

$$|\eta|(\{x\}) \cdot \begin{cases} T_{\max} & (\text{for even } k), \\ (c+k-1)2^{k-1} + T_{\max} - T_{\min} & (\text{for odd } k), \end{cases} \tag{5.28c}$$

---

[2]For notational simplicity, we do not indicate explicitly the dependence of $T_{\max}$ and $T_{\min}$ on the integer n. This convention should be kept in mind when reading the proof of Theorems 5.10 and 5.14 below.

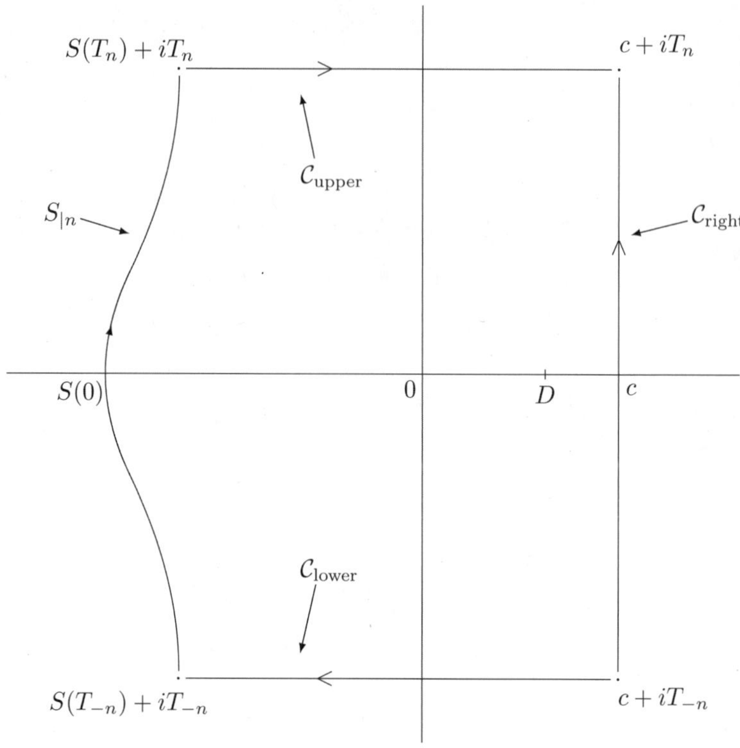

Figure 5.1: The contour $\mathcal{C}$.

and

$$CT_{\max}^{\kappa}(c - \inf S) \max\{x^c, x^{\inf S}\}. \tag{5.28d}$$

Further, for each point $s = S(t)+it$ ($|t| \geq 1$, $t \in \mathbb{R}$) *and for all* $x > 0$, *the integrand in the integral over the truncated screen* $S_{|n}$ *occurring in* (5.26) (*namely,* $x^{s+k-1}\zeta_\eta(s)/(s)_k$) *is bounded in absolute value by*

$$Cx^{k-1} \max\{x^{\sup S}, x^{\inf S}\}|t|^{\kappa-k}, \tag{5.29}$$

when hypothesis **L2** *holds, and by*

$$CA^{-\inf S}x^{k-1} \max\{x^{\sup S}, x^{\inf S}\}|t|^{\kappa-k}, \tag{5.29$'$}$$

when the stronger hypothesis **L2**$'$ *holds.* (*See Equations* (5.23) *and* (5.24) *above.*)

Proof. The proof is given in two steps. The first step consists of deriving an approximate expression for $N_\eta^{[k]}(x)$. For this, we consider the line integral

$$J(x,n) = \frac{1}{2\pi i}\int_{c+iT_{-n}}^{c+iT_n} x^{s+k-1}\zeta_\eta(s)\frac{ds}{(s)_k},$$

for some $c > D$. (See the contour $\mathcal{C}$ in Figure 5.1.) We substitute the expression of formula (4.4),

$$\zeta_\eta(s) = \int_0^\infty y^{-s}\eta(dy),$$

and interchange the order of integration. This interchange is justified since the integral is bounded by[3]

$$x^{c+k-1}\frac{1}{2\pi}\int_{T_{-n}}^{T_n}\int_0^\infty y^{-c}|\eta|(dy)\frac{dt}{c^k} \leq \frac{T_n - T_{-n}}{2\pi c^k}x^{c+k-1}\zeta_{|\eta|}(c).$$

We find that

$$J(x,n) = \int_0^\infty \frac{1}{2\pi i}\int_{c+iT_{-n}}^{c+iT_n} x^{s+k-1}y^{-s}\frac{ds}{(s)_k}\,\eta(dy). \tag{5.30}$$

In view of (5.11) and (5.30), the difference $N_\eta^{[k]}(x) - J(x,n)$ can now be bounded via an application of Lemma 5.1 (with $T_- = T_{-n}$ and $T_+ = T_n$). The absolute value of this difference is bounded by

$$\begin{aligned}
&\int_0^\infty \left| H^{[k]}(x-y) - \frac{1}{2\pi i}\int_{c+iT_{-n}}^{c+iT_n} x^{s+k-1}y^{-s}\frac{ds}{(s)_k}\right| |\eta|(dy) \\
&\leq \int_{y\neq x} x^{c+k-1}y^{-c}T_{\min}^{-k}\cdot\min\left\{T_{\max}, \frac{1}{|\log x - \log y|}\right\}|\eta|(dy) \qquad (5.31)\\
&\quad + x^{k-1}T_{\min}^{-k}|\eta|(\{x\})\cdot\begin{cases} T_{\max} & (k \text{ even}),\\ T_{\max} - T_{\min} + 2^{k-1}(c+k-1) & (k \text{ odd}).\end{cases}
\end{aligned}$$

To obtain a good bound for the last integral in this inequality, we split it as a sum of two integrals: one integral over y such that $y \leq x(1-T_{\min}^{-1/2})$ or $y \geq x(1+T_{\min}^{-1/2})$, where $|\log x - \log y| \gg T_{\min}^{-1/2}$, and another integral over the two open intervals in between, namely, $(x - xT_{\min}^{-1/2}, x)$ and $(x, x+xT_{\min}^{-1/2})$. The first integral is bounded by

$$x^{c+k-1}\zeta_{|\eta|}(c)T_{\min}^{-k}T_{\min}^{1/2}, \tag{5.32a}$$

and the second by

$$2x^{c+k-1}x^{-c}T_{\min}^{-k}T_{\max}\left(|\eta|(x - xT_{\min}^{-1/2}, x) + |\eta|(x, x+xT_{\min}^{-1/2})\right). \tag{5.32b}$$

---

[3]For simplicity, a reader unfamiliar with (local) complex-valued measures may wish to assume throughout the proofs given in this chapter that η is a positive measure on $(0,\infty)$ that is locally bounded; i.e., η is bounded for every bounded subinterval of $(0,\infty)$. Then, one may set $\eta = |\eta|$ in all the arguments presented here. See Remark 4.3.

Thus, the difference $|N_\eta^{[k]}(x) - J(x,n)|$ is bounded by the sum of three terms: (5.32a), (5.32b) and the term

$$x^{k-1}T_{\min}^{-k}\cdot|\eta|(\{x\})\cdot\begin{cases}T_{\max} & (k \text{ even}),\\ T_{\max}-T_{\min}+2^{k-1}(c+k-1) & (k \text{ odd}).\end{cases} \tag{5.32c}$$

Next we compute $J(x,n)$ by replacing the right contour in Figure 5.1 (that is, the line segment from $c+iT_{-n}$ to $c+iT_n$) by $\mathcal{C}_{\text{lower}}+S_{|n}+\mathcal{C}_{\text{upper}}$. Here, $S_{|n}$ is the truncated screen; i.e., the part of the screen for t going from T_{-n} to T_n, and the upper and lower parts of the contour are the horizontal lines $s=\sigma+iT_{\pm n}$, for $S(T_{\pm n})\le\sigma\le c$. (See Figure 5.1.) By the Theorem of Residues [Ahl, Theorem 17, p. 150], we obtain the following expression for $J(x,n)$,

$$\sum_{\omega\in\mathcal{D}(W_{|n})}\operatorname{res}\left(\frac{x^{s+k-1}\zeta_\eta(s)}{(s)_k};\omega\right)+\sum_{\substack{j=0\\ -j\in W\setminus\mathcal{D}}}^{k-1}\operatorname{res}\left(\frac{x^{s+k-1}\zeta_\eta(s)}{(s)_k};-j\right)$$
$$+\frac{1}{2\pi i}\int_{S_{|n}}x^{s+k-1}\zeta_\eta(s)\frac{ds}{(s)_k}+\frac{1}{2\pi i}\int_{\mathcal{C}_{\text{lower}}+\mathcal{C}_{\text{upper}}}x^{s+k-1}\zeta_\eta(s)\frac{ds}{(s)_k}.$$

A computation similar to that performed in the proof of Lemma 5.1 above shows that the residue at $-j$ is equal to

$$\frac{1}{(k-1)!}\binom{k-1}{j}(-1)^j x^{k-1-j}\zeta_\eta(-j),$$

provided that $-j$ is not a pole of ζ_η; i.e., provided that $-j\notin\mathcal{D}$. The integral over the upper side $\mathcal{C}_{\text{upper}}$ is bounded by

$$\begin{aligned}\left|\frac{1}{2\pi i}\int_{S(T_n)}^{c}x^{\sigma+iT_n+k-1}\zeta_\eta(\sigma+iT_n)\frac{d\sigma}{(\sigma+iT_n)_k}\right|&\\ \le\frac{1}{2\pi}x^{k-1}CT_n^{\kappa}T_n^{-k}\int_{S(T_n)}^{c}x^\sigma\,d\sigma&\\ \le\frac{C}{2\pi}T_n^{\kappa-k}x^{k-1}(c-\inf S)\max\{x^c,x^{\inf S}\},&\end{aligned} \tag{5.32d}$$

by hypothesis **L1**. The integral over the lower side $\mathcal{C}_{\text{lower}}$ is bounded similarly.

In summary, we have now obtained the desired estimate:

$$\left| N_\eta^{[k]}(x) - \sum_{\omega\in\mathcal{D}(W_{|n})} \operatorname{res}\left(\frac{x^{s+k-1}\zeta_\eta(s)}{(s)_k};\omega\right) - \frac{1}{(k-1)!}\sum_{\substack{j=0\\ -j\in W\setminus\mathcal{D}}}^{k-1}\binom{k-1}{j}(-1)^j x^{k-1-j}\zeta_\eta(-j) - \frac{1}{2\pi i}\int_{S_{|n}} x^{s+k-1}\zeta_\eta(s)\frac{ds}{(s)_k}\right| \leq d(x,n), \tag{5.33}$$

where $d(x,n)$ is the sum of the four terms (5.32a)–(5.32d).

Finally, at $s = S(t)+it$, we have $|(s)_k| \geq |t|^k$. Hence the integrand in (5.33) at $s = S(t)+it$ ($|t|\geq 1$) is bounded by a constant times

$$x^{k-1+S(t)}|t|^{\kappa-k} \leq x^{k-1}\max\{x^{\sup S}, x^{\inf S}\}|t|^{\kappa-k}.$$

This constant is either C, when we assume **L2**, or $C\cdot A^{-S(t)} \leq C\cdot A^{-\inf S}$, when we assume **L2′**. This completes the proof of Lemma 5.9. □

We can now state the main result of this section, which will be used in most situations (in this and in the later chapters) to obtain pointwise explicit formulas for geometric or spectral counting functions. In the next section, we will obtain its distributional analogue, Theorem 5.18. (See also Theorem 5.26 along with Theorem 5.30.)

Theorem 5.10 (The pointwise explicit formula, with error term). *Let η be a languid*[4] *generalized fractal string, and let k be an integer such that $k > \max\{1, \kappa+1\}$, where κ is the exponent occurring in the statement of* **L1** *and* **L2**. *Then, for all $x>0$, the pointwise explicit formula is given by the following equality:*[5]

$$N_\eta^{[k]}(x) = \sum_{\omega\in\mathcal{D}_\eta(W)} \operatorname{res}\left(\frac{x^{s+k-1}\zeta_\eta(s)}{(s)_k};\omega\right) + \frac{1}{(k-1)!}\sum_{\substack{j=0\\ -j\in W\setminus\mathcal{D}_\eta}}^{k-1}\binom{k-1}{j}(-1)^j x^{k-1-j}\zeta_\eta(-j) + R_\eta^{[k]}(x). \tag{5.34}$$

[4]That is, it satisfies hypotheses **L1** and **L2** of Definition 5.2, Equations (5.19) and (5.20) above.

[5]Recall that $\mathcal{D}_\eta = \mathcal{D}_\eta(W)(=\mathcal{D})$ denotes the set of visible complex dimensions of η (for the window W), and that $(s)_k$ is defined by (5.12). The complement of $\mathcal{D}_\eta(W)$ in W is denoted by $W\setminus\mathcal{D}_\eta$.

Here, for $x > 0$, $R(x) = R_\eta^{[k]}(x)$ is the error term, given by the absolutely convergent integral

$$R(x) = R_\eta^{[k]}(x) = \frac{1}{2\pi i} \int_S x^{s+k-1} \zeta_\eta(s) \frac{ds}{(s)_k}. \tag{5.35}$$

Further, for all $x > 0$, we have

$$R(x) = R_\eta^{[k]}(x) \leq C(1 + \|S\|_{\mathrm{Lip}}) \frac{x^{k-1}}{k - \kappa - 1} \max\{x^{\sup S}, x^{\inf S}\} + C', \tag{5.36}$$

where C is the positive constant occurring in **L1** *and* **L2** *and C' is some suitable positive constant. The constants $C(1 + \|S\|_{\mathrm{Lip}})$ and C' depend only on η and the screen, but not on k. (Here, $\inf S$ and $\sup S$ are given by (5.17).)*

In particular, we have the following pointwise error estimate:

$$R(x) = R_\eta^{[k]}(x) = O\big(x^{\sup S + k - 1}\big), \tag{5.37}$$

as $x \to \infty$. Moreover, if $S(t) < \sup S$ for all $t \in \mathbb{R}$ (i.e., if the screen lies strictly to the left of the line $\operatorname{Re} s = \sup S$), then $R(x)$ is of order less than $x^{\sup S + k - 1}$ as $x \to \infty$:

$$R(x) = R_\eta^{[k]}(x) = o\big(x^{\sup S + k - 1}\big), \tag{5.38}$$

as $x \to \infty$.

Remark 5.11. The proof of Theorem 5.10 does not give information about the nature of the convergence of the sum over the complex dimensions ω in (5.34). Therefore, we need to specify the meaning of $\sum_{\omega \in \mathcal{D}_\eta(W)} \cdots$ as the limit $\lim_{n \to +\infty} \sum_{\omega \in \mathcal{D}_\eta(W_{|n})} \cdots$, where $W_{|n}$ is the truncated window, given by Definition 5.8. (The fact that this limit exists follows from the proof of the theorem.) The same remark applies to the sum occurring in formula (5.40) of Theorem 5.14 below.

Further, the precise form of the term corresponding to ω in this sum will change depending on the multiplicity of the pole of ζ_η at $s = \omega$. In particular, a multiple pole will give rise to logarithmic terms. On the other hand, if $\omega \in \mathcal{D}_\eta(W)$ is a simple pole, then the corresponding term becomes

$$\operatorname{res}\left(\frac{x^{s+k-1} \zeta_\eta(s)}{(s)_k}; \omega \right) = \operatorname{res}\left(\zeta_\eta(s); \omega\right) \frac{x^{\omega+k-1}}{(\omega)_k}. \tag{5.39}$$

See Section 6.1 for a more elaborate discussion of the local term associated with ω.

Remark 5.12. In most applications, we have a lot of freedom for choosing the sequence $\{T_n\}_{n \in \mathbb{Z}}$ satisfying hypothesis **L1**. If η is a real-valued (rather than a complex-valued) measure, we can always choose W to be symmetric and set $T_{-n} = -T_n$ for all $n \geq 1$. Then, in the sum over the complex dimensions ω in (5.34), we can collect the terms in complex conjugate pairs $\omega, \overline{\omega}$ (as is done in classical number-theoretic explicit formulas involving the Riemann zeta function [Da, Edw, In, Pat]).

Finally, we note that in practice we can often choose the screen S to be a vertical line (with equation $\operatorname{Re} s = \sigma$, say). Then $S(t) \equiv \sigma$ is clearly Lipschitz continuous and bounded, and we have $\sup S = \inf S = \sigma$.

Similar remarks apply to all explicit formulas in this chapter.

Remark 5.13. Note that our hypotheses **L1** and **L2** imply, in particular, that ζ_η has an analytic continuation to a neighborhood of $[\operatorname{Re} s \geq D]$. It is in general hard to determine whether this condition is satisfied. Indeed, this condition is essentially equivalent to the existence of a suitable explicit formula for the counting function N_η. Establishing such a result would entail developing a theory of almost periodic functions with amplitude of polynomial growth (rather than with bounded amplitude as in [Bohr]). This theory would be of independent interest. It would be naturally motivated by our explicit formulas, in which the real parts of the underlying complex dimensions give rise to generalized Fourier series with variable amplitudes. See also Problem 4.11 at the end of the previous chapter.

Proof of Theorem 5.10. For a given $n \geq 1$, we apply Lemma 5.9, the truncated pointwise formula, with $T_+ = T_n$, $T_- = T_{-n}$. Since $k \geq 2$, the first term (5.32c) of $d(x,n)$ above tends to zero as $n \to \infty$. Also the second term (5.32a) and the middle term (5.32b) go to 0 as $n \to \infty$. Finally, since $k > \kappa$, the last term tends to zero. All these terms tend to zero at the rate of some negative power of $T_{\min}$.[6]

Finally, the error term is absolutely convergent since $k > \kappa + 1$. Note that $S(t)$ is differentiable almost everywhere since it is assumed to be Lipschitz. Furthermore, since for almost every $t \in \mathbb{R}$, the derivative of $t \mapsto S(t) + it$ is bounded by $1 + \|S\|_{\mathrm{Lip}}$, where $\|S\|_{\mathrm{Lip}}$ denotes the Lipschitz norm of $S(t)$, and since

$$\int_1^\infty t^{\kappa - k}\, dt = \frac{1}{k - \kappa - 1}$$

(because $\kappa - k < -1$, by assumption), the estimate (5.36) of the error term follows from (5.29). The constant C' on the right-hand side of (5.36) comes from a bound for the integral over the part of the screen for $-1 \leq t \leq 1$.

To obtain the better estimate (5.38) when the screen S approximates the vertical line $\operatorname{Re} s = \sup S$, but stays to the left of it, we use a well-known method to estimate (5.35); see, e.g., [In, pp. 33–34]. Given $\varepsilon > 0$, we want

[6] Recall from Lemma 5.1 that $T_{\max} = \max\{T_n, |T_{-n}|\}$ and $T_{\min} = \min\{T_n, |T_{-n}|\}$ both depend on the integer n.

to show that (5.35) is bounded by $\varepsilon \cdot x^{\sup S+k-1}$. Write this integral as the sum of the integral over the part of S contained in $-T \leq \operatorname{Im} s \leq T$, and the part of S contained in $|\operatorname{Im} s| > T$. Since the second integral is absolutely convergent, it is bounded by $\frac{1}{2}\varepsilon \cdot x^{\sup S+k-1}$, provided we choose T sufficiently large. Then, since $S(t)$ has a maximum strictly less than $\sup S$ on the compact interval $[-T, T]$, we can find $\delta > 0$ such that $S(t) \leq \sup S - \delta$ for $|t| \leq T$. It follows that the integral over this part of the screen is of order $O(x^{\sup S-\delta+k-1})$ as $x \to \infty$. Hence for large x, it is bounded by $\frac{1}{2}\varepsilon \cdot x^{\sup S+k-1}$. This proves that $R(x) = o\big(x^{\sup S+k-1}\big)$ as $x \to \infty$. □

We assume in Theorem 5.10 that $k \geq 2$. However, in certain situations—namely, when ζ_η satisfies the stronger hypothesis **L2′** rather than **L2**—we can push the screen S to $-\infty$ and thereby obtain a pointwise explicit formula without error term, valid for a lower level $k > \kappa$ (rather than $k > \kappa + 1$, as in Theorem 5.10).

We will apply this result, in particular, to the geometric counting function (at level 1) of a self-similar string, as defined in Section 2.2 (see Section 6.4). We will see that we can set $\kappa = 0$, so that $k = 1$ will indeed be allowed in applying formula (5.40) below.

Theorem 5.14 (The pointwise formula, without error term). *Let η be a generalized fractal string satisfying hypotheses* **L1** *and* **L2′**; *i.e., η is strongly languid* (*see Definition 5.3 and Equations* (5.19) *and* (5.21) *above*). *Let k be a positive integer such that $k > \kappa$. Then, for all $x > A$, the pointwise explicit formula is given by the following equality*:

$$N_\eta^{[k]}(x) = \sum_{\omega \in \mathcal{D}_\eta(W)} \operatorname{res}\left(\frac{x^{s+k-1}\zeta_\eta(s)}{(s)_k}; \omega\right) + \frac{1}{(k-1)!} \sum_{\substack{j=0 \\ -j \in W \setminus \mathcal{D}_\eta}}^{k-1} \binom{k-1}{j} (-1)^j x^{k-1-j} \zeta_\eta(-j). \tag{5.40}$$

Here, A is the positive number given by hypothesis **L2′** (*cf. Equation* (5.21)) *and κ is the exponent occurring in the statement of hypotheses* **L1** *and* **L2′**.

Proof. For a fixed integer $n \geq 1$, we apply Lemma 5.9 with the screen S_m given by hypothesis **L2′**. We now assume that $k > \kappa$ (instead of $k > \kappa + 1$ as in the proof of Theorem 5.10). We first let m tend to ∞, while keeping n fixed. Since the functions $S_m(t)$ have a uniform Lipschitz bound, the sequence of integrals over the screens tends to 0 as $m \to \infty$, provided $x > A$.

Then we let $n \to \infty$. For $k \geq 2$, in Equation (5.27), the first three terms of $d(x, n)$ tend to zero as $n \to \infty$, as we have seen in the proof of Theorem 5.10, and for $k > \kappa$, the last term also tends to zero.

Further, if $k = 1$, the first term (5.32c) of $d(x, n)$ tends to 0, by our assumption that $T_{\max}/T_{\min} \to 1$ as $n \to \infty$. Note that this assumption implies that $(T_{\max} - T_{\min})/T_{\min} \to 0$ as $n \to \infty$. The situation is more complicated for the middle term (5.32b), in case $k = 1$, since η can have a large portion of its mass close to x. However, by writing the interval $(x, 2x)$ as a disjoint countable union

$$(x, 2x) = \bigcup_{j=1}^{\infty} \left[x + \frac{x}{j+1}, x + \frac{x}{j}\right),$$

and noting that $|\eta|(x, 2x)$ is finite, we see that

$$|\eta| \left(x, x + \frac{x}{m}\right) = \sum_{j=m}^{\infty} |\eta| \left[x + \frac{x}{j+1}, x + \frac{x}{j}\right)$$

goes to 0 when $m \to \infty$, at some rate depending on the positive measure $|\eta|$ (on the given interval $(x, 2x)$). Thus also $|\eta|(x, x + xT_{\min}^{-1/2}) \to 0$ as $n \to \infty$. Similarly, we check that $|\eta|(x - xT_{\min}^{-1/2}, x) \to 0$ as $n \to \infty$. This completes the proof of Theorem 5.14. $\square$

Remark 5.15. As can be seen from the above proof, the full strength of the asymptotic symmetry condition $\lim_{n\to+\infty} T_n/|T_{-n}| = 1$ is needed only when $k = 1$, which is only allowed in Theorem 5.14.

Remark 5.16. There is another situation in which we can push the screen to $-\infty$ and obtain an explicit formula without error term. Indeed, sometimes the geometric zeta function satisfies a functional equation (in the sense of [JorLan3]). Then the theory of [JorLan3] (see also [JorLan1–2]) applies to yield an explicit formula without error term, but with a Weil term [Wei4, 6], coming from the fudge factor involved in the statement of the functional equation in [JorLan3].

5.3.1 The Order of the Sum over the Complex Dimensions

In the applications given in Chapters 7, 8, 9 and 11 below, we often want to single out the sum over the complex dimensions on the line $\operatorname{Re} s = D$, and estimate the sum over those to the left of $\operatorname{Re} s = D$. We also want the resulting error term to be of lower order. If we can choose a screen that makes all complex dimensions to the left of this line invisible, then we apply the estimate (5.38). But as Examples 5.32 and 5.33 show, we cannot always choose such a screen. In that case, we use the following argument.

Theorem 5.17. *Let*

$$\sum_{\operatorname{Re} \omega < D} a_\omega x^\omega (\log x)^{m_\omega} \tag{5.41}$$

be an absolutely convergent sum over the visible complex dimensions with real part less than D, arising from Theorem 5.10 or 5.14 (hence $a_\omega \in \mathbb{C}$

and $m_\omega \in \mathbb{N}$ for each ω). Then this sum is of order $o(x^D)$, as $x \to \infty$, and the corresponding sum

$$\sum_{\operatorname{Re}\omega = D} a_\omega x^\omega (\log x)^{m_\omega}$$

converges.

Proof. The (conditional) convergence of the sum over the complex dimensions with real part D follows from Theorem 5.10 or 5.14, and the assumption that (5.41) converges absolutely.

It remains to estimate the sum in (5.41). This is done by adapting the method of [In, pp. 82 and 33] as follows. Let $\varepsilon > 0$. Since (5.41) is absolutely convergent, the sum over ω with $|\operatorname{Im}\omega| > T$ is less than εx^D for T large enough. Next, since there are only finitely many visible complex dimensions ω with $|\operatorname{Im}\omega| < T$, their real parts attain a maximum $M < D$. Then (5.41) is bounded by $\varepsilon x^D + C x^M < 2\varepsilon x^D$ for large enough x. This shows that the expression in (5.41) is $o\left(x^D\right)$, as $x \to \infty$. □

5.4 Distributional Explicit Formulas

In this section, we give a distributional counterpart of the pointwise explicit formulas obtained in the previous section. We also obtain several refinements of the distributional formula that will be needed in the rest of this book; see Sections 5.4.1 and 5.4.2.

We view η as a distribution, acting on a test function φ by $\int_0^\infty \varphi\, d\eta$. More generally, for $k \geq 1$, the *k-th primitive* $\mathcal{P}_\eta^{[k]}$ of η is the distribution given by Equation (5.5) or, equivalently, (5.4). For $k \leq 0$, we extend this definition by differentiating $|k| + 1$ times the distribution $\mathcal{P}_\eta^{[1]}$. Thus, in particular, $\mathcal{P}_\eta^{[0]} = \eta$.

In the following,

$$(s)_k = \frac{\Gamma(s+k)}{\Gamma(s)}, \tag{5.42}$$

for $k \in \mathbb{Z}$. This extends the definition of the Pochhammer symbol (5.12) to $k \leq 0$. Thus, $(s)_0 = 1$ and, for $k \geq 1$, $(s)_k = s(s+1)\cdots(s+k-1)$.

We denote by $\widetilde{\varphi}$ the *Mellin transform* of a (suitable) function φ on $(0, \infty)$; it is defined by

$$\widetilde{\varphi}(s) = \int_0^\infty \varphi(x) x^{s-1}\, dx. \tag{5.43}$$

As before, we denote by $\operatorname{res}(g(s); \omega)$ the residue of a meromorphic function $g = g(s)$ at ω. It vanishes unless ω is a pole of g. Since $\operatorname{res}(g(s); \omega)$ is linear in g, we have that

$$\int_0^\infty \varphi(x) \operatorname{res}\left(x^{s+k-1} g(s); \omega\right) dx = \operatorname{res}\left(\widetilde{\varphi}(s+k) g(s); \omega\right), \tag{5.44}$$

for φ in any of the classes of test functions considered in this chapter.

We first formulate our distributional explicit formulas, Theorems 5.18 and 5.22, with error term and without error term respectively, for (complex-valued) test functions φ in the class $C^\infty(0,\infty)$ (or $C^\infty(A,\infty)$, respectively),[7] with $\varphi^{(q)}(t)t^m \to 0$ for all $m \in \mathbb{Z}$ and $q \in \mathbb{N}$, as $t \to 0^+$ (respectively, as $t \to A^+$, for Theorem 5.22) or as $t \to \infty$; see also Remark 5.21. We then formulate the most general conditions under which our explicit formula applies; see Theorems 5.26 and 5.27. In particular, we want to weaken the decay condition on φ at 0 and ∞ and the differentiability assumption on φ. (See especially Section 5.4.1 below, the results of which will be applied, in particular, in Section 6.2.3 and in Chapter 8.)

The following result provides a distributional analogue of the pointwise formula obtained in Theorem 5.10. This result will be complemented below by Theorem 5.30, which will provide an estimate for the distributional error term.

Theorem 5.18 (The distributional formula, with error term). *Let η be a languid generalized fractal string; i.e., it satisfies hypotheses* **L1** *and* **L2** *(see Equations* (5.19) *and* (5.20) *above). Then, for every $k \in \mathbb{Z}$, the distribution $\mathcal{P}_\eta^{[k]}$ is given by*

$$\mathcal{P}_\eta^{[k]}(x) = \sum_{\omega \in \mathcal{D}_\eta(W)} \operatorname{res}\left(\frac{x^{s+k-1}\zeta_\eta(s)}{(s)_k}; \omega\right) + \frac{1}{(k-1)!} \sum_{\substack{j=0 \\ -j \in W \setminus \mathcal{D}_\eta}}^{k-1} \binom{k-1}{j} (-1)^j x^{k-1-j} \zeta_\eta(-j) + \mathcal{R}_\eta^{[k]}(x). \tag{5.45a}$$

That is, the action of $\mathcal{P}_\eta^{[k]}$ on a test function φ is given by

$$\langle \mathcal{P}_\eta^{[k]}, \varphi \rangle = \sum_{\omega \in \mathcal{D}_\eta(W)} \operatorname{res}\left(\frac{\zeta_\eta(s)\widetilde{\varphi}(s+k)}{(s)_k}; \omega\right) + \frac{1}{(k-1)!} \sum_{\substack{j=0 \\ -j \in W \setminus \mathcal{D}_\eta}}^{k-1} \binom{k-1}{j} (-1)^j \zeta_\eta(-j) \widetilde{\varphi}(k-j) + \langle \mathcal{R}_\eta^{[k]}, \varphi \rangle. \tag{5.45b}$$

[7]Given an open interval J, $C^\infty(J)$ denotes the space of infinitely differentiable functions on J.

Here, the distribution $\mathcal{R} = \mathcal{R}_\eta^{[k]}$ *is the error term, given by*

$$\langle \mathcal{R}, \varphi \rangle = \langle \mathcal{R}_\eta^{[k]}, \varphi \rangle = \frac{1}{2\pi i} \int_S \zeta_\eta(s) \widetilde{\varphi}(s+k) \frac{ds}{(s)_k}. \tag{5.46}$$

Remark 5.19. In (5.45), the sum over j is interpreted as being equal to 0 if $k \leq 0$. The same comment applies to the corresponding sum in Equation (5.48) of Theorem 5.22 below.

Proof of Theorem 5.18. First, given an integer $n \geq 1$, we apply Lemma 5.9 for $k > \kappa + 1$. Let φ be a test function. Then

$$\langle \mathcal{P}_\eta^{[k]}, \varphi \rangle = \int_0^\infty N_\eta^{[k]}(x) \varphi(x)\, dx,$$

which is approximated by

$$\sum_{\omega \in \mathcal{D}_\eta(W_{|n})} \operatorname{res}\left(\frac{\zeta_\eta(s)\widetilde{\varphi}(s+k)}{(s)_k}; \omega \right) + \frac{1}{(k-1)!} \sum_{\substack{j=0 \\ -j \in W \setminus \mathcal{D}_\eta}}^{k-1} \binom{k-1}{j} (-1)^j \zeta_\eta(-j) \widetilde{\varphi}(k-j) + \frac{1}{2\pi i} \int_{S_{|n}} \zeta_\eta(s) \widetilde{\varphi}(s+k) \frac{ds}{(s)_k}.$$

(Here, we use the notation $\mathcal{D}_\eta(W_{|n})$ and $S_{|n}$ introduced in Definition 5.8.) Since $k > \kappa + 1$, the integral converges and the error of the approximation vanishes as $n \to \infty$, by the same argument as in the proof of Theorem 5.10. Then we derive formula (5.45a) for every $k \in \mathbb{Z}$ by differentiating, as a distribution, the above formula sufficiently many times. □

Remark 5.20. The method used above (which may be coined the *descent method*) shows that from a pointwise formula applied at a high enough level (so as to avoid any problem of convergence in the sum or in the integrals involved), we can deduce a corresponding distributional formula on every level. This is reminiscent of the method used to establish the convergence of the Fourier series associated to a periodic distribution. (See, for example, [Schw1, §VII, I, esp. p. 226] or [Schw2, Chapter IV].)

We now provide an alternate proof of this theorem. It is a direct proof, in the sense that by contrast to that given above, it does not involve the descent method of Remark 5.20.

Alternate proof of Theorem 5.18. Note that

$$\begin{aligned} \langle \mathcal{P}_\eta^{[k]}, \varphi \rangle &= (-1)^q \langle \mathcal{P}_\eta^{[k+q]}, \varphi^{(q)} \rangle \\ &= (-1)^q \int_0^\infty \varphi^{(q)}(x) N_\eta^{[k+q]}(x)\, dx, \end{aligned} \tag{5.47}$$

when $k+q>\kappa+1$. We then apply the pointwise explicit formula, Theorem 5.10. Using the fact that $\widetilde{\varphi}(s)=\frac{(-1)^q}{(s)_q}\widetilde{\varphi^{(q)}}(s+q)$ (see Equation (5.57) below), we obtain the explicit formula of Theorem 5.18. □

Remark 5.21. The (infinite) sum over the visible complex dimensions of η appearing on the right-hand side of (5.45a) defines a distribution. Hence, formula (5.45a) is not only an equality between distributions, but each term on the right-hand side of (5.45a) also defines a distribution. Indeed, this is a simple consequence of the following well-known fact about the convergence of distributions. (See, for example, [Schw1–2] or [Hö2, Theorem 2.18, p. 39]; this property was also used in a related context in [DenSchr, p. 50].) For definiteness, we will work with the space of test functions

$$\mathbf{D}=\mathbf{D}(0,\infty),$$

consisting of all infinitely differentiable functions with compact support contained in $(0,\infty)$, and the associated classical space of distributions

$$\mathbf{D}'=\mathbf{D}'(0,\infty),$$

the topological dual of $\mathbf{D}$; see [Schw1, Hö2]. (Note that $\mathbf{D}$ is contained in our space of admissible test functions.) Let $\{\mathcal{T}_n\}_{n=1}^{\infty}$ be a sequence of distributions in $\mathbf{D}'$ such that

$$\langle\mathcal{T},\varphi\rangle:=\lim_{n\to\infty}\langle\mathcal{T}_n,\varphi\rangle$$

exists for every test function $\varphi\in\mathbf{D}$. Then $\mathcal{T}$ is a distribution in $\mathbf{D}'$. (This follows from a well-known extension from Banach spaces to suitable topological vector spaces of the Uniform Boundedness Principle [Ru3, Theorems 2.5 and 2.8, pp. 44 and 46], also called the Banach–Steinhauss Theorem.)

In our present setting, we can apply this result to an appropriate sequence of partial sums (the sequence of partial sums mentioned in Remark 5.11, applied to φ) to deduce that the sum over $\mathcal{D}_\eta(W)$ on the right-hand side of (5.45a) is really a distribution in $\mathbf{D}'$, as stated above, and hence that each term taken separately on the right-hand side of (5.45a) defines a distribution in $\mathbf{D}'$.[8]

An entirely analogous comment applies to Theorem 5.22 below, the distributional explicit formula without error term, provided that we work instead with $\mathbf{D}=\mathbf{D}(A,\infty)$ and $\mathbf{D}'=\mathbf{D}'(A,\infty)$.

---

[8]Note that, for every $k\geq 1$, $N_\eta^{[k]}(x)$ defines $\mathcal{P}_\eta^{[k]}$ as a distribution in $\mathbf{D}'$ because $N_\eta^{[k]}(x)$ is a locally integrable function. Also, since η is a (local) measure, $\mathcal{P}_\eta^{[0]}=\eta$ is a distribution in $\mathbf{D}'$ having $\mathcal{P}_\eta^{[k]}$ as its k-th primitive.

Next, we obtain the distributional analogue of the pointwise formula without error term stated in Theorem 5.14. This is an asymptotic distributional formula; i.e., it applies to test functions that are supported on the right of $x = A$, where $A > 0$ is as in hypothesis **L2**$'$.

Theorem 5.22 (The distributional formula, without error term). *Let η be a generalized fractal string that is strongly languid; i.e., it satisfies hypotheses* **L1** *and* **L2**$'$ *(see Equations* (5.19) *and* (5.21) *above). Then, for every $k \in \mathbb{Z}$ and for test functions with compact support contained in (A, ∞), the distribution $\mathcal{P}_\eta^{[k]}$ in $\mathbf{D}'(A, \infty)$ is given by*

$$\mathcal{P}_\eta^{[k]}(x) = \sum_{\omega \in \mathcal{D}_\eta(W)} \operatorname{res}\left(\frac{x^{s+k-1}\zeta_\eta(s)}{(s)_k}; \omega\right) + \frac{1}{(k-1)!} \sum_{\substack{j=0 \\ -j \in W \setminus \mathcal{D}_\eta}}^{k-1} \binom{k-1}{j} (-1)^j x^{k-1-j} \zeta_\eta(-j). \quad (5.48a)$$

That is, the action of $\mathcal{P}_\eta^{[k]}$ on a test function φ with compact support contained in (A, ∞) (i.e., on $\varphi \in \mathbf{D}(A, \infty)$) is given by

$$\langle \mathcal{P}_\eta^{[k]}, \varphi \rangle = \sum_{\omega \in \mathcal{D}_\eta(W)} \operatorname{res}\left(\frac{\zeta_\eta(s)\widetilde{\varphi}(s+k)}{(s)_k}; \omega\right) + \frac{1}{(k-1)!} \sum_{\substack{j=0 \\ -j \in W \setminus \mathcal{D}_\eta}}^{k-1} \binom{k-1}{j} (-1)^j \zeta_\eta(-j) \widetilde{\varphi}(k-j). \quad (5.48b)$$

Proof. Again, given $n \geq 1$, we apply Lemma 5.9, but now for $k > \kappa$ (instead of $k > \kappa + 1$ as in the proof of Theorem 5.18). Let φ be a test function whose support is contained in $[A + \delta, \infty)$, for some $\delta > 0$. Then $\widetilde{\varphi}(s)$ is bounded by

$$(A + \delta)^\sigma \int_0^\infty |\varphi(x)| \frac{dx}{x}.$$

It follows that the integral over the screen S_m tends to 0 as $m \to \infty$. The rest of the argument is exactly the same as in the proof of Theorem 5.18 above. □

By applying Theorem 5.18 (respectively, 5.22) at level $k = 0$, we obtain the following result, which is central to our theory and will be used repeatedly in the rest of this book. (See Sections 6.3.1, 6.3.2 and 5.1.1 for further discussion and interpretation of (5.49) and, for example, Section 5.5 as well as Chapters 9 and 11 for applications of this result.)

Corollary 5.23 (The density of states formula). *Under the same hypotheses as in Theorem 5.18 (respectively, Theorem 5.22), we have the following distributional explicit formula for* $\mathcal{P}_\eta^{[0]} = \eta$*:*

$$\eta = \sum_{\omega \in \mathcal{D}_\eta(W)} \operatorname{res}\left(\frac{x^{s+k-1}\zeta_\eta(s)}{(s)_k}; \omega \right) + \mathcal{R}_\eta^{[0]}, \tag{5.49}$$

where $\mathcal{R}_\eta^{[0]}$ *is the distribution given by formula* (5.46) *with* $k = 0$ *(respectively, if* **L2′** *is satisfied,* $\mathcal{R}_\eta^{[0]} \equiv 0$*).*

5.4.1 Extension to More General Test Functions

We extend our distributional explicit formulas (Theorems 5.18 and 5.22) to a broader class of test functions φ, not necessarily C^∞ and decaying less rapidly near 0 and ∞ than considered previously. In particular, in Theorem 5.26, we allow test functions that have a suitable asymptotic expansion at $x = 0$. In the corresponding distributional explicit formula without error term, Theorem 5.27, we only allow test functions having a very special expansion on $(0, A + \delta)$, for some $\delta > 0$.

We will use this extension to obtain an explicit formula for the volume of the tubular neighborhoods of the boundary of a fractal string; see especially the proof of Theorem 8.1 in Section 8.1. Moreover, although this aspect will not be stressed as much in this work, we mention that we can also use this extension to obtain explicit formulas for other geometric or spectral functions, such as the partition function (see Section 6.2.3).

Given $v \in \mathbb{R}$, we will say that a function φ on $(0, \infty)$ has an asymptotic expansion of order v at 0 if there are finitely many complex exponents α, with $\operatorname{Re}\alpha > -v$, and complex coefficients a_α such that

$$\varphi(x) = \sum_{\alpha} a_\alpha x^{-\alpha} + O(x^v), \qquad \text{as } x \to 0^+. \tag{5.50}$$

Clearly, the coefficients are uniquely determined by φ. For each such α, we write τ_α to denote the corresponding distribution,

$$\langle \tau_\alpha, \varphi \rangle = a_\alpha, \tag{5.51}$$

where φ is given by (5.50).

Remark 5.24. Observe that τ_0 is the Dirac delta distribution $\delta_{\{0\}}$ at $x = 0$. More generally, $\tau_{-n} = \delta_{\{0\}}^{(n)}/n!$ is related to the n-th derivative of this distribution.

Note that if $\varphi(x) = O(x^a)$ as $x \to 0^+$ and $\varphi(x) = O(x^b)$ as $x \to \infty$, with $a > b$, then $\widetilde{\varphi}(s)$ is defined and holomorphic in the strip $-a < \operatorname{Re} s < -b$.

Moreover, for any given integer $k \geq 1$,

$$x^{-\alpha} = x^{-\alpha}e^{-x}\left(1 + x + \frac{x^2}{2!} + \cdots + \frac{x^k}{k!}\right) + O\bigl(x^{-\alpha+k+1}\bigr), \qquad \text{as } x \to 0^+. \tag{5.52}$$

The Mellin transform of $x^{-\alpha}e^{-x}$ is $\Gamma(s-\alpha)$, with simple poles at the points $s = \alpha, \alpha - 1, \ldots$, and with residue $(-1)^k/k!$ at $s = \alpha - k$. Consequently, the Mellin transform of the first term on the right-hand side of (5.52) is meromorphic in $\operatorname{Re} s > \alpha - k - 1$, with a simple pole of residue 1 at $s = \alpha$, the residues at the other possible poles adding up to 0. Provided that $\varphi(x) = O(x^b)$ at ∞, it follows that the Mellin transform of the function φ in (5.50) is meromorphic in $-v < \operatorname{Re} s < -b$, with a simple pole at $s = \alpha$ of residue a_α for each α.

We will need the following lemma.

Lemma 5.25. *Let $f(s)$ and $g(s)$ be meromorphic in a disc around 0. Then the function $z \mapsto \operatorname{res}(f(s-z)g(s);0) + \operatorname{res}(f(s-z)g(s);z)$ is holomorphic in the same disc. Its value at $z = 0$ is $\operatorname{res}(f(s)g(s);0)$.*

Proof. This follows since the function can be written as the integral over a small circle around $s = 0$ and $s = z$ of $f(s-z)g(s)$, and this function is analytic in z. □

Theorem 5.26 (Extended distributional formula, with error term). *Let η be a languid generalized fractal string (see Definition 5.2). Let $k \in \mathbb{Z}$ and let $q \in \mathbb{N}$ be such that $k+q \geq \kappa+1$, where κ is given as in (5.19) and (5.20). Further, let φ be a test function that is q-times continuously differentiable on $(0,\infty)$, and assume that its j-th derivative satisfies, for each $0 \leq j \leq q$ and some $\delta > 0$,*

$$\varphi^{(j)}(x) = O(x^{-k-j-D-\delta}), \qquad \text{as } x \to \infty, \tag{5.53a}$$

and[9]

$$\varphi^{(j)}(x) = \sum_\alpha a_\alpha^{(j)} x^{-\alpha-j} + O(x^{-k-j-\inf S+\delta}), \qquad \text{as } x \to 0^+, \tag{5.53b}$$

[9]Note that $a_\alpha^{(j)} = -\alpha\cdots(-\alpha-j+1)a_\alpha^{(0)}$ and that φ has an asymptotic expansion of order $-k-\inf S+\delta$ at 0, where $\inf S$ is given by (5.17a).

for a finite sequence of complex exponents α as above. We then have the following distributional explicit formula with error term for $\mathcal{P}_\eta^{[k]}$:

$$\mathcal{P}_\eta^{[k]}(x) = \sum_{\omega \in \mathcal{D}_\eta(W)} \operatorname{res}\left(\frac{x^{s+k-1}\zeta_\eta(s)}{(s)_k};\omega\right) + \sum_{\substack{\alpha \in W\setminus\mathcal{D}_\eta \\ \alpha\notin\{1-k,\dots,0\}}} \tau_\alpha(x)\frac{\zeta_\eta(\alpha)}{(\alpha)_k}$$
$$+ \frac{1}{(k-1)!}\sum_{\substack{j=0 \\ -j\in W\setminus\mathcal{D}_\eta}}^{k-1} \binom{k-1}{j}(-1)^j x^{k-j-1}\zeta_\eta(-j) + \mathcal{R}_\eta^{[k]}(x), \tag{5.54a}$$

where $\mathcal{R}_\eta^{[k]}(x)$ and $\tau_\alpha(x)$ are the distributions given by (5.46) and (5.51), respectively. Note that the sum over α is finite by our assumption on the space of test functions. Applied to the test function φ, the distribution $\mathcal{P}_\eta^{[k]}$ is given by

$$\langle \mathcal{P}_\eta^{[k]}, \varphi\rangle = \sum_{\omega \in \mathcal{D}_\eta(W)} \operatorname{res}\left(\frac{\zeta_\eta(s)\widetilde{\varphi}(s+k)}{(s)_k};\omega\right) + \sum_{\substack{\alpha \in W\setminus\mathcal{D}_\eta \\ \alpha\notin\{1-k,\dots,0\}}} a_\alpha^{(0)}\frac{\zeta_\eta(\alpha)}{(\alpha)_k}$$
$$+ \frac{1}{(k-1)!}\sum_{\substack{j=0 \\ -j\in W\setminus\mathcal{D}_\eta}}^{k-1} \binom{k-1}{j}(-1)^j \zeta_\eta(-j)\widetilde{\varphi}(k-j) + \langle\mathcal{R}_\eta^{[k]}, \varphi\rangle. \tag{5.54b}$$

For the extended distributional formula without error term, we require that the test function is a finite linear combination of terms $x^{-\beta}e^{-c_\beta x}$ in a neighborhood of the entire interval $(0, A]$. Here, the constants c_β are complex numbers with positive real part. Note that the β's come from a subset of $\bigcup_\alpha\{\alpha, \alpha-1, \dots\}$.

Theorem 5.27 (Extended distributional formula, without error term). *Let η bea strongly languid generalized fractal string (see Definition 5.3). Let $k \in \mathbb{Z}$ and let $q \in \mathbb{N}$ be such that $k+q > \max\{1, \kappa\}$, where κ is given as in (5.19) and (5.21). Further, let φ be a test function that is q-times continuously differentiable on $(0,\infty)$. Assume that as $x \to \infty$, the j-th derivative $\varphi^{(j)}(x)$ satisfies (5.53a) and (5.53b), and that there exists a number $\delta > 0$ and constants $b_\beta^{(j)}$ and c_β with $\operatorname{Re} c_\beta > 0$, such that*

$$\varphi^{(j)}(x) = \sum_\beta b_\beta^{(j)} x^{-\beta}e^{-c_\beta x}, \quad \text{for } x \in (0, A+\delta) \text{ and } 0 \le j \le q. \tag{5.55}$$

Then formula (5.54), with $\mathcal{R}_\eta^{[k]} \equiv 0$, gives the distributional explicit formula without error term at level k for φ.

Remark 5.28. For $k \le 0$, the last sum over j in (5.54) equals 0 and the condition $\alpha \notin \{1-k, \dots, 0\}$ is vacuous, so that the explicit formula is given by a sum over the complex dimensions and an error term alone (or no error term at all, in the case of Theorem 5.27). It will be useful to keep this in mind when applying Theorem 5.26 or 5.27, especially in Chapter 8 in the proof of Theorem 8.1.

Proof of Theorems 5.26 and 5.27. First of all, the condition at infinity on φ implies that $\langle \mathcal{P}_\eta^{[k+j]}, \varphi^{(j)} \rangle$ is well defined. Furthermore, since

$$\langle \mathcal{P}_\eta^{[k]}, \varphi \rangle = (-1)^q \langle \mathcal{P}_\eta^{[k+q]}, \varphi^{(q)} \rangle, \tag{5.56}$$

and

$$\widetilde{\varphi}(s) = \frac{(-1)^q}{(s)_q} \widetilde{\varphi^{(q)}}(s+q) \tag{5.57}$$

is meromorphic in $k + \inf S \le \operatorname{Re} s \le k + D$, it suffices to establish the theorem when $q = 0$ and $k > \kappa + 1$ (respectively, $k > \max\{1, \kappa\}$ for Theorem 5.27).

We first assume that $\varphi(x) = O\left(x^{-k-\inf S+\delta}\right)$ as $x \to 0^+$ (or $\varphi(x) = 0$ for all $x \in (0, A+\delta)$, in the case of Theorem 5.27). Using Lemma 5.9, with $c = D + \delta/2$, we obtain a truncated explicit formula,

$$\begin{aligned}
\langle \mathcal{P}_\eta^{[k]}, \varphi \rangle \approx \sum_{\omega \in \mathcal{D}_\eta(W_{|n})} \operatorname{res}\left(\frac{\zeta_\eta(s)\widetilde{\varphi}(s+k)}{(s)_k}; \omega \right) + \sum_{\substack{\alpha \in W_{|n} \setminus \mathcal{D}_\eta \\ \alpha \notin \{1-k,\dots,0\}}} a_\alpha^{(0)} \frac{\zeta_\eta(\alpha)}{(\alpha)_k} \\
+ \frac{1}{(k-1)!} \sum_{\substack{j=0 \\ -j \in W \setminus \mathcal{D}_\eta}}^{k-1} \binom{k-1}{j} (-1)^j \zeta_\eta(-j)\widetilde{\varphi}(k-j) \\
+ \frac{1}{2\pi i} \int_{S_{|n}} \zeta_\eta(s)\widetilde{\varphi}(s+k) \frac{ds}{(s)_k},
\end{aligned} \tag{5.58}$$

up to an error not exceeding a constant times

$$\int_0^\infty |\varphi(x)| x^{k-1} T^{-k} \left(T x^{D+\delta/2} + T^\kappa \max\{x^{D+\delta/2}, x^{\inf S}\} \right) dx. \tag{5.59}$$

Here, we have used the fact that for fixed $0 < \beta_1 < \beta_2$,

$$|\eta| \, (\beta_1 x, \beta_2 x) = O\big(x^{D+\delta/2}\big), \qquad \text{as } x \to \infty,$$

since D is the abscissa of convergence of $\zeta_{|\eta|}$. Firstly, one checks that the integral (5.59) converges, due to the conditions we imposed on φ at 0 and ∞. Then we see that it vanishes as $T \to \infty$, provided $k > 1$ and $k > \kappa$. The integral over the truncated screen converges as $T \to \infty$ if $k > \kappa + 1$. If **L2**$'$ is satisfied, we first let $m \to \infty$; i.e., we derive a truncated formula without an integral over the truncated screen. Then we do not need the assumption that $k > \kappa + 1$ to ensure convergence. This establishes the formula when $\varphi(x) = O\big(x^{-k-\inf S+\delta}\big)$ as $x \to 0^+$.

We now prove the theorem for the special test function $\varphi(x) = x^{-\alpha}e^{-cx}$, with $\alpha \in \mathbb{C}$. Then, in light of Equation (5.52), we obtain the formula for general test functions by subtracting sufficiently many functions of this type, depending on the asymptotic expansion at 0 of the test function.

Let $\varphi(x) = x^{-\alpha}e^{-cx}$. For the formula with error term, varying c does not give any extra generality, and we simply put $c = 1$, but when **L2′** is satisfied, we let c be any complex number with positive real part. Then

$$\widetilde{\varphi}(s) = c^{-s+\alpha}\Gamma(s-\alpha).$$

This function has poles at the points $\alpha, \alpha - 1, \dots$, and it has the right decay as $\operatorname{Im} s \to \pm\infty$, so that the integral defining the error term in (5.58), namely,

$$\frac{1}{2\pi i}\int_{S_{|n}} \zeta_\eta(s)\widetilde{\varphi}(s+k)\,\frac{ds}{(s)_k},$$

converges as $n \to \infty$. Note in addition that by Stirling's formula [In, p. 57], we have

$$|\Gamma(s-\alpha)| \ll_a \exp(a \operatorname{Re} s)$$

for every $a > 0$, as $\operatorname{Re} s \to -\infty$ away from the poles $\alpha, \alpha - 1, \dots$. Thus we can first let $m \to \infty$ to obtain a formula without error term. In this case, as $m \to \infty$, we pick up a residue $\zeta_\eta(\alpha - l - k)c^l(-1)^l/\left(l!(\alpha - l - k)_k\right)$ at each point $\alpha - l - k$ where $\widetilde{\varphi}(s+k)$ has a pole. Since the integral over the screen converges and the expression for the error (5.59) vanishes as $T \to \infty$, the sum of these residues converges.

For the formula with error term, we first apply the explicit formula for $\operatorname{Re}\alpha$ small enough, so that none of the points $\alpha, \alpha - 1, \dots$ lies inside W. The left-hand side of the explicit formula is clearly an analytic function in α, taking into account the fact that η is supported away from 0. The proof will be complete when we have shown that the right-hand side is also analytic in α. For this, we need to show that the right-hand side changes analytically when one of the points $\alpha, \alpha - 1, \dots$ crosses the screen S or coincides with a complex dimension or with one of the points $-j$, for $j = 0, \dots, k-1$. Indeed, when one of the points $\alpha, \alpha - 1, \dots$ crosses S, the integral over the screen changes by minus the residue of the integrand at this point, and this cancels the corresponding term in the last sum. Secondly, when one of the points $\alpha, \alpha - 1, \dots$ coincides with a complex dimension or with one of the points $-j$, $j = 0, \dots, k-1$, the analyticity follows from Lemma 5.25. This completes the proof of Theorems 5.26 and 5.27. □

5.4.2 The Order of the Distributional Error Term

We now provide a more quantitative version of our distributional explicit formulas with error term (Theorem 5.18, or more generally, Theorem 5.26), which will play a key role in the remainder of this book. (See, for example, Chapters 7–9 and 11.)

Given $a > 0$ and a test function φ, we set

$$\varphi_a(x) = \frac{1}{a}\,\varphi\left(\frac{x}{a}\right). \tag{5.60}$$

In light of Equation (5.43), the Mellin transform of $\varphi_a(x)$ is given by

$$\widetilde{\varphi_a}(s) = a^{s-1}\widetilde{\varphi}(s). \tag{5.61}$$

Definition 5.29. We will say that a distribution $\mathcal{R}$ is of *asymptotic order* at most x^α (respectively, less than x^α)—and we will write $\mathcal{R}(x) = O(x^\alpha)$ (respectively, $\mathcal{R}(x) = o(x^\alpha)$), as $x \to \infty$—if applied to a test function φ, we have that[10]

$$\langle \mathcal{R}, \varphi_a \rangle = O\left(a^\alpha\right) \quad \text{(respectively, } \langle \mathcal{R}, \varphi_a \rangle = o\left(a^\alpha\right)\text{)}, \quad \text{as } a \to \infty. \tag{5.62}$$

When we apply Definition 5.29, we use an arbitrary test function φ of the type considered in Theorem 5.18 (respectively, Theorem 5.26).

The following theorem completes Theorems 5.18 and 5.26 by specifying the asymptotic order of the distributional error term $\mathcal{R}_\eta^{[k]}$ obtained in our distributional explicit formula (5.45). It shows that, in a suitable sense, Theorem 5.18 and its extension, Theorem 5.26, are as flexible as and more widely applicable than their pointwise counterpart, Theorem 5.10. As was mentioned above, we will take advantage of this fact on many occasions in the rest of this book.

Recall that the screen S is the parametrized curve $S\colon t \mapsto S(t) + it$, for some bounded Lipschitz continuous function $S(t)$, and that the least upper bound of S is denoted $\sup S$; see Section 4.1 and formula (5.17b).

Theorem 5.30 (Order of the distributional error term). *Fix $k \in \mathbb{Z}$. Assume that the hypotheses of Theorem 5.18 (or more generally, of Theorem 5.26, with $k + q > \kappa + 1$) are satisfied. Then the distribution $\mathcal{R}_\eta^{[k]}$, given by (5.46), is of asymptotic order at most $x^{\sup S + k - 1}$ as $x \to \infty$:*

$$\mathcal{R}_\eta^{[k]}(x) = O\big(x^{\sup S + k - 1}\big), \quad \text{as } x \to \infty, \tag{5.63}$$

in the sense of Definition 5.29.

Moreover, if $S(t) < \sup S$ for all $t \in \mathbb{R}$ (i.e., if the screen lies strictly to the left of the line $\operatorname{Re} s = \sup S$), then this distribution is of asymptotic order less than $x^{\sup S + k - 1}$ as $x \to \infty$:

$$\mathcal{R}_\eta^{[k]}(x) = o\big(x^{\sup S + k - 1}\big), \quad \text{as } x \to \infty. \tag{5.64}$$

---

[10] In this formula, the implicit constants depend on the given test function φ.

Proof. The integral (5.46) for $\langle \mathcal{R}_\eta^{[k]}, \varphi \rangle$ converges absolutely. Let φ be a test function with compact support. When we replace φ by φ_a in (5.46), we see, by formula (5.61), that the absolute value of the integrand is multiplied by $a^{\operatorname{Re} s+k-1} \leq a^{\sup S+k-1}$ for $s = S(t) + it$. Hence $\left|\langle \mathcal{R}_\eta^{[k]}, \varphi_a \rangle\right|$ is bounded by a constant times $a^{\sup S+k-1}$.

To obtain the better estimate (5.64) when S approximates the line $\operatorname{Re} s = \sup S$, but stays strictly to the left of it, we use an argument similar to the one used to derive estimate (5.38) of Theorem 5.10. □

The analysis of the error term given in Theorem 5.30 also allows us to estimate the sum over the complex dimensions occurring in our distributional explicit formulas.

Theorem 5.31. *Let $v \leq D$. Assume that the hypotheses of Theorem 5.18 (or of Theorem 5.26, with $k + q > \kappa + 1$) are satisfied, with a screen contained in the open half-plane* $\operatorname{Re} s < v$. *Assume, in addition, that there exists a screen S_0 contained in* $\operatorname{Re} s < v$, *satisfying* **L2** *and such that every complex dimension to the right of S_0 has real part $\geq v$. Then*

$$\sum_{\omega \in \mathcal{D}_\eta(W),\, \operatorname{Re}\omega < v} \operatorname{res}\left(\frac{x^{s+k-1}\zeta_\eta(s)}{(s)_k}; \omega\right) = o\left(x^{v+k-1}\right), \qquad \text{as } x \to \infty, \quad (5.65)$$

in the sense of Definition 5.29.

Proof. We write this distribution as $\mathcal{R}_{0,\eta}^{[k]}(x) - \mathcal{R}_\eta^{[k]}(x)$, where $\mathcal{R}_{0,\eta}^{[k]}$ is the error term associated with the screen S_0. The result then follows from Theorem 5.30 applied to $\mathcal{R}_{0,\eta}^{[k]}$ and $\mathcal{R}_\eta^{[k]}$. □

We will apply Theorem 5.31 in Chapters 7–9 and 11 with $v = D$. However, the hypotheses of this theorem are not always satisfied: it is not always possible to choose a screen S_0 passing between $\operatorname{Re} s = D$ and the complex dimensions to the left of this line. Nevertheless, in the case of self-similar strings, we can still obtain information from the complex dimensions; see Remark 6.15. In the following example, we construct a nonlattice string that does not satisfy **L2** for any such screen S_0 (hence it is neither languid nor a fortiori, strongly languid, along this screen).

Example 5.32. Recall the definition of the continued fraction of a real number $\alpha > 1$ given in Section 3.5.1. We construct a nonlattice self-similar string $\mathcal{L}$ with two scaling ratios $r_1 = e^{-1}$, $r_2 = e^{-\alpha}$, where α will be specified below.[11] Consider the function $f(s) = 1 - e^{-s} - e^{-\alpha s}$. Let D be the real zero of f; i.e., the dimension of $\mathcal{L}$. Let p_n/q_n be a convergent of α, so that $q_n\alpha - p_n = (-1)^{n+1}/q'_{n+1}$. Then $f(D + 2\pi i q_n)$ is very close to 0.

[11] Here, $e = \exp(1)$ is the base of the natural logarithm.

Indeed, we have $e^{-(D+2\pi i q_n)} = e^{-D}$ and

$$e^{-(D+2\pi i q_n)\alpha} = e^{-D\alpha} e^{-2\pi i q_n \alpha + 2\pi i p_n} = e^{-D\alpha} e^{2\pi i (-1)^n / q'_{n+1}}.$$

Hence, by the same techniques as those used in Chapter 3, Section 3.5.2,

$$\begin{aligned} f(D+2\pi i q_n) &= 1 - e^{-D} - e^{-D\alpha}\left(1 + O(2\pi/q'_{n+1})\right) \\ &= O(2\pi/q'_{n+1}), \qquad \text{as } n \to \infty. \end{aligned}$$

The reason for this near vanishing is that there is a zero of f close to $s = D + 2\pi i q_n$, but slightly to the left. It follows that

$$|\zeta_{\mathcal{L}}(D + 2\pi i q_n)| \gg q'_{n+1}.$$

Note that

$$q'_{n+1} = \alpha_{n+1} q_n + q_{n-1} > a_{n+1} q_n.$$

We next choose the integers a_j as follows: $a_0 = 1$, and if $a_0, a_1, \dots, a_n$ are already constructed, then we compute q_n and set $a_{n+1} = q_n^n$ (thus $a_1 = 1$, $a_2 = 1$, $a_3 = 4$, $a_4 = 9^3$, $a_5 \approx 1.8 \cdot 10^{15}$, ...). Then

$$|\zeta_{\mathcal{L}}(D + 2\pi i q_n)| \gg q_n^{n+1},$$

so that on the vertical line $\operatorname{Re} s = D$, condition **L2** is violated for every value of κ—see Equation (5.20).

Naturally, the value of $f(s)$ will be even smaller between $s = D + 2\pi i q_n$ and the nearby zero of this function. Thus **L2** will not be satisfied on any screen passing between $\operatorname{Re} s = D$ and the poles of $\zeta_{\mathcal{L}}$.

The reader sees that the integers a_j in this example, and hence the numbers q_n, grow extremely fast. Thus the set of points where $\zeta_{\mathcal{L}}$ is large is very sparse.

One could think of a way out by replacing hypothesis **L2** in (5.20) by the following integrability condition: The function

$$t \longmapsto |\zeta_{\mathcal{L}}(D + it)|(|t| + 1)^{-\kappa - 1} \tag{5.66}$$

is integrable on $\mathbb{R}$. Indeed, our theory easily goes through under this weaker condition.

However, we can even choose α such that $|\zeta_{\mathcal{L}}(D + it)|/g(|t|)$ is not integrable as a function of $t \in \mathbb{R}$ no matter how fast the positive function g grows. Namely, we choose α such that $\log q_{n+1}/g(q_n)$ grows unboundedly. This is the case, for example, if we choose the coefficients of α so as to satisfy

$$a_{n+1} \geq e^{ng(q_n)},$$

for $n = 0, 1, 2, \dots$.

We again give an example, now of a string that is not self-similar, showing that condition (5.66) is not satisfied by every generalized fractal string.

Example 5.33. Consider the measure μ on $[1, \infty)$ defined by

$$\mu = \sum_{n \in \mathbb{Z} \backslash \{0\}} \frac{x^{D-1-d_n+in}}{n^2} \, dx,$$

where the real numbers $d_n = d_{-n}$ are small, and will be specified below. The geometric zeta function of this generalized fractal string has poles at $D - d_n \pm in$, with residue $1/n^2$, for $n \in \mathbb{Z} - \{0\}$. Hence the value of the integral of $|\zeta_\mu(s)|$ over an interval from $D + i\left(n - \frac{1}{2}\right)$ to $D + i\left(n + \frac{1}{2}\right)$ is close to

$$\frac{1}{n^2} \int_{-1/2}^{1/2} \frac{dt}{|t + id_n|} \approx -\frac{\log d_n}{n^2},$$

as $|n| \to \infty$. Therefore, we have

$$\int_{D+i(n-1/2)}^{D+i(n+1/2)} |\zeta_\mu(D+it)| \, t^{-\kappa} \, dt \approx -\frac{\log d_n}{n^{2+\kappa}},$$

as $n \to \infty$.

We conclude that if $d_n = \exp(-n^n)$, then for every value of κ, the function $|\zeta_\mu(D + it)|(|t| + 1)^{-\kappa}$ is not integrable along the vertical line $\operatorname{Re} s = D$—and hence along any screen passing between $\operatorname{Re} s = D$ and the poles of ζ_μ. It follows that condition (5.66) (or its obvious analogue along a suitable screen S) is never satisfied for this string μ.

The following example explains rather clearly why, in general, our explicit formulas must have an error term. Indeed, in this extreme situation, there is only an error term because the set $\mathcal{D}_\eta$ of visible complex dimensions is empty and therefore the corresponding sum over $\mathcal{D}_\eta$ vanishes.

Example 5.34. The generalized fractal string

$$\eta = \sum_{n=1}^{\infty} (-1)^{n-1} \delta_{\{n\}} \tag{5.67}$$

has no complex dimensions. Indeed, by [Ti, Eq. (2.2.1), p. 16], the associated geometric zeta function is

$$\zeta_\eta(s) = \sum_{n=1}^{\infty} (-1)^{n-1} n^{-s} = \left(1 - 2^{1-s}\right) \zeta(s). \tag{5.68}$$

Since the pole of $\zeta(s)$ at $s = 1$ is canceled by the corresponding zero of $\left(1 - 2^{1-s}\right)$ in Equation (5.68), $\zeta_\eta(s)$ is holomorphic in all of $\mathbb{C}$ and hence $\mathcal{D}_\eta$ is empty, as claimed above. Thus the explicit formula for $N_\eta(x)$ (and for $N_\eta^{[k]}(x)$ at level k) has no sum over ω, only an error term and a constant term.

We now explain this in more detail in the case when $k = 1$. Choose a screen to the left of $\operatorname{Re} s = 0$. Since $\zeta(\sigma + it)$ grows like $|t|^{1/2-\sigma}$ (see Equation (6.21) in Chapter 6 below), we have to interpret $N_\eta(x)$ as a distribution. Also the more stringent assumptions of Theorem 5.22, needed to obtain a distributional explicit formula without error term, are not satisfied here. Theorem 5.18 yields

$$N_\eta(x) = \frac{1}{2} + \mathcal{R}^{[1]}(x). \tag{5.69}$$

By construction, we have

$$N_\eta(x) = \begin{cases} 0, & \text{for } 2n \leq x < 2n+1, \ n = 0, 1, \dots, \\ 1, & \text{for } 2n+1 \leq x < 2n+2, \ n = 0, 1, \dots. \end{cases} \tag{5.70}$$

Observe that $N_\eta(x)$ is an additively periodic function, with period 2. Its Fourier series is

$$N_\eta(x) = \frac{1}{2} + \frac{1}{\pi i} \sum_{n=-\infty}^{\infty} \frac{e^{\pi i(2n+1)x}}{2n+1}. \tag{5.71}$$

We point out that the reason why our explicit formula gives only the constant term $1/2$ and an error term in (5.69) is that this function is not multiplicatively periodic. On the other hand, our explicit formula does yield the Fourier series of a multiplicatively periodic counting function.

Remark 5.35. By (5.70), the average value of $N_\eta(x)$ in the previous example is equal to $1/2$, as expressed by the first term of the right-hand side of (5.69). Further, this function jumps by $+1$ at every odd integer, and by -1 at every even integer. Hence, as $x \to \infty$, $N_\eta(x) = 1/2 + O(1)$ as a function, and no better pointwise estimate holds as $x \to \infty$. On the other hand, as a distribution, $\mathcal{R}^{[1]}(x) = O\left(x^\sigma\right)$ for every $\sigma < 0$, since we can choose screens arbitrarily far to the left. Thus $N_\eta(x)$ does not oscillate. This seems to be a contradiction.

The resolution of this apparent paradox is important since in Chapters 9 and 11, we will make use of the fact that a certain type of oscillations, multiplicative oscillations, are reflected in our explicit formulas. We discuss this paradox here in the case of the harmonic string h, which will serve as a paradigm for this phenomenon.

Since by (4.11), $h = \sum_{n=1}^{\infty} \delta_{\{n\}}$, we have

$$N_h(x) = \#\{n \geq 1 : n \leq x\} = [x]. \tag{5.72}$$

Hence $N_h(x)$ jumps at every positive integer. On the other hand, according to Theorem 5.18 applied at level $k = 1$ and $k = 0$ respectively, we have

$$N_h(x) = x - \frac{1}{2} + \mathcal{R}_h^{[1]}(x), \tag{5.73a}$$

for the counting function of the lengths, and

$$h = 1 + \mathcal{R}_h^{[0]}(x), \tag{5.73b}$$

for the density of states. In the sense of Definition 5.29, the error terms are estimated by

$$\mathcal{R}_h^{[1]}(x) = O\big(x^{\sigma}\big), \text{ as } x \to \infty, \tag{5.74a}$$

and

$$\mathcal{R}_h^{[0]}(x) = O\big(x^{\sigma-1}\big), \text{ as } x \to \infty, \tag{5.74b}$$

for every $\sigma < 0$. As in Example 5.34, this seems to be in contradiction with the jumps of $N_h(x)$ and the point masses of h at the positive integers.

The meaning of the distributional equalities (5.73) and (5.74) becomes clear when we do not apply them to a localized test function, but to a test function having an asymptotic expansion near 0,

$$\varphi(x) = \sum_{\operatorname{Re}\alpha<1} a_\alpha x^{-\alpha}, \qquad \text{as } x \to 0^+,$$

as in (5.53b), with $a_\alpha = 0$ for $\operatorname{Re}\alpha \geq 1$. Write

$$\varphi_t(x) = \frac{1}{t}\varphi\left(x/t\right)$$

and $J = \int_0^\infty \varphi(x)\,dx$. Then, by Theorem 5.26, applied at level $k = 0$,

$$\sum_{n=1}^{\infty} \varphi_t(n) = J + \sum_{\operatorname{Re}\alpha\geq-\sigma} a_\alpha \zeta(\alpha) t^{\alpha-1} + \big\langle \mathcal{R}_h^{[0]}, \varphi_t \big\rangle,$$

and the error term is of order $O\big(t^{\sigma-1}\big)$ as $t \to \infty$, for every $\sigma < 0$.[12]

On a logarithmic scale, the points $1, 2, 3, \ldots$ become dense on the real line, which is why h is distributionally approximated by the density 1 up to every order; see Figures 5.2 and 5.3. On the other hand, multiplicative oscillations of η, like, for example, in the case of the Cantor string, do give rise to oscillatory terms in the explicit formula for η, as we will see in Section 8.4.2 and in Chapter 10.

Remark 5.36. A notation of the same type as the first part of Equation (5.62) of Definition 5.29, $\mathcal{R}(x) = O\left(x^\alpha\right)$, was introduced independently (and probably a little earlier) by Yves Meyer in [Mey, Definitions 1.2 and 1.3, p. 10] for different classes of test functions, and with a closely related meaning. (See also, e.g., the memoir by Stéphane Jaffard and Yves Meyer [JafMey].)[13]

---

[12]This very useful classical formula does not appear in the literature. It was explained heuristically to the second author by Don Zagier (private communication, 1994).

[13]There are some obvious but confusing notational differences between [Mey] and Definition 5.29, due to the fact that in our case, $x > 0$ and $x \to \infty$, whereas in [Mey], $x \in \mathbb{R}^d$ and $x \to 0$ (or more generally, $x \to x_0$ for some $x_0 \in \mathbb{R}^d$).

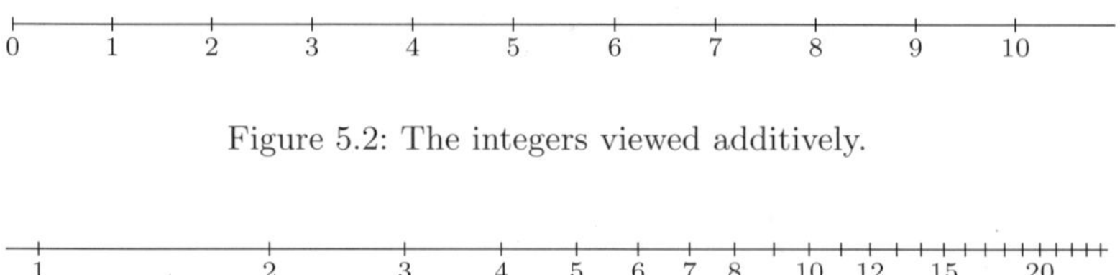

Figure 5.2: The integers viewed additively.

Figure 5.3: The integers viewed multiplicatively.

In [Lap-vF1–4] and [Lap-vF5, Definition 4.22, p. 99], we were led naturally to formulate Definition 5.29 in order to obtain suitable distributional explicit formulas with error term. In our theory, the latter were needed originally to establish the extension to other zeta functions in Section 11.2 of the results of Section 11.1. See especially Theorem 11.12 about the absence of infinite vertical arithmetic progressions of zeros. See also the corresponding result in [Lap-vF5, Theorem 9.5, pp. 184–185], announced in [Lap-vF2, 4].

5.5 Example: The Prime Number Theorem

In the next chapter, we will discuss a number of examples illustrating our pointwise and distributional explicit formulas. In the present section, we will concentrate on one particular application of these formulas, namely, to the Prime Number Theorem and issues surrounding it, particularly the Riemann–von Mangoldt formula. Our main purpose is to indicate by means of this example the flexibility provided by our distributional formula and by the introduction of a screen in the formulation of our explicit formulas.

This example is also included in order to clarify the relationships between our explicit formulas and the traditional ones from number theory. Our proof of the Prime Number Theorem with error term obtained in this manner is essentially the same as the usual one, except for the analysis of the error term. We note, however, that in the present context, the systematic use of the screen S in our theory makes particularly transparent the connection between a given zero-free region for $\zeta(s)$ and the corresponding error term $\mathcal{R}_\pi(x)$ in the asymptotic formula (5.77).

In Chapter 7, we shall use our explicit formulas to obtain an analogue of the Prime Number Theorem (with error term) for the primitive periodic orbits of a self-similar flow and other suspended flows. See especially Section 7.4.

The reader familiar with the classical explicit formulas from number theory—the first one of which was discovered by Riemann and described in his famous 1858 paper [Rie1]—may wonder why the theorems obtained in this chapter are called explicit formulas. Indeed, Riemann's explicit formula—as interpreted and proved thirty-six years later by von Mangoldt [vM1–2]—relates a suitable counting function associated with the prime numbers to an infinite sum involving the zeros of the Riemann zeta function (that is, the poles of its logarithmic derivative ζ'/ζ); see, for example, [Pat, esp. Chapters 1 and 3], [Da, Chapter 17], [In, Chapters II–IV], [Edw, Chapters 1 and 3]. See also the discussion on pages 3–4 of the introduction and the one surrounding Equation (5.8) in Section 5.1. (A similar statement holds for the various types of number-theoretic explicit formulas found in the literature; see, for instance, the references given in Section 5.6.)

However, we show in this section that we can easily recover the Riemann–von Mangoldt formula from our own explicit formulas. Actually, we can obtain different versions of it, either pointwise or distributional, at various levels k, the most basic of which is the distributional formula obtained when $k = 0$, corresponding to the density of prime powers viewed as the generalized fractal string $\mathfrak{P} = \sum_{m\geq 1,\, p} (\log p)\, \delta_{\{p^m\}}$, the prime string, introduced in (4.13); see Equation (5.82) in Section 5.5.1 below.

In turn, now standard arguments (see, e.g., [Da, Chapter 18], [Edw, Chapter 4], [Pat, Chapter 3] or [In, p. 5 and Chapter II])—first used in slightly different forms by Hadamard [Had2] and de la Vallée Poussin [dV1–2][14]—enable us to deduce from this explicit formula the Prime Number Theorem with error term; namely, if

$$\pi(x) = \#\{p \leq x \colon p \text{ is a prime number}\} \tag{5.76}$$

denotes the prime number counting function, then

$$\pi(x) = \mathrm{Li}(x) + \mathcal{R}_\pi(x), \quad \text{as } x \to \infty, \tag{5.77}$$

---

[14] Their results, obtained independently and published almost simultaneously in 1896, relied on von Mangoldt's work [vM1–2], as well as on Hadamard's product formula for entire functions [Had1]. In their proof, both Hadamard [Had2] and de la Vallée Poussin [dV1] showed (also independently) that $\zeta(s) \neq 0$ for all s on the vertical line $\operatorname{Re} s = 1$. Hadamard's classical proof of the Prime Number Theorem is the simplest of the two, but in his second paper [dV2], de la Vallée Poussin established the existence of a zero-free region to the left of $\operatorname{Re} s = 1$ and thus went further in his investigation of the approximation of $\pi(x)$, in the form of the error estimate (5.77). (See, e.g., [In, p. 5], [Edw, Chapter 5], [Ti, Chapter III] and [Pat, Chapter 3].) We note that the original form of the Prime Number Theorem obtained in [Had2, dV1] was as follows:

$$\pi(x) = \mathrm{Li}(x)\,(1 + o(1)) = \frac{x}{\log x}\,(1 + o(1)), \text{ as } x \to \infty. \tag{5.75}$$

where Li denotes the logarithmic integral

$$\mathrm{Li}(x) = \lim_{\varepsilon \to 0^+} \left(\int_0^{1-\varepsilon} + \int_{1+\varepsilon}^{x} \right) \frac{1}{\log t}\, dt, \qquad \text{for } x > 1, \tag{5.78}$$

and estimates for the error term $\mathcal{R}_\pi(x)$ depend on our current knowledge of the zero-free region for the Riemann zeta function. For example, in (5.77), we have

$$\mathcal{R}_\pi(x) = O\left(x e^{-c\sqrt{\log x}} \right), \qquad \text{as } x \to \infty, \tag{5.79}$$

for some explicit positive constant c. Better error estimates—following from a refined analysis of the zero-free regions of ζ—can be found, for instance, in Ivić's book [Ivi] (see also, e.g., [In, pp. xi–xii] and the references therein) and can also be deduced from our explicit formula, simply by choosing a different screen S and thus a different window W suitably adapted to the zero-free region.

5.5.1 The Riemann–von Mangoldt Formula

We now briefly explain how to interpret and derive from our explicit formula the Riemann–von Mangoldt formula. (Compare, for example, with the discussion in [Edw, Chapter 3, esp. §3.5] or else in [In, Pat].) As was seen in Section 4.1.1, the geometric zeta function of the prime string

$$\mathfrak{P} = \sum_{m \geq 1,\, p} (\log p) \delta_{\{p^m\}} \tag{5.80}$$

is the function

$$\zeta_{\mathfrak{P}}(s) = -\frac{\zeta'(s)}{\zeta(s)}. \tag{5.81}$$

(In (5.80) above and in (5.83) below, p runs over the prime numbers.) This function satisfies **L1** and **L2**$'$ for every $\kappa > 0$ and with $W = \mathbb{C}$. Recall that the poles of $\zeta_{\mathfrak{P}}$ coincide precisely with the zeros and the pole at 1 of ζ, except that they occur with multiplicity one.

By Theorem 5.22, applied at level $k = 0$, and with $A = 1$, the density of lengths formula without error term for the prime string reads as follows (see also Corollary 5.23 above and Section 6.3.1 below):

$$\mathfrak{P} = 1 - \sum_{\rho} x^{\rho - 1} - \sum_{n=1}^{\infty} x^{-1-2n}, \qquad \text{for } x > 1, \tag{5.82}$$

where ρ runs through the sequence of critical zeros of ζ. (See the last comment at the end of Section 4.1.1.) This formula should be interpreted distributionally on the open interval $(1,\infty)$, as in Theorem 5.22. Further, note that the last sum on the right-hand side of (5.82) converges to $\frac{1}{x(x^2-1)}$ and corresponds to the trivial zeros of ζ, located at $-2,-4,-6,\ldots$.

The corresponding formula for the measure

$$\eta = \sum_{m\geq 1,\, p} \frac{1}{m}\delta_{\{p^m\}}, \tag{5.83}$$

with geometric zeta function

$$\zeta_\eta(s) = \log\zeta(s) = \sum_p \sum_{m=1}^{\infty} \frac{1}{m} p^{-ms},$$

can be derived by means of the following artefact. One first establishes that the string $\frac{dx}{\log x}$ has a geometric zeta function with a logarithmic singularity at 1 that cancels that of $\log\zeta(s)$. Hence, our explicit formula applies to the string $\eta - \frac{dx}{\log x}$, with a screen to the left of $\operatorname{Re} s = 1$, but to the right of all zeros of ζ. We thus find (with $\sup S \leq 1$ given as in formula (5.17b)):

$$\eta = \left(\frac{1}{\log x} + o\big(x^{\sup S - 1}\big)\right) dx = \frac{1}{\log x}(1+o(1))dx, \qquad \text{as } x\to\infty. \tag{5.84}$$

This provides a distributional interpretation of (part of) Riemann's original explicit formula [Rie1].[15]

Remark 5.37. Using the explicit formula (5.82), together with a deep analysis of the location of the critical zeros ρ of the Riemann zeta function, one can prove the Prime Number Theorem with error term, stating that

$$\pi(x) = \operatorname{Li}(x) + O\Big(xe^{-c\sqrt{\log x}}\Big), \qquad \text{as } x\to\infty, \tag{5.85}$$

for some constant $c>0$. This result does not seem to be attainable by a Tauberian argument.

5.6 Notes

A proof of the Prime Number Theorem can be found in any of the books [Da, Edw, In, Pat, Ti].

For information on pointwise and distributional explicit formulas in analytic number theory see, for example, [Da, Edw, In, Pat, Wei4–6]. Distributional-type explicit formulas (in a sense somewhat different from this book) can be found in [Wei4–6, Bar, Haran1, Den1–3, DenSchr].

[15] The latter is difficult to establish pointwise (see, e.g., [Edw, §1.18, p. 36]) and was not properly justified in Riemann's original paper [Rie1].

The first number-theoretic explicit formula is due to Riemann in his classic work [Rie1]. It was later extended and rigorously established by von Mangoldt in [vM1] and especially [vM2] (using, in particular, Hadamard's factorization for entire functions [Had1] and the Euler product representation (4.14) of the Riemann zeta function). A sample of additional references dealing with explicit formulas in number theory includes the works by Cramér [Cram], Guinand [Gui1–2], Delsarte [Del], Weil [Wei4–6], Barner [Bar], Haran [Haran1], Burnol [Bu1–3], Schröter and Soulé [SchrSo], Jorgenson and Lang [JorLan1, 3], Deninger [Den1–3], Deninger and Schröter [DenSchr], as well as Rudnick and Sarnak [RudSar].

Further, an excellent introduction to the classical number-theoretic formulas (with varying degrees of sophistication) can be found in the books by Ingham [In, esp. Chapter IV], Edwards [Edw, esp. Chapter 3] (and the Appendix [Rie1]), Lang [Lan], Davenport [Da], Patterson [Pat, esp. Chapter 3], as well as that written by Manin and Panchishkin (and edited by Parshin and Shafarevich) [ParSh1, esp. §2.5].

We refer to [Pos, Section 27, pp. 109–112] and [Shu, Theorem 14.1, p. 115] for a discussion of the Wiener–Ikehara Tauberian Theorem. See also [Va] for a concise exposition of the logic of the proof of the Prime Number Theorem by means of the Wiener–Ikehara Tauberian Theorem. We also recommend the exposition of J. Korevaar [Kor] on Tauberian theory.

6

The Geometry and the Spectrum of Fractal Strings

In this chapter, we give various examples of explicit formulas for the counting function of the lengths and frequencies of (generalized) fractal strings and sprays.

In Section 6.1, we give a detailed discussion of the oscillatory term associated with a single complex dimension of the (generalized) fractal string η. In Section 6.2, we then derive the explicit formulas for the geometric and spectral counting functions of η, and in Section 6.2.3, for the geometric and spectral partition functions, which are sometimes a useful substitute for the corresponding counting functions. In Section 6.3.1, we obtain the explicit formulas for the geometric and spectral density of states of η, and use them in Section 6.3.2 to define the spectral operator, which formalizes the direct spectral problem. We also obtain an Euler product representation for the spectral operator in that section, and an Euler sum representation for its additive counterpart.

In Section 6.4, we explore the important special case of self-similar strings studied in Chapters 2 and 3. We discuss, in particular, the geometry and the spectrum of both lattice and nonlattice strings. As an application, we shall prove a Prime Orbit Theorem for self-similar flows in Chapter 7. We study two examples of non-selfsimilar strings in Section 6.5. We close this chapter in Section 6.6 by explaining how our explicit formulas apply to the study of the geometry and the spectrum of fractal sprays, a higher-dimensional analogue of fractal strings.

6.1 The Local Terms in the Explicit Formulas

Our explicit formulas[1] give expansions of various functions associated with a fractal string as a sum over the complex dimensions of this string. The term corresponding to the complex dimension ω of multiplicity one is of the form

Cx^ω, where C is a constant depending on ω. If ω is real, the function x^ω simply has a certain asymptotic behavior as $x \to \infty$. If, on the other hand, $\omega = \sigma + it$ has a nonzero imaginary part t, then $x^\omega = x^\sigma \cdot x^{it}$ is of order $O(x^\sigma)$ as $x \to \infty$, with a multiplicatively periodic behavior: The function $x^{it} = \exp(it \log x)$ takes the same value at the points $e^{2\pi n/t}x$ $(n \in \mathbb{Z})$. Thus, the term corresponding to ω will be called an oscillatory term. If there are complex dimensions with higher multiplicity, there will also be terms of the form $Cx^\omega (\log x)^m$, $m \in \mathbb{N}^*$, which have a similar oscillatory behavior.

6.1.1 *The Geometric Local Terms*

Let η be a generalized fractal string and let ω be a complex dimension of η; i.e., $s = \omega$ is a pole of $\zeta_\eta(s)$. The term corresponding to ω in the explicit formula of Theorems 5.10, 5.14, 5.18, 5.22, 5.26 and 5.27 is

$$\operatorname{res}\left(\frac{x^{s+k-1}\zeta_\eta(s)}{(s)_k};\omega\right). \tag{6.1}$$

The nature of this term depends on the multiplicity of ζ_η at ω and on the expansion of ζ_η around ω. Let $m = m(\omega)$ be the multiplicity of ω, and let

$$\frac{a_m}{(s-\omega)^m} + \frac{a_{m-1}}{(s-\omega)^{m-1}} + \frac{a_{m-2}}{(s-\omega)^{m-2}} + \cdots + \frac{a_1}{s-\omega} \tag{6.2}$$

be the principal part of the Laurent series of $\zeta_\eta(s)$ at $s = \omega$. In particular,

$$a_1 = \operatorname{res}\left(\zeta_\eta(s);\omega\right). \tag{6.3}$$

The expansion of x^{s-1} around $s = \omega$ is given by

$$x^{s-1} = x^{\omega-1}\left(1 + (s-\omega)\log x + \frac{(s-\omega)^2}{2!}(\log x)^2 + \dots\right). \tag{6.4}$$

In general, for $k \geq 1$, the expansion of $x^{s+k-1}/(s)_k$ around $s = \omega$ is given by

$$x^{\omega+k-1}\sum_{n=0}^{\infty}(s-\omega)^n \sum_{\mu=0}^{n}\frac{(\log x)^\mu}{\mu!}\sum_{\iota=0}^{k-1}\frac{(-1)^{n+\iota-\mu}}{\iota!(k-1-\iota)!}\frac{1}{(\omega+\iota)^{n+1-\mu}}. \tag{6.5}$$

---

[1]Throughout this discussion, we assume that ζ_η satisfies the appropriate growth conditions; namely, hypotheses **L1** and **L2** (respectively, **L1** and **L2′**) if the explicit formula of Theorem 5.10 or 5.18 (respectively, 5.14 or 5.22) is applied.

To derive this formula, we first compute the partial fraction expansion of $1/(s)_k$ (see formula (5.12) for the Pochhammer symbol $(s)_k$). The residue at $s = -\iota$ is determined by a computation similar to that in the proof of Lemma 5.1, and equals $(-1)^\iota/(\iota!(k-\iota-1)!)$. Hence, for $k \geq 1$,

$$\frac{1}{(s)_k} = \sum_{\iota=0}^{k-1} \frac{(-1)^\iota}{\iota!(k-1-\iota)!(s+\iota)}. \tag{6.6}$$

Next, we substitute the power series

$$\frac{1}{s+\iota} = \sum_{n=0}^{\infty} (-1)^n \frac{(s-\omega)^n}{(\omega+\iota)^{n+1}}$$

and multiply by the power series of x^{s+k-1} to obtain formula (6.5).

In the pointwise explicit formulas (Theorems 5.10 and 5.14), k is always positive. On the other hand, we can choose k to be negative in the distributional explicit formulas of Theorems 5.18, 5.22, 5.26 and 5.27. In that case, $(s)_k$ is defined by (5.42). Thus $1/(s)_k$ is a polynomial,

$$\frac{1}{(s)_k} = (s-1)(s-2)\dots(s+k).$$

For negative values of k, it is much harder to compute the expansion of $x^{s+k-1}/(s)_k$ around $s = \omega$. In the next theorem, we only discuss $k = -1$ and arbitrary negative values of k if ω is a simple pole of ζ_η.

Theorem 6.1. *Let ω be a complex dimension of multiplicity m and let the principal part of $\zeta_\eta(s)$ at $s = \omega$ be given by formula (6.2). Then the local term (6.1) in the explicit formula at ω is given by*

$$x^{\omega-1} \sum_{j=1}^{m} a_j \frac{(\log x)^{j-1}}{(j-1)!}, \tag{6.7}$$

for $k = 0$, and by

$$x^{\omega+k-1} \sum_{j=1}^{m} a_j \sum_{\iota=0}^{k-1} \sum_{\mu=0}^{j-1} \frac{(\log x)^\mu}{\mu!} \frac{(-1)^{j-1+\iota-\mu}}{\iota!(k-1-\iota)!} \frac{1}{(\omega+\iota)^{j-\mu}}, \tag{6.8}$$

for $k \geq 1$. For $k = -1$, this local term is given by

$$x^{\omega-2} \sum_{j=1}^{m} a_j \frac{(\log x)^{j-2}}{(j-1)!} (j-1+(\omega-1)\log x). \tag{6.9}$$

In particular, by (6.3), if ω is a simple pole of ζ_η, then the local term is given by these formulas for $m = 1$:

$$x^{\omega-1} \operatorname{res}\left(\zeta_\eta(s); \omega\right), \tag{6.10}$$

if $k = 0$, *and by*

$$\operatorname{res}\left(\zeta_\eta(s);\omega\right)\frac{x^{\omega+k-1}}{(\omega)_k} = \operatorname{res}\left(\zeta_\eta(s);\omega\right)x^{\omega+k-1}\sum_{\iota=0}^{k-1}\frac{(-1)^\iota}{\iota!(k-1-\iota)!}\frac{1}{\omega+\iota}, \tag{6.11}$$

if $k \geq 1$. *For* $k < 0$, *the local term is given by*

$$\operatorname{res}\left(\zeta_\eta(s);\omega\right)x^{\omega+k-1}(\omega-1)(\omega-2)\dots(\omega+k). \tag{6.12}$$

6.1.2 The Spectral Local Terms

Let ν be the spectral measure of the generalized fractal string η, as in Section 4.2, and let ω be a *spectral complex dimension* of η (i.e., ω is a pole of the spectral zeta function ζ_ν, and therefore, a complex dimension of ν). Since, by formula (4.17), $\zeta_\nu(s) = \zeta_\eta(s)\zeta(s)$, either $\omega = 1$, the pole of the Riemann zeta function $\zeta(s)$, or ω is a pole of the geometric zeta function $\zeta_\eta(s)$ (i.e., ω is a *geometric complex dimension* of η) that is not canceled by a zero of $\zeta(s)$ at ω. See the end of this subsection and also Chapter 9 for a discussion of the possibility of cancellation of complex dimensions.

The spectral dimension $\omega = 1$ gives the Weyl term and is discussed in the next section.

In order to study the corresponding local terms, we need to compute the principal part of the Laurent series of $\zeta_\nu(s) = \zeta_\eta(s)\zeta(s)$ at $s = \omega$. For $\omega \neq 1$, let this principal part be

$$\frac{b_m}{(s-\omega)^m} + \frac{b_{m-1}}{(s-\omega)^{m-1}} + \frac{b_{m-2}}{(s-\omega)^{m-2}} + \dots + \frac{b_1}{s-\omega}. \tag{6.13}$$

The coefficients b_j can be expressed in terms of the a_j's of formula (6.2):

$$b_j = \sum_{q=1}^{j} a_q \frac{\zeta^{(j-q)}(\omega)}{(j-q)!}, \tag{6.14}$$

where $\zeta^{(n)}(\omega)$ denotes the n-th derivative of the Riemann zeta function at $s = \omega$. In particular, note that we have $b_m = a_m\zeta(\omega)$. (Indeed, since in our applications, $\operatorname{Re}\omega < 1$, ω is not a pole of the Riemann zeta function.) But if $\zeta(s)$ vanishes at $s = \omega$, then the multiplicity of the complex dimension ω diminishes in the spectrum. The complex dimension ω could even disappear altogether, as was mentioned above. This is the subject of Chapter 9.

6.1.3 The Weyl Term

There is one special complex dimension associated with the spectrum, namely $s = 1$. We call it the spectral complex dimension of the Bernoulli string.[2] It is never a pole of ζ_η, so that we can easily compute the corresponding term in the explicit formulas.

Definition 6.2. The *Weyl term* associated with the Bernoulli string is the term at $s = 1$ in the explicit formula for the spectrum. It is given by

$$W^{[k]}(x) = \zeta_\eta(1)\frac{x^k}{k!}, \tag{6.15}$$

for $k \geq 0$.

Remark 6.3. Recall that if η is associated to an ordinary fractal string $\mathcal{L}$, then $\zeta_\eta(1) = \mathrm{vol}_1(\mathcal{L})$ is the total length of the string. Note that for a smooth, compact and connected manifold, the leading term of the spectral asymptotics of the Laplacian is often referred to as the Weyl term in the literature. According to Weyl's asymptotic formula (see Equations (B.2) and (B.3) in Section B.1 of Appendix B), the Weyl term is proportional to the volume of the manifold.

6.1.4 The Distribution $x^\omega \log^m x$

The local terms are finite sums of terms proportional to $x^{\omega+k-1}(\log x)^m$. In Theorems 5.18 and 5.22, these terms have to be interpreted as distributions, which act on a test function φ in the following way:

$$\left\langle x^{\omega+k-1}(\log x)^m, \varphi\right\rangle = \int_0^\infty x^{\omega+k-1}(\log x)^m \varphi(x)\,dx. \tag{6.16}$$

Now, $\widetilde{\varphi}(s) = \int_0^\infty x^{s-1}\varphi(x)\,dx$ is an analytic function (it is meromorphic in a strip with finitely many poles under the more general assumptions of Theorems 5.26 and 5.27). Since this integral is absolutely convergent, we can differentiate under the integral sign. The m-th derivative with respect to s is

$$\widetilde{\varphi}^{(m)}(s) = \int_0^\infty x^{s-1}(\log x)^m \varphi(x)\,dx. \tag{6.17}$$

Combining (6.16) and (6.17), putting $s = \omega + k$, we find that

$$\left\langle x^{\omega+k-1}(\log x)^m, \varphi\right\rangle = \widetilde{\varphi}^{(m)}(\omega + k), \tag{6.18}$$

where $\widetilde{\varphi}^{(m)}$ is the m-th derivative of the holomorphic function $\widetilde{\varphi}$.

---

[2] The Bernoulli string $\mathcal{B}$ (called the Sturm–Liouville string in [Lap2, Remark 2.5, p. 144]) is defined as $\Omega = (0, 1)$, a single open interval of length 1. Its spectral zeta function is the Riemann zeta function, $\zeta_{\nu,\mathcal{B}}(s) = \zeta(s)$.

6.2 Explicit Formulas for Lengths and Frequencies

In this section, we give the explicit formulas for the geometric and spectral counting functions $N_\eta(x)$ and $N_\nu(x)$. We also give these formulas on the k-th level, for $k \geq 2$.

6.2.1 The Geometric Counting Function of a Fractal String

Choose a screen S such that **L1** and **L2** are satisfied for some κ, so that η is languid in the corresponding window W. When $k > \max\{1, \kappa+1\}$, the pointwise explicit formula (Theorem 5.10) for the counting function is valid. Otherwise (that is, if $k \leq \kappa + 1$ or $k \leq 1$, $k \in \mathbb{Z}$), we have to interpret $N_\eta^{[k]}(x)$ as a distribution. Note that for the usual counting function $N_\eta(x) = N_\eta^{[1]}(x)$, we do not have a pointwise formula with error term. But as a distribution, according to Theorem 5.18, it is given by

$$N_\eta(x) = \sum_{\omega \in \mathcal{D}_\eta(W)} \operatorname{res}\left(\frac{x^s \zeta_\eta(s)}{s}; \omega\right) + \{\zeta_\eta(0)\} + R_\eta^{[1]}(x), \tag{6.19a}$$

and, if the poles are simple, by

$$N_\eta(x) = \sum_{\omega \in \mathcal{D}_\eta(W)} \frac{x^\omega}{\omega} \operatorname{res}\left(\zeta_\eta(s); \omega\right) + \{\zeta_\eta(0)\} + R_\eta^{[1]}(x), \tag{6.19b}$$

where the term in braces is included only if $0 \in W \backslash \mathcal{D}_\eta$.

These are special cases of the following more general formulas, for $k \geq 1$:

$$N_\eta^{[k]}(x) = \sum_{\omega \in \mathcal{D}_\eta(W)} \operatorname{res}\left(\frac{x^{s+k-1}\zeta_\eta(s)}{(s)_k}; \omega\right) + \frac{1}{(k-1)!} \sum_{\substack{j=0 \\ -j \in W \backslash \mathcal{D}_\eta}}^{k-1} \binom{k-1}{j} (-1)^j x^{k-1-j} \zeta_\eta(-j) + R_\eta^{[k]}(x), \tag{6.20a}$$

and, if the complex dimensions are simple,

$$N_\eta^{[k]}(x) = \sum_{\omega \in \mathcal{D}_\eta(W)} \frac{x^{\omega+k-1}}{(\omega)_k} \operatorname{res}\left(\zeta_\eta(s); \omega\right) + \frac{1}{(k-1)!} \sum_{\substack{j=0 \\ -j \in W \backslash \mathcal{D}_\eta}}^{k-1} \binom{k-1}{j} (-1)^j x^{k-1-j} \zeta_\eta(-j) + R_\eta^{[k]}(x). \tag{6.20b}$$

We note that the distributional error term occurring in formulas (6.19a)–(6.20b) can be estimated by means of Theorem 5.30.

Remark 6.4. From now on, for notational simplicity, we will no longer distinguish between the distributional error terms at level k (previously denoted by $\mathcal{R}_\eta^{[k]}(x)$) and their pointwise counterparts, $R_\eta^{[k]}(x)$. In other words, both a pointwise and distributional error term at level k will be denoted by $R_\eta^{[k]}(x)$, as we have done in formulas (6.19a)–(6.20b) above. As usual, the case $k \leq 0$ is allowed in the distributional formula.

6.2.2 The Spectral Counting Function of a Fractal String

For the spectral zeta function $\zeta_\nu(s)$, hypothesis **L2′** is never satisfied. Hence only Theorems 5.10 and 5.18 apply; i.e., our explicit formulas for the spectrum will always contain an error term. Indeed, by [Edw, §9.2, p. 185], the Riemann zeta function satisfies

$$|\zeta(\sigma + it)| \leq C_\sigma \left(|t| + 1\right)^{\frac{1}{2}-\sigma} \tag{6.21}$$

for $\sigma < 0$, and it does not satisfy a better estimate in this half-plane. (The constant C_σ decreases like $(2\pi e)^{\sigma-1/2}$ as $\sigma \to -\infty$.) Moreover, inside the critical strip $0 \leq \sigma \leq 1$, one has the estimate

$$|\zeta(\sigma + it)| \leq K \cdot (|t| + 1)^{(1-\sigma)/2} \log |t| \tag{6.22}$$

for some constant K. It follows from these estimates that if $\mathcal{L}$ is a fractal string satisfying hypotheses **L1** and **L2** for some value of κ, then, by Equation (1.38), its spectral zeta function $\zeta_\nu(s) = \zeta_{\mathcal{L}}(s)\zeta(s)$ satisfies the same assumptions for the same screen, except with κ replaced by $\kappa + \frac{1}{2} - \inf S$ if $\inf S \leq 0$, or by $\kappa + \frac{1}{2}$ if $\inf S > 0$.

Given $\sigma_0 \leq 0$, we apply Theorem 5.10 to find the following pointwise explicit formula, valid for all positive integers $k > \frac{3}{2} - \sigma_0 + \kappa$:

$$\begin{aligned} N_\nu^{[k]}(x) = W_\eta^{[k]}(x) &+ \sum_{\omega \in \mathcal{D}_\eta(W)} \operatorname{res}\left(\frac{x^{s+k-1}\zeta_\nu(s)}{(s)_k}; \omega\right) \\ &+ \frac{1}{(k-1)!} \sum_{\substack{j=0 \\ -j \in W \setminus \mathcal{D}_\eta}}^{k-1} \binom{k-1}{j} (-1)^j x^{k-1-j} \zeta_\nu(-j) + R_\nu^{[k]}(x), \end{aligned} \tag{6.23a}$$

where $W_\eta^{[k]}(x) = \zeta_\eta(1)x^k/k!$ is the Weyl term (at level k), introduced in Section 6.1.3. If the poles are simple, this becomes

$$\begin{aligned} N_\nu^{[k]}(x) = W_\eta^{[k]}(x) &+ \sum_{\omega \in \mathcal{D}_\eta(W)} \frac{x^{\omega+k-1}}{(\omega)_k} \zeta(\omega) \operatorname{res}\left(\zeta_\eta(s); \omega\right) \\ &+ \frac{1}{(k-1)!} \sum_{\substack{j=0 \\ -j \in W \setminus \mathcal{D}_\eta}}^{k-1} \binom{k-1}{j} (-1)^j x^{k-1-j} \zeta_\nu(-j) + R_\nu^{[k]}(x). \end{aligned} \tag{6.23b}$$

Alternatively, if we apply Theorem 5.18, we obtain a distributional explicit formula (with error term) valid for all $k \in \mathbb{Z}$ (and in particular for $k = 1$), which is still given by Equation (6.23), but which is now interpreted distributionally (see Remark 6.4 above). This distributional formula will be very useful later in the book; see, in particular, Chapter 9 and especially Section 11.2.

Remark 6.5. If we set $k = 1$ in (6.23a) and (6.23b) (which is only possible provided these formulas are interpreted distributionally), then Equation (6.23a) becomes, using that $\zeta(0) = -1/2$,

$$N_\nu(x) = \zeta_\eta(1)x + \sum_{\omega \in \mathcal{D}_\eta(W)} \operatorname{res}\left(\frac{x^s \zeta_\nu(s)}{s}; \omega\right) - \{\zeta_\eta(0)/2\} + R_\nu(x). \tag{6.24a}$$

Moreover, if all the complex dimensions of the fractal string $\mathcal{L}$ are simple, then Equation (6.23b) becomes

$$N_\nu(x) = \zeta_\eta(1)x + \sum_{\omega \in \mathcal{D}_\eta(W)} \frac{x^\omega}{\omega} \zeta(\omega) \operatorname{res}\left(\zeta_\eta(s); \omega\right) - \{\zeta_\eta(0)/2\} + R_\nu(x). \tag{6.24b}$$

In both formulas, the term in braces is included only if $0 \in W \backslash \mathcal{D}_\eta$.

6.2.3 *The Geometric and Spectral Partition Functions*

The *geometric partition function* of an ordinary fractal string $\mathcal{L} = (l_j)_{j=1}^\infty$ is given by

$$\theta_{\mathcal{L}}(t) = \sum_{j=1}^{\infty} e^{-tl_j^{-1}}, \qquad \text{for } t > 0. \tag{6.25}$$

More generally, for a generalized fractal string η, it is given by

$$\theta_\eta(t) = \int_0^\infty e^{-xt} \eta(dx) = \langle \mathcal{P}_\eta^{[0]}, \varphi_t \rangle, \tag{6.26}$$

where $\varphi_t(x) = e^{-xt}$ for $t > 0$. Hence, if η satisfies **L1** and **L2** (that is, η is languid, in the sense of Definition 5.2), the explicit formula of Theorem 5.26 applies, and we obtain the following result (note that by (5.43), we have $\widetilde{\varphi_t}(s) = \Gamma(s)t^{-s}$, where Γ is the gamma function):

$$\begin{aligned}\theta_\eta(t) = &\sum_{\omega \in \mathcal{D}_\eta(W)} \operatorname{res}\left(\zeta_\eta(s)\Gamma(s)t^{-s}; \omega\right) \\ &+ \sum_{\substack{m=0 \\ -m \in W \backslash \mathcal{D}_\eta}}^{\infty} \frac{(-1)^m}{m!} t^m \zeta_\eta(-m) + \langle R_\eta^{[0]}, \varphi_t \rangle.\end{aligned} \tag{6.27a}$$

If the complex dimensions of η are simple and do not overlap with the poles $0, -1, -2, \ldots$ of the gamma function, this yields

$$\theta_\eta(t) = \sum_{\omega \in \mathcal{D}_\eta(W)} \operatorname{res}\left(\zeta_\eta(s); \omega\right) \Gamma(\omega) t^{-\omega} + \sum_{\substack{m=0 \\ -m \in W \setminus \mathcal{D}_\eta}}^{\infty} \frac{(-1)^m}{m!} t^m \zeta_\eta(-m) + \left\langle R_\eta^{[0]}, \varphi_t \right\rangle. \tag{6.27b}$$

Note that by Theorem 5.27, if η satisfies the stronger hypothesis **L2**$'$ instead of **L2**, so that η is strongly languid, in the sense of Definition 5.3, we can choose $W = \mathbb{C}$ and set $\left\langle R_\eta^{[0]}, \varphi_t \right\rangle = 0$ in both (6.27a) and (6.27b).

Recall from (4.21) that the spectral measure of the ordinary fractal string $\mathcal{L}$ is given by

$$\nu = \sum_f w_f^{(\nu)} \delta_{\{f\}} = \sum_{n,j=1}^{\infty} \delta_{\{n l_j^{-1}\}}, \tag{6.28}$$

where f runs over the sequence of (distinct) frequencies of $\mathcal{L}$ and $w_f^{(\nu)}$ denotes the multiplicity of f, as in (1.33). Therefore, according to definition (6.26), the *spectral partition function* of $\mathcal{L}$ is given by

$$\theta_\nu(t) = \sum_f w_f^{(\nu)} e^{-ft} = \sum_{n,j=1}^{\infty} e^{-n l_j^{-1} t}, \quad \text{for } t > 0. \tag{6.29}$$

In general, the spectral measure ν associated with a generalized fractal string η (that is, the measure ν representing the frequency spectrum of η) is defined by (4.16) or (4.20). Then, the spectral partition function of η is given by (6.26), with η replaced by ν. When we apply formula (6.27a) to ν, we thus obtain the following formula for the spectral partition function, provided 1 is not a complex dimension of η:

$$\theta_\nu(t) = \zeta_\eta(1)\frac{1}{t} + \sum_{\omega \in \mathcal{D}_\eta(W)} \operatorname{res}\left(\zeta_\eta(s)\zeta(s)\Gamma(s)t^{-s}; \omega\right) + \sum_{\substack{m=0 \\ -m \in W \setminus \mathcal{D}_\eta}}^{\infty} \frac{(-1)^m}{m!} t^m \zeta_\eta(-m)\zeta(-m) + \left\langle R_\nu^{[0]}, \varphi_t \right\rangle. \tag{6.30a}$$

If the complex dimensions of η are simple and do not overlap with the poles of Γ, this yields

$$\theta_\nu(t) = \zeta_\eta(1)\frac{1}{t} + \sum_{\omega \in \mathcal{D}_\eta(W)} \operatorname{res}\left(\zeta_\eta(s); \omega\right) \zeta(\omega)\Gamma(\omega) t^{-\omega} + \sum_{\substack{m=0 \\ -m \in W \setminus \mathcal{D}_\eta}}^{\infty} \frac{(-1)^m}{m!} t^m \zeta_\eta(-m)\zeta(-m) + \left\langle R_\nu^{[0]}, \varphi_t \right\rangle. \tag{6.30b}$$

Remark 6.6. In the literature on spectral geometry or on mathematical physics (see, for example, [Gi] or [BaltHi, Sim]), the spectral partition function $\theta_\nu(t)$ is frequently used as a substitute for the spectral counting function $N_\nu(x)$. It usually has a clear physical interpretation. For example, in quantum statistical mechanics, the eigenvalues correspond to the possible energy levels E_n of the underlying quantum system (with possible interactions). Further, the variable t is replaced by $\beta = 1/k_B T$, where k_B is the Boltzmann constant and T stands for the absolute temperature. (Note that physically, $\beta\hbar$ has the dimension of a time, where $\hbar$ is the Planck constant divided by 2π.) In this notation, the (quantum) partition function, defined as the trace of the Schrödinger semigroup,[3] equals $\sum_{j=1}^{\infty} e^{-\beta E_j}$. For each $n \geq 1$, $e^{-\beta E_n} / \sum_j e^{-\beta E_j}$ is then interpreted as the probability (or statistical weight) attached to the n-th state of the system (counting multiplicities in the spectrum). See [BaltHi, Sim] along with [Fey], and in a broader context, [JohLap, Section 20.2].

The interested reader can find in Section B.2 of Appendix B (and the relevant references therein) some information about the short time asymptotics of $\theta_\nu(t)$ in the classical case of a smooth manifold, along with their relationship with the spectral zeta function $\zeta_\nu(s)$. We note that for the spectral problems of this kind, one usually needs information about the asymptotic behavior of $\theta_\nu(t)$ as $t \to 0^+$, which corresponds to the asymptotic behavior of $N_\nu(x)$ as $x \to \infty$.

6.3 The Direct Spectral Problem for Fractal Strings

6.3.1 The Density of Geometric and Spectral States

Of special interest are the explicit formulas at level $k = 0$, the density of states formulas. They only have a distributional interpretation and they express the measure of the generalized fractal string, or the spectral measure, as a sum over the complex dimensions.

We apply our distributional explicit formula with error term (Theorem 5.18) to $\eta = \mathcal{P}_\eta^{[0]}$, where η is a generalized fractal string satisfying hypotheses **L1** and **L2**. We then obtain that, as a distribution, the measure η is given by the following *density of geometric states* (or *density of lengths*) formula:

$$\eta = \sum_{\omega \in \mathcal{D}_\eta(W)} \operatorname{res}\left(\zeta_\eta(s) x^{s-1}; \omega\right) + R_\eta^{[0]}(x). \tag{6.31a}$$

[3]It is often given and calculated via a path integral.

If the complex dimensions of η are simple, this becomes

$$\eta = \sum_{\omega \in \mathcal{D}_\eta(W)} \operatorname{res}\left(\zeta_\eta(s); \omega\right) x^{\omega-1} + R_\eta^{[0]}(x). \tag{6.31b}$$

The error term is given by the integral over the screen,

$$R_\eta^{[0]}(x) = \frac{1}{2\pi i} \int_S \zeta_\eta(s) x^{s-1}\, ds. \tag{6.32}$$

Further, by Theorem 5.22, $R_\eta^{[0]}(x)$ vanishes identically if condition **L2**$'$, rather than **L2**, is satisfied and $W = \mathbb{C}$.

Next, let ν be the generalized fractal string representing the frequencies of η. Note that we can use the same window W for ν, since ζ grows polynomially in vertical strips (see Equation (6.21) above or [In]). If 1 is not a complex dimension of η, we obtain the following *density of spectral states* (or *density of frequencies*) formula, by Theorem 5.18 applied to $\nu = \mathcal{P}_\nu^{[0]}$:

$$\nu = \zeta_\eta(1) + \sum_{\omega \in \mathcal{D}_\eta(W)} \operatorname{res}\left(\zeta_\eta(s)\zeta(s)x^{s-1}; \omega\right) + R_\nu^{[0]}(x). \tag{6.33a}$$

If, in addition, the complex dimensions of η are simple, we find by evaluating the residues in the sum that

$$\nu = \zeta_\eta(1) + \sum_{\omega \in \mathcal{D}_\eta(W)} \operatorname{res}\left(\zeta_\eta(s); \omega\right) \zeta(\omega) x^{\omega-1} + R_\nu^{[0]}(x). \tag{6.33b}$$

The error term is given by the integral over the screen:

$$R_\nu^{[0]}(x) = \frac{1}{2\pi i} \int_S \zeta_\eta(s)\zeta(s) x^{s-1}\, ds. \tag{6.34}$$

Note that in this case, **L2**$'$ is never satisfied, as is explained at the beginning of Section 6.2.2.

Heuristically, this distributional formula describes what physicists call the density of states of the string η (or, quantum-mechanically, the density of energy levels of η); see, e.g., [Berr3,BogKe,Ke] and the relevant references therein.[4]

[4]The authors of these physical references study the spectrum of suitable Hamiltonians and not that of (generalized) fractal strings. Moreover, from the mathematical point of view, it is clear that their formal treatment of the density of spectral states should be interpreted distributionally (as could be done using, for example, our explicit formulas). We do not claim, however, to be able to recover from our explicit formulas all the results contained in those papers. See Section 12.5.3 for a related discussion.

6.3.2 The Spectral Operator and its Euler Product

Our explicit formulas for the density of geometric and spectral states obtained in the previous section—along with previous work on related subjects (especially [LapPo1–2], [LapMa1–2] and Part II of [Lap3])—suggest introducing the following definition, which fits naturally within the present framework of generalized fractal strings. In turn, it helps conceptualize aspects of the aforementioned work and will become useful in our own investigations (see Chapters 9 and 11). In particular, it formalizes the direct spectral problem which consists of deducing spectral information from the geometry.

Definition 6.7. The *spectral operator* maps the density of geometric states onto the density of spectral states. Specifically, it adds the Weyl term $\zeta_\eta(1)$ to the explicit formula for η, and locally, if the complex dimensions are simple, it multiplies each $x^{\omega-1}$-term by $\zeta(\omega)$. Further, the integrand of the error term is multiplied by $\zeta(s)$.

Remark 6.8. We could use the geometric and spectral partition function of Section 6.2.3, instead of the density of geometric and spectral states, to formulate the definition of the spectral operator.

Remark 6.9. The *inverse spectral operator* maps ν back onto η. Since it involves dividing by $\zeta(\omega)$ at the complex dimensions and by $\zeta(s)$ in the error term, the extent to which it exists depends on the location of the critical zeros of ζ. This sheds new light on the work of the first author with H. Maier [LapMa1–2], to be revisited and extended in Chapter 9.

It follows from our results in Section 10.2 below that in some sense the inverse spectral operator is well defined on the class of generalized Cantor strings studied in Chapter 10 (see Remark 10.11). This is the guiding principle underlying our proof in Chapter 11 that many number-theoretic zeta functions and other Dirichlet series do not have an infinite vertical sequence of critical zeros forming
an arithmetic progression.

The Euler Product for the Spectral Operator

Let $N(x) = N_\eta(x)$ be the counting function of a (generalized) fractal string η. We define the *spectral operator* as the counting function of $\eta * h$ (see Section 4.2, along with Theorem 1.19 of Section 1.3),

$$\nu(N)(x) = \sum_{k=1}^{\infty} N(x/k).$$

Note that this sum is finite, since $N(x) = 0$ for x close to 0. To fix ideas, let us suppose that $N(x) = 0$ for $0 \leq x \leq 1$. Then k runs up to $[x]$ in the sum. Define also the p-factor of ν, for every prime p, by

$$\nu_p(N)(x) = \sum_{k=0}^{\infty} N(xp^{-k}),$$

where the terms in the sum vanish when $p^k \geq x$. This is the counting function of $\eta * h_p$.[5] The operators ν_p commute with each other, and their composition gives the *Euler product* for ν:

$$\nu(N) = \left(\prod\nolimits_p \nu_p\right)(N).$$

Making a change of variables $x = e^t$, and writing $f(t) = N(x)$, we define the *additive spectral operator*

$$a(f)(t) = \sum_{k=1}^{\infty} f(t - \log k),$$

and the p-factors

$$a_p(f)(t) = \sum_{k=0}^{\infty} f(t - k \log p).$$

Both these sums have finitely many terms for each t provided f is supported on a right half-line $[b, \infty)$, for some real number b. These operators are again related by an Euler product,

$$a(f) = \left(\prod_p a_p\right)(f),$$

where the product means composition of operators.

To study the continuity (boundedness) of these operators, we work in the Hilbert space of functions f on $[0, \infty)$ such that the norm

$$\|f\|_c^2 = \int_0^{\infty} |f(t)|^2 e^{-2ct}\, dt$$

is finite. The Hilbert space depends on a positive parameter c which needs to be suitably chosen. Note that if $N(x) = O(x^D)$ then $f(t) = O(e^{Dt})$ and f has finite norm for $c > D$. In particular, the spectral operator is bounded for $c > 1$, and unbounded for $c \leq 1$. For $c = 1$, the p-factors are bounded on the space of functions f for which there exists a $D < 1$ (depending on f) such that $f(t) = O(e^{Dt})$, which corresponds to the space of counting functions of fractal strings of dimension D. Such strings have

---

[5] Here, $h_p = \sum_{j=0}^{\infty} \delta_{\{p^j\}}$ denotes the p-th elementary prime string, as in Equation (4.25) of Section 4.2.1.

complex dimensions ω with $\operatorname{Re}\omega \leq D$. In this sense, for $c = 1$, the spectral operator converges *inside* the critical strip $0 < \operatorname{Re} s < 1$.

Let $\partial = \frac{d}{dt}$ be the differentiation operator. The Taylor series for a smooth function f can be written as

$$f(t+h) = f(t) + \frac{f'(t)}{1!}h + \frac{f''(t)}{2!}h^2 + \frac{f'''(t)}{3!}h^3 + \cdots = e^{h\partial}(f)(t).$$

That is, the derivative is the infinitesimal generator of the group of shifts on the real line. Using this operator, we see the analogy with the usual Euler product for $\zeta(s)$ (see Equations (4.14) and (4.27)):

$$a_p(f) = \sum_{k=0}^{\infty} e^{-k(\log p)\partial}(f) = \left(1 - p^{-\partial}\right)^{-1}(f) = \zeta_{h_p}(\partial)(f)$$

and

$$a(f) = \sum_{k=1}^{\infty} e^{-(\log k)\partial}(f) = \zeta(\partial)(f).$$

To justify this formalism, we need to verify that the spectral operator and its factors are normal; i.e., $aa^* = a^*a$ and $a_p a_p^* = a_p^* a_p$. The adjoint of a shift is the shift in the opposite direction. Since here, our Hilbert space depends on a parameter c, we obtain

$$a_p^*(f)(t) = \sum_{k=0}^{\infty} f(t + k\log p)p^{-2kc},$$

and

$$a^*(f)(t) = \sum_{k=1}^{\infty} f(t + \log k)k^{-2c}.$$

We compute that these operators are indeed normal:

$$a_p a_p^*\,(f)\,(t) = a_p^* a_p\,(f)\,(t) = \frac{1}{1 - p^{-2c}} \sum_{k=-\infty}^{\infty} f(t - k\log p)p^{2c\min\{0,k\}},$$

and

$$aa^*\,(f)\,(t) = a^*a\,(f)\,(t) = \zeta(2c) \sum_{(m,n)=1} f\left(t - \log\tfrac{m}{n}\right)n^{-2c},$$

where the last sum is taken over all pairs of integers $m, n \geq 1$ without common factor.

Note that

$$a_p^{-1}(f)(t) = f(t) - f(t - \log p).$$

To compute the spectrum of a_p, we solve $a_p(f) - \lambda f = g$ for f, and consider for which complex numbers λ this does not define a bounded operator. Applying a_p^{-1} first, we obtain

$$(1 - \lambda)f(t) + \lambda f(t - \log p) = g(t) - g(t - \log p).$$

From this we readily deduce that

$$f(t)=\frac{1}{1-\lambda}g(t)-\sum_{k=1}^{\infty}g(t-k\log p)\frac{\lambda^{k-1}}{(\lambda-1)^{k+1}}.$$

Taking $g=0$, we see first of all that a_p has no eigenfunctions. The function f has bounded norm exactly when $|\frac{\lambda}{\lambda-1}|<p^c$. Hence, the spectrum $\sigma(a_p)$ of the operator a_p consists of those values for which $|\frac{\lambda}{\lambda-1}|\geq p^c$. We find

$$\sigma(a_p)=\left\{z\in\mathbb{C}\colon \left|z-\tfrac{1}{1-p^{-2c}}\right|\leq\tfrac{p^{-c}}{1-p^{-2c}}\right\}.$$

We invite the reader to further explore the material contained in this section. In particular, we suggest the following problem, which may be related to the Riemann hypothesis:

Problem 6.10. Determine the spectrum of the operator a.

6.4 Self-Similar Strings

We now consider the case of self-similar strings studied in Chapter 2. The reader may first wish to briefly review some of the main results and definitions of Sections 2.2–2.5, particularly Theorems 2.4 and 2.17, as well as Definition 2.14.

According to Theorem 2.4, the geometric zeta function of a self-similar string $\mathcal{L}$ of length L with scaling ratios $r_1,r_2,\ldots,r_N$ and gaps scaled by $g_1,\ldots,g_K$ is given by

$$\zeta_{\mathcal{L}}(s)=L^s\frac{g_1^s+\cdots+g_K^s}{1-r_1^s-\cdots-r_N^s}. \tag{6.35}$$

We deduce that (recall from Section 2.1 that g_K is the smallest gap),

$$|\zeta_{\mathcal{L}}(s)|\ll\left(Lg_Kr_N^{-1}\right)^{-|\sigma|}\qquad\text{as } \sigma=\operatorname{Re}s\to-\infty. \tag{6.36}$$

It follows that we can let $W=\mathbb{C}$ and that $\zeta_{\mathcal{L}}(s)$ satisfies **L1** and **L2′** with $\kappa=0$ and $A=L^{-1}g_K^{-1}r_N$, as will be further explained just below. Furthermore, Theorem 3.26 enables us to find a suitable sequence $\{T_n\}_{n\in\mathbb{Z}}$, with $T_n\to\pm\infty$ as $n\to\pm\infty$, such that $\zeta_{\mathcal{L}}(s)$ is uniformly bounded on the horizontal lines $\operatorname{Im}s=T_n$ (for every $n\in\mathbb{Z}$), and hence such that hypothesis **L1** given by (5.19) is satisfied with $\kappa=0$. Secondly, in view of estimate (6.36) above, we can simply choose the screen S_m to be the vertical line $\operatorname{Re}s=-m$ for $m\geq1$ to verify that hypothesis **L2′** given by (5.21) holds with $\kappa=0$ and $A=L^{-1}g_K^{-1}r_N$. It follows that geometrically, a self-similar string is always strongly languid, for $\kappa=0$ (and hence, by Remark 5.4, for every $\kappa\geq0$). Thus Theorems 5.14 and 5.22 apply to obtain asymptotic pointwise and distributional formulas without error term, valid for all $x > L^{-1}g_K^{-1}r_N$. See also Theorem 3.25, along with its application given in Section 8.4.4 below.

Remark 6.11. In what follows, it will be useful to recall from Theorem 2.17 that for a self-similar string, D is always simple, where D denotes the dimension of $\mathcal{L}$.[6] Moreover, all the other complex dimensions of $\mathcal{L}$ are located to the left of the vertical line $\operatorname{Re} s = D$. According to the basic dichotomy of Theorem 2.17, for a nonlattice string, they all lie strictly to the left of this line, and a subsequence of complex dimensions lies arbitrarily close to it, whereas if $\mathcal{L}$ is a lattice string, there is an infinite sequence of complex dimensions ω on the line $\operatorname{Re} s = D$, namely $\omega = D + in\mathbf{p}$ $(n \in \mathbb{Z})$, where $\mathbf{p}$ denotes the oscillatory period of $\mathcal{L}$.

6.4.1 Lattice Strings

The explicit formulas for lattice self-similar strings are particularly simple, since by Theorem 2.17, the complex dimensions of $\mathcal{L}$ are located on finitely many vertical lines. Moreover, if the gaps are integral powers of r as well, as in Equation (2.40), then the residues of $\zeta_{\mathcal{L}}$ at the complex dimensions on a given line are all equal. We assume this in this section and the lattice part of Section 6.4.3, leaving the more general case to the reader (see, for example, Remark 2.19). This assumption is satisfied in all our lattice examples, since they all have a single gap, normalized as in Remark 2.6. We do not make (and indeed, cannot make) this assumption in Section 6.4.2 or the part of Section 6.4.3 dealing with nonlattice strings.

If the complex dimensions $\omega + in\mathbf{p}$ $(n \in \mathbb{Z})$ on the vertical line $\operatorname{Re} s = \operatorname{Re}\omega$ are simple (i.e., if ω is a simple pole of $\zeta_{\mathcal{L}}$), then the corresponding sum in the pointwise formula for $N_{\mathcal{L}}(x)$ gives the multiplicatively periodic function

$$\operatorname{res}\left(\zeta_{\mathcal{L}}(s);\omega\right) \sum_{n=-\infty}^{\infty} \frac{x^{\omega+in\mathbf{p}}}{\omega+in\mathbf{p}}. \tag{6.37}$$

Using the Fourier series (1.13), with $\log b = 2\pi\omega/\mathbf{p}$ and $u = \mathbf{p}\log x/2\pi$, we see that this sum is equal to

$$\operatorname{res}\left(\zeta_{\mathcal{L}}(s);\omega\right) \sum_{n=-\infty}^{\infty} \frac{x^{\omega+in\mathbf{p}}}{\omega+in\mathbf{p}} = \operatorname{res}\left(\zeta_{\mathcal{L}}(s);\omega\right) \frac{b}{b-1} b^{-\{u\}} x^{\omega} \frac{2\pi}{\mathbf{p}}. \tag{6.38}$$

In particular, since D is simple (see Remark 6.11 above), the local term corresponding to D in the explicit formula for $N_{\mathcal{L}}(x)$ is given by (6.37) (with the infinite sum expressed as in (6.38)) with $\omega = D$.

---

[6] The number D is the Minkowski dimension (or the only real dimension) of $\mathcal{L}$, also called the similarity dimension of $\mathcal{L}$. See Remark 2.22.

Likewise, these complex dimensions contribute the quantity

$$\operatorname{res}\left(\zeta_{\mathcal{L}}(s);\omega\right)\sum_{n=-\infty}^{\infty}\Gamma\left(\omega+in\mathbf{p}\right)t^{-\omega-in\mathbf{p}} \tag{6.39}$$

to the asymptotic expansion of the geometric partition function $\theta_{\mathcal{L}}(t)$, as defined by Equation (6.25).

For example, for the Cantor string defined in Section 2.3.1, we have $\omega = D = \log_3 2$ and $\mathbf{p} = 2\pi/\log 3$. Hence we can set $b = 2$ in the Fourier expansion (1.13), and we recover formula (1.31) stated at the end of Section 1.2.2:[7]

$$N_{\mathrm{CS}}(x) = 2^{1-\{\log_3 x\}}x^D - 1 = \frac{1}{\log 3}\sum_{n=-\infty}^{\infty}\frac{x^{D+in\mathbf{p}}}{D+in\mathbf{p}} - 1. \tag{6.40}$$

In the same way, the geometric partition function of the Cantor string,

$$\theta_{\mathrm{CS}}(t) = \sum_{n=0}^{\infty} 2^n e^{-3^n t},$$

is given by, in view of formula (6.27b) and the comment following it,

$$\begin{aligned}\frac{1}{\log 3}\sum_{n=-\infty}^{\infty}\Gamma\left(D+in\mathbf{p}\right)t^{-D-in\mathbf{p}} &+ \sum_{m=0}^{\infty}\frac{(-1)^m}{m!}t^m\frac{1}{1-2\cdot 3^m}\\ &= t^{-D}G_{\mathrm{CS}}\left(\log_3 t^{-1}\right) + \sum_{m=0}^{\infty}\frac{(-1)^m}{m!}t^m\frac{1}{1-2\cdot 3^m},\end{aligned} \tag{6.41}$$

where G_{CS} is the nonconstant periodic function (of period 1) given by

$$G_{\mathrm{CS}}(u) = \frac{1}{\log 3}\sum_{n=-\infty}^{\infty}\Gamma\left(D+in\mathbf{p}\right)e^{2\pi inu}. \tag{6.42}$$

Remark 6.12. The periodic function G_{CS}, defined by (6.42) and occurring in the explicit formula (6.41) for $\theta_{\mathrm{CS}}(t)$, is smooth on $(-\infty,+\infty)$. (In light of a well-known theorem about Fourier series, this follows from the fast decay of $\Gamma(D+in\mathbf{p})$ as $|n|\to\infty$; see, e.g., [Schw1, §VII.1] along with Stirling's formula [In, p. 57].) In contrast, the corresponding periodic function occurring in the explicit formula (6.40) (or (1.31)) for the geometric counting function $N_{\mathrm{CS}}(x)$ has infinitely many discontinuities.

An analogous comment applies to the Fibonacci string studied below—and, more generally, to every lattice string.

---

[7] Recall that our definition of the Cantor string in Chapter 1 was slightly different from the present one in that the first length was equal to 1/3 instead of 1.

Remark 6.13. It is instructive to see to what extent formula (6.41) could have been derived by a direct computation (even formally). Indeed, writing

$$\frac{1}{1-2\cdot 3^m} = -\sum_{l=1}^{\infty} 2^{-l}3^{-ml},$$

we find that the sum over m in (6.41) is equal to $-\sum_{n=-\infty}^{-1} 2^n e^{-3^n t}$. Hence, it remains to establish the formula

$$\sum_{n=-\infty}^{-\infty} 2^n e^{-3^n t} = t^{-D} G_{\mathrm{CS}}\left(\log_3 t^{-1}\right).$$

Now, since the left-hand side changes by a factor of 2 when t is divided by 3, it is immediate that G_{CS} is periodic with period 1. But to compute the Fourier series of G_{CS}, as is done in (6.41) and (6.42), is not an easy task.

For the Fibonacci string, introduced in Section 2.3.2, there are two lines of complex dimensions, see Figure 2.5. Also, we have $\mathbf{p} = 2\pi/\log 2$. For the sum corresponding to the line of complex dimensions above $D = \log_2\phi$, we have $b = \phi$, and for the other line, above $-D + \frac{1}{2}i\mathbf{p}$, we have $b = -\phi^{-1}$ in formula (1.13). Thus we find that

$$\begin{aligned} N_{\mathrm{Fib}}(x) &= \frac{3+4\phi}{5}\phi^{-\{\log_2 x\}}x^D - 1 + \frac{7-4\phi}{5}(-\phi)^{\{\log_2 x\}}x^{-D+\frac{1}{2}i\mathbf{p}} \\ &= \frac{3+4\phi}{5}\phi^{-\{\log_2 x\}}x^D - 1 + \frac{7-4\phi}{5}\phi^{\{\log_2 x\}}x^{-D}(-1)^{[\log_2 x]}. \end{aligned} \tag{6.43}$$

In view of formula (6.27b) and the comment following it, the geometric partition function of the Fibonacci string,

$$\theta_{\mathrm{Fib}}(t) = \sum_{n=0}^{\infty} F_{n+1}e^{-2^n t}$$

(where the Fibonacci numbers F_{n+1} are defined by the recursive equation (2.19)), has an asymptotic expansion given by

$$\begin{aligned} \theta_{\mathrm{Fib}}(t) = \frac{\phi+2}{5\log 2}\sum_{n=-\infty}^{\infty}\Gamma(D+in\mathbf{p})t^{-D-in\mathbf{p}} + \sum_{m=0}^{\infty}\frac{(-1)^m}{m!}t^m\frac{1}{1-2^m-4^m} \\ + \frac{3-\phi}{5\log 2}\sum_{n=-\infty}^{\infty}\Gamma(-D+in\mathbf{p})t^{D-in\mathbf{p}}. \end{aligned} \tag{6.44}$$

If the complex dimensions of the lattice string $\mathcal{L}$ on a line are not simple, then formula (6.37) has to be changed. Moreover, the resulting function is no longer multiplicatively periodic in that case. For example, if the complex dimensions on the (discrete) line $\omega + in\mathbf{p}$ are double poles of $\zeta_{\mathcal{L}}$, and the Laurent series of $\zeta_{\mathcal{L}}$ around the pole at ω is given by

$$\zeta_{\mathcal{L}}(s) = \alpha_{-2}(s-\omega)^{-2} + \alpha_{-1}(s-\omega)^{-1} + \dots,$$

then the sum over these complex dimensions in the explicit formula for $N_{\mathcal{L}}(x)$ is the function

$$\alpha_{-2}\frac{2\pi}{\mathbf{p}}\frac{b}{b-1}b^{-\{u\}}x^{\omega}\log x + \alpha_{-1}\frac{2\pi}{\mathbf{p}}\frac{b}{b-1}b^{-\{u\}}x^{\omega}, \tag{6.45}$$

where, as above, we write $u = (\mathbf{p}/2\pi)\log x$ and $b = \exp(2\pi\omega/\mathbf{p})$.

In particular, for the lattice string $\mathcal{L}$ with multiple poles introduced in Section 2.3.4, we find

$$N_{\mathcal{L}}(x) = \frac{4}{9}2^{1-\{u\}}x^{D} + (-1)^{[\log_3 x]}\frac{3\log_3 x + 5}{18} - \frac{1}{4}. \tag{6.46}$$

The asymptotic expansion of the geometric partition function of that string is given by

$$\begin{aligned}\theta_{\mathcal{L}}(t) = {} & \frac{4}{9\log 3}\sum_{n=-\infty}^{\infty}\Gamma(D+in\mathbf{p})\,t^{-D-in\mathbf{p}} \qquad (6.47)\\ & + \sum_{m=0}^{\infty}\frac{(-1)^m}{m!}t^m\frac{1}{1-3\cdot 9^m - 2\cdot 27^m} + \frac{1}{9\log 3}\sum_{n=-\infty}^{\infty}t^{-\frac{1}{2}i\mathbf{p}-in\mathbf{p}}\\ & \cdot\Big((5-3\log_3 t)\,\Gamma\left(\tfrac{1}{2}i\mathbf{p}+in\mathbf{p}\right) + \frac{3}{\log 3}\Gamma'\left(\tfrac{1}{2}i\mathbf{p}+in\mathbf{p}\right)\Big),\end{aligned}$$

where $\Gamma'(s)$ denotes the derivative of the gamma function.

6.4.2 *Nonlattice Strings*

As was recalled in Remark 6.11, according to the nonlattice case of Theorem 2.17, there is only one complex dimension with real part $\geq D$, namely D itself. Further, D is always simple. Consequently, by estimate (5.64) of Theorem 5.30, if there exists a screen passing between $\operatorname{Re} s = D$ and the complex dimensions to the left of this line (for the general case, see Remark 6.15 below), we have for $k \geq 1$,

$$N_{\mathcal{L}}^{[k]}(x) = \operatorname{res}\left(\zeta_{\mathcal{L}}(s); D\right)\frac{x^{D+k-1}}{(D)_k} + o\big(x^{D+k-1}\big), \quad \text{as } x \to \infty. \tag{6.48}$$

In particular, for $k=1$, we have, again provided a suitable screen exists,

$$N_{\mathcal{L}}(x) = \operatorname{res}\left(\zeta_{\mathcal{L}}(s); D\right)\frac{x^{D}}{D} + o\big(x^{D}\big), \quad \text{as } x \to \infty. \tag{6.49}$$

With the same restriction, we find for the partition function that

$$\theta_{\mathcal{L}}(t) = \operatorname{res}\left(\zeta_{\mathcal{L}}(s); D\right) \Gamma\left(D\right) t^{-D} + o\big(t^{-D}\big), \quad \text{as } t \to 0^+. \tag{6.50}$$

Similarly, nonlattice strings are Minkowski measurable, since (as will follow from Theorem 8.36 below),

$$V(\varepsilon) = \operatorname{res}\left(\zeta_{\mathcal{L}}(s); D\right) \frac{(2\varepsilon)^{1-D}}{D(1-D)} + o\big(\varepsilon^{1-D}\big), \quad \text{as } \varepsilon \to 0^+. \tag{6.51}$$

Remark 6.14. We point out that the exponents of ε in the estimates in (6.48)–(6.51) are the best possible, since by the nonlattice case of Theorem 2.17, there always exist complex dimensions of $\mathcal{L}$ arbitrarily close to, but strictly to the left of, the vertical line $\operatorname{Re} s = D$. However, the results of Section 3.6 allow us to improve these estimates with a factor of lower order, depending on the size of the dimension-free region for $\mathcal{L}$.

Remark 6.15. Recall that we cannot always choose a screen passing between the line $\operatorname{Re} s = D$ and all complex dimensions strictly to the left of this line; see Example 5.32. Hence the above analysis is valid only for nonlattice strings that allow the choice of such a screen. However, if $\mathcal{L}$ does not allow such a screen, we apply Theorem 3.25 to write, for all small positive numbers δ,

$$N_{\mathcal{L}}^{[k]}(x) = \operatorname{res}\left(\zeta_{\mathcal{L}}(s); D\right) \frac{x^{D+k-1}}{(D)_k} + \sum_{D-\delta/2<\operatorname{Re}\omega<D} \operatorname{res}\left(\zeta_{\mathcal{L}}(s); \omega\right) \frac{x^{\omega+k-1}}{(\omega)_k} + o\big(x^{D-\delta/4+k-1}\big), \quad \text{as } x \to \infty, \tag{6.52}$$

and

$$V(\varepsilon) = \operatorname{res}\left(\zeta_{\mathcal{L}}(s); D\right) \frac{(2\varepsilon)^{1-D}}{D(1-D)} + \sum_{D-\delta/2<\operatorname{Re}\omega<D} \operatorname{res}\left(\zeta_{\mathcal{L}}(s); \omega\right) \frac{(2\varepsilon)^{1-\omega}}{\omega(1-\omega)} + o\big(\varepsilon^{1-D+\delta/4}\big), \quad \text{as } \varepsilon \to 0^+. \tag{6.53}$$

The sum in (6.53) converges absolutely, since the residues are bounded, by Theorem 3.25. For the same reason, the sum in (6.52) converges absolutely provided that $k \geq 2$. Then we apply the estimate of Theorem 5.17 to deduce (6.48) and (6.51) above. In Section 8.4.4, we give more details of this analysis. The same argument also applies to the geometric partition function.

Remark 6.16. The estimate for the volume of the tubular neighborhood $V(\varepsilon)$ will be shown in Section 8.4.4 to hold pointwise. Further, the estimates for $N_{\mathcal{L}}^{[k]}(x)$ can be understood pointwise for $k \geq 1$, except when we are in the situation of the previous remark, in which case one needs $k \geq 2$. In the latter situation, the estimate for $N_{\mathcal{L}}^{[1]}(x)$ can still be interpreted distributionally.

6.4.3 The Spectrum of a Self-Similar String

We now study the spectral asymptotics of lattice and nonlattice self-similar strings.

Lattice Case

Let $\mathcal{L}$ be a lattice self-similar string with oscillatory period $\mathbf{p}$ and multiplicative generator r. Suppose that ω is a simple complex dimension and that the gaps are integral powers of r, as in Equation (2.40).[8] In the explicit formula for the spectral counting function, the sum over the complex dimensions on the line $\omega + in\mathbf{p}$ gives the periodic distribution

$$\operatorname{res}\left(\zeta_{\mathcal{L}}(s);\omega\right) \sum_{n=-\infty}^{\infty} \frac{x^{\omega+in\mathbf{p}}}{\omega+in\mathbf{p}} \zeta(\omega+in\mathbf{p}). \tag{6.54}$$

In contrast to the geometric formula (6.37), there is no nice closed formula for this expression, because $0 < \operatorname{Re}\omega < 1$ and hence $\sum_{j=1}^{\infty} j^{-\omega}$ does not converge.

Again, in view of Remark 6.11 above, we can apply (6.54) to $\omega = D$ itself. We find that the spectrum of a lattice self-similar string is, as a distribution, given by its counting function,

$$N_\nu(x) = \operatorname{vol}_1(\mathcal{L}) \cdot x + \operatorname{res}\left(\zeta_{\mathcal{L}}(s);D\right) \sum_{n=-\infty}^{\infty} \zeta(D+in\mathbf{p}) \frac{x^{D+in\mathbf{p}}}{D+in\mathbf{p}} + O\left(x^{\Theta}(\log x)^{m-1}\right) + O(1), \tag{6.55}$$

as $x \to \infty$, where $\operatorname{Re} s = \Theta$ is the first line of complex dimensions to the left of D, and m is the maximal multiplicity of the complex dimensions of $\mathcal{L}$ on this line.[9] Note that $\operatorname{vol}_1(\mathcal{L}) = \zeta_{\mathcal{L}}(1)$ is the one-dimensional total length of $\mathcal{L}$, and $\operatorname{res}\left(\zeta_{\mathcal{L}}(s);D\right)$ is related to the D-dimensional volume of the boundary of the string; see Chapter 8, Theorem 8.15. Further note that the error term $O(1)$ in (6.55) (or in (6.56) below) is needed if $\Theta < 0$, as is the case, for example, for the Fibonacci string.

Similarly, the spectral partition function has an asymptotic expansion

$$\theta_\nu(t) = \operatorname{vol}_1(\mathcal{L}) \cdot t^{-1} + \operatorname{res}\left(\zeta_{\mathcal{L}}(s);D\right) \sum_{n=-\infty}^{\infty} \Gamma\left(D+in\mathbf{p}\right) \zeta(D+in\mathbf{p}) t^{-D-in\mathbf{p}} + O\left(t^{-\Theta}\left(\log t^{-1}\right)^{m-1}\right) + O(1), \tag{6.56}$$

as $t \to 0^+$, where Θ and m are as above.

---

[8] As in Section 6.4.1, we leave the general case to the reader; see Remark 2.19.

[9] Observe that in the notation of Theorem 2.17, different ω_u's can have the same real part.

Remark 6.17. We note that $\zeta(D + in\mathbf{p})$ does not vanish for infinitely many values of n, by Theorem 11.1 of Chapter 11. Hence the sum over the line of complex dimensions $D + in\mathbf{p}$ contains infinitely many oscillatory terms. In other words, the spectral counting function and the spectral partition function of every lattice string have oscillations of order D. This additional information cannot be obtained by means of the Renewal Theorem or by means of a Tauberian-type argument.

For the Cantor string of Section 2.3.1 we obtain

$$N_\nu(x) = 3x + \frac{1}{\log 3} \sum_{n=-\infty}^{\infty} \zeta(D + in\mathbf{p}) \frac{x^{D+in\mathbf{p}}}{D + in\mathbf{p}} + O(1) \tag{6.57}$$

as $x \to \infty$, where $D = \log_3 2$ and $\mathbf{p} = 2\pi/\log 3$, and

$$\theta_\nu(t) = 3t^{-1} + \frac{1}{\log 3} \sum_{n=-\infty}^{\infty} \Gamma(D + in\mathbf{p})\zeta(D + in\mathbf{p})t^{-D-in\mathbf{p}} + O(1), \tag{6.58}$$

as $t \to 0^+$. In light of Remark 6.17, the (multiplicatively) periodic functions represented by the series in (6.57) and (6.58) are both nonconstant. The same comment would hold for arbitrary lattice strings not necessarily satisfying Equation (2.40).

A detailed analysis of the spectral counting function $N_\nu(x)$ of the Cantor string (and of other integral Cantor strings) is provided in Section 10.2.1 below. (Set $a = 3$ and $b = 2$ in Theorem 10.7 and Corollary 10.8; see also [LapPo2, Theorem 4.6, p. 65] for an earlier, although less precise, study of this particular case.)

As was pointed out above, each line of complex dimensions of a lattice string yields a corresponding sum in our explicit formulas. For example, for the Fibonacci string of Section 2.3.2, we obtain the following expression (with $D = \log_2 \phi$ and $\mathbf{p} = 2\pi/\log 2$, where $\phi = (1 + \sqrt{5})/2$ is the golden ratio):

$$N_\nu(x) = 4x + \frac{\phi + 2}{5 \log 2} \sum_{n=-\infty}^{\infty} \zeta(D + in\mathbf{p}) \frac{x^{D+in\mathbf{p}}}{D + in\mathbf{p}} + \frac{1}{2}$$
$$+ \frac{3 - \phi}{5 \log 2} \sum_{n=-\infty}^{\infty} \zeta(-D + i(n + 1/2)\mathbf{p}) \frac{x^{-D+i(n+1/2)\mathbf{p}}}{-D + i(n + 1/2)\mathbf{p}} + O(x^\rho) \tag{6.59}$$

as $x \to \infty$, and

$$\theta_\nu(t) = 4t^{-1} + \frac{\phi+2}{5\log 2}\sum_{n=-\infty}^{\infty} \Gamma\left(D + in\mathbf{p}\right)\zeta(D+in\mathbf{p})t^{-D-in\mathbf{p}} + \frac{1}{2}$$
$$+\frac{3-\phi}{5\log 2}\sum_{n=-\infty}^{\infty} \Gamma\left(-D + i(n+1/2)\mathbf{p}\right)\zeta(-D+i(n+1/2)\mathbf{p})t^{D-i(n+1/2)\mathbf{p}}$$
$$+ O\left(t^{-\rho}\right), \quad (6.60)$$

as $t \to 0^+$, for every $\rho < -D$.

Remark 6.18. We leave it as an exercise for the interested reader to write down, possibly with the help of a symbolic computation package, the explicit formulas for the spectral counting and partition functions of the lattice string with multiple poles introduced in Section 2.3.4.

Nonlattice Case

Next, let $\mathcal{L}$ be a nonlattice self-similar string. After separating the term corresponding to the dimension of the string, we find

$$N_\nu(x) = \mathrm{vol}_1(\mathcal{L}) \cdot x - \mathrm{res}\left(\zeta_{\mathcal{L}}(s); D\right)\left(-\zeta(D)\right)\frac{x^D}{D} + o\left(x^D\right), \quad \text{as } x \to \infty, \tag{6.61}$$

interpreted distributionally. Note that by formula (8.25) and since $\zeta(D) < 0$, the coefficient of x^D is negative. Again, when there is no screen passing between $\mathrm{Re}\, s = D$ and the complex dimensions to the left of this line, one has to apply the technique of Remark 6.15 to derive this formula, still interpreted distributionally.

For the spectral partition function, one obtains, as $t \to 0^+$,

$$\theta_\nu(t) = \mathrm{vol}_1(\mathcal{L}) \cdot t^{-1} - \mathrm{res}\left(\zeta_{\mathcal{L}}(s); D\right)\left(-\zeta(D)\right)\Gamma(D)t^{-D} + o\left(t^{-D}\right). \tag{6.62}$$

We leave it as an exercise for the interested reader to apply the formulas obtained for the geometric and spectral density of states in Section 6.3.1.

Remark 6.19. As in Remark 6.14, we point out that in (6.61) and (6.62), the exponent in the error term is the best possible. Indeed, by [Ti, Theorem 9.19(C), p. 204], the density of the zeros off the critical line $\mathrm{Re}\, s = 1/2$ of the Riemann zeta function is less than linear. On the other hand, by Theorem 2.17, every nonlattice string has complex dimensions with real part arbitrary close to D (from the left) with linear density. Hence these complex dimensions cannot all be canceled by zeros of $\zeta(s)$. A moment's reflection shows that this argument is valid for any D in $(0, 1)$, including $D = 1/2$. However, using the dimension-free regions obtained in Section 3.6, we can improve the error term by a factor of lower order.

6.5 Examples of Non-Self-Similar Strings

Let η be a generalized fractal string with geometric zeta function ζ_η. In general, $\zeta_\eta(s)$ does not need to have a continuation as a meromorphic function beyond the line $\operatorname{Re} s = D$ (see Example 1.17), or the meromorphic continuation does not need to satisfy **L1** and **L2** beyond a certain screen. Thus, the analysis of ζ_η and the choice of a good screen become the key to understanding the geometry and the spectrum of a fractal string. We give here an example of such an analysis.

6.5.1 The a-String

Given $a > 0$, an arbitrary positive real number, we consider the ordinary fractal string $\mathcal{L}$ with lengths

$$l_j = j^{-a} - (j+1)^{-a}, \quad j = 1, 2, \dots. \tag{6.63}$$

This string (which we call the a-string) has already been discussed in a related context in [Lap1, Example 5.1, pp. 512–513] and was later on revisited in [LapPo2, pp. 64–65]. However, thanks to Theorem 6.20 below and our explicit formulas, we will be able to obtain much more precise results than in these earlier papers.

Recall from [Lap1, Example 5.1] that $\mathcal{L}$ can be realized as the open set $\Omega \subset \mathbb{R}$ obtained by removing the points j^{-a} ($j = 1, 2, \dots$) from the interval $(0, 1)$; namely,

$$\Omega = \bigcup_{j=1}^{\infty} \left((j+1)^{-a}, j^{-a}\right). \tag{6.64}$$

Hence, its boundary is the countable compact subset of $\mathbb{R}$ given by

$$\partial\Omega = \left\{j^{-a} : j = 1, 2, \dots\right\} \cup \{0\}. \tag{6.65}$$

We begin by determining the possible complex dimensions of $\mathcal{L}$, that is, the poles of $\zeta_{\mathcal{L}}$.[10]

Theorem 6.20. *Let $a > 0$ and let $\mathcal{L}$ be the ordinary fractal string with lengths l_j given by (6.63). Then $\zeta_{\mathcal{L}}(s) = \sum_{j=1}^{\infty} l_j^s$ has a meromorphic continuation to all of $\mathbb{C}$. The poles of $\zeta_{\mathcal{L}}$ are located at $\frac{1}{a+1}$ and at (a subset of) the points $-\frac{1}{a+1}, -\frac{2}{a+1}, -\frac{3}{a+1}, \dots$, and they are all simple. In particular, the dimension of $\mathcal{L}$ is $D = \frac{1}{a+1}$, and this is the only pole of $\zeta_{\mathcal{L}}$ with positive real part. The residue of $\zeta_{\mathcal{L}}$ at this pole is equal to a^D.*

[10] We are grateful to Driss Essouabri [Es1–2] for providing us with the main idea of the proof of Theorem 6.20.

Further, for any screen S not passing through a pole, $\zeta_{\mathcal{L}}$ satisfies **L1** *and* **L2** *(i.e., $\mathcal{L}$ is languid) with $\kappa = \frac{1}{2} - (a+1)\inf S$ if $\inf S \leq 0$ and $\kappa = \frac{1}{2}$ if $\inf S \geq 0$. (Here, $\inf S$ is defined as in* (5.17a).)

Proof. We compute the first term of an asymptotic expansion of l_j:

$$l_j = j^{-a} - (j+1)^{-a} = a\int_j^{j+1} x^{-a-1}\,dx = aj^{-a-1} + H(j),$$

where $H(j) = a\int_j^{j+1}\left(x^{-a-1} - j^{-a-1}\right)\,dx$. It follows that

$$h_j := a^{-1}j^{a+1}H(j) = j\int_0^{1/j}\left((1+t)^{-a-1} - 1\right)\,dt. \tag{6.66}$$

Note that $h_j = O(1/j)$, as $j \to \infty$. Choose an integer $M \geq 0$. Then

$$\begin{aligned} l_j^s &= \left(aj^{-a-1}(1+h_j)\right)^s \\ &= a^s j^{-s(a+1)}\left(\sum_{n=0}^{M}\binom{s}{n}h_j^n + O\left(\frac{(|s|+1)^{M+1}}{j^{M+1}}\right)\right), \end{aligned}$$

where we have set

$$\binom{s}{n} = \frac{s(s-1)\dots(s-n+1)}{n!}, \qquad \text{for } s \in \mathbb{C}. \tag{6.67}$$

We thus obtain

$$\zeta_{\mathcal{L}}(s) = \sum_{n=0}^{M} a^s\binom{s}{n}\sum_{j=1}^{\infty} h_j^n j^{-s(a+1)} + f(s), \tag{6.68}$$

where $f(s)$ is defined and holomorphic for $\operatorname{Re} s > -\frac{M}{a+1}$. The first term of this sum, for $n = 0$, is $a^s\zeta((a+1)s)$. Thus we find the first pole at $s = \frac{1}{a+1}$. Note that the first term grows as $(|t|+1)^{\frac{1}{2}-\sigma(a+1)}$ on vertical lines $\operatorname{Re} s = \sigma$ with $\sigma < 0$.

It remains to analyze the functions

$$\sum_{j=1}^{\infty} h_j^n j^{-s(a+1)}, \tag{6.69}$$

for $n \geq 1$. We will show that these functions are meromorphic with simple poles at the points $0, -\frac{1}{a+1}, -\frac{2}{a+1}, \dots$.

Using the asymptotic expansion $(1+t)^{-a-1} = \sum_{m=0}^{M}\binom{-a-1}{m}t^m + O\left(t^{M+1}\right)$ as $t \to 0$, we obtain in view of (6.66) that

$$\begin{aligned} h_j &= j\int_0^{1/j}\sum_{m=1}^{M}\binom{-a-1}{m}t^m\,dt + O\left(j^{-M-1}\right) \\ &= -\frac{1}{a}\sum_{m=1}^{M}\binom{-a}{m+1}j^{-m} + O\left(j^{-M-1}\right), \qquad \text{as } j \to \infty. \end{aligned}$$

By taking the n-th power of this expansion, we find an asymptotic expansion for h_j^n. Substituting this expansion, we write each of the functions in (6.69) as a sum of constant multiples of $\zeta(m+(a+1)s)$ (which has a pole at $s=\frac{1-m}{a+1}$), for $n \leq m \leq M$. In view of Equation (6.68), we thus deduce that $\zeta_{\mathcal{L}}$ has a meromorphic continuation to $\operatorname{Re} s > -\frac{M}{a+1}$, with simple poles at $s=\frac{1-m}{a+1}$, $m=0, 2, 3, 4, \ldots, M$ (note that 0 is not a pole of $\zeta_{\mathcal{L}}$, due to the factor $\binom{s}{1}=s$ on the right-hand side of (6.68)). Since M is arbitrary, it follows that $\zeta_{\mathcal{L}}$ has a meromorphic continuation to all of $\mathbb{C}$. A direct computation shows that the residue of $\zeta_{\mathcal{L}}$ at $D=1/(a+1)$ is equal to a^D.

Finally, for $m \geq 1$, the growth of $\zeta(m+(a+1)s)$ is superseded by the growth of the first term $a^s\zeta((a+1)s)$. Thus for any screen $\operatorname{Re} s = \inf S$ with $\inf S < 0$, we can choose $\kappa = \frac{1}{2} - (a+1)\inf S$. □

It follows from Theorem 6.20 (as well as from Theorems 5.18 and 5.30) that the geometric counting function of $\mathcal{L}$ satisfies

$$N_{\mathcal{L}}(x) = a^D x^D + \zeta_{\mathcal{L}}(0) + O\left(x^{-D}\right), \qquad \text{as } x \to \infty, \tag{6.70}$$

while its spectral counting function satisfies

$$N_\nu(x) = x + \zeta(D)a^D x^D - \frac{1}{2}\zeta_{\mathcal{L}}(0) + O\left(x^{-D}\right), \qquad \text{as } x \to \infty. \tag{6.71}$$

Recall from [Ti] that $\zeta(D) < 0$ since $D=\frac{1}{a+1} \in (0,1)$. Also note that $\zeta_{\mathcal{L}}(0) = -1/2$ because the first term in (6.68) is $a^s\zeta((a+1)s)$, $\zeta(0)=-1/2$, and all other terms vanish at $s=0$ since $\binom{0}{n}=0$ for $n \geq 1$.

We note that in [LapPo2, Example 4.3], the error term is only $o(x^D)$ in the counterpart of (6.71) (or of (6.70)). See, in particular, [LapPo2, Equation (4.25), p. 65].

Actually, since the poles of $\zeta_{\mathcal{L}}$ are all simple, it follows from Theorem 6.20 and our explicit formulas that Equations (6.70) and (6.71) can be replaced respectively by the much more precise expressions

$$N_{\mathcal{L}}(x) = a^D x^D + \zeta_{\mathcal{L}}(0) + \sum_{m=1}^{M} \operatorname{res}\left(\zeta_{\mathcal{L}}(s); -mD\right) x^{-mD} + O\left(x^{-(M+1)D}\right), \qquad \text{as } x \to \infty, \tag{6.70$'$}$$

and

$$N_\nu(x) = x + \zeta(D)a^D x^D - \frac{1}{2}\zeta_{\mathcal{L}}(0) + \sum_{m=1}^{M} \operatorname{res}\left(\zeta_{\mathcal{L}}(s); -mD\right)\zeta(-mD) x^{-mD} + O\left(x^{-(M+1)D}\right), \qquad \text{as } x \to \infty, \tag{6.71$'$}$$

valid for every $M = 0, 1, 2, \ldots$.

Remark 6.21. Depending on the arithmetic properties of the parameter a, the residue of $\zeta_{\mathcal{L}}$ at $s = -mD$ may vanish for some values of $m \in \mathbb{N}^*$. This is why one cannot in general specify the exact set of complex dimensions of the a-string $\mathcal{L}$ in the statement of Theorem 6.20.

6.5.2 *The Spectrum of the Harmonic String*

An interesting example of a non-self-similar fractal string is provided by the harmonic string, introduced and studied in [Lap2, Example 5.4(ii), pp. 171–172] or [Lap3, Remark 2.5, pp. 144–145]. Recall from Equation (4.11) above that it is given by the measure $h = \sum_{j=1}^{\infty} \delta_{\{j\}}$ and hence has lengths $l_j = \frac{1}{j}$ for $j = 1, 2, \ldots$ (note that this string has infinite total length $\sum_{j=1}^{\infty} \frac{1}{j} = \infty$). Thus

$$\zeta_h(s) = \zeta(s) \tag{6.72}$$

and

$$\zeta_\nu(s) = \zeta_h(s) \cdot \zeta(s) = (\zeta(s))^2 . \tag{6.73}$$

It follows from (6.73) that at the geometric level, $s = 1$ is the only pole of $\zeta_h(s)$, and that it is simple, while at the spectral level, $s = 1$ is the only pole of $\zeta_\nu(s)$ and has multiplicity two.

Since we have, as was noted in [Lap2, p. 171] or [Lap3, p. 144],[11]

$$N_\nu(x) = \sum_{1 \le k \le x} \tau(k), \tag{6.74}$$

where $\tau(k)$ denotes the number of divisors of the integer k, it follows from Equation (6.73) above (and our error estimates) that

$$N_\nu(x) = \sum_{1 \le k \le x} \tau(k) = x \log x + (2\gamma - 1)x + o(x), \tag{6.75}$$

as $x \to \infty$, where γ denotes Euler's constant. This well-known formula was recovered in a similar manner in [Lap2–3] and is often referred to as the Dirichlet divisor formula. A long standing open problem—called the Dirichlet divisor problem in the literature [Ti, §12.1, pp. 312–314]—consists of obtaining a much sharper form for the error term on the right-hand side of (6.75).

---

[11] Indeed, as is observed in [Lap2–3], the frequencies of the harmonic string consist of the sequence of positive integers $1, 2, 3, \ldots$, with multiplicity respectively equal to $\tau(1), \tau(2), \tau(3), \ldots$. With our notation from Section 4.1 and in view of formulas (4.11) and (4.20), this can be seen as follows: $\nu = h * h = \sum_{m,n=1}^{\infty} \delta_{\{mn\}} = \sum_{k=1}^{\infty} \tau(k) \delta_{\{k\}}$.

6.6 Fractal Sprays

We consider fractal sprays, defined in Section 1.4 (following [LapPo3]) and, more generally, in Section 4.3. For simplicity, we assume that

$$\zeta_B(s) = \sum_f w_f f^{-s},$$

the spectral zeta function of the basic shape B, has at most finitely many (visible) poles, called the *spectral complex dimensions of* B.[12] Moreover, we restrict our attention to those fractal sprays of a (possibly generalized) fractal string η on B for which no complex dimension of η coincides with a spectral complex dimension of B.

The *Weyl term* at level k associated with the fractal spray of η on B is

$$W_{B,\eta}^{[k]}(x) = \sum_{u:\,\text{pole of }\zeta_B} \operatorname{res}\left(\frac{\zeta_B(s)\zeta_\eta(s)}{(s)_k} x^{s+k-1}; u\right). \tag{6.76}$$

The terms in this sum can be analyzed in the same way as in Section 6.1.1. The result depends on the multiplicity of the spectral complex dimensions. In particular, if these complex dimensions are simple and avoid the zeros $0, -1, -2, \dots, 1-k$ of $(u)_k$, we obtain the formula

$$W_{B,\eta}^{[k]}(x) = \sum_{u:\,\text{pole of }\zeta_B} \operatorname{res}\left(\zeta_B(s); u\right) \zeta_\eta(u) \frac{x^{u+k-1}}{(u)_k}. \tag{6.77}$$

Recall from Equation (4.35) that the spectral zeta function of this fractal spray is given by

$$\zeta_\nu(s) = \zeta_\eta(s) \cdot \zeta_B(s). \tag{6.78}$$

Therefore, much as was done for fractal strings in Section 6.3.2, we state the following definition:

Definition 6.22. The *spectral operator* (for fractal sprays) is the operator that adds the Weyl term and multiplies each term corresponding to a complex dimension of η (assumed to be simple) by $\zeta_B(\omega)$, and also multiplies the integrand in the error term by $\zeta_B(s)$.

We will focus here on examples of self-similar sprays. However, it should be clear to the reader that non-self-similar fractal sprays—such as those studied, for example, in [LapPo3]—can be treated as well, once their complex dimensions have been analyzed.

---

[12]For example, 1 is the only spectral complex dimension of the Bernoulli string, as discussed in Section 6.1.3.

Before considering general self-similar sprays in Section 6.6.2, we illustrate by the following example our methods applied to fractal sprays. See also Section 11.5, formulas (11.57), (11.58) and (11.60) for a further example of a fractal spray and of the corresponding Weyl term (in a case where $\zeta_B(s)$ has infinitely many poles). Further, see formulas (11.12) and (11.13) for the simpler Cantor sprays considered in Section 11.2.

6.6.1 *The Sierpinski Drum*

The Dirichlet Laplacian on the equilateral triangle $\mathcal{T}$ with sides 1 has for eigenvalue spectrum (see [Pin, Note on p. 820 and footnote 1] or [Bér]),

$$\lambda_{m,n} = \frac{16\pi^2}{9}\left(m^2+mn+n^2\right), \quad m,n=1,2,3,\ldots,$$

and (in view of our convention defining the frequencies as $\sqrt{\lambda_{m,n}/\pi}$) the corresponding spectral zeta function is equal to[13]

$$\zeta_{\mathcal{T}}(s) = \left(\frac{3}{4}\right)^s \sum_{m,n=1}^{\infty} \left(m^2+mn+n^2\right)^{-s/2}. \tag{6.79}$$

We find the poles and the corresponding residues of this function as follows. The zeta function of the cyclotomic field $\mathbb{Q}[\rho]$, obtained by adjoining a cubic root of unity to the rationals, is given by

$$\zeta_{\mathbb{Q}[\rho]}(s) = \sum_{m=0}^{\infty}\sum_{n=1}^{\infty} \left(m^2+mn+n^2\right)^{-s}. \tag{6.80}$$

It has a simple pole at $s=1$ with residue $\pi/(3\sqrt{3})$; see Appendix A, Equation (A.4). Further, it is related to $\zeta_{\mathcal{T}}(s)$ by the equation

$$\zeta_{\mathcal{T}}(s) = \left(\frac{3}{4}\right)^s \zeta_{\mathbb{Q}[\rho]}(s/2) - \left(\frac{3}{4}\right)^s \zeta(s). \tag{6.81}$$

Thus $\zeta_{\mathcal{T}}(s)$ has a simple pole at $s=2$, with residue $\pi\sqrt{3}/8$, and one at $s=1$, with residue $-3/4$.

Consider the spray of $\mathcal{L}$ on $\mathcal{T}$ obtained by scaling the middle triangle in Figure 6.1 by the normalized[14] lattice string $\mathcal{L}$ with one gap and scaling ratios $r_1=r_2=r_3=\frac{1}{2}$. The boundary of this spray is the classical Sierpinski gasket; see Figure 6.1. We will call this spray the *Sierpinski drum.* Hence, it corresponds to the Dirichlet Laplacian on the infinitely connected bounded open subset of $\mathbb{R}^2$ with boundary the Sierpinski gasket.

In light of Equation (6.78), the spectral zeta function of the Sierpinski drum is given by

$$\zeta_\nu(s) = \zeta_{\mathcal{L}}(s)\zeta_{\mathcal{T}}(s) = \sum_{n=0}^{\infty} 3^n 2^{-ns}\zeta_{\mathcal{T}}(s) = \frac{1}{1-3\cdot 2^{-s}}\zeta_{\mathcal{T}}(s), \tag{6.82}$$

[13]Note that the quadratic form $m^2+mn+n^2=(m+n/2)^2+3n^2/4$ is positive definite.

[14]Meaning that $\zeta_{\mathcal{L}}(s)=\frac{1}{1-3\cdot 2^{-s}}$.

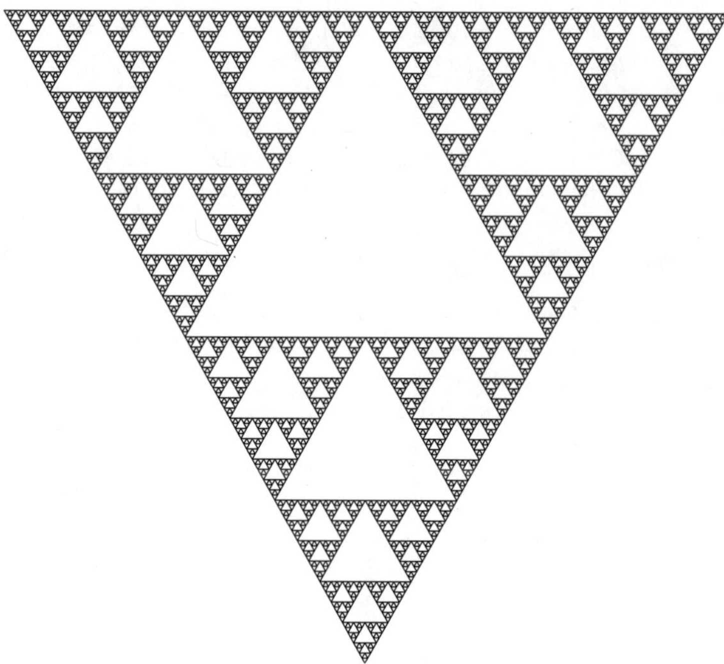

Figure 6.1: The Sierpinski gasket, the boundary of the Sierpinski drum.

and its Weyl term is, by (6.77) along with the comment following (6.81),

$$W_{\mathcal{T},\mathcal{L}}^{[1]}(x) = \frac{\pi\sqrt{3}}{4}x^2 + \frac{3}{2}x. \tag{6.83}$$

From the explicit formula for the geometric counting function of $\mathcal{L}$ (see Section 6.4.1),

$$N_{\mathcal{L}}(x) = \frac{1}{\log 2}\sum_{n=-\infty}^{\infty} \frac{x^{D+in\mathbf{p}}}{D+in\mathbf{p}} - \frac{1}{2}, \tag{6.84}$$

where $D = \log_2 3$ and $\mathbf{p} = 2\pi/\log 2$, we deduce the explicit formula for the counting function of the frequencies of the Sierpinski drum,

$$N_{\mathcal{T},\mathcal{L}}(x) = \frac{\pi\sqrt{3}}{4}x^2 + x^D G(\log x) + \frac{3}{2}x - \frac{1}{2}\zeta_{\mathcal{T}}(0) + o(1), \tag{6.85}$$

as $x \to \infty$. Here, G is the periodic function, of period $2\pi/\mathbf{p} = \log 2$, given by the Fourier series

$$G(u) = \frac{1}{\log 2}\sum_{n=-\infty}^{\infty} \frac{e^{in\mathbf{p}u}}{D+in\mathbf{p}}\zeta_{\mathcal{T}}(D+in\mathbf{p}), \tag{6.86}$$

where $\zeta_{\mathcal{T}}(s)$ is given by (6.79) or (6.81).

Remark 6.23. Note that $\mathcal{L}$ has dimension $\log_2 3$, which lies between 1 and 2. Hence, strictly speaking, it is not a string in the sense of Section 1.1. Thus, the triangles constituting the Sierpinski gasket would not fit on a line of finite length if they were all aligned, but they do fit inside a bounded region in the plane. Note that in Chapter 3, we did not make any restrictive assumptions on the dimension of the generalized self-similar strings involved, so that our theory can be applied without any problem.

Remark 6.24. It is noteworthy that the spectral zeta function $\zeta_\nu(s)$ of the Sierpinski drum has poles at nonreal values of s. Specifically, according to (6.82) and the comment following (6.81), the spectral complex dimensions of this fractal spray are located at $s = 2$ and at $s = D + in\mathbf{p}$, $n \in \mathbb{Z}$, with D and $\mathbf{p}$ as above. This is in contrast to the spectral zeta function of a smooth manifold, which has only real poles; see Theorem B.4 and Remark B.5 of Appendix B.

Theorem 6.25. *The Sierpinski drum has oscillations of order D in its spectrum, where $D = \log_2 3$ is the Minkowski dimension of the Sierpinski gasket.*

More precisely, the asymptotic behavior as $x \to \infty$ of the spectral counting function of the Sierpinski drum is given by the formula (6.85), *with G given by* (6.86). *Further, G is a* nonconstant *periodic function, with period* $\log 2$.

Proof. The fact that G is nonconstant follows from Theorem 11.12 below, according to which the Dirichlet series $\zeta_{\mathcal{T}}(s)$ does not have an infinite vertical sequence of zeros in arithmetic progression. It follows that the expansion (6.86), interpreted distributionally (as, for example, in [Schw1, Section VII.1]), has infinitely many nonzero terms. □

Remark 6.26. In view of [Pin], an entirely analogous result can be obtained for Neumann rather than for Dirichlet boundary conditions. Moreover, in light of [Bér], fractal sprays with more general basic shapes associated with crystallographic groups can be analyzed in the same manner. The necessary computations, however, will be significantly more complicated.

Remark 6.27. One can also obtain the counterpart of Theorem 6.25 for the drum with fractal boundary the Sierpinski carpet, or its higher-dimensional analogue, the Menger sponge. See, for instance, [Man1, Plate 145, pp. 144–145] for a picture of these classical fractals. In particular, the basic shape of this fractal spray is the unit square or the unit cube, respectively. It is also interesting to study similarly the main example of Brossard and Carmona [BroCa], revisited and extended by Fleckinger and Vassiliev in [FlVa]. Finally, as an exercise, the reader may wish to consider the Cantor spray introduced in Section 1.4; see Figure 1.7. We note that more general Cantor sprays will be studied in Section 11.2 below.

The Sierpinski drum, as well as the examples mentioned in Remark 6.27, is an example of a lattice self-similar spray, in the sense of Theorem 6.28 below (its oscillatory period is $\mathbf{p} = 2\pi/\log 2$). Such a self-similar drum has been studied, in particular, by the first author in [Lap2; Lap3, Section 4.4.1] and, in more detail, by Gerling [Ger] and Gerling and Schmidt [GerSc1–2]. We note that the fact that the periodic function G in (6.86) is nonconstant (and thus that the Sierpinski drum has oscillations of order D in its spectrum, as stated in Theorem 6.25 above) was not established in these references. Within our framework, it is a direct consequence of Theorem 11.12, which establishes the nonexistence of infinite arithmetic progressions of zeros for a suitable class of Dirichlet series (including the spectral zeta function $\zeta_{\mathcal{T}}$) and provides us with a very useful tool for proving such results. See Remark 6.30 below for further references and extensions, which can also be dealt with by our methods.

6.6.2 *The Spectrum of a Self-Similar Spray*

By entirely analogous methods—based on our explicit formulas and on Theorem 11.12 in Section 11.2 below—we can establish the analogue for self-similar sprays of the results obtained in Section 6.4 for self-similar strings.

Theorem 6.28. *The exact counterpart of the results of Section 6.4 holds for self-similar sprays (see Remark 6.29 below for more precision). In particular, a self-similar spray $\mathbb{S}$ with basic shape B, a connected manifold with piecewise smooth boundary (much as in Section 6.6.1), has oscillations of order D, the dimension of the self-similar string used to define the spray, if and only if $\mathbb{S}$ is a lattice spray. In that case, the oscillations are necessarily multiplicatively periodic.*

On the other hand, if B has a fractal boundary, then it can have oscillations of order D in its spectrum, and hence the corresponding spray may have such oscillations, even if $\mathbb{S}$ is a nonlattice spray.

This is in agreement with [Lap3, Conjecture 3, p. 163] stated for more general self-similar drums.

Remark 6.29. By definition, a *self-similar spray* is a fractal spray with basic shape B scaled by a self-similar string $\mathcal{L}$.[15] Moreover, in Theorem 6.28, we assume that $\zeta_B(s)$, the spectral zeta function of B, is languid (i.e., satisfies the growth conditions **L1** and **L2** along some screen). We further suppose that no complex dimension of $\mathcal{L}$ coincides with any of the poles of $\zeta_B(s)$. These growth conditions will be satisfied, for instance, when $\zeta_B(s)$ is the spectral zeta function of the Dirichlet (or Neumann) Laplacian on

[15] which is allowed to have dimension greater than 1, as was the case for the Sierpinski drum (see Remark 6.23).

a piecewise smooth bounded domain of $\mathbb{R}^d$ (and the window W is not all of $\mathbb{C}$), provided that the dimension of $\mathcal{L}$ is less than d (in that case, D is also the dimension of the boundary of the spray). In particular, it is so for all the examples mentioned in Section 6.6.1 and in the next remark.

Remark 6.30. Theorem 6.28 extends and specifies the corresponding results obtained earlier in [Ger, GerSc1–2, Lap3, Section 4.4.1b; LeVa, vB-Le]. In particular, the statement about the presence of oscillations of order D in the spectrum of lattice sprays seems to be new and cannot be established by means of the Renewal Theorem [Fel, Theorem 2, p. 39], which is used in [Lap3, LeVa]. As was noted earlier, to obtain this result, we make use of Theorem 11.12 about the nonexistence of infinite arithmetic progressions of zeros for certain zeta functions. See Remark 8.40 for a similar comment in a related situation. It has been verified, however, in special cases, such as for the main example of [FlVa], extended in [vB-Le].

7

Periodic Orbits of Self-Similar Flows

In this chapter, we apply our explicit formulas to obtain an asymptotic expansion for the prime orbit counting function of suspended flows. The resulting formula involves a sum of oscillatory terms associated with the dynamical complex dimensions of the flow. We then focus in Section 7.3 on the special case of self-similar flows and deduce from our explicit formulas a Prime Orbit Theorem with error term. For a self-similar flow, we define the lattice and the nonlattice case in Definition 7.27. In the lattice case, the counting function of the prime orbits has oscillatory leading asymptotics. The explicit formula for this counting function enables us to give a very precise expression for this function in terms of multiplicatively periodic functions. In the nonlattice case (which is the generic case), the leading term does not have oscillations, and we provide a detailed analysis of the error term. The precise order of the error term depends on the dimension-free region of the dynamical zeta function, as in the classical Prime Number Theorem. Applying the results of Chapter 3, we find that this region, and hence the error term, depends on properties of Diophantine approximation of the weights of the flow.

The dynamical complex dimensions of a suspended flow are defined as the poles of the logarithmic derivative of the dynamical zeta function. On the other hand, the geometric complex dimensions of a fractal string have been defined in Section 1.2.1 as the poles of the geometric zeta function itself, which coincides with the dynamical zeta function when the string and the flow are both self-similar (see Theorem 7.15). Thus the geometric complex dimensions of a self-similar flow only depend on the poles of the corresponding zeta function, and they are counted with a multiplicity, whereas the dynamical complex dimensions of a flow depend on the zeros and the poles of the dynamical zeta function, and they have no multiplicity. Due to the fact that the dynamical zeta function of a self-similar flow has no zeros, since a self-similar flow always corresponds to a self-similar set with a single gap, the two sets of complex dimensions coincide (as sets of points without multiplicity) in the case of self-similar flows.

7.1 Suspended Flows

Let $N \geq 0$ be an integer and let $\Sigma = \{0, \dots, N-1\}^{\mathbb{N}}$ be the space of infinite sequences over the alphabet $\{0, \dots, N-1\}$, called the space of infinite *words*. Let $\mathfrak{w} \colon \Sigma \to (0, \infty]$ be a function, called the *weight*. On Σ, we have the left shift σ, given on a word (a_n) by $(\sigma a)_n = a_{n+1}$. We define the *suspended flow* $\mathcal{F}_{\mathfrak{w}}$ on the space $[0, \infty) \times \Sigma$ as the following dynamical system (or time evolution, see [PaPol2, Chapter 6]):

$$\mathcal{F}_{\mathfrak{w}}(t, a) = \begin{cases} (t, a) & \text{if } 0 \leq t < \mathfrak{w}(a), \\ \mathcal{F}_{\mathfrak{w}}(t - \mathfrak{w}(a), \sigma a) & \text{if } t \geq \mathfrak{w}(a). \end{cases} \tag{7.1}$$

We think of it as the path of a particle, starting at the word $a = (a_n)$. On each word $b \in \Sigma$, an interval of length $\mathfrak{w}(b)$ is suspended, along which the particle moves up, with unit velocity. When it reaches the end of the interval suspended at a (at time $t = \mathfrak{w}(a)$), it jumps to σa. Thus the position at time t, starting at a, is the value $\mathcal{F}_{\mathfrak{w}}(t, a)$. If the series

$$\mathfrak{w}(a) + \mathfrak{w}(\sigma a) + \mathfrak{w}(\sigma^2 a) + \dots$$

converges, then this position is not defined for t larger than its sum. However, $\mathcal{F}_{\mathfrak{w}}(t, a)$ is always defined on periodic words.

Note that the number of words in Σ is uncountable if $N \geq 2$. See Figure 7.1 for an example where the length of the interval suspended on a word depends only on the first element of the word.

Remark 7.1. This formalism is seemingly less general than the one introduced in [PaPol2, Chapter 1]. However, defining $\mathfrak{w}(a) = \infty$ when the word a contains a prohibited word of length 2, and $e^{-s\infty} = 0$, allows us to deal with the general case.

Remark 7.2. In the literature on symbolic dynamics (see, e.g., [PaPol2] and the references therein), Σ is also referred to as the *code space*. See the notes to the present chapter (Section 7.6) for references on a variety of dynamical systems that can be viewed as (possibly more general) suspended flows. These include subshifts of finite type and interval maps.

7.1.1 The Zeta Function of a Dynamical System

Given a finite word $\mathfrak{x} = a_1, a_2, \ldots, a_l$ of length $l = l(\mathfrak{x})$, we let

$$a = a_1, a_2, \ldots, a_l, a_1, a_2, \ldots, a_l, \ldots$$

be the corresponding periodic word, and we define $\sigma\mathfrak{x} = a_2, \ldots, a_l, a_1$. The *total weight* of the orbit of σ on $\mathfrak{x}$ is

$$\mathfrak{w}_{\mathrm{tot}}(\mathfrak{x}) = \mathfrak{w}(a) + \mathfrak{w}(\sigma a) + \cdots + \mathfrak{w}(\sigma^{l-1} a). \tag{7.2}$$

Definition 7.3 ([Bow1–2, Rue2] and [PaPol2, Chapter 5]). The *dynamical zeta function* of $\mathcal{F}_{\mathfrak{w}}$ is defined by

$$\zeta_{\mathfrak{w}}(s) = \exp\left(\sum_{\mathfrak{x}} \frac{1}{l(\mathfrak{x})} e^{-s\mathfrak{w}_{\mathrm{tot}}(\mathfrak{x})} \right), \tag{7.3}$$

for $\operatorname{Re} s$ sufficiently large, where the sum extends over all finite words $\mathfrak{x}$ of positive length.

In (7.4) below, we introduce the logarithmic derivative of $\zeta_{\mathfrak{w}}$. The abscissa of convergence of this function will be denoted D, which is also the abscissa of convergence of the sum in (7.3).

Remark 7.4. For $N = 0$, the alphabet is empty, and we interpret $\mathcal{F}_{\mathfrak{w}}$ as the static flow on a point, and $\zeta_{\mathfrak{w}}(s) = 1$. For $N = 1$, there is only one periodic orbit $(0, 0, 0, \ldots)$, and we have the dynamical system of a point moving around a circle of length $\mathfrak{w}_{\mathrm{tot}}(0) = \mathfrak{w}(0, 0, \ldots)$. Then

$$\zeta_{\mathfrak{w}}(s) = \frac{1}{1 - e^{-s\mathfrak{w}_{\mathrm{tot}}(0)}}.$$

The logarithmic derivative of the dynamical zeta function is

$$-\frac{\zeta_{\mathfrak{w}}'}{\zeta_{\mathfrak{w}}}(s) = \sum_{\mathfrak{x}} \frac{\mathfrak{w}_{\mathrm{tot}}(\mathfrak{x})}{l(\mathfrak{x})} e^{-s\mathfrak{w}_{\mathrm{tot}}(\mathfrak{x})}. \tag{7.4}$$

For $N \geq 1$, this series does not converge at $s = 0$. We assume that (7.4) converges for some value of $s > 0$, and the abscissa of convergence of this series will be denoted by D, the *dimension* of $\mathcal{F}_{\mathfrak{w}}$. Clearly, $D \geq 0$. Then the series on the right-hand side of (7.4) is absolutely convergent for $\operatorname{Re} s > D$. Moreover, we assume that there exists a screen

$$S\colon t \mapsto S(t) + it,$$

satisfying $S(t) < D$ for every $t \in \mathbb{R}$, such that $-\zeta_{\mathfrak{w}}'/\zeta_{\mathfrak{w}}$ has a meromorphic continuation to an open neighborhood of the corresponding window (see Section 1.2.1, Equations (1.22) and (1.23), along with the beginning of Section 5.3),

$$W = \{ s = \sigma + it\colon \sigma \geq S(t) \}.$$

In Section 7.4, we will also assume that $-\zeta_{\mathfrak{w}}'/\zeta_{\mathfrak{w}}$ satisfies the growth conditions **L1** and **L2**, as introduced in Section 5.3. We will then say that $\mathcal{F}_{\mathfrak{w}}$ is *languid,* as in Definition 5.2.

Remark 7.5. The dimension of a suspended flow often coincides with the topological entropy of the flow; see [PaPol2, Chapter 5] and the references therein. It would be interesting to be able to interpret the poles of $-\zeta_{\mathfrak{w}}'/\zeta_{\mathfrak{w}}$ as "complex entropies" in a way that is meaningful for these dynamical systems.

Definition 7.6. The poles of $-\zeta_{\mathfrak{w}}'/\zeta_{\mathfrak{w}}$ in W are called the *complex dimensions* (or dynamical complex dimensions) of the flow $\mathcal{F}_{\mathfrak{w}}$. The *set of complex dimensions* of $\mathcal{F}_{\mathfrak{w}}$ in W is denoted by $\mathcal{D}_{\mathfrak{w}}(W)$ or $\mathcal{D}_{\mathfrak{w}}$ for short.

The nonreal complex dimensions of a flow come in complex conjugate pairs ω, $\overline{\omega}$ (provided that W is symmetric about the real axis). If $\zeta_{\mathfrak{w}}$ has a meromorphic continuation to an open neighborhood of W as well, then the complex dimensions of $\mathcal{F}_{\mathfrak{w}}$ are simple and they are located at the zeros and poles of $\zeta_{\mathfrak{w}}$,

$$\mathcal{D}_{\mathfrak{w}}(W) = \{\omega \in W \colon \zeta_{\mathfrak{w}}(\omega) = 0 \text{ or } \infty\}.$$

The residue at a complex dimension ω is then equal to $-\operatorname{ord}(\zeta_{\mathfrak{w}};\omega)$, where $\operatorname{ord}(\zeta_{\mathfrak{w}};\omega)$ is the order of $\zeta_{\mathfrak{w}}$ at ω:

$$\zeta_{\mathfrak{w}}(s) = C(s-\omega)^n + O((s-\omega)^{n+1}), \qquad C \neq 0, \quad n = \operatorname{ord}(\zeta_{\mathfrak{w}};\omega).$$

In general, the complex dimensions of $\mathcal{F}_{\mathfrak{w}}$ in W are not simple, and the residues are not necessarily integers. We extend the notation, so that we write $\operatorname{ord}(\zeta_{\mathfrak{w}};\omega) = \operatorname{res}(\zeta_{\mathfrak{w}}'/\zeta_{\mathfrak{w}};\omega)$ if $\zeta_{\mathfrak{w}}'/\zeta_{\mathfrak{w}}$ has a meromorphic continuation with a simple pole at ω, even if the residue is not an integer. In that case, $\zeta_{\mathfrak{w}}$ is not analytic at ω but has a logarithmic singularity.

7.2 Periodic Orbits, Euler Product

A periodic word a in Σ with period l, $a = a_1, \dots, a_l, a_1, \dots, a_l, \dots$, gives rise to the finite orbit $\{a, \sigma a, \dots, \sigma^{l-1}a\}$ of the shift σ. Clearly, l is a multiple of the cardinality $\#\{a, \sigma a, \dots, \sigma^{l-1}a\}$ of this orbit.

Definition 7.7. A finite word $\mathfrak{x}$ is *primitive* if its length $l(\mathfrak{x})$ coincides with the length of the corresponding periodic orbit of σ.

We denote by $\sigma\backslash\Sigma$ the space of primitive periodic orbits of σ. Thus

$$\sigma\backslash\Sigma = \big\{\{\sigma^k\mathfrak{x} \colon k \in \mathbb{N}\} \colon \mathfrak{x} \text{ is a finite word}\big\}. \tag{7.5}$$

We reserve the letter $\mathfrak{p}$ for elements of $\sigma\backslash\Sigma$. Thus $\mathfrak{p}$ denotes the set of points in a primitive periodic orbit of σ, and we write $\#\mathfrak{p}$ for its length.

The *total weight* of a primitive orbit $\mathfrak{p}$ is

$$\mathfrak{w}_{\mathrm{tot}}(\mathfrak{p}) = \sum_{a \in \mathfrak{p}} \mathfrak{w}(a). \tag{7.6}$$

Theorem 7.8 (Euler sum). *For* $\operatorname{Re} s > D$, *we have the following expression for the logarithmic derivative of* $\zeta_{\mathfrak{w}}$:

$$-\frac{\zeta'_{\mathfrak{w}}}{\zeta_{\mathfrak{w}}}(s) = \sum_{\mathfrak{p} \in \sigma \backslash \Sigma} \sum_{k=1}^{\infty} \mathfrak{w}_{\mathrm{tot}}(\mathfrak{p}) e^{-sk\mathfrak{w}_{\mathrm{tot}}(\mathfrak{p})}, \tag{7.7}$$

where $\mathfrak{p}$ *runs through all primitive periodic orbits of* $\mathcal{F}_{\mathfrak{w}}$ *(i.e., through all elements of* $\sigma \backslash \Sigma$*).*

Proof. We write the sum in (7.4) over the finite words $\mathfrak{x}$ as a sum over the primitive words and repetitions of these. An orbit $\mathfrak{p}$ contains $\#\mathfrak{p}$ different primitive words of length $\#\mathfrak{p}$, hence we obtain

$$\begin{aligned} \sum_{\mathfrak{x}} \frac{\mathfrak{w}_{\mathrm{tot}}(\mathfrak{x})}{l(\mathfrak{x})} e^{-s\mathfrak{w}_{\mathrm{tot}}(\mathfrak{x})} &= \sum_{\mathfrak{x}:\text{primitive}} \sum_{k=1}^{\infty} \frac{k\mathfrak{w}_{\mathrm{tot}}(\mathfrak{x})}{kl(\mathfrak{x})} e^{-ks\mathfrak{w}_{\mathrm{tot}}(\mathfrak{x})} \\ &= \sum_{\mathfrak{p} \in \sigma \backslash \Sigma} \#\mathfrak{p} \sum_{k=1}^{\infty} \frac{k\mathfrak{w}_{\mathrm{tot}}(\mathfrak{p})}{k\#\mathfrak{p}} e^{-ks\mathfrak{w}_{\mathrm{tot}}(\mathfrak{p})}. \end{aligned}$$

The theorem follows after cancelling k and $\#\mathfrak{p}$. □

Corollary 7.9 (Euler product). *The function* $\zeta_{\mathfrak{w}}(s)$ *has the following expansion as a product over all primitive periodic orbits* $\mathfrak{p}$ *of* $\mathcal{F}_{\mathfrak{w}}$:

$$\zeta_{\mathfrak{w}}(s) = \prod_{\mathfrak{p} \in \sigma \backslash \Sigma} \frac{1}{1 - e^{-s\mathfrak{w}_{\mathrm{tot}}(\mathfrak{p})}}. \tag{7.8}$$

The product converges for $\operatorname{Re} s > D$.

Proof. In (7.7), we sum over k to obtain

$$\frac{\zeta'_{\mathfrak{w}}}{\zeta_{\mathfrak{w}}}(s) = -\sum_{\mathfrak{p} \in \sigma \backslash \Sigma} \frac{\mathfrak{w}_{\mathrm{tot}}(\mathfrak{p}) e^{-s\mathfrak{w}_{\mathrm{tot}}(\mathfrak{p})}}{1 - e^{-s\mathfrak{w}_{\mathrm{tot}}(\mathfrak{p})}} = -\sum_{\mathfrak{p} \in \sigma \backslash \Sigma} \frac{d}{ds} \log\left(1 - e^{-s\mathfrak{w}_{\mathrm{tot}}(\mathfrak{p})}\right). \tag{7.9}$$

The theorem then follows upon integrating and taking exponentials. The constant of integration is determined by $\lim_{s \to \infty} \zeta_{\mathfrak{w}}(s) = 1$. □

Since the Euler product converges for $\operatorname{Re} s > D$, we also obtain the following corollary:

Corollary 7.10. *The dynamical zeta function $\zeta_{\mathfrak{w}}(s)$ is holomorphic and does not have any zeros for* $\operatorname{Re} s > D$.

On the other hand, the meromorphic continuation of $\zeta_{\mathfrak{w}}$ (to an open neighborhood of W) may have zeros or poles for $\operatorname{Re} s \leq D$, $s \in W$.

Definition 7.11. The following function counts the primitive periodic orbits and their multiples by their total weight:

$$\psi_{\mathfrak{w}}(x) = \sum_{k\mathfrak{w}_{\text{tot}}(\mathfrak{p}) \leq \log x} \mathfrak{w}_{\text{tot}}(\mathfrak{p}), \tag{7.10}$$

where k ranges through all positive integers and $\mathfrak{p}$ through all primitive periodic orbits of $\mathcal{F}_{\mathfrak{w}}$ (i.e., $\mathfrak{p} \in \sigma \backslash \Sigma$).

The function $\psi_{\mathfrak{w}}(x)$ is the counterpart of $\psi(x) = \sum_{p^k \leq x} \log p$, which counts prime powers p^k with a weight $\log p$. See also Section 5.5.1.

In Section 7.4 below, we combine the above Euler sum representation of $-\zeta_{\mathfrak{w}}'/\zeta_{\mathfrak{w}}$ with our explicit formulas of Chapter 5 to derive a Prime Orbit Theorem for primitive periodic orbits, using the next corollary.

Corollary 7.12. *We have the following relation between $\zeta_{\mathfrak{w}}'/\zeta_{\mathfrak{w}}$ and $\psi_{\mathfrak{w}}$:*

$$-\frac{\zeta_{\mathfrak{w}}'}{\zeta_{\mathfrak{w}}}(s) = \int_0^\infty x^{-s}\, d\psi_{\mathfrak{w}}(x), \tag{7.11}$$

for $\operatorname{Re} s > D$.

The integral on the right-hand side of (7.11) is a Riemann–Stieltjes integral associated with the monotonic function $\psi_{\mathfrak{w}}$.

Remark 7.13. We use $\psi_{\mathfrak{w}}$ instead of the more direct counting function

$$\pi_{\mathfrak{w}}(x) = \sum_{\mathfrak{w}_{\text{tot}}(\mathfrak{p}) \leq \log x} 1,$$

where $\mathfrak{p} \in \sigma \backslash \Sigma$. However, setting $\theta_{\mathfrak{w}}(x) = \sum_{\mathfrak{w}_{\text{tot}}(\mathfrak{p}) \leq \log x} \mathfrak{w}_{\text{tot}}(\mathfrak{p})$, so that

$$\psi_{\mathfrak{w}}(x) = \theta_{\mathfrak{w}}(x) + \theta_{\mathfrak{w}}(x^{1/2}) + \theta_{\mathfrak{w}}(x^{1/3}) + \dots$$

and $\theta_{\mathfrak{w}}(x) = \psi_{\mathfrak{w}}(x) - O\left(\sqrt{\psi_{\mathfrak{w}}(x)}\right)$, as $x \to \infty$, we find that

$$\pi_{\mathfrak{w}}(x) = \int_0^x \frac{1}{\log t}\, d\theta_{\mathfrak{w}}(t) = \frac{\theta_{\mathfrak{w}}(x)}{\log x} + \int_0^x \frac{\theta_{\mathfrak{w}}(t)}{\log^2 t} \frac{dt}{t},$$

from which it is easy to derive the corresponding theorems for $\pi_{\mathfrak{w}}$ from those obtained for $\psi_{\mathfrak{w}}$ in Sections 7.4 and 7.5 below.

7.3 Self-Similar Flows

If $\mathfrak{w}(a)$ depends only on the first element of the word, then the flow is self-similar in the following sense. We count the periodic orbits with a weight $\mathfrak{w}_{\mathrm{tot}}(\mathfrak{p})$. Thus, for example, the empty orbit is not counted. If we append the element j in front of an orbit ($j = 0, \ldots, N-1$), the total weight of this orbit becomes larger by $\mathfrak{w}(j)$, and the exponential weight (7.14) below is scaled by the factor r_{j+1}. The total collection of periodic orbits is the union of these copies with a scaled weight function. Hence, we recover the intuitive notion of a self-similar object—in this case, a self-similar flow—as the union of N scaled copies of itself.

Definition 7.14. A flow $\mathcal{F}_{\mathfrak{w}}$ is *self-similar* if $N \geq 2$ and the weight function $\mathfrak{w}$ depends only on the first digit of the word on which it is evaluated. We then put

$$w_j = \mathfrak{w}(j-1, j-1, j-1, \ldots) \tag{7.12}$$

and

$$r_j = e^{-w_j} = r(j-1, j-1, j-1, \ldots), \tag{7.13}$$

for $j = 1, \ldots, N$, and define

$$r(\mathfrak{x}) = e^{-\mathfrak{w}_{\mathrm{tot}}(\mathfrak{x})}, \tag{7.14}$$

where $\mathfrak{w}_{\mathrm{tot}}(\mathfrak{x})$ is given by (7.2) for a finite word $\mathfrak{x}$.

The numbers r_j and w_j (for $j = 1, \ldots, N$) are called respectively the *scaling ratios* and *weights* of the self-similar flow $\mathcal{F}_{\mathfrak{w}}$.

Note that $0 < r_j < 1$ for $j = 1, \ldots, N$. We will assume that the weights $w_j = \log r_j^{-1}$ are ordered in nondecreasing order,

$$0 < w_1 \leq w_2 \leq \cdots \leq w_N,$$

so that $1 > r_1 \geq r_2 \geq \cdots \geq r_N > 0$.

When $N = 2$, the flow is called a *Bernoulli flow*. Such flows play an important role in ergodic theory (see [BedKS, Chapters 2, 6 and 8]).

A self-similar flow is best viewed as the following dynamics on the region of Figure 7.1. A point $x = x_1 N^{-1} + x_2 N^{-2} + \cdots = .x_1 x_2 \ldots$ on the unit interval moves vertically upward with unit speed until it reaches the graph, at which moment it jumps to $\{Nx\} = Nx - [Nx] = .x_2 x_3 \ldots$, the fractional part of Nx, and continues from there. In Figure 7.1, $N = 5$, and the expansions of A, B and C in base 5 are $A = 17/124 = .\overline{032}$, $B = 85/124 = .\overline{320}$, $C = 53/124 = .\overline{203}$.

Theorem 7.15. *The dynamical zeta function associated with a self-similar flow has a meromorphic continuation to the whole complex plane, given by*

$$\zeta_{\mathfrak{w}}(s) = \frac{1}{1 - \sum_{j=1}^{N} r_j^s}. \tag{7.15}$$

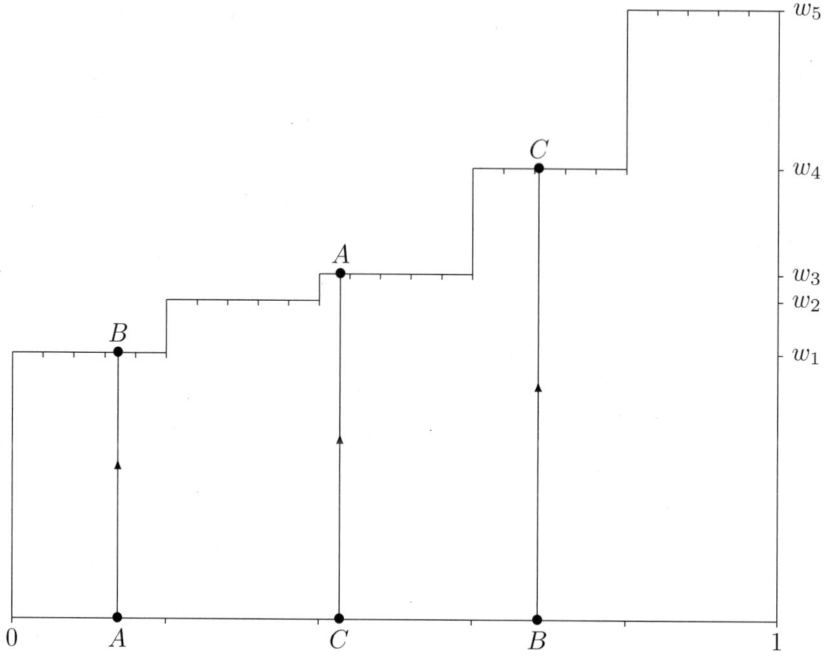

Figure 7.1: A self-similar flow, $N = 5$, with the orbit of $17/124$.

Its logarithmic derivative is given by

$$-\frac{\zeta'_{\mathfrak{w}}}{\zeta_{\mathfrak{w}}}(s) = \frac{\sum_{j=1}^{N} w_j r_j^s}{1 - \sum_{j=1}^{N} r_j^s}. \tag{7.16}$$

The dimension $D > 0$ of the flow is the unique real solution of the equation

$$\sum_{j=1}^{N} r_j^s = 1. \tag{7.17}$$

Proof. The sum over periodic words of fixed length l can be computed as follows:

$$\begin{aligned}\sum_{\mathfrak{r}:\, l(\mathfrak{r})=l} r(\mathfrak{r})^s &= \sum_{a_1=1}^{N} \sum_{a_2=1}^{N} \cdots \sum_{a_l=1}^{N} r_{a_1}^s \dots r_{a_l}^s \\ &= \left(r_1^s + \cdots + r_N^s\right)^l.\end{aligned}$$

Hence, for $\operatorname{Re} s > D$, the sum over all periodic words is equal to

$$\sum_{l=1}^{\infty} \frac{1}{l} \sum_{\mathfrak{r}:\, l(\mathfrak{r})=l} r(\mathfrak{r})^s = \sum_{l=1}^{\infty} \frac{1}{l} \left(r_1^s + \cdots + r_N^s\right)^l = -\log\Bigl(1 - \sum\nolimits_{j=1}^{N} r_j^s\Bigr).$$

The theorem follows upon exponentiation and analytic continuation. Since the function $1 - \sum_{j=1}^{N} r_j^s$ is holomorphic, $\zeta_{\mathfrak{w}}$ is meromorphic. □

Remark 7.16. Thanks to Theorem 7.15, for a self-similar flow we can take the full complex plane for the window: $W = \mathbb{C}$. In that case, there is no screen. However, in applying our explicit formulas, we sometimes choose a screen to obtain information about the error of an approximation; see, e.g., Section 7.5 and the discussion above Remark 6.11.

Remark 7.17. A self-similar flow corresponds to a self-similar string with one gap g_1, and with a total length L normalized by $g_1 L = 1$, or equivalently, such that the first length of $\mathcal{L}$ is 1. To make the connection, we must assume that $g_1 = 1 - \sum_{j=1}^{N} r_j > 0$, which corresponds to a lower bound on the weights $w_j = -\log r_j$. Note that a general suspended flow does not always satisfy this condition.

On the other hand, the logarithmic derivative of $\zeta_{\mathfrak{w}}$ for the self-similar flow $\mathcal{F}_{\mathfrak{w}}$ is

$$\frac{\zeta_{\mathfrak{w}}'}{\zeta_{\mathfrak{w}}}(s) = \frac{\sum_{j=1}^{N} w_j e^{-w_j s}}{1 - \sum_{j=1}^{N} e^{-w_j s}}.$$

This corresponds to a self-similar string with gaps e^{-w_j}, with (in general, noninteger) multiplicity w_j, as in Chapter 3.

Remark 7.18. By the previous remark and Corollary 7.9, the geometric zeta function of any self-similar string with a single gap has an Euler product. This Euler product does not seem to have a clear geometric interpretation in the language of fractal strings. Also, the connection with self-similar fractal strings with multiple gaps remains to be clarified.

Note that this Euler product is different from that for the spectral operator, discussed in the previous chapter (within Section 6.3.2), which is related to the Euler product for the Riemann zeta function.

The following result is a counterpart for self-similar flows of Corollary 2.5 for self-similar strings. It is an immediate consequence of Equation (7.16) of Theorem 7.15.

Corollary 7.19. *The set of (dynamical) complex dimensions $\mathcal{D}_{\mathfrak{w}} = \mathcal{D}_{\mathfrak{w}}(\mathbb{C})$ of the self-similar flow $\mathcal{F}_{\mathfrak{w}}$ is the set of solutions of the equation*

$$\sum_{j=1}^{N} r_j^{\omega} = 1, \qquad \omega \in \mathbb{C}. \tag{7.18}$$

Moreover, the complex dimensions are simple (that is, the pole of $-\zeta_{\mathfrak{w}}'/\zeta_{\mathfrak{w}}$ at ω is simple). The residue at ω equals $-\operatorname{ord}(\zeta_{\mathfrak{w}};\omega)$.

Remark 7.20 (Geometric and dynamical complex dimensions). In Chapter 1, the geometric complex dimensions of a fractal string are defined as the poles of its geometric zeta function. Thus the complex dimensions are counted with a multiplicity, and the zeros of the geometric zeta function are unimportant. On the other hand, the dynamical complex dimensions are defined as the poles of the logarithmic derivative of the dynamical zeta function. Thus the complex dimensions are simple, and both the zeros and the poles of the dynamical zeta function are counted. For self-similar flows, the dynamical zeta function and the geometric zeta function of the corresponding string coincide (up to normalization), and this zeta function has no zeros. Hence, as sets (without multiplicity), the geometric and dynamical complex dimensions coincide for self-similar flows and self-similar fractal strings with a single gap.

Remark 7.21. It would be interesting to extend the formalism presented in this chapter to include self-similar flows corresponding to self-similar strings with multiple gaps (rather than with a single gap), as defined in Section 2.1. Even in the present deterministic (as opposed to random) setting, this may possibly be done within the broader context of random fractal trees and recursive constructions studied by Ben Hambly and the first author in [HamLap]. The latter work is discussed in Section 12.4.1.

Remark 7.22 (Higher-dimensional case). It is clear that our results can be applied to higher-dimensional self-similar fractals [Fa3, Man1] as well. This allows us to obtain information about the symbolic dynamics of self-similar fractals. On the other hand, it does not give information about the actual geometry of such fractals.

Remark 7.23. Let $\mathcal{L} = l_1, l_2, \dots$ be a fractal string. There is a natural suspended flow on $\mathcal{L}$, namely, the flow

$$\mathcal{F}_{\mathcal{L}}(t, j, x) = \begin{cases} (0, j, xe^t) & \text{if } xe^t < l_j, \\ \mathcal{F}_{\mathcal{L}}(t - \log l_j, j, 1) & \text{otherwise.} \end{cases} \tag{7.19}$$

The lengths l_j correspond to the periodic words $\mathfrak{x}$ of the flow $\mathcal{F}_{\mathfrak{w}}$ via the formula

$$l_j = \prod_{k=0}^{l(\mathfrak{x})-1} r(\sigma^k \mathfrak{x}). \tag{7.20}$$

This construction can be carried out for any fractal string. The resulting flow is self-similar in the present sense if and only if the associated fractal string is self-similar.

7.3.1 Examples of Self-Similar Flows

Example 7.24 (The Cantor flow). This is the self-similar flow on the alphabet $\{0, 1\}$, with two equal weights $w_1 = w_2 = \log 3$. It has 2^n periodic words[1] of weight $n \log 3$, for $n = 1, 2, \ldots$. The dynamical zeta function of this flow is given by

$$\zeta_{\mathrm{CF}}(s) = \frac{1}{1 - 2 \cdot 3^{-s}}. \tag{7.21}$$

The logarithmic derivative equals

$$-\frac{\zeta'_{\mathrm{CF}}}{\zeta_{\mathrm{CF}}}(s) = 2 \log 3 \cdot \frac{3^{-s}}{1 - 2 \cdot 3^{-s}}. \tag{7.22}$$

Example 7.25 (The Fibonacci flow). Next we consider a self-similar flow with two lines of complex dimensions. The *Fibonacci flow* is the flow Fib on the alphabet $\{0, 1\}$ with weights $w_1 = \log 2$, $w_2 = 2 \log 2$. Its periodic words have weight $\log 2, 2 \log 2, \ldots, n \log 2, \ldots$, with multiplicity respectively $1, 2, \ldots, F_{n+1}, \ldots$, the Fibonacci numbers defined by (2.19). The dynamical zeta function of the Fibonacci flow is

$$\zeta_{\mathrm{Fib}}(s) = \frac{1}{1 - 2^{-s} - 4^{-s}}, \tag{7.23}$$

with logarithmic derivative

$$-\frac{\zeta'_{\mathrm{Fib}}}{\zeta_{\mathrm{Fib}}}(s) = \log 2 \cdot \frac{2^{-s} + 2 \cdot 4^{-s}}{1 - 2^{-s} - 4^{-s}}. \tag{7.24}$$

Example 7.26 (The golden flow). We consider the nonlattice flow GF with weights $w_1 = \log 2$ and $w_2 = \phi \log 2$, where $\phi = (1 + \sqrt{5})/2$ is the golden ratio. We call it the *golden flow*. Its dynamical zeta function is

$$\zeta_{\mathrm{GF}}(s) = \frac{1}{1 - 2^{-s} - 2^{-\phi s}}, \tag{7.25}$$

and its complex dimensions are the solutions of the transcendental equation

$$2^{-\omega} + 2^{-\phi\omega} = 1, \qquad \omega \in \mathbb{C}. \tag{7.26}$$

A diagram of the dynamical complex dimensions of the golden flow is given in Figures 2.12 and 3.9. By Theorem 3.30 and the comment preceding it, we even know that the complex dimensions of the corresponding self-similar string are all simple, hence the dynamical and geometric complex dimensions correspond exactly.

---

[1] The number of *primitive* periodic words of weight $n \log 3$ is much harder to determine.

7.3.2 The Lattice and Nonlattice Case

Let $\mathcal{F}_{\mathfrak{w}}$ be a self-similar flow. Recall that $\mathfrak{w}$ depends only on the first symbol and $w_j = \mathfrak{w}(j,j,\dots)$ for $j = 1,\dots,N$. Consider the subgroup A of $\mathbb{R}$ generated by these weights,

$$A = \sum_{j=1}^{N} \mathbb{Z} w_j.$$

As in Chapter 3, we define the lattice and the (generic and nongeneric) nonlattice case.

Definition 7.27. We say that $\mathcal{F}_{\mathfrak{w}}$ is a *nonlattice flow* if A is dense in $\mathbb{R}$. This is the case when the rank of A is at least 2. The *generic nonlattice case* is when the rank of A equals $\#\{w_j \colon j = 1,\dots,N\}$, the number of different weights of the flow.

We call $\mathcal{F}_{\mathfrak{w}}$ a *lattice flow* if the group A is not dense (and hence discrete) in $\mathbb{R}$. In this situation there exists a unique positive real real number w, called the *generator* of the flow, and positive integers $k_1,\dots,k_N$ without common divisor, such that $1 \le k_1 \le \dots \le k_N$ and

$$w_j = k_j w, \tag{7.27}$$

for $j = 1,\dots,N$.

The generator of a lattice flow generates the flow in the sense that the weight of every periodic orbit is an integer multiple of w.

To determine the number D_l as introduced in Section 3.3 (see Equation (3.8b)), we need to count the multiplicity of r_N. Let m be the number of integers j in $\{1,\dots,N\}$ such that $r_j = r_N$. Then D_l is given as the unique real solution of[2]

$$1 + \sum_{j=1}^{N-m} r_j^{D_l} = m r_N^{D_l}. \tag{7.28}$$

We then have the counterpart of Theorems 2.17 and 3.6, of which we recall the main features in the present context.

Theorem 7.28. *Let $\mathcal{F}_{\mathfrak{w}}$ be a self-similar flow of dimension D and with scaling ratios $1 > r_1 \ge \dots \ge r_N > 0$. Then $s = D$ is the only complex dimension of $\mathcal{F}_{\mathfrak{w}}$ on the real line. All complex dimensions are simple, and the residue at a complex dimension (i.e., $\operatorname{res}(-\zeta_{\mathfrak{w}}'/\zeta_{\mathfrak{w}};\omega)$) is a positive integer. The set of complex dimensions in $\mathbb{C}$ (see Remark 7.16) of $\mathcal{F}_{\mathfrak{w}}$ is*

[2]This definition coincides with the one given by formula (3.8b) for D_l, since in that formula, the numbers r_j ($j = 1,\dots,M-1$) are all distinct and counted with multiplicity $m_j = |m_j|$, and r_M corresponds to r_N in (7.28).

contained in the bounded strip $D_l \leq \operatorname{Re} s \leq D$, *where* D_l *is given in* (7.28) *above:*

$$\mathcal{D}_{\mathfrak{w}} = \mathcal{D}_{\mathfrak{w}}(\mathbb{C}) \subset \{s \in \mathbb{C} \colon D_l \leq \operatorname{Re} s \leq D\}. \tag{7.29}$$

It is symmetric with respect to the real axis and infinite, with density

$$\#\left(\mathcal{D}_{\mathfrak{w}} \cap \{\omega \in \mathbb{C} \colon |\operatorname{Im} \omega| \leq T\}\right) \leq \frac{w_N}{\pi} T + O(1), \tag{7.30}$$

as $T \to \infty$.

In the lattice case, $\zeta_{\mathfrak{w}}(s)$ *is a rational function of* e^{-ws}, *where* w *is the generator of* $\mathcal{F}_{\mathfrak{w}}$. *So, as a function of* s, *it is periodic with period* $2\pi i/w$. *The complex dimensions* ω *are obtained as the complex solutions* z *of the polynomial equation* (*of degree* k_N)

$$\sum_{j=1}^{N} z^{k_j} = 1, \quad \textit{with } e^{-w\omega} = z. \tag{7.31}$$

The solutions of this equation are described in Theorem 2.17, following Equation (2.38).

In the nonlattice case, D *is the unique pole of* $\zeta_{\mathfrak{w}}$ *on the line* $\operatorname{Re} s = D$. *Further, there is an infinite sequence of complex dimensions of* $\mathcal{F}_{\mathfrak{w}}$ *coming arbitrarily close* (*from the left*) *to the line* $\operatorname{Re} s = D$. *There exists a screen* S *to the left of the line* $\operatorname{Re} s = D$, *such that* $-\zeta_{\mathfrak{w}}'/\zeta_{\mathfrak{w}}$ *satisfies* **L1** *and* **L2** *with* $\kappa = 0$ (*see Equations* (5.19) *and* (5.20)), *and the residue of* $-\zeta_{\mathfrak{w}}'/\zeta_{\mathfrak{w}}$ *at the pole* ω *in the associated window* W *is equal to* 1. *Finally, the complex dimensions of* $\mathcal{F}_{\mathfrak{w}}$ *can be approximated* (*via an explicit procedure, as described in Chapter 3*) *by the complex dimensions of a sequence of lattice flows, with smaller and smaller generators. Hence the complex dimensions of a nonlattice flow have a quasiperiodic structure.*

Corollary 7.29. *Every self-similar flow has infinitely many complex dimensions with positive real part.*

Proof of Theorem 7.28. For a proof of most these facts, see Theorems 2.17 and 3.6. The density estimate (7.30) follows from the fact that, according to Theorem 3.6, the right-hand side of (7.30) gives the asymptotic density of the number of poles of $\zeta_{\mathfrak{w}}$, counted with multiplicity, whereas the dynamical complex dimensions have multiplicity one (i.e., they are simple). □

7.4 The Prime Orbit Theorem for Suspended Flows

Let $\mathcal{F}_{\mathfrak{w}}$ be a suspended flow as in Section 7.1. In Corollary 7.12, we have written the logarithmic derivative of $\zeta_{\mathfrak{w}}(s)$ as the Mellin transform of the counting function $\psi_{\mathfrak{w}}$ of the weighted periodic orbits of σ, as defined in (7.10). Put $\eta = d\psi_{\mathfrak{w}}$, so that $\zeta_\eta = -\zeta_{\mathfrak{w}}'/\zeta_{\mathfrak{w}}$. The poles of $-\zeta_{\mathfrak{w}}'/\zeta_{\mathfrak{w}}$ are the complex dimensions of $\mathcal{F}_{\mathfrak{w}}$ and the residue at ω is $-\operatorname{ord}(\zeta_{\mathfrak{w}};\omega)$. By Theorems 5.18 and 5.30, applied to the generalized fractal string η with window W, we obtain the following distributional explicit formula for the counting function of the weighted periodic orbits of σ.

Theorem 7.30 (The Prime Orbit Theorem with Error Term). *Let $\mathcal{F}_{\mathfrak{w}}$ be a languid suspended flow (i.e., $\zeta_{\mathfrak{w}}'/\zeta_{\mathfrak{w}}$ satisfies conditions* **L1** *and* **L2** *of Definition* 5.2*). Then we have the following equality between distributions:*

$$\psi_{\mathfrak{w}}(x) = \frac{x^D}{D} + \sum_{\omega \in \mathcal{D}_{\mathfrak{w}} \setminus \{D,0\}} -\operatorname{ord}(\zeta_{\mathfrak{w}};\omega)\,\frac{x^\omega}{\omega} + \operatorname{res}\left(-\frac{x^s \zeta_{\mathfrak{w}}'(s)}{s\zeta_{\mathfrak{w}}(s)};0\right) + R(x), \tag{7.32}$$

where $\operatorname{ord}(\zeta_{\mathfrak{w}};\omega) < 0$ *denotes the order of $\zeta_{\mathfrak{w}}$ at ω,*[3] *and*

$$R(x) = -\int_S \frac{\zeta_{\mathfrak{w}}'}{\zeta_{\mathfrak{w}}}(s)x^s\,\frac{ds}{s} = O\left(x^{\sup S}\right), \tag{7.33}$$

as $x \to \infty$.

If 0 is not a complex dimension of the flow, then the third term on the right-hand side of (7.32) *simplifies to $-\zeta_{\mathfrak{w}}'(0)/\zeta_{\mathfrak{w}}(0)$. In general, this term is of the form $\alpha + \beta\log x$, for some constants α and β.*

If D is the only complex dimension on the line $\operatorname{Re} s = D$, *then the error term,*

$$\sum_{\omega \in \mathcal{D}_{\mathfrak{w}} \setminus \{D,0\}} -\operatorname{ord}(\zeta_{\mathfrak{w}};\omega)\,\frac{x^\omega}{\omega} + \operatorname{res}\left(-\frac{x^s \zeta_{\mathfrak{w}}'(s)}{s\zeta_{\mathfrak{w}}(s)};0\right) + R(x), \tag{7.34}$$

is estimated by $o(x^D)$, as $x \to \infty$. If this is the case, then we obtain a Prime Orbit Theorem for $\mathcal{F}_{\mathfrak{w}}$ as follows:

$$\psi_{\mathfrak{w}}(x) = \frac{x^D}{D} + o\left(x^D\right), \tag{7.35}$$

as $x \to \infty$.

Proof. The first part of the theorem follows from the distributional explicit formula with error term (Theorem 5.18) and from the first part of Theorem 5.30, while the second part follows from the second part of Theorem 5.30. See also Theorem 5.17, in case there is no screen such that only D is visible. ∎

---

[3] See the paragraph following Definition 7.6—or, for more details, the text surrounding Equation (3.19)—for the precise definition of the order, $\operatorname{ord}(\zeta_{\mathfrak{w}};\omega)$.

Remark 7.31. In exactly the same way, one deduces from Theorem 5.22, the distributional explicit formula without error term, that if $\zeta'_{\mathfrak{w}}/\zeta_{\mathfrak{w}}$ satisfies hypotheses **L1** and **L2**$'$ (i.e., if $\mathcal{F}_{\mathfrak{w}}$ is strongly languid, in the sense of Definition 5.3), then we may put $R(x) \equiv 0$ in Equations (7.32) and (7.34).

Remark 7.32. Lalley considers in [Lal2–3] the (approximately) self-similar case. Using a nonlinear extension of the Renewal Theorem, he shows that in the nonlattice case, the leading asymptotics are nonoscillatory. In the lattice case, the leading asymptotics are periodic, and it becomes a natural question whether they are constant or nontrivially periodic. For self-similar flows, this problem is resolved in Section 7.4.2 below.

Remark 7.33. We can apply our results to obtain explicit formulas for the more general dynamical systems considered, for example, in [PaPol2] or [Lal2–3] and the relevant references therein. (See also the notes to this chapter.) The precise form of the resulting expressions and error terms, however, still remains to be fully worked out; see Remark 7.34. Even in the situation of self-similar flows studied in Sections 7.4.1–7.4.3 and 7.5 below, our results are significantly more precise than those previously available in the literature because the explicit formula (7.32)—combined with our Diophantine approximation techniques of Chapter 3—not only yields a Prime Orbit Theorem with error term (as, e.g., in Section 7.5), but also a full expansion of the prime powers counting function in terms of the complex dimensions of the flow (that is, the poles of the associated dynamical zeta function); see, e.g., Equations (7.32), (7.34), and for self-similar flows, Equations (7.39)–(7.40) and (7.43).

7.4.1 The Prime Orbit Theorem for Self-Similar Flows

For self-similar flows, the function $\zeta_{\mathfrak{w}}$ does not have any zeros (see (7.15)); i.e., such a flow corresponds to a self-similar string with a single gap. Hence every contribution to (7.32) comes from a pole of $\zeta_{\mathfrak{w}}$, and each coefficient $-\operatorname{ord}(\zeta_{\mathfrak{w}};\omega)$ is positive. Furthermore, 0 is never a complex dimension, so the third term on the right-hand side of (7.32) in the explicit formula is

$$-\frac{\zeta'_{\mathfrak{w}}(0)}{\zeta_{\mathfrak{w}}(0)} = \frac{1}{N-1}\sum_{j=1}^{N} w_j. \tag{7.36}$$

We can obtain information about $\psi_{\mathfrak{w}}$ by choosing a suitable screen. Note that the hypotheses **L1** and **L2** (or **L2**$'$ for a distributional formula without error term) are satisfied in view of Section 6.4, in particular estimate (6.36), and the second part of Remark 7.17.

7.4.2 Lattice Flows

In the lattice case, we obtain the Prime Orbit Theorem for lattice self-similar flows:

$$\psi_{\mathfrak{w}}(x) = G_1(\log x)x^D + \frac{1}{N-1}\sum_{j=1}^{N} w_j + O\left(x^{D-\alpha}\right), \tag{7.37}$$

as $x \to \infty$. Here, $D - \alpha$ (with $\alpha > 0$) is the abscissa of the first vertical line of complex dimensions next to D, and the periodic function G_1, of period w, is given by

$$G_1(y) = \sum_{n=-\infty}^{\infty} \frac{e^{2\pi i n y/w}}{D + 2\pi i n/w} = \frac{w}{1-e^{-wD}}\, e^{-wD\{y/w\}}. \tag{7.38}$$

By the pointwise explicit formula without error term (Theorem 5.14), we can even obtain more precise information about $\psi_{\mathfrak{w}}$,

$$\begin{aligned}
\psi_{\mathfrak{w}}(x) &= \sum_{u=1}^{q} -\operatorname{ord}\left(\zeta_{\mathfrak{w}}; \omega_u\right) \sum_{n\in\mathbb{Z}} \frac{x^{\omega_u + 2\pi i n/w}}{\omega_u + 2\pi i n/w} + \frac{1}{N-1}\sum_{j=1}^{N} w_j \\
&= \sum_{u=1}^{q} -\operatorname{ord}\left(\zeta_{\mathfrak{w}}; \omega_u\right) G_u(\log x) x^{\omega_u} + \frac{1}{N-1}\sum_{j=1}^{N} w_j,
\end{aligned} \tag{7.39}$$

where for each $u = 1, \dots, q$, the function G_u is periodic of period w and is given by

$$G_u(y) = \sum_{n\in\mathbb{Z}} \frac{e^{2\pi i n y/w}}{\omega_u + 2\pi i n/w} = \frac{w}{1-e^{-w\omega_u}}\, e^{-w\omega_u\{y/w\}}. \tag{7.40}$$

Here, $\omega_1 (= D), \omega_2, \dots, \omega_q$ are given as in the lattice case of Theorem 2.17 (just after Equation (2.38)), and $\operatorname{ord}(\zeta_{\mathfrak{w}}; \omega_1) = -1$.

For instance, for the Cantor flow of Example 7.24 (with $D = \log_3 2$ and $w_1 = w_2 = w = \log 3$), we have

$$\psi_{\mathrm{CF}}(x) = G_1(\log x)x^D + 2\log 3, \tag{7.41}$$

with $G_1(y) = w 2^{1-\{y/w\}}$. As another example, for the Fibonacci flow[4] of Example 7.25 (with $D = \log_2 \phi$ and $w_1 = w = \log 2$, $w_2 = 2w$), we have

$$\psi_{\mathrm{Fib}}(x) = G_1(\log x)x^D + G_2(\log x)x^{\pi i/w}x^{-D} - 3\log 2, \tag{7.42}$$

[4] Also called the golden mean flow in the literature (see, e.g., [BedKS, p. 59]), but not to be confused with the golden flow in this book, which is a nonlattice self-similar flow.

where $G_1(y) = w\phi^{2-\{y/w\}}$ and

$$G_2(y) = \sum_{n\in\mathbb{Z}} \frac{e^{2\pi iny/w}}{-D + 2\pi i(n+1/2)/w} = w\phi^{\{y/w\}-2}e^{-\pi i\{y/w\}}.$$

In the second term of Equation (7.42), the product

$$e^{-\pi i\{(\log x)/w\}}x^{\pi i/w}$$

combines to give the sign $(-1)^{[(\log x)/w]}$.

7.4.3 Nonlattice Flows

In the nonlattice case, we use Theorem 3.25 according to which there exists $\delta > 0$ and a screen S lying to the left of the vertical line $\operatorname{Re} s = D - \delta$ such that $-\zeta_{\mathfrak{w}}'/\zeta_{\mathfrak{w}}$ is bounded on S and all the complex dimensions ω to the right of S have residue $\operatorname{res}(-\zeta_{\mathfrak{w}}'/\zeta_{\mathfrak{w}};\omega) = 1$. Then $R(x) = O(x^{D-\delta})$, as $x \to \infty$. There are no complex dimensions with $\operatorname{Re}\omega = D$ except for D itself. Hence, the assumptions of Theorem 7.30 are satisfied. Therefore, in view of Theorem 7.30, we deduce by a classical argument (see the proof of Theorems 7.37 and 7.41) the Prime Orbit Theorem for nonlattice suspended flows:

$$\psi_{\mathfrak{w}}(x) = \frac{x^D}{D} + \sum_{\omega\in\mathcal{D}_{\mathfrak{w}}\setminus\{D\}} \frac{x^\omega}{\omega} + O\left(x^{D-\delta}\right) = \frac{x^D}{D} + o\left(x^D\right), \tag{7.43}$$

as $x \to \infty$ (see also Theorem 7.41, and when $N = 2$, Theorem 7.37 below for a better estimate of the error). We note that the exponent of x in this estimate is best possible, since by the nonlattice case of Theorem 2.17, there always exist complex dimensions of $\mathfrak{w}$ arbitrarily close to the vertical line $\operatorname{Re} s = D$. However, the dimension-free regions obtained in Section 3.6 allow us to improve the error term by a factor of lower order than a power of x. See Section 7.5 below.

Remark 7.34. It would be interesting to apply Theorem 7.30 to suspended flows that are more general than self-similar flows: For example, (i) the hyperbolic flows associated with subshifts of finite type studied by Parry and Pollicott in [PaPol2]; and (ii) the suspended flows considered by Lalley in [Lal2–3], such as the approximately self-similar flows naturally associated with limit sets of suitable Kleinian groups. This would require a more detailed study of the dynamical zeta function of each of these flows. (In case (i), some of the required information is already available in [PaPol2] and the relevant references therein.) We note that, in case (ii), the lattice–nonlattice dichotomy applies in a suitable extended sense, via the nonlinear renewal theory developed in [Lal2–3].

7.5 The Error Term in the Nonlattice Case

In this section, we provide a detailed analysis of the error term in the counting function for the periodic orbits in the nonlattice case, using the information on the dimension-free region for a nonlattice string obtained in Chapter 3, Section 3.6.

7.5.1 Two Generators

Definition 7.35. A domain in the complex plane containing the vertical line $\operatorname{Re} s = D$ is a *dimension-free region* for the flow $\mathcal{F}_{\mathfrak{w}}$ if the only pole of $-\zeta'_{\mathfrak{w}}/\zeta_{\mathfrak{w}}$ in that region is $s = D$.

The following result is the direct analogue of Corollary 3.44, which is a corollary of Theorem 3.34. Recall from (7.15) and Sections 3.5.1 and 3.5.2 that

$$\zeta_{\mathfrak{w}}(s) = \frac{1}{f(s)},$$

where $f(s) = 1 - m_1 e^{-w_1 s} - m_2 e^{-w_2 s}$ and m_1 and m_2 are the multiplicities (necessarily positive and integral) of the weights w_1 and w_2, respectively. In the nonlattice case, we have that $w_2/w_1 = \alpha > 1$ is irrational, with continued fraction $\alpha = [a_0, a_1, a_2, \dots]$. Further recall form Section 3.5.1 that we denote by q_k the denominators of the convergents of α; see Equations (3.24) and (3.25). These numbers q_k grow roughly like $(a_1+1)(a_2+1)\dots(a_k+1)$, and hence exponentially if the partial quotients of α are bounded, and otherwise faster than exponentially. The dimension of $\mathcal{F}_{\mathfrak{w}}$ satisfies $f(D) = 0$, and clearly,

$$f'(D) = m_1 w_1 e^{-w_1 D} + m_2 w_2 e^{-w_2 D}$$

is a positive number.

Corollary 7.36. *Assume that the partial quotients $a_0, a_1, \dots$ of α are bounded by b. Put $B = \pi^4 e^{-(w_1+w_2)D}/(2f'(D)^3)$ (see also Equation (3.43) of Section 3.6). Then $\mathcal{F}_{\mathfrak{w}}$ has a dimension-free region of the form*

$$\left\{\sigma + it \in \mathbb{C} \colon \sigma > D - \frac{B}{b^2 t^2}\right\}. \tag{7.44}$$

Further, the function $-\zeta'_{\mathfrak{w}}/\zeta_{\mathfrak{w}}$ satisfies hypotheses **L1** *and* **L2** *(i.e., the flow $\mathcal{F}_{\mathfrak{w}}$ is languid) with $\kappa = 2$ on the screen $\operatorname{Re} s = D - Bb^{-2}/(\operatorname{Im} s)^2$.*

More generally, let $b \colon \mathbb{R}^+ \to [1, \infty)$ be a function such that the partial quotients $\{a_k\}_{k=0}^{\infty}$ of the continued fraction of α satisfy $a_{k+1} \le b(q_k)$ for every $k \ge 0$. Then $\mathcal{F}_{\mathfrak{w}}$ has a dimension-free region of the form

$$\left\{\sigma + it \in \mathbb{C} \colon \sigma > D - \frac{B}{t^2 b^2(tw_1/2\pi)}\right\}. \tag{7.45}$$

If b grows at most polynomially, then $-\zeta_{\mathfrak{w}}'/\zeta_{\mathfrak{w}}$ satisfies hypotheses **L1** *and* **L2** *with κ such that $t^\kappa \geq t^2 b^2(tw_1 / 2\pi)$.*

Proof. This follows from Theorem 3.34, if we note that for $t = 2\pi q_k/w_1$, we have $q'_{k+1} = \alpha_{k+1} q'_k \leq 2b(q_k)q'_k \leq 4b(q_k)q_k$. So the complex dimension close to $D+it$ is located at $D+i(t+O(q'^{-1}_{k+1})) - (w_1^2/\pi^2)Bq'^{-2}_{k+1} + O(q'^{-4}_{k+1})$, where the big-$O$ terms denote real-valued functions. The real part of this complex dimension is less than $D - Bt^{-2}b^{-2}(tw_1/(2\pi))$. □

This has the following consequence for the Prime Orbit Theorem.

Theorem 7.37 (Prime Orbit Theorem with Error Term, for Bernoulli Flows). *Let $\alpha = w_2/w_1$ have bounded partial quotients in its continued fraction. Then*

$$\psi_{\mathfrak{w}}(x) = \frac{x^D}{D} + O\left(x^D \left(\frac{\log\log x}{\log x}\right)^{1/4}\right), \tag{7.46}$$

as $x \to \infty$.

If α is polynomially approximable, with partial quotients in its continued fraction satisfying $a_{k+1} \leq b(q_k)$, for some increasing function b such that $b(x) = O(x^l)$, as $x \to \infty$, then

$$\psi_{\mathfrak{w}}(x) = \frac{x^D}{D} + O\left(x^D \left(\frac{\log\log x}{\log x}\right)^{\frac{1}{4l+4}}\right), \tag{7.47}$$

as $x \to \infty$.

The proof of Theorem 7.37 will be given in the next section, following the statement of Theorem 7.41.

Remark 7.38. The estimates for the error terms in (7.46), (7.47), (7.50) and (7.51) are not best possible. Indeed, as is clear from the proof, these bounds are obtained by using a Tauberian argument, which makes the exponents worse by a factor of two. See also Remark 7.42.

7.5.2 More Than Two Generators

We formulate the following counterpart of Theorem 3.41. The proof is the same as that of Theorem 3.41.

Recall that M denotes the number of different weights $w_1 < \cdots < w_M$, where w_j is counted with the positive integral multiplicity m_j. The total number of weights is $N = m_1 + \cdots + m_M$.

Theorem 7.39. *Let $M \geq 2$ and let $w_1, \ldots, w_M$ be weights. Let Q and q be as in Lemma 3.16. Then $\mathcal{F}_{\mathfrak{w}}$ has a complex dimension close to $D+2\pi iq/w_1$ at a distance of at most $O(Q^{-2})$ from the line $\operatorname{Re} s = D$, as $Q \to \infty$. On the line $\operatorname{Re} s = D$, the function $|\zeta_{\mathfrak{w}}'/\zeta_{\mathfrak{w}}|$ reaches a maximum of order Q^2 around the point $s = D + 2\pi iq/w_1$.*

Corollary 7.40. *The best dimension-free region that $\mathcal{F}_{\mathfrak{w}}$ can have is of size*

$$\left\{\sigma + it \colon \sigma \geq D - O\left(t^{-2/(M-1)}\right)\right\}. \tag{7.48}$$

The implied constant is positive and depends only on $w_1, \dots, w_M$.

Let $w_1, \dots, w_M$ be b-approximable, where $b \colon [1, \infty) \to \mathbb{R}^+$ is an increasing function such that for every integer $q \geq 1$,

$$|qw_j - p_j w_1| \geq \frac{w_1}{b(q)} q^{-1/(M-1)}$$

for $j = 1, \dots, M$. Then the dimension-free region has the form

$$\left\{\sigma + it \colon \sigma \geq D - O\left(b^{-2}(w_1 t / 2\pi) t^{-2/(M-1)}\right)\right\}. \tag{7.49}$$

The O-terms in (7.48) and (7.49) are positive functions, bounded as indicated.

This has the following consequence for the Prime Orbit Theorem.

Theorem 7.41 (Prime Orbit Theorem with Error Term). *Suppose the weights $w_1, \dots, w_M$ are badly approximable, in the sense that*

$$|qw_j - p_j w_1| \gg q^{-1/(M-1)}$$

for $j = 1, \dots, M$ and every $q \geq 1$. Then

$$\psi_{\mathfrak{w}}(x) = \frac{x^D}{D} + O\left(x^D \left(\frac{\log\log x}{\log x}\right)^{\frac{M-1}{4}}\right), \tag{7.50}$$

as $x \to \infty$.

If $w_1, \dots, w_M$ is polynomially approximable, in the sense that

$$|qw_j - p_j w_1| \geq \frac{w_1}{b(q)} q^{-1/(M-1)}$$

for $j = 1, \dots, M$ and every $q \geq 1$, for some increasing function b on $[1, \infty)$ such that $b(x) = O(x^l)$ as $x \to \infty$, then

$$\psi_{\mathfrak{w}}(x) = \frac{x^D}{D} + O\left(x^D \left(\frac{\log\log x}{\log x}\right)^{\frac{M-1}{4l(M-1)+4}}\right), \tag{7.51}$$

as $x \to \infty$.

Proof of Theorems 7.37 and 7.41. We apply the pointwise explicit formula at level $k = 2$ (see Theorem 5.10) to obtain

$$\psi_{\mathfrak{w}}^{[2]}(x) = \frac{x^{D+1}}{D(D+1)} + \sum_{\omega \in \mathcal{D}_{\mathfrak{w}} \setminus \{D\}} \frac{x^{\omega+1}}{\omega(\omega+1)} + R^{[2]}(x).$$

The error term is estimated by $R^{[2]}(x) = O(x^{D+1-c})$ for some positive c. We will estimate the sum by using an argument which is classical in the theory of the Riemann zeta function and the Prime Number Theorem, under the assumptions that the complex dimensions ω have a linear density, and that every $\omega = \sigma + it$ satisfies

$$\sigma \leq D - Ct^{-\rho} \tag{7.52}$$

for some positive number ρ. Taking $\rho = 2/(M-1) + 2l$, we obtain Theorem 7.41, and Theorem 7.37 corresponds to the case when $M = 2$.

The sum $\sum_\omega \frac{x^{\omega+1}}{\omega(\omega+1)}$ is absolutely convergent. We split this sum into the parts with $|\operatorname{Im}\omega| > T$ and with $|\operatorname{Im}\omega| \leq T$. Put

$$U = \sum_\omega \frac{1}{|\omega(\omega+1)|}.$$

From the fact that the complex dimensions have a linear density (see Theorem 3.6), it follows that there exists a positive constant V such that

$$\sum_{|\operatorname{Im}\omega|\geq T} \frac{1}{|\omega(\omega+1)|} \leq \frac{V}{T}$$

for every $T > 0$. Then

$$\left| \sum_\omega \frac{x^{\omega+1}}{\omega(\omega+1)} \right| \leq U x^{D+1-CT^{-\rho}} + V \frac{x^{D+1}}{T},$$

where ρ is the constant introduced in (7.52) above. We now choose T so that the two terms on the right-hand side have the same order of magnitude, namely, $T = (\rho C \log x / \log\log x)^{1/\rho}$. Thus we find

$$\left| \sum_\omega \frac{x^{\omega+1}}{\omega(\omega+1)} \right| = O\left(x^{D+1} \left(\frac{\log\log x}{\log x} \right)^{1/\rho} \right).$$

We now apply a Tauberian argument to deduce a similar error estimate for $\psi_{\mathfrak{w}}(x)$; see [In, p. 64]. Let $h = x(\log\log x/\log x)^{1/(2\rho)}$. Thus

$$\psi_{\mathfrak{w}}(x) \leq \frac{1}{h} \int_x^{x+h} \psi_{\mathfrak{w}}(t)\, dt = \frac{\psi_{\mathfrak{w}}^{[2]}(x+h) - \psi_{\mathfrak{w}}^{[2]}(x)}{h}.$$

Observe that

$$\begin{aligned}
\frac{(x+h)^{D+1} - x^{D+1}}{hD(D+1)} &= \frac{x^D}{D} + O(x^{D-1}h) \\
&= \frac{x^D}{D} + x^D O\Big((\log\log x/\log x)^{1/(2\rho)} \Big).
\end{aligned}$$

Further,

$$O\Big(x^{D+1}(\log\log x/\log x)^{1/\rho}/h\Big) = x^D O\Big((\log\log x/\log x)^{1/(2\rho)}\Big),$$

from which the desired estimate follows. □

A nice exposition of Diophantine approximation in a related context and, in particular, of the notion of well approximable and badly approximable irrational numbers, can be found in the article [DodKr].

Remark 7.42. Note that by using the Tauberian argument, we lose a factor two in the exponent. Indeed, the estimate

$$\sum_{\omega\in\mathcal{D}_{\mathfrak{w}}\setminus\{D\}} \frac{x^\omega}{\omega} + R(x) = O\left(x^D\left(\frac{\log\log x}{\log x}\right)^{\frac{M-1}{2l(M-1)+2}}\right)$$

holds distributionally (instead of pointwise).

Remark 7.43. If $b(q)$ grows faster than polynomially, we obtain a bound of the form $x^D/b^{\mathrm{inv}}(\log x)$ for the error in the Prime Orbit Theorem, where b^{inv} is the inverse function of b.

Remark 7.44. If $l>0$ in the exponent $(M-1)/(4l(M-1)+4)$ of $\log x$ in the error term of Theorems 7.37 and 7.41, then the error term does not depend much on M, and is essentially of order $x^D(\log x)^{-1/4l}$ (ignoring the factor $\log\log x$). Thus, if the weights are well approximable, the error term is never better than x^D divided by a fixed power of the logarithm of x. On the other hand, when $l=0$—that is, roughly speaking, when the weights are never close to rational numbers—the error term is essentially of order $x^D(\log x)^{-(M-1)/4}$. Hence, the larger M, the smaller the error term in that case.

We may compare this—somewhat superficially in view of the Riemann hypothesis—with the situation of the Riemann zeta function. In view of Section 5.5.1, the weights are $w_p=\mathfrak{w}_{\mathrm{tot}}(p)=\log p$, for each prime number p, and there are infinitely many of them. Since it is expected that the infinite sequence $\{\log p\}_{p\colon\mathrm{prime}}$ is badly approximable, one expects an error term of order "$x^D(\log x)^{-\infty}$". Indeed, in (5.85), $e^{-c\sqrt{\log x}}=O\left((\log x)^{-n}\right)$ for every $n>0$. The corresponding pole-free region has width of order $O(1/\log t)$ at height t (see [In, Theorem 19]), which is "$t^{-1/\infty}$". This lends credibility to the conjecture that $\{\log p\}_{p:\mathrm{prime}}$ is badly approximable by rational numbers.

7.6 Notes

Definition 7.3: the dynamical zeta function (with weight) is often called in the literature the Ruelle (or Bowen–Ruelle) zeta function of the suspended flow. In the present form, it was introduced by Ruelle in [Rue1, 2]. Motivations for the study of such objects included number theory [ArMazu], dynamical systems [Sma, Bow1–2] and statistical physics [Rue3]. (See [Lag2] for a much more detailed discussion.) We refer the interested reader to the monographs by Parry and Pollicott [PaPol2] and by Ruelle [Rue4], as well as to the research expository articles by Baladi [Bal1–2] and Lagarias [Lag2], for a number of references on this subject and a detailed account of the theory of such zeta functions from several different points of view. See also, for example, [Bal3, BalKel, BedKS, Gou, Lal2–3] and the relevant references therein.

Theorem 7.30 was first published in [Lap-vF6]. In [PaPol1], Parry and Pollicott obtain a Prime Orbit Theorem (without error term) for suspended flows (see also [PaPol2, Chapter 6]). The first results of this kind were obtained in special cases by Huber [Hub], Sinai [Sin], and Margulis [Marg], among others (see [PaPol1, 2] and the relevant references therein, as well as the historical note in [BedKS, p. 154]). Parry and Pollicott derive the first term in the asymptotic expansion of the counting function of prime orbits, by applying the Wiener–Ikehara Tauberian Theorem to the logarithmic derivative of the dynamical zeta function. An alternate approach, based on a suitable nonlinear extension of the Renewal Theorem obtained in [Lal3], is taken by Lalley in [Lal2–3] for a class of approximately self-similar flows. See also Remarks 7.32–7.34. In our setting, thanks to the explicit formulas developed in Chapter 5, we obtain a full expansion over the complex dimensions of the flow, and using the Diophantine approximation techniques of Chapter 3, a Prime Orbit Theorem with error term.

8

Tubular Neighborhoods and Minkowski Measurability

In this chapter, we obtain (in Section 8.1) a distributional formula for the volume of the tubular neighborhoods of the boundary of a fractal string, called a *tube formula*. In Section 8.1.1, under more restrictive assumptions, we also derive a tube formula that holds pointwise. In Section 8.3, we then deduce from these formulas a new criterion for the Minkowski measurability of a fractal string, in terms of its complex dimensions. Namely, under suitable assumptions, we show that a fractal string is Minkowski measurable if and only if it does not have any nonreal complex dimensions of real part D, its Minkowski dimension. This completes and extends the earlier criterion obtained in [LapPo1–2].

In Section 8.4, we illustrate our results by extensively discussing the class of self-similar strings. In particular, we obtain specific tube formulas in the lattice case and carefully study the error term in the nonlattice case.

The results obtained in this chapter provide further insight into the geometric meaning of the notion of complex dimension and suggest new analogies between aspects of fractal and Riemannian geometry. See, in particular, Section 8.2.

This chapter makes use of the notions of Minkowski dimension and measurability, which are introduced at the beginning of Chapter 1 in Section 1.1. Also, in Sections 1.1.2 and 2.3.2 we directly computed the volume of the tubular neighborhood of the Cantor string and the Fibonacci string. We recover the resulting tube formulas from our general results in Sections 8.4.1 and 8.4.2.

8.1 Explicit Formulas for the Volume of Tubular Neighborhoods

Let $\mathcal{L}$ be a standard fractal string given by the sequence of lengths $(l_j)_{j=1}^{\infty}$ and of Minkowski dimension $D \in (0,1)$. Let $\eta = \sum_{j=1}^{\infty} \delta_{\{l_j^{-1}\}}$ be the associated measure, as explained in Chapter 4.

Given $\varepsilon > 0$, let $V(\varepsilon)$ denote the volume of the ε-neighborhood of the boundary of $\mathcal{L}$, as defined by (1.3). By formula (1.9), the volume is given by

$$V(\varepsilon) = \sum_{j\,:\, l_j \geq 2\varepsilon} 2\varepsilon + \sum_{j\,:\, l_j < 2\varepsilon} l_j \tag{8.1}$$

(see also [LapPo1], [LapPo2, Eq. (3.2)]). This formula defines the volume in a very straightforward manner, and can be directly used to obtain a closed form for $V(\varepsilon)$ in such simple examples as the Cantor string and the Fibonacci string, as we did in Sections 1.1.2 and 2.3.2, respectively. In general, however, this formula does not reveal oscillations in the tubular neighborhoods of a fractal string as ε varies. Such oscillations are revealed by the explicit formulas for $V(\varepsilon)$ to be obtained in this chapter. Therefore, we rewrite (8.1) as

$$\begin{aligned} V(\varepsilon) &= \int_0^{\frac{1}{2\varepsilon}} 2\varepsilon\, \eta(dx) + \int_{\frac{1}{2\varepsilon}}^{\infty} \frac{1}{x}\, \eta(dx) \\ &= \left\langle \mathcal{P}_\eta^{[0]}, v_\varepsilon \right\rangle, \end{aligned} \tag{8.2}$$

where $v_\varepsilon(x)$ is the function defined on $(0,\infty)$ by

$$v_\varepsilon(x) = \begin{cases} 2\varepsilon & \text{for } x \leq (2\varepsilon)^{-1}, \\ 1/x & \text{for } x > (2\varepsilon)^{-1}. \end{cases} \tag{8.3}$$

We assume that $\zeta_{\mathcal{L}}$ has a meromorphic continuation to some neighborhood of the closed half-plane $\{s\colon \operatorname{Re} s \geq D\}$ and that the resulting function satisfies **L1** and **L2** for some real exponent κ; i.e., the fractal string $\mathcal{L}$ is languid, in the sense of Definition 5.2. Our extended distributional formula (Theorem 5.26) does not directly apply to the present situation since the test function v_ε is not sufficiently differentiable. Indeed, in view of (8.3), the function v_ε is clearly continuous but not differentiable. Therefore we interpret $V(\varepsilon)$ itself as a distribution. However, a more careful analysis in Section 8.1.1 allows us to derive an explicit formula for $V(\varepsilon)$ that holds pointwise (under somewhat stronger hypotheses, see Theorem 8.7).

Recall that a fractal string is languid if its geometric zeta function satisfies certain growth conditions along a screen S. Thus a screen will not pass through any of the complex dimensions of the string. To be useful, S has to lie to the left of the line $\operatorname{Re} s = D$. For the next theorem, it must also avoid the point $s = 0$.

Theorem 8.1 (The distributional tube formula). *Let $\mathcal{L}$ be languid for some real exponent κ and a screen that does not pass through* 0. *Then the volume of the (one-sided) tubular neighborhood of radius ε of the boundary of $\mathcal{L}$ is given by the following distributional explicit formula, on test functions in* $\mathbf{D}(0,\infty)$:[1]

$$V(\varepsilon)=\sum_{\omega\in\mathcal{D}_{\mathcal{L}}(W)}\operatorname{res}\left(\frac{\zeta_{\mathcal{L}}(s)(2\varepsilon)^{1-s}}{s(1-s)};\omega\right)+\{2\varepsilon\zeta_{\mathcal{L}}(0)\}+\mathcal{R}(\varepsilon), \tag{8.4}$$

where the term in braces is only included if $0\in W\backslash\mathcal{D}_{\mathcal{L}}(W)$, and $\mathcal{R}(\varepsilon)$ is the distributional error term, given by

$$\mathcal{R}(\varepsilon)=\frac{1}{2\pi i}\int_S(2\varepsilon)^{1-s}\zeta_{\mathcal{L}}(s)\,\frac{ds}{s(1-s)}. \tag{8.5}$$

It is estimated, in the sense of Definition 5.29 of Section 5.4.2, by

$$\mathcal{R}(\varepsilon)=O\big(\varepsilon^{1-\sup S}\big), \tag{8.6}$$

as $\varepsilon\to0^+$ (here, $\sup S$ is given by Equation (5.17b)).

Moreover, if $\mathcal{L}$ is strongly languid[2] *then we may choose $W=\mathbb{C}$ and then we have no error term (i.e., $\mathcal{R}(\varepsilon)\equiv0$), provided we apply this formula to a test function supported on a compact subset of $[0,1/(2A))$. Again, the term $2\varepsilon\zeta_{\mathcal{L}}(0)$ is included if* 0 *belongs to W and is not a complex dimension of $\mathcal{L}$.*

Proof. Let $\varphi(\varepsilon)$ be a smooth, compactly supported test function on $(0,\infty)$; i.e., $\varphi\in\mathbf{D}(0,\infty)$. Then, by (8.3),

$$\begin{aligned}\int_0^\infty\varphi(\varepsilon)v_\varepsilon(x)\,d\varepsilon&=\int_0^{\frac{1}{2x}}2\varepsilon\varphi(\varepsilon)\,d\varepsilon+\frac{1}{x}\int_{\frac{1}{2x}}^\infty\varphi(\varepsilon)\,d\varepsilon\\&=\varphi_1(x)+\varphi_2(x),\end{aligned} \tag{8.7}$$

where φ_1 and φ_2 are smooth (but not compactly supported) test functions, given by

$$\varphi_1(x):=\int_0^{\frac{1}{2x}}2\varepsilon\varphi(\varepsilon)\,d\varepsilon \tag{8.8a}$$

and

$$\varphi_2(x):=\frac{1}{x}\int_{\frac{1}{2x}}^\infty\varphi(\varepsilon)\,d\varepsilon. \tag{8.8b}$$

[1] $\mathbf{D}(0,\infty)$ is the space of C^∞ functions with compact support contained in $(0,\infty)$; see also Remark 5.21.

[2] i.e., it satisfies (for some real exponent κ) the stronger hypotheses **L1** and **L2′** of Definition 5.3 with $W=\mathbb{C}$ and some constant $A>0$ in Equation (5.21), rather than **L1** and **L2**.

Thus, in light of (8.2) and (8.7),

$$\begin{aligned}\langle V(\varepsilon), \varphi\rangle &= \int_0^\infty \varphi(\varepsilon) \int_0^\infty v_\varepsilon(x)\,\eta(dx)\,d\varepsilon \\ &= \left\langle \mathcal{P}_\eta^{[0]}, \varphi_1 + \varphi_2 \right\rangle.\end{aligned}$$

Since φ is compactly supported, the function φ_1 is constant near 0 (with value $\int_0^\infty 2\varepsilon\varphi(\varepsilon)\,d\varepsilon$), and it vanishes for $x \gg 0$. Similarly, the function φ_2 vanishes near 0 and is equal to $\frac{1}{x}\int_0^\infty \varphi(\varepsilon)\,d\varepsilon$ for $x \gg 0$.[3] Therefore, the asymptotic expansions (5.53a) and (5.53b) are satisfied, with $k = 0$ and for any fixed integer $q > \kappa + 1$ (e.g., $q = [\kappa + 2]$). Hence the explicit formula of Theorem 5.26, the extended distributional formula with error term, applies to φ_1 and φ_2, with $k = 0$ and $q > \kappa + 1$.

Using Definition (5.43) of the Mellin transform, we find, for $\operatorname{Re} s > 0$, that

$$\widetilde{\varphi_1}(s) = \frac{2^{1-s}}{s}\,\widetilde{\varphi}(2 - s) \tag{8.9a}$$

and, for $\operatorname{Re} s < 1$,

$$\widetilde{\varphi_2}(s) = \frac{2^{1-s}}{1-s}\,\widetilde{\varphi}(2 - s). \tag{8.9b}$$

Thus we obtain a priori for $0 < \operatorname{Re} s < 1$,

$$\widetilde{(\varphi_1 + \varphi_2)}(s) = \widetilde{\varphi_1}(s) + \widetilde{\varphi_2}(s) = \frac{2^{1-s}}{s(1-s)}\,\widetilde{\varphi}(2 - s). \tag{8.10}$$

After analytic continuation, formulas (8.9) and (8.10) are valid for all $s \in \mathbb{C}$. Since the function $\widetilde{\varphi}(z)$ is entire (because φ is compactly supported), the Mellin transform of $\varphi_1 + \varphi_2$ is meromorphic on $\mathbb{C}$ with simple poles at 0 and 1. Also $\widetilde{\varphi_1}$ and $\widetilde{\varphi_2}$ are meromorphic on $\mathbb{C}$, with simple poles at 0 and 1, respectively.

In particular, by formula (8.9a), $\widetilde{\varphi_1}(s)$ has a pole at $s = 0$, and no other poles. If 0 is not a complex dimension, this pole gives the term $2\varepsilon\zeta_{\mathcal{L}}(0)$, to be included if $0 \in W$. Applying the distribution $\mathcal{P}_\eta^{[0]} = \eta$ to $\varphi_1 + \varphi_2$ in Equation (5.54a), we deduce from Theorem 5.26 (applied at level $k = 0$,

---

[3]Indeed, if the support of φ is contained in the compact interval $[a, b]$, with $a > 0$, then $\varphi_1(x) = 0$ and $\varphi_2(x) = \frac{1}{x}\int_0^\infty \varphi(\varepsilon)\,d\varepsilon$ for $x \geq 1/2a$, and $\varphi_1(x) = \int_0^\infty 2\varepsilon\varphi(\varepsilon)\,d\varepsilon$ and $\varphi_2(x) = 0$ for $x \in (0, 1/2b)$.

see Remark 5.28),

$$\begin{aligned}\langle V(\varepsilon),\varphi\rangle &= \sum_{\omega\in\mathcal{D}_{\mathcal{L}}(W)\backslash\{0\}} \operatorname{res}\left(\zeta_{\mathcal{L}}(s)\widetilde{(\varphi_1+\varphi_2)}(s);\omega\right) \\ &\quad + \left\{\operatorname{res}\left(\zeta_{\mathcal{L}}(s)\widetilde{(\varphi_1+\varphi_2)}(s);0\right)\right\} + \mathcal{R}_{\eta}^{[0]}(\varphi_1+\varphi_2) \\ &= \int_0^\infty \sum_{\omega\in\mathcal{D}_{\mathcal{L}}(W)\backslash\{0\}} \operatorname{res}\left(\zeta_{\S}tring(s)\frac{(2\varepsilon)^{1-s}}{s(1-s)};\omega\right)\varphi(\varepsilon)d\varepsilon \qquad (8.11) \\ &\quad + \left\{\int_0^\infty \operatorname{res}\left(\zeta_{\mathcal{L}}(s)\frac{(2\varepsilon)^{1-s}}{s(1-s)};0\right)\varphi(\varepsilon)d\varepsilon\right\} + \int_0^\infty \mathcal{R}(\varepsilon)\varphi(\varepsilon)d\varepsilon,\end{aligned}$$

where the second equality follows from (8.10) and the linearity of the residue. Finally, observe that (in view of (8.3) and (5.43))

$$\widetilde{v_\varepsilon}(s) = \frac{1}{s(1-s)}(2\varepsilon)^{1-s}, \tag{8.12}$$

a priori for $0 < \operatorname{Re} s < 1$, but after meromorphic continuation for all $s \in \mathbb{C}$. Comparing (5.45a) and (5.45b) of Theorem 5.18, using formula (5.44), we see that Equation (8.11) is equivalent to the equality of distributions (8.4), as desired. This completes the proof of the distributional tube formula, with the error term $\mathcal{R}(\varepsilon)$ given by (8.5). The error estimate (8.6) follows from the first part of Theorem 5.30 (applied with $k = 0$ and the variable $x = 1/\varepsilon$).

Finally, the fact that $\mathcal{R}(\varepsilon)$ vanishes identically when $\mathcal{L}$ is strongly languid (i.e., $\mathcal{L}$ satisfies the stronger hypothesis **L2′**) follows in an entirely analogous fashion from the corresponding extended distributional explicit formula without error term given in Theorem 5.27 (and applied with $k = 0$ and $q > \max\{1, \kappa\}$). This concludes the proof of Theorem 8.1. □

Remark 8.2. Theorem 8.1 is valid without change for a generalized fractal string η (rather than just an ordinary fractal string $\mathcal{L}$), in which case $V(\varepsilon)$ is defined by formula (8.2),[4] which was the starting point of the proof of the theorem. In formulating the counterpart of Theorem 8.1 and Corollary 8.3, one must then take into account the fact that 1 may also be a complex dimension.

If the complex dimension ω (and hence also $\overline{\omega}$) is simple and different from 0,[5] then the associated local term in formula (8.4) is equal to

$$\operatorname{res}\left(\zeta_{\mathcal{L}}(s);\omega\right)\frac{(2\varepsilon)^{1-\omega}}{\omega(1-\omega)}. \tag{8.13}$$

We thus obtain the following corollary of Theorem 8.1.

---

[4]This is used, in particular, to study the truncated Cantor string in Section 10.3.

[5]Note that $\omega \neq 1$ since $\operatorname{Re}\omega \leq D < 1$ for an ordinary fractal string.

Corollary 8.3. *If in Theorem 8.1, we assume in addition that all the visible complex dimensions of $\mathcal{L}$ are simple, then the distributional tube formula* (8.4) *becomes*

$$V(\varepsilon) = \sum_{\omega \in \mathcal{D}_{\mathcal{L}}(W) \setminus \{0\}} \operatorname{res}\left(\zeta_{\mathcal{L}}(s); \omega\right) \frac{(2\varepsilon)^{1-\omega}}{\omega(1-\omega)} + \left\{2\varepsilon(1 - \log(2\varepsilon)) \operatorname{res}\left(\zeta_{\mathcal{L}}(s); 0\right) + 2\varepsilon\zeta_{\mathcal{L}}(0)\right\}_{0 \in W} + \mathcal{R}(\varepsilon), \quad (8.14)$$

where the distributional error term $\mathcal{R}(\varepsilon)$, in case $\mathcal{L}$ is languid, is given by formula (8.5) *and estimated by* (8.6). *Here, in case $\zeta_{\mathcal{L}}$ has a (simple) pole at $s = 0$, the notation $\zeta_{\mathcal{L}}(0)$ means the constant term in the Laurent expansion of $\zeta_{\mathcal{L}}$ around $s = 0$, and if $\zeta_{\mathcal{L}}$ does not have a pole at $s = 0$, then* $\operatorname{res}(\zeta_{\mathcal{L}}(s); 0) = 0$. *The braces indicate that these terms are only included if* 0 *is visible (i.e., if $0 \in W$). (See Remark 8.11 for further discussion.)*

Moreover if $\mathcal{L}$ is assumed to be strongly languid rather than languid, then we may take $W = \mathbb{C}$, and the error term $\mathcal{R}(\varepsilon)$ vanishes identically. As in the statement of the second part of Theorem 8.1, this formula applies to test functions supported on a compact subset of $[0, 1/2A)$.

For example, this corollary applies to (lattice and nonlattice) self-similar strings, choosing a suitable screen close enough to the line $\operatorname{Re} s = D$, since by Theorem 2.17, all complex dimensions on or sufficiently close to this line are simple. In view of Theorem 3.30, it also applies to all nonlattice self-similar strings with exactly two distinct scaling ratios (with multiplicities). Indeed, in that case, all the complex dimensions of the string are simple.

Remark 8.4. Note that the sum over the complex dimensions in (8.14) can be written as

$$\sum_{\omega \in \mathcal{D}_{\mathcal{L}}(W) \setminus \{0\}} \operatorname{res}\left(\zeta_{\mathcal{L}}(s); \omega\right) \frac{(2\varepsilon)^{1-\omega}}{\omega} + \sum_{\omega \in \mathcal{D}_{\mathcal{L}}(W) \setminus \{0\}} \operatorname{res}\left(\zeta_{\mathcal{L}}(s); \omega\right) \frac{(2\varepsilon)^{1-\omega}}{1-\omega}.$$

Both these sums must be interpreted distributionally because they are usually pointwise divergent.

Remark 8.5. In view of (5.17b) and (5.24), we have $\sup S \leq D$, and so estimate (8.6) implies that $\mathcal{R}(\varepsilon) = O\left(\varepsilon^{1-D}\right)$, as $\varepsilon \to 0^+$, both in Theorem 8.1 and Corollary 8.3. We point out that under the stronger assumptions of Theorem 8.15 below, we actually have the better estimate $\mathcal{R}(\varepsilon) = o\left(\varepsilon^{1-D}\right)$, as $\varepsilon \to 0^+$. This follows from Theorem 5.31 and from the second part of Theorem 5.30.

Remark 8.6. Since a self-similar string satisfies hypothesis **L2′**, we can let $W = \mathbb{C}$, as was seen at the beginning of Section 6.4. We refer to Sections 8.4.2 and 8.4.4 below for the precise form of the corresponding tube formula (8.4) in the lattice and nonlattice cases.

8.1.1 The Pointwise Tube Formula

In this section, we obtain a formula for the volume of the tubular neighborhood of a fractal string that holds pointwise. This pointwise formula holds under somewhat more restrictive assumptions than those for Theorem 8.1, the distributional formula for the volume of the tubular neighborhood.

Our starting point is Lemma 5.1 with $k = 1$, i.e., the identity

$$H^{[1]}(x-y) = \lim_{T\to\infty} \int_{c-iT}^{c+iT} x^s y^{-s} \frac{ds}{2\pi i s}, \tag{8.15}$$

for $c > 0$. Recall from the defining formula (5.10) that the Heaviside function is given by

$$H^{[1]}(x-y) = \begin{cases} 0 & \text{if } x < y, \\ 1/2 & \text{if } x = y, \\ 1 & \text{if } x > y. \end{cases} \tag{8.16}$$

The pointwise counterpart of Theorem 8.1 above is the following result (again, the screen cannot pass through 0):

Theorem 8.7 (The pointwise tube formula). *Let $\mathcal{L}$ be a languid fractal string of dimension $D < 1$ and with exponent $\kappa < 1$ in hypotheses* **L1** *and* **L2**, *for a screen that does not pass through* 0. *Then, for every $\varepsilon > 0$, the volume of the (one-sided) tubular neighborhood of radius ε of the boundary of $\mathcal{L}$ is given by the pointwise explicit formula with error term*

$$V(\varepsilon) = \sum_{\omega\in\mathcal{D}_{\mathcal{L}}(W)} \operatorname{res}\left(\frac{\zeta_{\mathcal{L}}(s)(2\varepsilon)^{1-s}}{s(1-s)};\omega\right) + \{2\varepsilon\zeta_{\mathcal{L}}(0)\} + R(\varepsilon), \tag{8.17}$$

where the term in braces is only included if $0 \in W\backslash\mathcal{D}_{\mathcal{L}}(W)$. The pointwise error term, $R(\varepsilon)$, is given by the following absolutely convergent contour integral:

$$R(\varepsilon) = \frac{1}{2\pi i}\int_S (2\varepsilon)^{1-s}\zeta_{\mathcal{L}}(s)\,\frac{ds}{s(1-s)}. \tag{8.18}$$

It is estimated pointwise by

$$R(\varepsilon) = O\left(\varepsilon^{1-\sup S}\right), \tag{8.19}$$

as $\varepsilon \to 0^+$ (here, $\sup S$ is given by Equation (5.17b)*).*

Moreover, if $\mathcal{L}$ is strongly languid[6] *with $\kappa < 2$, then we can take $W = \mathbb{C}$ and the explicit formula* (8.17) *holds for all positive $\varepsilon < 1/2A$, and it is exact (i.e., $R(\varepsilon) \equiv 0$). Further, the term $2\varepsilon\zeta_{\mathcal{L}}(0)$ is included if 0 is not a complex dimension of $\mathcal{L}$.*

---

[6] i.e., $\zeta_{\mathcal{L}}$ satisfies **L1** of Definition 5.2 and **L2**$'$ of Definition 5.3, for some constant $A > 0$ and real exponent κ.

Proof. Let the fractal string $\mathcal{L}$ be given by $\mathcal{L} = \{l_j\}_{j=1}^{\infty}$. Then, by (8.16), we have that $H^{[1]}(l_j - 2\varepsilon)$ equals 1 for $l_j > 2\varepsilon$ and $1/2$ for $l_j = 2\varepsilon$. Likewise, $H^{[1]}(2\varepsilon - l_j)$ equals 1 for $l_j < 2\varepsilon$ and $1/2$ for $l_j = 2\varepsilon$. By formula (8.1) and Equation (8.15), we obtain for $0 < c < 1$ that

$$\begin{aligned} V(\varepsilon) &= \sum_{j=1}^{\infty} \left(2\varepsilon H^{[1]}\left(l_j - 2\varepsilon\right) + l_j H^{[1]}\left(2\varepsilon - l_j\right) \right) \\ &= \sum_{j=1}^{\infty} \lim_{T \to \infty} \left(\int_{c-iT}^{c+iT} l_j^s (2\varepsilon)^{1-s} \frac{ds}{2\pi i s} + \int_{1-c-iT}^{1-c+iT} l_j^{1-t} (2\varepsilon)^t \frac{dt}{2\pi i t} \right). \end{aligned}$$

In the second integral, we substitute $t = 1 - s$ and then combine the two integrals into one that converges absolutely to obtain

$$V(\varepsilon) = \sum_{j=1}^{\infty} \int_{c-i\infty}^{c+i\infty} l_j^s (2\varepsilon)^{1-s} \frac{ds}{2\pi i s(1-s)}.$$

For $c > D$, the integrand is bounded by a constant multiple of

$$\sum_{j=1}^{\infty} \int_{-\infty}^{\infty} l_j^c (2\varepsilon)^{1-c} \frac{dt}{1+t^2} = \pi \zeta_{\mathcal{L}}(c) (2\varepsilon)^{1-c}.$$

Therefore, we can interchange the sum and the integral. For $D < c < 1$, we thus obtain the following expression for $V(\varepsilon)$ as an absolutely convergent integral,

$$V(\varepsilon) = \int_{c-i\infty}^{c+i\infty} \zeta_{\mathcal{L}}(s) (2\varepsilon)^{1-s} \frac{ds}{2\pi i s(1-s)}.$$

Using the sequence $\{T_n\}_{n \in \mathbb{Z}}$ of (5.18) and of hypothesis **L1** in (5.19), we truncate this integral to obtain

$$V(\varepsilon) = \lim_{n \to \infty} \int_{c-iT_{-n}}^{c+iT_n} \zeta_{\mathcal{L}}(s) (2\varepsilon)^{1-s} \frac{ds}{2\pi i s(1-s)}.$$

By the Residue Theorem, with the notation of Lemma 5.9 (the truncated pointwise formula), we obtain, after moving the line of integration to the truncated screen $S_{|n}$, that the integral equals

$$\sum_{\omega \in \mathcal{D}_{\mathcal{L}}(W_{|n})} \operatorname{res} \left(\frac{\zeta_{\mathcal{L}}(s)(2\varepsilon)^{1-s}}{s(1-s)}; \omega \right) + \{2\varepsilon \zeta_{\mathcal{L}}(0)\} + R_{|n}(\varepsilon) + U_n + L_n,$$

where $R_{|n}(\varepsilon)$ is given by the integral over the truncated screen $S_{|n}$, and U_n and L_n are given by integrals over the upper and lower parts of $S_{|n}$ (see Figure 5.1). Note that as in formula (8.17), the term between braces is included only if $0 \in W \backslash \mathcal{D}_{\mathcal{L}}(W)$.

If **L2** is satisfied with $\kappa < 1$, then $R_{|n}(\varepsilon)$ converges as $n \to \infty$ to $R(\varepsilon)$, the integral over the screen. Moreover, since $\kappa < 2$, the upper and lower terms U_n and L_n vanish as $n \to \infty$.

Finally, if **L2**$'$ is satisfied with $\kappa < 2$, then we first take the limit as $m \to \infty$ over the truncated screens $S_{m|n}$, keeping n fixed. Here, $\{S_m\}_{m=1}^{\infty}$ is the sequence of screens occurring in the statement of hypothesis **L2**$'$ (see Definition 5.3), and $S_{m|n}$ is the screen S_m, truncated between T_{-n} and T_n, as in Lemma 5.9. Provided that $\varepsilon < 1/(2A)$, the limit as $m \to \infty$ of the integrals over $S_{m|n}$ vanishes. Then the limit as $n \to \infty$ of U_n and L_n vanishes since $\kappa < 2$. □

Remark 8.8. Recall from Remark 5.4 that if $\mathcal{L}$ is strongly languid for some value of κ, then it is strongly languid for every larger value of κ, but not necessarily for any smaller value. Consequently, the assumptions of the second part of Theorem 8.7 (the pointwise tube formula without error term) do not imply those of the first part (i.e., its counterpart with error term).

This is related to the fact that for the geometric counting function $N^{[1]}(x)$ or for $V(\varepsilon)$ itself, we cannot apply the pointwise explicit formula with error term to the Cantor string (since the Cantor string is languid with $\kappa = 0$, the conditions of Theorem 5.10 are not satisfied for $k = 1$), but we do have a pointwise formula without error term (because the conditions of Theorem 5.14 are satisfied for $k = 1$ and $\kappa = 0$; see Sections 1.1–1.2, 6.4.1 and 8.4.1). Indeed, to prove the second part of Theorem 8.7, we first take the limit for $m \to \infty$; i.e., we push the truncated screens away to the left, which works provided $\kappa < 2$, and we then take the limit for $n \to \infty$; that is, we remove the truncation. Since the truncation at this stage of the proof only applies to the complex dimensions and not to the screen, the condition that the integral over the screen converges, $\kappa < 1$, is not needed.

Remark 8.9. Theorem 8.1 holds without any condition on κ. Therefore, the conditions of (the first or second part of) Theorem 8.7 imply the respective conditions of Theorem 8.1.

In view of (8.13) and the sentence preceding it, we can deduce from Theorem 8.7 the following exact pointwise counterpart of Corollary 8.3 (which is itself a consequence of Theorem 8.1).

Corollary 8.10. *If in Theorem 8.7, we assume in addition that all the visible complex dimensions of $\mathcal{L}$ are simple, then the sum over the complex dimensions in the pointwise tube formula* (8.17) *becomes*

$$\begin{aligned} V(\varepsilon) = &\sum_{\omega \in \mathcal{D}_{\mathcal{L}}(W) \setminus \{0\}} \operatorname{res}(\zeta_{\mathcal{L}}(s); \omega) \frac{(2\varepsilon)^{1-\omega}}{\omega(1-\omega)} \\ &+ \{2\varepsilon(1 - \log(2\varepsilon)) \operatorname{res}(\zeta_{\mathcal{L}}(s); 0) + 2\varepsilon\zeta_{\mathcal{L}}(0)\}_{0 \in W} + R(\varepsilon), \qquad (8.20) \end{aligned}$$

where the pointwise error term $R(\varepsilon)$, in case $\mathcal{L}$ is languid, is given by formula (8.18) *and is estimated by* (8.19)*. As in Corollary 8.3, the term in braces is included only if* 0 *is visible. Here, if $\zeta_{\mathcal{L}}$ has a pole at $s = 0$, the notation $\zeta_{\mathcal{L}}(0)$ denotes the constant term in the Laurent expansion around $s = 0$.*

Moreover, if $\mathcal{L}$ is assumed to be strongly languid with $\kappa < 2$ rather than languid, as in the second part of Theorem 8.7, then we may take $W = \mathbb{C}$, and the error term $R(\varepsilon)$ vanishes identically in formula (8.20).

Remark 8.11. Note that in Corollaries 8.3 and 8.10, the condition for the inclusion of the term in braces is different from the one in Theorems 8.1 and 8.7: it is included if 0 is visible (that is, if $0 \in W$). The reason is that by assumption, the pole of $\zeta_{\mathcal{L}}(s)/s$ at $s = 0$ is at most of the second order, and the expression between the braces $\{\dots\}_{0 \in W}$ gives an explicit determination of the contribution of this pole to the explicit formula, if 0 is visible. On the other hand, in the formulas (8.4) and (8.17), the pole of $\zeta_{\mathcal{L}}(s)/s$ at $s = 0$ could be of any order, and we only include the explicit determination of this term in the explicit formula in case this pole is simple; i.e., in case 0 is visible but not a complex dimension.

8.1.2 Example: The a-String

Let $\mathcal{L}$ be the a-string studied in Section 6.5.1. Recall from Theorem 6.20 that $\mathcal{L}$ has Minkowski dimension $D = \frac{1}{a+1}$ and that its complex dimensions are all simple, real, and located at D and possibly at $-D, -2D, \dots$. Also, the residue of $\zeta_{\mathcal{L}}$ at D is equal to a^D. Hence, by Corollary 8.3, we have the distributional formula, for every fixed integer $J \geq 0$,

$$V(\varepsilon) = \frac{(2\varepsilon)^{1-D}}{D(1-D)} a^D - \sum_{j=1}^{J} \frac{(2\varepsilon)^{1+jD}}{jD(1+jD)} \operatorname{res}(\zeta_{\mathcal{L}}; -jD) + O\big(\varepsilon^{1+(J+1)D}\big), \tag{8.21}$$

as $\varepsilon \to 0^+$. (For $J = 0$, the sum is interpreted as 0.) Using Corollary 8.10, but only for $J = 0$, we derive the same formula interpreted pointwise, and with an error term $O\big(\varepsilon^{1+D/2-\delta}\big)$ for every $\delta > 0$, since $\kappa < 1$ is satisfied only for screens to the right of the line $\operatorname{Re} s = -\frac{D}{2} = -\frac{1}{2(a+1)}$.

In particular, it follows that $\mathcal{L}$ is Minkowski measurable (this will be shown in Section 8.3 below) with Minkowski content

$$\mathcal{M} = \frac{2^{1-D}}{D(1-D)} a^D. \tag{8.22}$$

Since D is simple and is the only pole of $\zeta_{\mathcal{L}}$ above D, this is in agreement with Theorem 8.15 below. We stress, however, that our tube formula (8.21) gives much more precise information about $V(\varepsilon)$.

8.2 Analogy with Riemannian Geometry

There is an interesting analogy between formula (8.4) (or (8.17)) and H. Weyl's formula [Wey3] for the volume of the tubular ε-neighborhood of a compact, n-dimensional Riemannian submanifold of Euclidean space $\mathbb{R}^d$. Indeed, by [BergGo, Theorem 6.9.9, p. 235], this volume is given by the formula

$$V_M(\varepsilon) = a_0\varepsilon^{d-n} + a_1\varepsilon^{d-n+1} + \cdots + a_n\varepsilon^d. \tag{8.23}$$

When $V_M(\varepsilon)$ is the volume of the two-sided ε-neighborhood, the odd-numbered coefficients $a_1, a_3, \ldots$ all vanish, and for $j = 0, 1, 2, \ldots$, the even numbered coefficient a_{2j} is given by

$$a_{2j} = \frac{1}{d-n+2j}\int_M K_{2j}\delta, \tag{8.24}$$

where K_{2j} is a universal j-th degree polynomial in the curvature tensor of M, and δ is the canonical density on M (see [BergGo, Proposition 6.6.1, p. 214]). Thus, formula (8.23) is a polynomial in ε whose coefficients (8.24) are expressed in terms of the Weyl curvatures in different (integer) dimensions.

If $V_M(\varepsilon)$ is the volume of the one-sided ε-neighborhood (which corresponds more closely to our definition of $V(\varepsilon)$), then the counterpart of (8.23) still holds. In that case, the odd coefficients do not necessarily vanish. We refer to [BergGo], Sections 6.6–6.9, and especially Section 6.9.8 and Theorem 6.9.9, and also to [Gra], for a much more detailed discussion and various helpful examples. For additional references, see also the endnotes in Section 8.5.

In the explicit formula (8.4) (or (8.17)), the volume $V(\varepsilon)$ is expressed as a sum of terms $\varepsilon^{1-\omega}$, where ω ranges over the visible complex dimensions. This suggests that the complex dimensions of fractal strings and the associated residues[7] could have a direct geometric interpretation. The work in progress [Pe] and [LapPe2–4], developed while this book was almost completed, indicates a close relationship to some aspects of geometric measure theory [Fed1, 2] and a possible new notion of fractal curvature measures indexed by the underlying complex dimensions, both in the present one-dimensional context of fractal strings and in higher dimensions. See Section 12.3.2 for a brief discussion of [Pe] and [LapPe2–4].

Remark 8.12. The Weyl curvatures occurring in Weyl's tube formula were shown in [Wey3] to be intrinsic to the submanifold. One of them is closely tied to the notion of Euler characteristic and hence to the Gauss–Bonnet formula (or its higher-dimensional counterpart, due to Chern).

---

[7] Or, more generally, the Laurent expansions at these poles, when they have higher multiplicities.

8.3 Minkowski Measurability and Complex Dimensions

In Theorem 8.15 of this section, we obtain a new criterion for the Minkowski measurability of the boundary of a fractal string $\mathcal{L}$, expressed in terms of the complex dimensions of $\mathcal{L}$. This completes and extends (under our present assumptions) the earlier criterion obtained by the first author and C. Pomerance in [LapPo1–2]. In Remarks 8.19 and 8.20, we will also comment on the relationship between our new criterion and the one obtained previously in [LapPo1–2].

Before stating and proving our main result, Theorem 8.15, we need to establish a technical lemma, which is an extension of Theorem 1.16 of Chapter 1.

Lemma 8.13. *Let $\mathcal{L}$ be a generalized fractal string, given by a (local) positive measure η. If the pole of $\zeta_{\mathcal{L}}$ at D is of order $m \geq 1$, then any pole at $D + it$ (with $t \in \mathbb{R}$) is of order at most m.*

Proof. Let $\operatorname{Re} s = \sigma > D$. Since $|\zeta_{\mathcal{L}}(s)| \leq \zeta_{\mathcal{L}}(\sigma)$, we deduce that the function $(\sigma - D)^m \, \zeta_{\mathcal{L}}(s)$ is bounded as $\sigma \to D^+$. □

Remark 8.14. It follows from the $m = 1$ case of Lemma 8.13 that if D is a simple complex dimension of $\mathcal{L}$, then every other complex dimension of $\mathcal{L}$ on the vertical line $\operatorname{Re} s = D$ is also simple. For example, this is consistent with the fact—established in Theorem 2.17—that for a lattice self-similar string $\mathcal{L}$, its dimension D, together with all the complex dimensions above D—namely, $D + in\mathbf{p}$, with $n \in \mathbb{Z}$, where $\mathbf{p}$ is the oscillatory period of $\mathcal{L}$—are simple.

The next result—which follows in particular from the tube formula given in Theorem 8.1 (in conjunction with Theorem 5.31)—extends and puts in a more conceptual framework the criterion for Minkowski measurability of fractal strings obtained by M. L. Lapidus and C. Pomerance in [LapPo1] and [LapPo2, Theorem 2.2, p. 46]. Note that statement (i) of Theorem 8.15 does not appear in [LapPo2].[8] We refer to Remarks 8.18–8.21 below for further comments about our Minkowski measurability criterion and about its relationship with earlier work, in particular [LapPo1–2].

Recall from Section 1.1 that $\mathcal{L}$ is Minkowski measurable if the limit

$$\lim_{\varepsilon \to 0^+} V(\varepsilon)\varepsilon^{D-1}$$

exists and lies in $(0, \infty)$. Then, necessarily, D coincides with the Minkowski dimension of $\mathcal{L}$.

---

[8] On the other hand, the hypotheses about $\zeta_{\mathcal{L}}$ made here are not assumed in [LapPo2]. See Remark 8.19.

Theorem 8.15 (Criterion for Minkowski measurability). *Let $\mathcal{L}$ be an ordinary fractal string that is languid for a screen passing between the vertical line $\operatorname{Re} s = D$ and all the complex dimensions of $\mathcal{L}$ with real part strictly less than D, and not passing through 0. Then the following statements are equivalent:*

(i) *D is the only complex dimension with real part D, and it is simple.*

(ii) *$N_{\mathcal{L}}(x) = E \cdot x^D + o(x^D)$ for some positive constant E.*

(iii) *The boundary of $\mathcal{L}$ is Minkowski measurable.*

Moreover, if any of these conditions is satisfied, then

$$\mathcal{M} = 2^{1-D}\frac{E}{1-D} = 2^{1-D}\frac{\operatorname{res}\left(\zeta_{\mathcal{L}}(s); D\right)}{D(1-D)} \tag{8.25}$$

is the Minkowski content of the boundary of $\mathcal{L}$.

Proof. Assume (i) and choose a screen such that only D is visible. By Theorem 5.18, the distributional explicit formula with error term, applied to $\mathcal{L}$, and using Theorems 5.30 and 5.31, we obtain for $\mathcal{P}_\eta^{[1]}(x) = N_{\mathcal{L}}(x)$,

$$\mathcal{P}_\eta^{[1]}(x) = \operatorname{res}\left(\zeta_{\mathcal{L}}(s)\frac{x^s}{s}; D\right) + \mathcal{R}_\eta^{[1]}(x) = \frac{x^D}{D}\operatorname{res}(\zeta_{\mathcal{L}}(s); D) + o\left(x^D\right),$$

as $x \to \infty$. Hence, (ii) holds with $E = D^{-1}\operatorname{res}\left(\zeta_{\mathcal{L}}(s); D\right)$. Since D is assumed to be a simple pole of $\zeta_{\mathcal{L}}(s)$ (by (i)) and $\zeta_{\mathcal{L}}(\sigma)$ is positive for $\sigma > D$, the number E, given by $D^{-1}\lim_{\sigma\to D^+}(\sigma - D)\zeta_{\mathcal{L}}(\sigma)$, is positive. Similarly, we deduce (iii) using Theorem 8.1.

Assume (ii). Then $\zeta_{\mathcal{L}}$ has only simple poles on the line $\operatorname{Re} s = D$, by Theorem 1.16 and Lemma 8.13. Let $\{D + i\gamma_n\}$ be the (finite or infinite) sequence of these poles. By Theorem 5.18, we have that

$$N_{\mathcal{L}}(x) = \sum_n a_n x^{D+i\gamma_n} + o\left(x^D\right), \qquad \text{as } x \to \infty,$$

where $a_n = \operatorname{res}\left(s^{-1}\zeta_{\mathcal{L}}(s); D + i\gamma_n\right)$. Hence

$$\sum_n a_n x^{D+i\gamma_n} - E \cdot x^D \to 0, \qquad \text{as } x \to \infty.$$

By the Uniqueness Theorem for almost periodic functions (see [Schw1, Section VI.9.6, p. 208]), we conclude that $a_n = 0$ for $\gamma_n \neq 0$. This implies (i).

To deduce (i) from (iii), we reason similarly, using $V(\varepsilon)$ instead of $N_{\mathcal{L}}(x)$, and letting $a_n = \operatorname{res}\left((s(1-s))^{-1}\zeta_{\mathcal{L}}(s); D + i\gamma_n\right)$. This concludes the proof of the Minkowski measurability criterion, Theorem 8.15. □

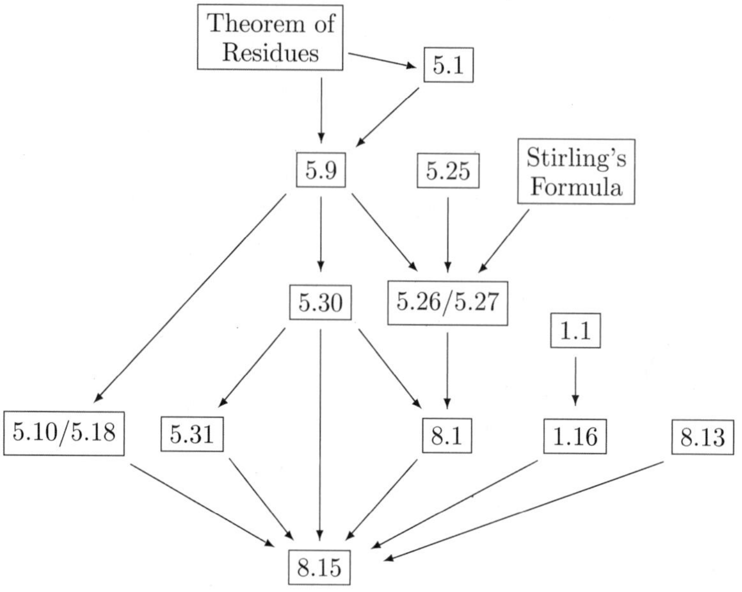

Figure 8.1: The structure of the proof of Theorem 8.15.

We refer the reader to Figure 8.1 for a diagrammatic illustration[9] of both the structure of the proof of Theorem 8.15 and the interdependence of many results obtained in Chapter 5 (on explicit formulas) and so far in this chapter on tube formulas.

In the following comment, we discuss a subtle point in the proof of Theorem 8.15. It explains, in particular, how the preliminary results of this section are used to circumvent an apparent difficulty (for example, in deriving (ii) from (i)).

Remark 8.16. Proposition 1.1, Theorem 1.16 and Lemma 8.13 above are needed to ensure (under the hypotheses of Theorem 8.15) that

$$N_{\mathcal{L}}(x) = x^D \cdot G(x) + o\big(x^D\big), \qquad \text{as } x \to \infty,$$

or

$$V(\varepsilon) = \varepsilon^{1-D} G_1(\varepsilon) + o\big(\varepsilon^{1-D}\big), \qquad \text{as } \varepsilon \to 0^+,$$

where G and G_1 are multiplicatively almost periodic functions. The proof would be simpler if we could split the sum

$$\sum_{\operatorname{Re}\omega = D} \operatorname{res}\left(\frac{\zeta_{\mathcal{L}}(s) x^{s+k-1}}{(s)_k}; \omega\right) \tag{8.26}$$

[9]We thank Erin Pearse for providing us with this figure relating to the corresponding theorem in [Lap-vF5].

into the different subsums of the type

$$(\log x)^n \sum_{\operatorname{Re}\omega=D} a_{\omega,n} x^{\omega+k-1},$$

for fixed n, which would arise when $\mathcal{L}$ has multiple complex dimensions. Then we would simply apply the Uniqueness Theorem for almost periodic functions to each of these sums. Even though we think that such a decomposition of the sum in the explicit formula is possible, we have not been able to prove this, since the series (8.26) is only conditionally convergent.

Remark 8.17. As was pointed out in Example 5.32 (see also Remark 6.15), for certain fractal strings, and in particular, for certain nonlattice strings, we cannot choose a screen as in Theorem 8.15, and the above proof of our criterion for Minkowski measurability does not apply to such strings. However, by the work of M. L. Lapidus and K. J. Falconer ([Lap3, Section 4.4.1b] and [Fa4]), nonlattice strings are always Minkowski measurable (see Remark 8.40 below). We show in Section 8.4.4 how we can also recover this result within our framework. Actually, the results of that section imply significantly more. It follows, in particular, from Theorems 8.23 and 8.36 that a self-similar string is Minkowski measurable if and only if it is nonlattice and that, in view of Theorem 2.17 or 3.6, for any (lattice or nonlattice string), the main conclusion of Theorem 8.15 (namely, the equivalence of conditions (i)–(iii)) still holds. Moreover, we will show in Section 8.4.3, Theorem 8.30, that even in the lattice case, formula (8.25) remains valid provided $\mathcal{M}$ is replaced by the average Minkowski content $\mathcal{M}_{\mathrm{av}}$, as defined by (8.55). See formula (8.56).

Remark 8.18. Let $\{l_j\}_{j=1}^{\infty}$ denote the sequence of lengths of the fractal string $\mathcal{L}$. Condition (ii) of Theorem 8.15 is then equivalent to the following condition:

(ii′) $l_j \sim M \cdot j^{-1/D}$ (i.e., $j^{1/D} l_j \to M$) as $j \to \infty$, for some positive constant $M > 0$.

Further, the constants E in (ii) and M in (ii′) are connected by $E = M^D$.

Our next remark explains the connections and the differences between our present Minkowski measurability criterion (Theorem 8.15) and the criterion previously obtained in [LapPo2].

Remark 8.19. The criterion for Minkowski measurability that was obtained in [LapPo2, Theorem 2.2, p. 46] is the following:

> Let $\mathcal{L} = \{l_j\}_{j=1}^{\infty}$ be an (arbitrary) ordinary fractal string of Minkowski dimension $D \in (0,1)$. Then the boundary of $\mathcal{L}$ is Minkowski measurable if and only if (ii) above holds for some $E > 0$ (or, equivalently, if and only if condition (ii′) from Remark 8.18 holds for some $M > 0$).

Further, in that case, the Minkowski content of $\mathcal{L}$ is given by

$$\mathcal{M} = 2^{1-D}\frac{E}{1-D} = 2^{1-D}\frac{M^D}{1-D}. \tag{8.27}$$

Note that in [LapPo1–2], $\zeta_{\mathcal{L}}$ is not required to admit a meromorphic extension to a neighborhood of $\operatorname{Re} s = D$ or to satisfy suitable growth conditions. On the other hand, under the hypotheses of Theorem 8.15, our present theory enables us to introduce the new criterion (i), expressed in terms of the notion of complex dimension. The latter criterion gives a rather clear and intuitive geometric meaning to the notion of Minkowski measurability in the present context of fractal strings. It also provides one more geometric interpretation of the notion of complex dimension.

The following comment may help the reader, as it did the authors, to develop further intuition for the notion of complex dimension and the associated oscillatory phenomena (in the geometry). It will also be very useful in Chapter 9, when we reformulate and extend the inverse spectral problem for fractal strings studied in [LapMa1–2].

Remark 8.20 (Dimensions above D and geometric oscillations). The Minkowski measurability of $\mathcal{L}$, condition (iii) in Theorem 8.15, means heuristically that the leading term of the volume of small tubular neighborhoods does not oscillate. Similarly, condition (ii) (or, equivalently, (ii′), by Remark 8.18) says that the sequence of lengths of $\mathcal{L}$ does not oscillate (asymptotically) either. That is, (ii) and (iii) can be interpreted as corresponding to the absence of oscillations of order D in the geometry of $\mathcal{L}$. Therefore, in some sense (provided that D is simple), Theorem 8.15 says that the absence of geometric oscillations of order D in $\mathcal{L}$ is equivalent to the absence of nonreal complex dimensions of $\mathcal{L}$ above D. Note that $\mathcal{L}$ could still have oscillations of lower order.

As will be discussed in more detail in Sections 8.4.2 and 8.4.4 below, for a self-similar string $\mathcal{L}$, this fact is illustrated rather clearly by the lattice vs. nonlattice dichotomy which corresponds precisely to the existence vs. the absence of (nonreal) complex dimensions above D and hence, by Theorems 8.23 and 8.36, to the non-Minkowski measurability vs. the Minkowski measurability of $\mathcal{L}$.

Remark 8.21. Our present approach enables us to analyze in more detail the effect on the geometry (or the spectrum) of fractal strings due to certain of the gauge functions (other than the usual power functions) involved in the definition of the generalized Minkowski content studied by C. Q. He and the first author in [HeLap1–2]. It also enables us to deal with, for example, gauge functions of the form of a power function times a multiplicatively periodic function, as for lattice self-similar strings, which are not within the scope of the theory developed in [HeLap2], where all the gauge functions were assumed to be monotonic. This is worked out in Section 8.4.2. (See also the notes to this chapter for further discussion.)

8.4 Tube Formulas for Self-Similar Strings

In this section, we discuss three classes of examples of self-similar strings, namely, the generalized Cantor strings and lattice strings (in Sections 8.4.1 and 8.4.2–8.4.3), and the nonlattice strings (in Section 8.4.4), in order to illustrate and complete our results of Sections 8.1 and 8.3.

8.4.1 Generalized Cantor Strings

A generalized Cantor string is a generalized fractal string (i.e., a measure) with a single line of complex dimensions, of the form $\{D+in\mathbf{p}\}_{n\in\mathbb{Z}}$. Such a string has lengths 1, a^{-1}, $a^{-2}, \dots$, with multiplicities given by 1, b, $b^2, \dots$, where $b=a^D$ and $a=e^{2\pi/\mathbf{p}}$ is the reciprocal of the multiplicative generator r. We assume $\mathbf{p}>0$ so that $a>1$. Such a string is lattice, except that we allow D to be an arbitrary complex number. Thus, $a=r^{-1}$ is real and positive, and b is a complex number. The parameters a and b are related to D and $\mathbf{p}$ by

$$\begin{aligned} a &= r^{-1} = e^{2\pi/\mathbf{p}}, & \mathbf{p} &= \frac{2\pi}{\log a}, \\ b &= a^D, & D &= \log_a b. \end{aligned} \tag{8.28}$$

Thus, for example, we allow $D=0$ and $D=1$, corresponding to $b=1$ and $b=a$, respectively.

The geometric zeta function of the generalized Cantor string of dimension D and oscillatory period $\mathbf{p}$ is given by

$$\zeta_{D,\mathbf{p}}(s) = \frac{1}{1-b\cdot a^{-s}},$$

with residue $1/\log a$ at $s=D$. We denote this string by $\mathcal{L}_{D,\mathbf{p}}$ (see also Definition 10.1). Clearly, $\mathcal{L}_{D,\mathbf{p}}$ has a single line of complex dimensions,

$$\mathcal{D} = \{D+in\mathbf{p}\colon n\in\mathbb{Z}\},$$

with the oscillatory period $\mathbf{p}=2\pi/\log a$ as in (8.28), and the residue at each complex dimension is equal to $1/\log a$.

If $a>b>1$ and b is an integer, then this string is an ordinary fractal string, and $V(\varepsilon)=V_{D,\mathbf{p}}(\varepsilon)$ has its standard geometric meaning. However, our computation is formally valid without these hypotheses, and it fits both in the framework of generalized fractal strings of Chapter 4 and of Dirichlet polynomials of Chapter 3. See also Chapter 10 for an extensive study of generalized Cantor strings.

We next compute the volume of the tubular neighborhood of $\mathcal{L}_{D,\mathbf{p}}$. In view of the second part of Corollary 8.10, we have that (provided $D\neq 0$, i.e., $b \neq 1$),

$$V_{D,\mathbf{p}}(\varepsilon) = \frac{1}{\log a} \sum_{n\in\mathbb{Z}} \frac{(2\varepsilon)^{1-D-in\mathbf{p}}}{(D+in\mathbf{p})(1-D-in\mathbf{p})} - \frac{2\varepsilon}{b-1}, \tag{8.29}$$

as a pointwise equality between functions, for $0 < \varepsilon < a/2$. In case $D = 0$, this formula needs to be replaced by

$$V_{0,\mathbf{p}}(\varepsilon) = \frac{1}{\log a} \sum_{n\neq 0} \frac{(2\varepsilon)^{1-in\mathbf{p}}}{in\mathbf{p}(1-in\mathbf{p})} - 2\varepsilon\left(\frac{1}{2} + \log_a(2\varepsilon)\right), \tag{8.30}$$

since then, 0 is a visible pole of $\zeta_{0,\mathbf{p}}$.

We can also obtain these formulas by a direct computation, starting with formula (1.9) and taking Definition 10.1 into account,

$$V_{D,\mathbf{p}}(\varepsilon) = \sum_{k=0}^{[\log_a \frac{1}{2\varepsilon}]} 2\varepsilon \cdot b^k + \sum_{k=1+[\log_a \frac{1}{2\varepsilon}]}^{\infty} b^k a^{-k}. \tag{8.31}$$

The second sum only converges if $\operatorname{Re} D < 1$, which we assume in the following. Using formula (1.13), we obtain (for $D \neq 0$)

$$V_{D,\mathbf{p}}(\varepsilon) = (2\varepsilon)^{1-D} G\big(\log_a(2\varepsilon)^{-1}\big) - \frac{2\varepsilon}{b-1}, \tag{8.32}$$

for $0 < \varepsilon < a/2$, where G is the periodic function (of period 1) given for all $x > -1$ by

$$\begin{aligned} G(x) &= \frac{1}{\log a} \sum_{n\in\mathbb{Z}} \frac{e^{2\pi i n x}}{(D+in\mathbf{p})(1-D-in\mathbf{p})} \\ &= \frac{r^{D\{x\}}}{1-r^D} + \frac{r^{(D-1)\{x\}}}{r^{D-1}-1}, \end{aligned} \tag{8.33}$$

thus recovering (8.29). (Formula (8.30) is recovered in a similar manner, using in addition the Fourier series for $\{u\}$.) Clearly, for $D \neq 0$, the function G is nonconstant since all its Fourier coefficients are nonzero. Further, G is a positive function that is bounded away from zero and infinity. In fact, $G(x)$ attains its maximum

$$\frac{r^{D-1} - r^D}{(1-r^D)(r^{D-1}-1)} \tag{8.34a}$$

at the endpoints $x = 0$ and $x = 1$, and its minimum

$$\frac{(1-D)^{D-1}(1-r^D)^{D-1}}{D^D(r^{D-1}-1)^D} \tag{8.34b}$$

at the point $x > 0$ such that

$$r^x = \frac{1-D}{D}\frac{1-r^D}{r^{D-1}-1}.$$

In view of (8.32) and Definitions (1.6a) and (1.6b), it follows from (8.34a) and (8.34b) that the upper and lower Minkowski contents of $\mathcal{L}_{D,\mathbf{p}}$ are given by

$$\mathcal{M}^* = 2^{1-D}\frac{r^{D-1}-r^D}{(1-r^D)(r^{D-1}-1)} = 2^{1-D}\frac{b(a-1)}{(a-b)(b-1)} \tag{8.35a}$$

and

$$\begin{aligned}\mathcal{M}_* &= 2^{1-D}\frac{(1-D)^{D-1}(1-r^D)^{D-1}}{D^D(r^{D-1}-1)^D} \\ &= 2^{1-D}D^{-D}(1-D)^{-(1-D)}\frac{b(b-1)^{D-1}}{(a-b)^D},\end{aligned} \tag{8.35b}$$

respectively, where $r = 1/a$ and $D = \log_a b$ (so that $r^D = 1/b$ and hence $r^{D-1} = a/b$). Note that $0 < \mathcal{M}_* < \mathcal{M}^* < \infty$ since G is nonconstant and bounded away from zero and infinity. It follows that $\mathcal{L}$ is not Minkowski measurable. By contrast, in case $D = 0$, both the upper and lower Minkowski dimension of $\mathcal{L}_{0,\mathbf{p}}$ are infinite.

Example 8.22 (The Cantor string). The Cantor string is a special case of the generalized Cantor string, with parameters $a = 3$ and $b = 2$. This lattice string was studied in Chapter 1 and in Section 2.3.1. Note that the first length of the Cantor string equals $\frac{1}{3}$ in Chapter 1 (so that the Cantor string fits in the unit interval), whereas in Section 2.3.1 and in this example, we adopt the usual convention that the first length is normalized to be unity, so that the total length of the Cantor string equals three. Thus formula (1.12) and the infinite sum in formula (1.14) of Chapter 1 need to be multiplied by $3^D = 2$ to obtain respectively (8.37) and (8.36) below.

In view of (8.32) and the discussion in Section 1.1.2, we have

$$V(\varepsilon) = \frac{1}{\log 3}(2\varepsilon)^{1-D}G\left(\log_3(2\varepsilon)^{-1}\right) - 2\varepsilon, \tag{8.36}$$

where $D = \log_3 2$, $r = 1/3$, and G is the (nonconstant) periodic function given by (8.33), with $\mathbf{p} = 2\pi/\log 3$. Further, G is bounded away from zero and infinity, as explained in (8.34a) and (8.34b). Moreover, by specializing (8.35) to $a = 3$ and $b = 2$, we have

$$\mathcal{M}^* = 2^{3-D} \quad \text{and} \quad \mathcal{M}_* = 2^{2-D}D^{-D}(1-D)^{-(1-D)}, \tag{8.37}$$

in agreement with the result of [LapPo2, Theorem 4.6, p. 65], recalled in Section 1.1.2. In particular, we recover the fact that the Cantor string is not Minkowski measurable.

8.4.2 Lattice Self-Similar Strings

Let $\mathcal{L}$ be a self-similar string, with boundary $\partial\mathcal{L}$ of Minkowski dimension D, as studied in Chapter 2. Thus, $\mathcal{L}$ has scaling ratios $r_1, \dots, r_N$ (with $N \geq 2$) and gaps scaled by $g_1, \dots, g_K$ (with $K \geq 1$). Furthermore, recall from Section 2.1 that $0 < r_N \leq \cdots \leq r_1 < 1$, $0 < g_K \leq \cdots \leq g_1 < 1$, and that Equation (2.4) holds:

$$\sum_{j=1}^{N} r_j + \sum_{k=1}^{K} g_k = 1. \tag{8.38}$$

In addition,

$$L = \zeta_{\mathcal{L}}(1) = \mathrm{vol}_1(\mathcal{L}) \tag{8.39}$$

denotes the total length of $\mathcal{L}$.

Recall that the geometric zeta function of $\mathcal{L}$ is given by Equation (2.10) of Theorem 2.4. In particular, 0 is not a pole of $\zeta_{\mathcal{L}}$. Also, $\mathcal{L}$ is called a lattice string if there exist a multiplicative generator $r \in (0,1)$ and positive integers $k_1, \dots, k_N$ without common factor such that $r_j = r^{k_j}$ for every $j = 1, \dots, N$. Otherwise, $\mathcal{L}$ is said to be a nonlattice string.

If $\mathcal{L}$ is a lattice string, then the complex dimensions located on the vertical line $\mathrm{Re}\, s = D$ are all simple and of the form $D + in\mathbf{p}$ (for $n \in \mathbb{Z}$), where $\mathbf{p} = 2\pi/\log r^{-1}$ is the oscillatory period of $\mathcal{L}$ and $r = e^{-2\pi/\mathbf{p}}$ is its multiplicative generator. (See Theorem 2.17 and Definition 2.14.) The generalized Cantor string is an example of a lattice self-similar string with a single line of complex dimensions.

In this section, we will prove the following theorem, along with additional more precise results (see Theorem 8.25 and Corollary 8.27).

Theorem 8.23. *A lattice string is never Minkowski measurable and always has multiplicatively periodic oscillations of order D, its dimension, in its geometry.*

Proof (*in the case of a single gap*). Let $\mathcal{L}$ be a self-similar lattice string with a single gap, normalized as in Remark 2.6.[10] Choose a number Θ with $0 \leq \Theta < D$ such that the first line of complex dimensions to the left of D lies to the left of the line $\mathrm{Re}\, s = \Theta$ (if there is no such line, then we take $\Theta = 0$). Then, by the computation for the generalized Cantor string

[10]The general case of a lattice string with multiple gaps will be treated in Theorem 8.25 and Corollary 8.27.

in Section 8.4.1,

$$\begin{aligned} V(\varepsilon) &= \operatorname{res}(\zeta_{\mathcal{L}}(s); D) \sum_{n\in\mathbb{Z}} \frac{(2\varepsilon)^{1-D-in\mathbf{p}}}{(D+in\mathbf{p})(1-D-in\mathbf{p})} \\ &\quad + \sum_{\operatorname{Re}\omega<D} \operatorname{res}\left(\frac{\zeta_{\mathcal{L}}(s)(2\varepsilon)^{1-s}}{s(1-s)}; \omega\right) + 2\varepsilon\zeta_{\mathcal{L}}(0) \qquad (8.40) \\ &= (2\varepsilon)^{1-D} G\left(\log_{r^{-1}}(2\varepsilon)^{-1}\right) + O\left(\varepsilon^{1-\Theta}\right), \qquad \text{as } \varepsilon \to 0^+, \end{aligned}$$

where

$$G(x) = \operatorname{res}\left(\zeta_{\mathcal{L}}(s); D\right) \sum_{n\in\mathbb{Z}} \frac{e^{2\pi i n x}}{(D+in\mathbf{p})(1-D-in\mathbf{p})}, \tag{8.41}$$

and $\operatorname{res}\left(\zeta_{\mathcal{L}}(s); D\right)$ is given by (2.43) (with $K = 1$ and $g_1 L = 1$):[11]

$$\operatorname{res}\left(\zeta_{\mathcal{L}}(s); D\right) = \frac{1}{\sum_{j=1}^N r_j^D \log r_j^{-1}} = \frac{1}{\log r^{-1} \sum_{j=1}^N k_j r^{k_j D}}. \tag{8.42}$$

Recall that as in Chapter 2, the positive integers $k_1, \dots, k_N$ are defined by $r_j = r^{k_j}$ for $j = 1, \dots, N$. Since the periodic function G is nonconstant (because it has nonzero Fourier coefficients for $n \neq 0$), it follows that $\mathcal{L}$ is not Minkowski measurable, in agreement with Theorems 8.15 and 2.17. Moreover, the geometric oscillations are multiplicatively periodic since G is periodic. □

Note that formula (8.40) converges pointwise since there are only finitely many lines of complex dimensions, the complex dimensions have bounded residues, and the denominators of the terms are quadratic in ω.

Given the special structure of the complex dimensions of a lattice string (see Theorem 2.17), we can rewrite the first equality of Equation (8.40) in a much more explicit manner. Indeed, continue to assume for simplicity that $\mathcal{L}$ has a single gap and is normalized as in Remark 2.6, so that it has first length 1, and the last statement of the lattice case of Theorem 2.17 applies (see Remark 8.24 below for the general case). Then, in the notation of Theorem 2.17 and in the case where the complex dimensions ω_u are all simple (as in Examples 8.22 above and 8.32 below), we have for $0 < \varepsilon < 1/(2r_N)$ (see Theorem 8.25 below)

$$V(\varepsilon) = \sum_{u=1}^{q} (2\varepsilon)^{1-\omega_u} G_u \left(\log_{r^{-1}}(2\varepsilon)^{-1}\right) + 2\varepsilon\zeta_{\mathcal{L}}(0), \tag{8.43}$$

---

[11]Note that for the generalized Cantor string of Section 8.4.1, we have

$$\operatorname{res}\left(\zeta_{\mathcal{L}}(s); D\right) = \frac{1}{\log a} = \frac{1}{\log r^{-1}},$$

so that formula (8.41) reduces to (8.33) in that case.

where $\omega_1 = D$, $G_1 = G$, as defined in (8.41), and for $u = 1, \ldots, q$, the function G_u is periodic (of period 1) given by[12]

$$G_u(x) = \operatorname{res}\left(\zeta_{\mathcal{L}}(s); \omega_u\right) \sum_{n\in\mathbb{Z}} \frac{e^{2\pi i n x}}{(\omega_u + in\mathbf{p})(1 - \omega_u - in\mathbf{p})}. \tag{8.44}$$

Clearly, G_u is nonconstant and, as above, it is bounded away from zero and from infinity. Further, according to Remark 2.19, $\operatorname{res}\left(\zeta_{\mathcal{L}}(s); \omega_u\right)$ is given by (2.50) with $n = 0$.

Remark 8.24. If the lattice string $\mathcal{L}$ has one gap but is not normalized, then an additional phase factor is introduced. More precisely, $V(\varepsilon)$ is still given by (8.43), with G_u as in (8.44) (and, in particular, $G = G_1$ as in (8.41)), except with $\operatorname{res}(\zeta_{\mathcal{L}}(s); \omega_u)$ multiplied by $(g_1 L)^{in\mathbf{p}}$ (inside the sum), where $g_1 L$ is the first length of $\mathcal{L}$. Therefore, $\operatorname{res}\left(\zeta_{\mathcal{L}}(s); \omega_u\right)$ is given by (2.50) with $K = 1$ and $n = 0$, for $u = 1, \ldots, q$. (Recall that $\omega_1 = D$.) See the comment following Equation (2.49), along with Remark 2.19. (Also, see Theorem 8.25 below according to which the counterpart of (8.43) is valid for all $0 < \varepsilon < \frac{1}{2} L g_1 r_N^{-1}$.)

The following result—which completes Theorem 8.23 in the case of multiple gaps—is really a corollary of the method of proof of that theorem. See also Corollary 8.27 below for the case when $\mathcal{L}$ may have complex dimensions of higher multiplicity.

Theorem 8.25 (Lattice strings with multiple gaps). *Let $\mathcal{L}$ be a lattice self-similar string with multiplicative generator r. Assume that the complex dimensions of $\mathcal{L}$ are all simple. Then for all ε with $0 < \varepsilon < \frac{1}{2} L g_K r_N^{-1}$, the volume $V(\varepsilon)$ is given by the following pointwise tube formula:*

$$V(\varepsilon) = \sum_{u=1}^{q} (2\varepsilon)^{1-\omega_u} G_u\left(\log_{r^{-1}}(2\varepsilon)^{-1}\right) + \frac{2K}{1-N}\varepsilon, \tag{8.45}$$

where for each $u = 1, \ldots, q$, the function G_u is the nonconstant, real-valued periodic function of period 1 on $\mathbb{R}$ corresponding to the line of complex dimensions through ω_u ($\omega_1 = D > \operatorname{Re}\omega_2 \geq \cdots \geq \operatorname{Re}\omega_q$), and given by the absolutely convergent Fourier series

$$G_u(x) = \sum_{n\in\mathbb{Z}} \frac{\operatorname{res}(\zeta_{\mathcal{L}}(s); \omega_u + in\mathbf{p})}{(\omega_u + in\mathbf{p})(1 - \omega_u - in\mathbf{p})} e^{2\pi i n x}, \tag{8.46}$$

---

[12] Recall from formula (2.39) that $\mathcal{D} = \{\omega_u + in\mathbf{p} \colon n \in \mathbb{Z}, u = 1, \ldots, q\}$ and from the last statement of the lattice case of Theorem 2.17 that for each fixed $u = 1, \ldots, q$, we have $\operatorname{res}\left(\zeta_{\mathcal{L}}(s); \omega_u + in\mathbf{p}\right) = \operatorname{res}\left(\zeta_{\mathcal{L}}(s); \omega_u\right)$, for all $n \in \mathbb{Z}$. Since $\mathcal{L}$ is normalized, this common value is given in Equation (2.50), with the numerator on the right-hand side of this equation set equal to one.

where

$$\operatorname{res}(\zeta_{\mathcal{L}}(s);\omega_u + in\mathbf{p}) = \frac{\sum_{\mu=1}^{K} (g_\mu L)^{\omega_u + in\mathbf{p}}}{\log r^{-1} \sum_{j=1}^{N} k_j r^{k_j \omega_u}}. \tag{8.47}$$

In particular, since $\omega_1 := D$, *we have that* $G_1 = G$ *is given by* (8.52) *of Corollary 8.27 below.*

Furthermore, if $\mathcal{L}$ *has a single gap and is normalized* (*i.e.*, $K = 1$ *and* $g_1 L = 1$), *then* (8.45) *reduces to, and specifies, formula* (8.43) (*which is therefore valid pointwise for all* $0 < \varepsilon < 1/2r_N$), *while* (8.46) *and* (8.47) *together yield* (8.44).

Proof. The theorem follows from the second part of Theorem 8.7 (the pointwise tube formula with error term). In light of the comment at the beginning of Section 6.4 (just above Remark 6.11), according to which $\mathcal{L}$ is strongly languid with $A = L^{-1} g_K^{-1} r_N$ and $\kappa = 0$ (thus also with any $\kappa \geq 0$), and since 0 is not a complex dimension of $\mathcal{L}$, the latter theorem can be applied to yield for all positive $\varepsilon < \frac{1}{2} L g_K r_N^{-1}$:

$$\begin{aligned} V(\varepsilon) &= \sum_{\omega \in \mathcal{D}_{\mathcal{L}}(\mathbb{C})} \operatorname{res}\left(\frac{(2\varepsilon)^{1-s}}{s(1-s)} \zeta_{\mathcal{L}}(s);\omega\right) + 2\varepsilon \zeta_{\mathcal{L}}(0) \\ &= \sum_{u=1}^{q} \sum_{n \in \mathbb{Z}} \operatorname{res}\left(\frac{(2\varepsilon)^{1-s}}{s(1-s)} \zeta_{\mathcal{L}}(s);\omega_u + in\mathbf{p}\right) + 2\varepsilon \frac{K}{1-N}. \end{aligned} \tag{8.48}$$

Here, the second equality results from the periodic structure of the complex dimensions in the lattice case (see Theorem 2.17).

If ω_u (and hence, $\omega_u + in\mathbf{p}$, for all $n \in \mathbb{Z}$) is simple, then

$$\operatorname{res}\left(\frac{(2\varepsilon)^{1-s}}{s(1-s)} \zeta_{\mathcal{L}}(s);\omega_u + in\mathbf{p}\right) = (2\varepsilon)^{1-\omega_u - in\mathbf{p}} \frac{\operatorname{res}(\zeta_{\mathcal{L}}(s);\omega_u + in\mathbf{p})}{(\omega_u + in\mathbf{p})(1 - \omega_u - in\mathbf{p})},$$

and according to formula (2.50) of Remark 2.19,

$$\operatorname{res}(\zeta_{\mathcal{L}};\omega_u + in\mathbf{p}) = \frac{\sum_{\mu=1}^{K} (g_\mu L)^{\omega_u + in\mathbf{p}}}{\log r^{-1} \sum_{j=1}^{N} k_j r^{k_j \omega_u}}. \tag{8.49}$$

□

Remark 8.26. Assume, in addition, that the gap sizes $g_\mu L$ are also integral powers of the multiplicative generator r of $\mathcal{L}$; i.e., $g_\mu L = r^{k'_\mu}$ for certain integers k'_μ, $\mu = 1, \dots, K$ (as in (2.40) of Theorem 2.17). Then for each $u = 1, \dots, q$, the residue $\operatorname{res}(\zeta_{\mathcal{L}}(s);\omega_u + in\mathbf{p})$ is independent of $n \in \mathbb{Z}$, as was observed in Remark 2.19. More precisely, since then $(g_\mu L)^{in\mathbf{p}} = 1$, formula (8.49) simplifies to (as in Equation (2.50'))

$$\operatorname{res}(\zeta_{\mathcal{L}};\omega_u + in\mathbf{p}) = \frac{\sum_{\mu=1}^{K} r^{k'_\mu \omega_u}}{\log r^{-1} \sum_{j=1}^{N} k_j r^{k_j \omega_u}}, \quad \text{for } n \in \mathbb{Z}. \tag{8.50}$$

A similar comment applies to formula (8.53) in Corollary 8.27 below.

The next corollary (of the proof of the foregoing theorem) applies to an arbitrary lattice string and establishes in particular Theorem 8.23 in full generality. It also completes the proof of Theorems 2.17 and 3.6 in the lattice case.

Corollary 8.27. *Let $\mathcal{L}$ be a lattice self-similar string, with scaling ratios $r_1, \dots, r_N$ $(N \geq 2)$ and gaps $g_1, \dots, g_K$ $(K \geq 1)$, as in Section 2.1. Let $\operatorname{Re} s = \Theta$ be the rightmost vertical line to the left of $\operatorname{Re} s = D$ containing complex dimensions of $\mathcal{L}$,*[13] *and let $m \geq 1$ be the maximal multiplicity of a complex dimension on this line. Then*

$$V(\varepsilon) = (2\varepsilon)^{1-D} G\left(\log_{r^{-1}}(2\varepsilon)^{-1}\right) + E(\varepsilon), \tag{8.51}$$

where $G = G_1$ is the nonconstant, real-valued periodic function of period 1 on $\mathbb{R}$ given by the absolutely convergent Fourier expansion

$$G(x) = \sum_{n \in \mathbb{Z}} \frac{\operatorname{res}(\zeta_{\mathcal{L}}(s); D + in\mathbf{p})}{(D + in\mathbf{p})(1 - D - in\mathbf{p})} e^{2\pi i n x}, \tag{8.52}$$

where the residues of the geometric zeta function of $\mathcal{L}$ are given by

$$\operatorname{res}(\zeta_{\mathcal{L}}(s); D + in\mathbf{p}) = \frac{\sum_{\mu=1}^{K} (g_\mu L)^{D+in\mathbf{p}}}{\log r^{-1} \sum_{j=1}^{N} k_j r^{k_j D}}, \tag{8.53}$$

infinitely many of which are nonvanishing.

In formula (8.51), the error $E(\varepsilon)$ is estimated as follows, as $\varepsilon \to 0^+$:

(i) *If $\Theta > 0$ (i.e., there are complex dimensions with real part strictly between 0 and D) or if $\Theta = 0$ and 0 is a complex dimension of $\mathcal{L}$ of multiplicity at most $m - 1$, then*

$$E(\varepsilon) = O\left(\varepsilon^{1-\Theta} \left|\log \varepsilon\right|^{m-1}\right). \tag{8.54a}$$

(ii) *If $\Theta < 0$, then*

$$E(\varepsilon) = O(\varepsilon). \tag{8.54b}$$

(iii) *If $\Theta = 0$ and 0 is a complex dimension of $\mathcal{L}$ of multiplicity m, then*

$$E(\varepsilon) = O\left(\varepsilon \left|\log \varepsilon\right|^{m}\right). \tag{8.54c}$$

It follows that $\mathcal{L}$ is not Minkowski measurable. In other words, a lattice string does not have a Minkowski content.

---

[13]Note that the meaning of Θ is slightly different from that in the proof of Theorem 8.23, where $\Theta \geq 0$.

Proof. We deduce formula (8.51)—with $G = G_1$ and the coefficients of G given by (8.52) and (8.53)—from the first part of the pointwise formula for the volume of the tubular neighborhoods, Theorem 8.7. To obtain the error estimate as given, we choose for the screen a vertical line $\text{Re}\, s = \sigma < \Theta$ if $\Theta \geq 0$, or a vertical line $\text{Re}\, s = \sigma$, where $\Theta < \sigma < 0$, if $\Theta < 0$. With this choice of screen, the poles of $\zeta(s)/s$ that contribute to the explicit formula for $V(\varepsilon)$ lie only on the two lines $\text{Re}\, s = D$ and either on $\text{Re}\, s = 0$ or on $\text{Re}\, s = \Theta$, whichever of these lines lies furthest to the right. By the choice of screen, $\zeta_{\mathcal{L}}$ is bounded on the screen, hence **L1** and **L2** are satisfied with $\kappa = 0$. Therefore, we can apply the pointwise tube formula (8.17) of the first part of Theorem 8.7.

From the computation of the local terms in Section 6.1.1, we deduce that the resulting terms in the explicit formula are of the given order. Note that the resulting periodic function $G_1(x)$ is nonconstant because as was noted in the course of the proof of Theorem 2.17, above Equation (2.49), not all[14] of the (genuinely complex) dimensions $D + in\mathbf{p}$ ($n \in \mathbb{Z}$, $n \neq 0$) are canceled by the zeros of the numerator of $\zeta_{\mathcal{L}}$ in formula (2.10). It follows that $\mathcal{L}$ always has multiplicatively periodic oscillations of leading order D and hence is not Minkowski measurable. □

Remark 8.28. In the above corollary, we only used the sum over the complex dimensions on the line $\text{Re}\, s = D$ in the explicit formula for $V(\varepsilon)$, choosing a screen that shields all the other complex dimensions. This way we avoided having to determine the terms in the explicit formula corresponding to complex dimensions of higher multiplicity. We leave it as an exercise for the interested reader to work out the general explicit formula for $V(\varepsilon)$, with a complete determination of all the terms, corresponding to complex dimensions of any multiplicity. (As a guide, see the computation of the local terms in Section 6.1, along with Example 8.34 below.)

8.4.3 The Average Minkowski Content

We supplement our study of lattice self-similar strings (Section 8.4.2, where it was established that lattice strings do not have a Minkowski content) by considering the notion of average Minkowski content.

Definition 8.29. Let $\mathcal{L}$ be a fractal string of dimension D. The *average Minkowski content*, $\mathcal{M}_{\text{av}}$, is defined by the logarithmic Cesaro average

$$\mathcal{M}_{\text{av}} = \lim_{T\to\infty} \frac{1}{\log T} \int_{1/T}^{1} \varepsilon^{-(1-D)} V(\varepsilon)\, \frac{d\varepsilon}{\varepsilon}, \tag{8.55}$$

provided this limit exists and is a finite positive real number.

---

[14] Indeed, infinitely many of these complex dimensions are not canceled.

Theorem 8.30. *Let $\mathcal{L}$ be a lattice self-similar string of total length L, with scaling ratios $r_1 = r^{k_1}, \ldots, r_N = r^{k_N}$ and gaps $g_1, \ldots, g_K$. Then the average Minkowski content of $\mathcal{L}$ exists and is given by the finite positive number*

$$\mathcal{M}_{\mathrm{av}} = \frac{2^{1-D} \sum_{\mu=1}^{K} (g_\mu L)^D}{D(1-D) \log r^{-1} \sum_{j=1}^{N} k_j r^{k_j D}} = \frac{2^{1-D}}{D(1-D)} \operatorname{res}\bigl(\zeta_{\mathcal{L}}(s); D\bigr). \quad (8.56)$$

Proof. The existence and the computation of the average Minkowski content results from an application of Corollary 8.27. More precisely, according to (8.51) and (8.54), we have for all $0 \leq \varepsilon \leq 1$ and some $\Theta < D$,

$$\varepsilon^{-(1-D)} V(\varepsilon) = 2^{1-D} G\left(\log_{r^{-1}} (2\varepsilon)^{-1}\right) + O\bigl(\varepsilon^{D-\Theta}\bigr),$$

where G is the (nonconstant) real-valued periodic function of period 1 given by (8.52) and (8.53).

After making the change of variable $x = \log_{r^{-1}} (2\varepsilon)^{-1}$, we obtain

$$\frac{1}{\log T} \int_{1/T}^{1} \varepsilon^{-(1-D)} V(\varepsilon) \frac{d\varepsilon}{\varepsilon} = 2^{1-D} \frac{1}{\log_{r^{-1}} T} \int_0^{\log_{r^{-1}} T} G(x)\, dx + o(1)$$
$$\to 2^{1-D} \int_0^1 G(x)\, dx$$

as $T \to \infty$. Hence, in view of (8.55), $\mathcal{M}_{\mathrm{av}}$ exists and we have

$$\mathcal{M}_{\mathrm{av}} = 2^{1-D} \int_0^1 G(x)\, dx = 2^{1-D} \frac{\operatorname{res}\left(\zeta_{\mathcal{L}}(s); D\right)}{D(1-D)} > 0.$$

By formula (8.53) in Corollary 8.27 (applied with $n = 0$), this yields the (finite and positive) expression (8.56) for $\mathcal{M}_{\mathrm{av}}$, as desired. □

Remark 8.31. Both the non-Minkowski measurability and the existence as well as the value (8.56) of the average Minkowski content hold for arbitrary lattice self-similar strings with multiple gaps. Indeed, the non-Minkowski measurability follows from Theorem 8.15, since we know from Theorem 2.17 that D is simple and that there are always nonreal complex dimensions with real part equal to D. It can also be verified directly by working with our general pointwise tube formula without error term (the second part of Theorem 8.7) which also yields the existence of $\mathcal{M}_{\mathrm{av}}$. Indeed, the main point is that the term in our tube formula corresponding to the pole at $s_n := D + in\mathbf{p}$ (with $n \in \mathbb{Z}$, $n \neq 0$) is an oscillatory function with (logarithmic Cesaro) average zero (since s_n is a simple pole of $\zeta_{\mathcal{L}}$). On the other hand, for $n = 0$ (i.e., for $s = s_0 = D$), the resulting function is constant, and hence its average is equal to $\operatorname{res}\left(\zeta_{\mathcal{L}}(s); D\right)$, which (by Theorem 2.17, Equation (2.43)) is equal to the right-hand side of (8.56), up to a positive multiplicative factor.

Example 8.32 (The Fibonacci string). Recall from Section 2.3.2 that this is a lattice string with two lines of complex dimensions (see Figure 2.5 on page 43). Since all these complex dimensions are simple, the tube formula (8.40) gives

$$\begin{aligned} V(\varepsilon) = \frac{2+\phi}{5\log 2} \sum_{n\in\mathbb{Z}} \frac{(2\varepsilon)^{1-D-in\mathbf{p}}}{(D+in\mathbf{p})(1-D-in\mathbf{p})} - 2\varepsilon \\ + \frac{3-\phi}{5\log 2} \sum_{n\in\mathbb{Z}} \frac{(2\varepsilon)^{1+D-i(n+\frac{1}{2})\mathbf{p}}}{(-D+i(n+\frac{1}{2})\mathbf{p})(1+D-i(n+\frac{1}{2})\mathbf{p})}, \end{aligned} \tag{8.57}$$

where $D = \log_2 \phi$, $\mathbf{p} = 2\pi/\log 2$ and $0 < \varepsilon < 1$. In particular, we deduce from Theorem 8.30 that the average Minkowski content of the Fibonacci string is given by

$$\mathcal{M}_{\mathrm{av}} = \frac{2^{1-D}(2+\phi)}{5(1-D)\log\phi}. \tag{8.58}$$

Note that formula (8.57) is in agreement with the result (2.25) of the direct computation of $V(\varepsilon)$ carried out in Section 2.3.2. Indeed, the two formulas are equivalent, since the golden ratio satisfies $(2+\phi)/5 = \phi/\sqrt{5}$ and $(3-\phi)/5 = (\phi-1)/\sqrt{5}$.

Remark 8.33. The volume of the tubular neighborhood of the modified Cantor and Fibonacci strings, introduced in Section 2.3.3, is given by the same explicit formulas as in Examples 8.22 of Section 8.4.1 and Example 8.32 above, since their geometric zeta functions are the same as those of their nonmodified counterparts.

Example 8.34 (A lattice string with multiple poles). An example of a lattice string $\mathcal{L}$ with multiple poles was considered in Section 2.3.4. Recall that $\mathcal{L}$ has one (discrete) line of complex dimensions above D, namely, $\omega = D + in\mathbf{p}$ (with $n \in \mathbb{Z}$, $D = \log_3 2$ and $\mathbf{p} = 2\pi/\log 3$), and another line of double poles, namely, $\omega = \frac{1}{2}i\mathbf{p} + in\mathbf{p}$ (with $n \in \mathbb{Z}$); see Figure 2.9 on page 48. Thus, the tube formula (8.40) becomes

$$\begin{aligned} V(\varepsilon) = \frac{4}{9\log 3} \sum_{n\in\mathbb{Z}} \frac{(2\varepsilon)^{1-D-in\mathbf{p}}}{(D+in\mathbf{p})(1-D-in\mathbf{p})} - \frac{\varepsilon}{2} \\ + \frac{1}{9(\log 3)^2} \sum_{n\in\mathbb{Z}} \frac{(2\varepsilon)^{1-i(n+\frac{1}{2})\mathbf{p}}}{i(n+\frac{1}{2})\mathbf{p}(1-i(n+\frac{1}{2})\mathbf{p})} \\ \cdot \left(5\log 3 + \frac{3i(2n+1)\mathbf{p}-1}{i(n+\frac{1}{2})\mathbf{p}(1-i(n+\frac{1}{2})\mathbf{p})}\right) \\ + \frac{1}{3\log 3} \sum_{n\in\mathbb{Z}} \frac{(2\varepsilon)^{1-i(n+\frac{1}{2})\mathbf{p}}\log_3(2\varepsilon)^{-1}}{i(n+\frac{1}{2})\mathbf{p}(1-i(n+\frac{1}{2})\mathbf{p})}, \end{aligned} \tag{8.59}$$

as $\varepsilon \to 0^+$. In view of Equation (8.56) of Theorem 8.30, its average Minkowski content is given by

$$\mathcal{M}_{\mathrm{av}} = \frac{8}{27(1-D)\log 2}. \tag{8.60}$$

Remark 8.35. It also follows from Theorem 8.30 that the average Minkowski content of the Cantor string (as defined in Section 2.3.1 and Example 8.22) is given by

$$\mathcal{M}_{\mathrm{av}} = \frac{2^{1-D}}{(1-D)\log 2}, \tag{8.61}$$

where $D = \log_3 2$.

8.4.4 Nonlattice Self-Similar Strings

A self-similar string, lattice or nonlattice, is always strongly languid. More precisely, as is explained at the beginning of Section 6.4, $\zeta_{\mathcal{L}}(s)$ satisfies **L1** and **L2$'$** with $\kappa = 0$ and $A = L^{-1} g_K^{-1} r_N$. Hence, by the second part of Theorem 8.7, $V(\varepsilon)$ is given by a pointwise formula without error term,

$$V(\varepsilon) = \sum_{\omega \in \mathcal{D}_{\mathcal{L}}(\mathbb{C})} \operatorname{res}\left(\frac{\zeta_{\mathcal{L}}(s)(2\varepsilon)^{1-s}}{s(1-s)}; \omega\right) + 2\varepsilon\zeta_{\mathcal{L}}(0), \tag{8.62}$$

for $0 \le \varepsilon < \frac{1}{2} L g_K r_N^{-1}$. Note that neither 0 nor 1 is a complex dimension of a self-similar string. Also, if ω is a simple complex dimension of $\mathcal{L}$, then

$$\operatorname{res}\left(\frac{\zeta_{\mathcal{L}}(s)(2\varepsilon)^{1-s}}{s(1-s)}; \omega\right) = \operatorname{res}(\zeta_{\mathcal{L}}(s); \omega)\,\frac{(2\varepsilon)^{1-\omega}}{\omega(1-\omega)}.$$

This is the case, for example, of all the complex dimensions of a nonlattice string with two scaling ratios and positive multiplicities. (See Theorem 3.30.)

For a nonlattice string we sometimes need an explicit formula with error term to obtain information about $V(\varepsilon)$. We then use a suitable screen and the resulting formula only involves the corresponding visible complex dimensions. By Theorem 2.17, D is the only complex dimension located on the line $\operatorname{Re} s = D$ and it is simple. Therefore, by Corollary 8.10, we have

$$\begin{aligned} V(\varepsilon) &= \frac{(2\varepsilon)^{1-D}}{D(1-D)} \operatorname{res}(\zeta_{\mathcal{L}}(s); D) + \sum_{\operatorname{Re}\omega < D} \operatorname{res}\left(\frac{\zeta_{\mathcal{L}}(s)(2\varepsilon)^{1-s}}{s(1-s)}; \omega\right) \qquad (8.63) \\ &\qquad + \{2\varepsilon\zeta_{\mathcal{L}}(0)\} + \mathcal{R}(\varepsilon) \\ &= \mathcal{M}\cdot\varepsilon^{1-D} + o\big(\varepsilon^{1-D}\big), \qquad (8.64) \end{aligned}$$

as $\varepsilon \to 0^+$, where ω ranges in $\mathcal{D}_{\mathcal{L}}(W)$ and, as usual, the term in braces is included if $0 \in W$. It follows that $\mathcal{L}$ is Minkowski measurable (in agreement with Theorem 8.15), with Minkowski content $\mathcal{M}$ given by (8.25):

$$\mathcal{M} = \operatorname{res}\left(\zeta_{\mathcal{L}}(s); D\right) \frac{2^{1-D}}{D(1-D)}. \tag{8.65}$$

For instance, if $\mathcal{L}$ is the golden string defined in Section 2.3.5, then (8.65) holds with $D \approx .77921$.

By Remark 6.15, this analysis is not valid if, as in Example 5.32, there is no screen passing between $\operatorname{Re} s = D$ and the complex dimensions strictly to the left of this line for which the nonlattice string $\mathcal{L}$ is languid. In that case, we apply formula (6.53) of Section 6.4.2, along with Theorem 3.25. It remains to estimate

$$\sum_{D-\delta/2<\operatorname{Re}\omega<D} \operatorname{res}\left(\zeta_{\mathcal{L}}(s); \omega\right) \frac{(2\varepsilon)^{1-\omega}}{\omega(1-\omega)}, \tag{8.66}$$

for small positive δ, and this is done in Theorem 5.17. The latter result implies that this sum is $o\big(x^D\big)$, as $x \to +\infty$, from which we deduce that (8.63) holds pointwise for an arbitrary nonlattice string.

Note that the hypotheses of Theorem 5.17 are satisfied because the sum (8.66) is absolutely convergent. Indeed, by Theorem 3.25, $\operatorname{res}\left(\zeta_{\mathcal{L}}(s); \omega\right)$ is uniformly bounded for all visible complex dimensions ω, so that (8.66) can be compared to

$$\sum_{D-\delta/2<\operatorname{Re}\omega<D} \frac{1}{\omega^2},$$

which converges by the density estimate (2.37).

We summarize this discussion in the following theorem:

Theorem 8.36. *Every nonlattice string is Minkowski measurable. Further, the volume $V(\varepsilon)$ of the tubular neighborhoods of $\mathcal{L}$ is given by the formulas in* (8.63) (*which hold pointwise for every $\varepsilon > 0$*), *where the Minkowski content $\mathcal{M}$ of $\mathcal{L}$ is given by* (8.65)*—or, more precisely, by* (8.69) *below.*

Using the dimension-free regions of Section 3.6, we also obtain the following estimate of the error of the approximation to $V(\varepsilon)$.

Theorem 8.37. *The volume $V(\varepsilon)$ of the tubular neighborhoods of the nonlattice string $\mathcal{L}$ is estimated by*

$$V(\varepsilon) = \mathcal{M}\varepsilon^{1-D} + O\left(\varepsilon^{1-D}\left(\frac{\log\log\varepsilon^{-1}}{\log\varepsilon^{-1}}\right)^{1/\rho}\right), \quad \text{as } \varepsilon \to 0^+, \tag{8.67}$$

where $\rho = \frac{2}{M-1} + 2l$ and M and l are as in Theorems 7.37 and 7.41.

Proof. The proof is similar to that of Theorems 7.37 and 7.41 on page 228. Note that for Theorem 8.37, the Tauberian argument of the last step of that proof is not needed, because Theorem 8.7 already gives $V(\varepsilon)$ as a pointwise formula. Consequently, the exponent in the error term is best possible in case $M = 2$, and may be best possible in case $M \geq 3$. □

Remark 8.38. The residue of $\zeta_{\mathcal{L}}$ at D can be computed explicitly. If $\mathcal{L}$ is a nonlattice self-similar string with scaling ratios $r_1, \ldots, r_N$ and gaps $g_1, \ldots, g_K$, we have (see Equation (2.43)),

$$\operatorname{res}\left(\zeta_{\mathcal{L}}(s); D\right) = \frac{L^D \sum_{\mu=1}^K g_\mu^D}{\sum_{j=1}^N r_j^D \log r_j^{-1}} \tag{8.68}$$

(as was shown in the proof of Theorem 2.17). Hence, in light of (8.65), the nonlattice string $\mathcal{L}$ has for Minkowski content

$$\mathcal{M} = \frac{2^{1-D} L^D \sum_{\mu=1}^K g_\mu^D}{D(1-D) \sum_{j=1}^N r_j^D \log r_j^{-1}}, \tag{8.69}$$

in agreement with Equation (2.44) of Theorem 2.17, for which we have now completed the proof.

Remark 8.39. Theorem 8.36 above and Theorem 8.23 of Section 8.4.2 establish completely the geometric part of [Lap3, Conjecture 3, p. 163] in the one-dimensional case (that is, for self-similar strings rather than for self-similar drums). When specialized to one dimension ($d = 1$), the latter stated, in particular, that nonlattice self-similar strings are Minkowski measurable whereas lattice strings are not (because they have oscillations of order D in their geometry). See especially the Minkowski measurability criterion in [LapPo1–2], recalled in Remark 8.19 above, and the example of the Cantor set studied in [LapPo2, Example 4.5]. See also the main example of [BroCa], revisited in [FlVa]. We note, however, that our results significantly supplement the geometric aspects of [Lap3, Conjecture 3], as is explained in the next remark.

Remark 8.40. The fact that nonlattice strings are Minkowski measurable, as stated in Theorem 8.36, was already known in the literature (at least in the case of a single gap). See the notes to this chapter. As was alluded to at the end of the previous remark, our present results go further (than in either [Lap3] or [Fa4]) in several directions (besides the extension to multiple gaps):

(i) We provide an explicit formula for the volume of the tubular neighborhoods, $V(\varepsilon)$, valid pointwise and expressed in terms of the complex dimensions of $\mathcal{L}$. In the lattice case, this formula shows particularly well the role of the finitely many lines of complex dimensions. (For the case of a single gap, see formula (8.40) and, when $\mathcal{L}$ has simple complex dimensions, formula (8.43). For more general lattice strings, see Corollary 8.27 and Theorem 8.25, along with Remark 8.28.)

(ii) In the lattice case, we show that the periodic function G is nonconstant and hence that a lattice string is never Minkowski measurable. We also obtain the Fourier series expansion of G. (See formula (8.41) for a single gap, and formula (8.52) for multiple gaps.) More generally, under suitable hypotheses, we obtain one periodic function for each line of complex dimensions (see Theorem 8.25).

(iii) Our results provide a new intuitive understanding of the dichotomy lattice vs. nonlattice and of its consequence for the Minkowski measurability of self-similar strings. Namely, the presence of nonreal complex dimensions of real part D characterizes lattice strings and explains the fact that such strings are not Minkowski measurable (since they have oscillations of order D in their geometry). Analogously, nonlattice strings are Minkowski measurable because they do not have nonreal complex dimensions above D. See Theorems 2.17, 8.1 and 8.15 above, along with the related discussion in Section 12.3 below.

The Error Term in the Volume of the Tubular Neighborhoods

The results of Chapter 3 allow us to obtain a considerable strengthening of formula (8.63) above. Recall from Section 3.6 that a *dimension-free region* is a subset of the complex plane such that its intersection with the set of complex dimensions is only D. If there exist positive numbers C and p such that

$$\left\{\omega \in \mathbb{C}\colon \operatorname{Re}\omega \geq D - C(1+|\operatorname{Im}\omega|)^{-p}\right\} \tag{8.70}$$

is a dimension-free region, then the pointwise explicit formula (8.20) allows us to deduce that for every $\delta > 0$,

$$V(\varepsilon) = \mathcal{M}\varepsilon^{1-D}\left(1 + O\left(|\log\varepsilon|^{\delta-(1/p)}\right)\right), \tag{8.71}$$

as $\varepsilon \to 0^+$. Here, $\mathcal{M} = \frac{2^{1-D}}{D(1-D)}\operatorname{res}(\zeta_{\mathcal{L}}; D)$ is the Minkowski content of the boundary of $\mathcal{L}$, as in formulas (8.65) and (8.69).

From the results of Sections 3.6 and 3.5, we know that a nonlattice string has a dimension-free region, which is of the form (8.70) if the nonlattice string is badly approximable by lattice strings. In general, the dimension-free region is much thinner, with a corresponding weaker form of (8.71).

Remark 8.41. For nonlattice strings with two scaling ratios, where the (irrational) ratio of the weights $w_2/w_1 = \alpha > 1$, the continued fraction expansion of α enables us to carry this analysis further. In particular, the error term depends on a function $b(q)$ with the property that $a_{n+1} \leq b(q_n)$, where $\alpha = [[a_0, a_1, a_2, \dots]]$ is the continued fraction of α, and q_n is the denominator of the n-th convergent of α.

The case when α is badly approximable by rationals corresponds to increasing functions b of polynomial growth. If $b(q)$ grows faster than polynomially, we obtain a bound of the form $\varepsilon^{1-D}/b^{\mathrm{inv}}(-\log\varepsilon)$ for the error in the explicit formula for $V(\varepsilon)$. Here, b^{inv} is the inverse function of b, which in this case grows slower than any positive power function.

8.5 Notes

Section 8.1.1: Theorem 8.7, the pointwise tube formula, is entirely new. In particular, it was not contained in [Lap-vF5].

Section 8.1.2: the fact that the a-string is Minkowski measurable and its Minkowski content is given by formula (8.22) was first established in [Lap1]. See [Lap1, Example 5.1, pp. 512–513], along with [Lap1, Appendix C, pp. 523–524], where it is proved by a direct computation. It was later reproved more conceptually in [LapPo2, Example 4.3, pp. 64–65] and then in [HeLap2].

Section 8.2: it is pointed out in [BergGo, 6.9.8, p. 235] that Chern's classical work on a higher-dimensional analogue of the Gauss–Bonnet formula and on characteristic classes, [Chern1–2], was influenced by the geometric interpretation of Weyl's tube formula given in [Wey3].[15] Additional references can be found in [CheeMüS1, Kow].

Further references closely related to or extending in various directions Weyl's tube formula (or the earlier Steiner's formula [Stein] for convex bodies in Euclidean space [Fed2, Theorem 3.2.35]) include [Mink, Bl, Fed1–2, Ban, CheeMüS2, Gra, Fu1–2, Mil, Schn], where a more detailed history and description of this formula can be found. We point out, in particular, the book by Gray [Gra], as well as the extension in the context of geometric measure theory [Fed2] of the notion of Weyl's curvatures obtained by Federer in his paper on curvature measures [Fed1]. The more recent paper [Fu2], and the relevant references therein, gives a further generalization and interpretation of [Fed1].

Section 8.3: a different proof of the Minkowski measurability criterion of [LapPo1–2] (recalled in Remark 8.19 above) was later obtained by Falconer in [Fa4]. Rather than being of a purely analytical nature, this proof is partly based on dynamical systems. Moreover, as is discussed in Remark 8.21 in relationship with our present work, the Minkowski measurability criterion of [LapPo1–2] (together with the notion of Minkowski measur-

[15] The fact that Weyl's tube formula influenced this aspect of his work in [Chern1–2] was confirmed by S.-S. Chern during a conversation with the first author on the occasion of the symposium given in his honor at the University of California, Berkeley (and at the Mathematical Sciences Research Institute (MSRI), September 2000). See also [Chern3].

ability) was extended to a large class of gauge functions (going beyond the traditional power functions) in [HeLap1; HeLap2, Theorem 2.5 and §4.1].

The idea that lattice strings are not Minkowski measurable and nonlattice strings are was motivated in part by the work of Lalley in [Lal1–3] on various geometric counting functions associated with self-similar sets and by works of the first author in [Lap1–2] as well as of Lapidus and Pomerance in [LapPo1–2]. It was formulated as a conjecture in [Lap3, Conjecture 3] in the more general context of self-similar drums, as is discussed elsewhere in these notes and this book.

Section 8.4.3: we refer to the interesting papers by Tim Bedford and Albert Fisher [BedFi], as well as of Richard Stone [Sto-r], for more information on Cesaro averaging in various contexts.

In [Fra1, 2]—following [BedFi,Man2,Gat] and motivated in part by [LapPo2,Lap3,Fa4] and the results in [Lap-vF5] for self-similar strings (with one gap)—the elusive notion of lacunarity (briefly discussed in Section 12.2.3; see also [Man1]) is defined as the reciprocal of the average Minkowski content. Hence, the lacunarity of any lattice self-similar string (with gaps) is well defined and is given by the reciprocal of formula (8.56),[16] in agreement with [Fra2, Theorem 2.1]. According to Remark 8.38, this is also true for nonlattice self-similar strings, in which case the average Minkowski content coincides with the Minkowski content, given by (8.69)—or, equivalently, by the last expression in (8.56). Motivated in part by [BedFi, KiLap1, Con], a similar type of averaging was used in [Lap5, 6] to study (in the lattice case) the average spectral volume in the counterpart of Weyl's asymptotic formula for drums with self-similar fractal membrane. See also the later paper [KiLap2] where this spectral volume is further studied.

Section 8.4.4: it was first observed by Lapidus in [Lap3, Section 4.4.1b] and later recovered independently by Falconer in [Fa4] that nonlattice strings are Minkowski measurable. The main goal of [Fa4] was to obtain an alternative proof of the Minkowski measurability criterion of [LapPo2, Theorem 2.1] for fractal strings. The method used in both [Lap3] and [Fa4] relies on a suitable use of the Renewal Theorem from probability theory[17] [Fel, Theorem 2, p. 39], much as was done earlier by Lalley [Lal1] in a related context. Shortly after the preliminary version of [Lap3] was written, a similar method—based on the Renewal Theorem—was used by Kigami and the first author in [KiLap1] to obtain a Weyl-type formula for the eigenvalue distribution of Laplacians on finitely ramified self-similar fractals, thereby proving (and specifying) for this class of fractals Conjec-

---

[16] In [Fra2], the expression involving the residue of $\zeta_{\mathcal{L}}(s)$ is not given since the approach used there is along the lines of [Lap3] and [Fa4] rather than of [Lap-vF5] or the present book.

[17] A more general version of the Renewal Theorem that is well suited for this or related situations was later obtained in [LeVa]. Other references concerning the Renewal Theorem include [Sto-c], [Ki2, Appendix B, §B.4], and [HamLap, §3].

ture 5 of [Lap3, p. 190] for self-similar drums with fractal membrane—rather than with fractal boundary, as in [Lap3, Conjecture 3]. (None of these references considered the case of multiple gaps.) In the lattice case, the results of [Lap3] and of [Fa4]—also based on the Renewal Theorem—yield the existence of a multiplicatively periodic function G in the leading term of $\varepsilon^{-(1-D)}V(\varepsilon)$ as $\varepsilon \to 0^+$. However, as was noted in [Lap3], one cannot expect to deduce from the Renewal Theorem alone (or from a Tauberian-type argument) that G is nonconstant (as was conjectured in [Lap3]) and hence that a lattice string is not Minkowski measurable.

Several results of this chapter can be extended to fractal sprays, in the sense of [LapPo3] and Section 1.4. The main difficulty consists in finding a formula for the volume of the (inner) tubular neighborhood of the underlying basic shape $B \subset \mathbb{R}^d$ and in suitably adapting the techniques of the proof of Theorem 8.1. (See [LapPe2–4] where a corresponding tube formula is obtained, applicable to self-similar systems and the associated tilings. In that work, briefly discussed in Section 12.3.2 below, finitely many basic shapes or generators are allowed.) We refer to formulas (8.23) and (8.24) for the case of smooth manifolds B, and to (12.2) and (12.3) for two examples of a fractal basic shape.

In the important special case of self-similar fractal sprays (in the sense of Section 6.6), under suitable assumptions a counterpart of Theorems 8.36 and 8.23 can also be obtained in the nonlattice and lattice case, respectively. For example, as was stated in [Lap2–3], the Sierpinski gasket (see Figure 6.1 on page 204) is not Minkowski measurable because it is a lattice spray, whereas (suitable) nonlattice self-similar sprays are always Minkowski measurable, as was shown in [Lap3] by means of the Renewal Theorem (see also [LeVa] and, in a significantly more general context, [LapPe2, 3]).

9
The Riemann Hypothesis and Inverse Spectral Problems

In this chapter, we provide an alternative formulation of the Riemann hypothesis in terms of a natural inverse spectral problem for fractal strings. After stating this inverse problem in Section 9.1, we show in Section 9.2 that its solution is equivalent to the nonexistence of critical zeros of the Riemann zeta function on a given vertical line. This modifies and extends the earlier work of [LapMa1–2], but now we use the point of view of complex dimensions and the explicit formulas of Chapter 5. In Section 9.3, we then extend this characterization to a large class of zeta functions, including all the number-theoretic zeta functions for which the extended Riemann hypothesis is expected to hold.

The reformulation of the Riemann hypothesis as an inverse spectral problem was obtained when the first author (and his collaborators, particularly Helmut Maier) considered the question of whether one can deduce geometric information about a fractal string from information about its spectrum. In other words, they considered the question

Can one hear the shape of a fractal drum?

In Section 6.3.2, we introduced the spectral operator, which allows us to formalize questions of this type as the question of the invertibility of this operator.

We give here two examples of such inverse spectral problems. Namely,

Can one hear the dimension of a fractal string?

and

Can one hear if a fractal string is Minkowski measurable?

In [LapPo2, Theorem 2.4, p. 47], the first question is answered in the affirmative, and in [LapMa2] (announced in [LapMa1]) it is shown that a positive answer to the second inverse problem is equivalent to the fact that $\zeta(s)$ has no zeros on the line $\operatorname{Re} s = D$. Thus, Lapidus and Maier obtained a reformulation of the Riemann hypothesis in terms of a natural inverse spectral problem for (standard) fractal strings. See Figure 9.1 on the opposite page.

9.1 The Inverse Spectral Problem

The inverse spectral problem studied in [LapMa2] is the following:

(S) Let a standard fractal string with Minkowski dimension $D \in (0,1)$ be given. If this string has no oscillations of order D in its spectrum, does it follow that it is Minkowski measurable?

See [LapMa2] and below for a precise formulation. In [LapMa2], condition (ii) of Theorem 8.15 was used to characterize the Minkowski measurability of a fractal string. In our present situation, we use the equivalent condition (i) of Theorem 8.15 involving the complex dimensions above D.

Remark 9.1. Intuitively, the inverse spectral problem (S) can be interpreted as follows (see Remark 8.20): If an ordinary fractal string $\mathcal{L}$ of Minkowski dimension D has no oscillations of order D in its frequency spectrum, does it follow that it has no oscillations of order D in its geometry (i.e., by Theorem 8.15, that D is simple and $\mathcal{L}$ has no other complex dimensions on the vertical line $\operatorname{Re} s = D$)? By contraposition, we obtain the equivalent formulation: If $\mathcal{L}$ has oscillations of order D in its geometry, does it follow that it has oscillations of order D in its spectrum?

The inverse spectral problem is in the spirit of questions raised in the spectral theory of smooth manifolds; see Appendix B. Let Ω be a bounded open subset of $\mathbb{R}$, with boundary $\partial\Omega$ of Minkowski dimension $D \in [0,1]$. That is, Ω is an ordinary fractal string $\mathcal{L} = (l_j)_{j=1}^{\infty}$. Then a counterpart of estimate (B.12) was obtained in [LapPo1–2], where the first connections between the vibrations of fractal strings and the Riemann zeta function $\zeta(s)$ were established. This provided, in particular, a resolution of the (modified) Weyl–Berry Conjecture for fractal strings (as formulated in [Lap1]). More precisely, [LapPo2, Corollary 2.3, p. 46] states that if $\partial\Omega$ is Minkowski measurable in dimension $D \in (0,1)$ with Minkowski content $\mathcal{M}(D;\partial\Omega)$, then an analogue of (B.12) holds for the Dirichlet Laplacian on Ω, with $d = 1$ and $d-1$ replaced by D. Namely, we have the following pointwise asymptotic formula:

$$N_\nu(x) = W(x) - c_D \mathcal{M}(D;\partial\Omega)x^D + o(x^D), \tag{9.1}$$

as $x \to \infty$, where

$$W(x) = \operatorname{vol}_1(\Omega)x$$

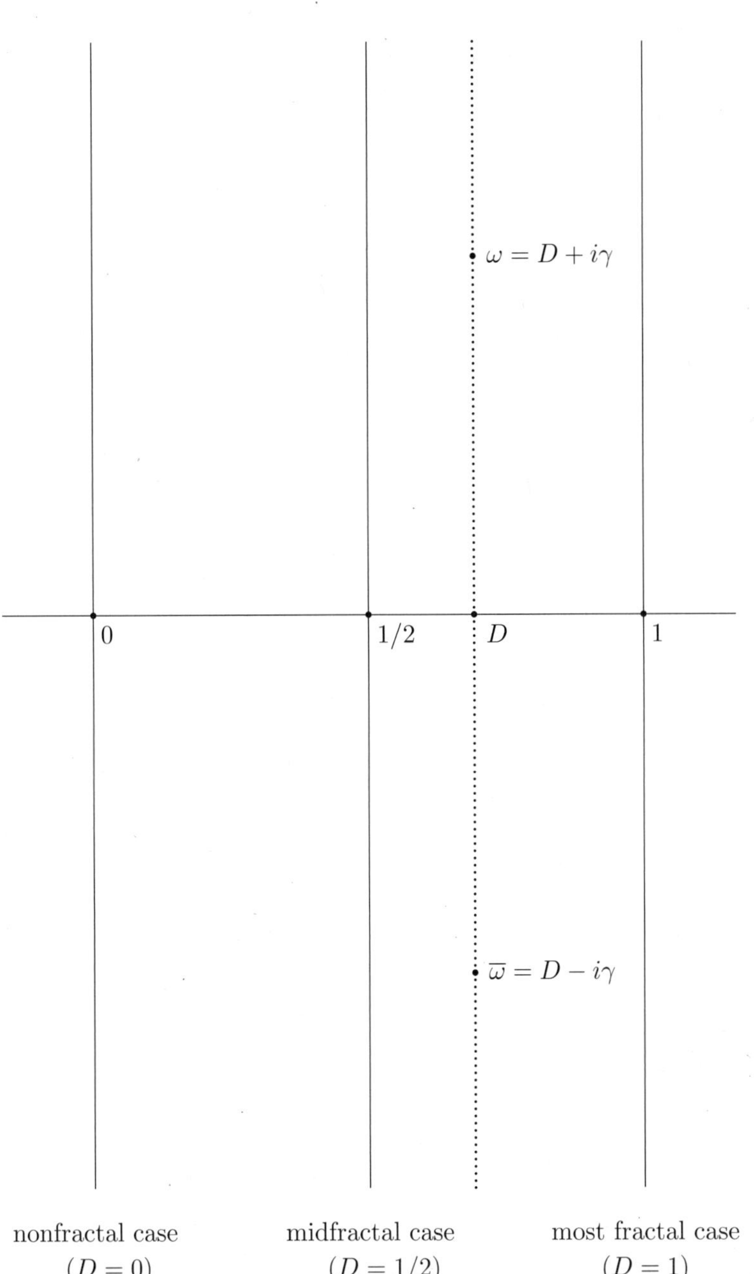

Figure 9.1: The critical strip for $\zeta(s)$: $0 \leq \operatorname{Re} s \leq 1$.

is the Weyl term and

$$c_D = 2^{-(1-D)}(1-D)\,(-\zeta(D))$$

(since $\zeta(D) < 0$ for D in the critical interval $(0,1)$, we have $c_D > 0$). Note that (9.1) implies that $N_\nu(x)$ admits a monotonic asymptotic second term.[1] The converse of this result (namely, the corresponding inverse spectral problem for fractal strings), was shown in [LapMa1–2] to be closely connected with the Riemann hypothesis. More precisely, Theorems 2.3 and 2.4 in [LapMa2, p. 20] show, in particular, that for a fixed $D \in (0,1)$, the existence of a monotonic asymptotic second term for $N_\nu(x)$ (with the pointwise error term $o(x^D)$ replaced by $O\big(x^D \log^{-(1+\varepsilon)} x\big)$, for some $\varepsilon > 0$) always implies that $\partial\Omega$ is Minkowski measurable if and only if the Riemann zeta function does not vanish on the vertical line $\operatorname{Re} s = D$.

Remark 9.2. For fractal strings, the following conditions were shown to be equivalent in [LapPo2, Theorem 2.4, p. 47]:

(i) $0 < \mathcal{M}_*(D;\partial\Omega) \le \mathcal{M}^*(D;\partial\Omega) < \infty$;

(ii) $0 < \liminf_{j\to\infty} l_j j^{1/D} \le \limsup_{j\to\infty} l_j j^{1/D} < \infty$;

(iii) $0 < \delta_* \le \delta^* < \infty$,

where δ_* (respectively, δ^*) denotes the lower (respectively, upper) limit of $x^{-D}\,(W(x) - N_\nu(x))$ as $x \to \infty$.[2] In conjunction with the aforementioned results [LapPo2, Theorems 2.1 and 2.2], this shows that the existence of an oscillatory second term of order D for $N_\nu(x)$ (i.e., by definition, $0 < \delta_* < \delta^* < \infty$) implies that (i) and (ii) hold with strict inequalities, and hence that $\partial\Omega$ is not Minkowski measurable.[3] On the other hand,

---

[1] The characterization of Minkowski measurability obtained in [LapPo2, Theorem 2.2, p. 46] played a key role in deriving this result. Recall that the criterion of [LapPo1–2] (namely, $l_j \sim Lj^{-1/D}$, for some $L > 0$) was extended in Chapter 8 and interpreted in terms of the notion of complex dimension; see especially Theorem 8.15 and Remark 8.19 above.

[2] Observe that by the Minkowski measurability criterion of [LapPo2, Theorem 2.2], condition (i) holds with strict inequalities if and only if (ii) does. Further observe that condition (i) (or, equivalently, (ii)) holds with strict inequalities exactly when the geometry of the boundary $\partial\Omega$ (i.e., of the fractal string $\mathcal{L}$) has oscillations of order D.

[3] For the easier part of the equivalence of (i) and (ii), namely,

$$\mathcal{M}^*(D;\partial\Omega) < \infty \Longleftrightarrow \limsup_{j\to\infty} l_j \cdot j^{1/D} < \infty,$$

compare also Proposition 1.1 and Theorem 1.16 of Chapter 1. On the other hand, by [LapPo2, Theorem 3.11, p. 55], only the implication

$$\liminf_{j\to\infty} l_j \cdot j^{1/D} > 0 \Longrightarrow \mathcal{M}_*(D;\partial\Omega) > 0$$

holds, whereas the converse is not true, by [LapPo2, Example 3.13, p. 56] the results of [LapMa1–2] (specifically, [LapMa2, Theorem 3.16, p. 28]) show that the converse is not true; namely, $\partial\Omega$ may not be Minkowski measurable (with strict inequalities in (i) and (ii)) whereas $N_\nu(x)$ admits a monotonic asymptotic second term of order D (i.e., $0 < \delta_* = \delta^* < \infty$). The fact that for Cantor-type sets (or strings), the spectral operator is invertible (in the terminology of Section 6.3.2) and hence this phenomenon cannot occur, is established in Chapter 10 and is used in Chapter 11 below. In turn, the results of Chapter 11 can be used to show that a lattice string always has oscillations of order D in its spectrum, and hence satisfies (iii) as well as conditions (i) and (ii) with strict inequalities; see Remark 6.17 in Section 6.4.3. Further, by using the results of Section 11.2, an entirely analogous statement can be shown to hold for a large class of (generalized) lattice self-similar sprays (as in Section 6.6).

9.2 Complex Dimensions of Fractal Strings and the Riemann Hypothesis

Our explicit formulas shed new light on the results and methods of M. L. Lapidus and H. Maier in [LapMa2]. We will use condition (i) of Theorem 8.15 to characterize the Minkowski measurability of a fractal string in order to recover (and extend) the results of [LapMa2] by focusing on the fractal string η introduced below and in [LapMa2, §3.3], as well as on its continuous analogue μ.

Consider the generalized fractal string

$$\mu(dx) = \left(x^{D-1} + \beta x^{D-1+i\gamma} + \overline{\beta} x^{D-1-i\gamma}\right) \mathbf{1}_{(1,\infty)}(x)\, dx, \tag{9.2}$$

where $\gamma \in \mathbb{R}$, $D \in (0,1)$ and $\beta \in \mathbb{C}$, $|\beta| \leq 1/2$, so that μ is a positive measure (here, $\mathbf{1}_{(1,\infty)}$ denotes the indicator function of the interval $(1,\infty)$). The geometric zeta function of μ is

$$\zeta_\mu(s) = \frac{1}{s-D} + \frac{\beta}{s-D-i\gamma} + \frac{\overline{\beta}}{s-D+i\gamma}, \tag{9.3}$$

and the complex dimensions of μ are D, $D+i\gamma$ and $D-i\gamma$. By Theorem 8.15, this string is not Minkowski measurable.[4] By Theorem 5.14, applied for $k=1$, we have the following pointwise equality, for all $x \geq 1$:

$$N_\mu(x) = \frac{x^D}{D} + \beta\frac{x^{D+i\gamma}}{D+i\gamma} + \overline{\beta}\frac{x^{D-i\gamma}}{D-i\gamma} + \zeta_\mu(0). \tag{9.4}$$

---

[4]Theorem 8.15 is formulated for ordinary fractal strings, but it clearly applies to this situation as well.

This formula can also be obtained by a direct computation.

For the frequency counting function, we cannot apply Theorem 5.14 to obtain a pointwise formula at level 1, hence we have to interpret $N_\nu(x)$ as a distribution. Thus, by Theorems 5.18 and 5.30, we have, in a distributional sense,

$$N_\nu(x) = \zeta_\mu(1)x + \zeta(D)\frac{x^D}{D} + \beta\zeta(D+i\gamma)\frac{x^{D+i\gamma}}{D+i\gamma} + \overline{\beta}\zeta(D-i\gamma)\frac{x^{D-i\gamma}}{D-i\gamma} + O(1), \quad (9.5)$$

as $x \to \infty$.

Now, if $\zeta(D+i\gamma)$, and hence also $\zeta(D-i\gamma)$, vanishes, then

$$N_\nu(x) = \zeta_\mu(1)x + \zeta(D)\frac{x^D}{D} + O(1). \quad (9.6)$$

We conclude that $N_\nu(x)$ has no oscillatory terms. Therefore, μ provides a counterexample to problem (S).

Remark 9.3. Instead of using the distributional explicit formula, we could apply the pointwise formula, Theorem 5.10, at level $k=2$. Then we obtain an integrated version of $N_\nu(x)$ (with $(s)_2 = s(s+1)$, as in Equation (5.12)):

$$N_\nu^{[2]}(x) = \zeta_\mu(1)\frac{x^2}{2} + \zeta(D)\frac{x^{D+1}}{(D)_2} + \beta\zeta(D+i\gamma)\frac{x^{D+1+i\gamma}}{(D+i\gamma)_2} + \overline{\beta}\zeta(D-i\gamma)\frac{x^{D+1-i\gamma}}{(D-i\gamma)_2} + O\left(x^{1+\varepsilon}\right), \quad (9.7)$$

as $x \to \infty$, for all $\varepsilon > 0$. In this formula, $O\left(x^{1+\varepsilon}\right)$ has the usual pointwise meaning, as opposed to $O(1)$ in formulas (9.5) and (9.6), which must be interpreted distributionally as in Section 6.3.1. Consequently, if $\zeta(s)$ vanishes at $s = D \pm i\gamma$, the spectrum has no oscillations of order $D+1$ at level 2.

Since the function (9.4) is strictly increasing, we can define an ordinary fractal string η by the property

$$N_\eta(x) = [N_\mu(x)], \text{ for every } x > 0. \quad (9.8)$$

Thus η is the ordinary fractal string $\eta = \sum_{j=1}^\infty \delta_{\{l_j^{-1}\}}$, with $l_j > 0$ defined uniquely by $N_\mu\left(l_j^{-1}\right) = j$, for each $j = 1, 2, \ldots$. Then we have

$$\zeta_\eta(s) = \zeta_\mu(s) + \zeta_{\eta-\mu}(s).$$

The function $\zeta_{\eta-\mu}(s)$ is holomorphic for $\operatorname{Re} s > 0$. Hence, η and μ have the same complex dimensions, namely D, $D+i\gamma$ and $D-i\gamma$, each of which is simple. Since η has nonreal complex dimensions above D, it follows from the above Minkowski measurability criterion (Theorem 8.15) that η is not Minkowski measurable: it has oscillations of order D in its geometry. Thus we recover the counterexample of [LapMa2] to the inverse spectral problem (S).

We now consider the converse of the above question. That is, we want to show that if the Riemann zeta function does not vanish on the vertical line $\operatorname{Re} s = D$, then the inverse spectral problem (S) in dimension D always has an affirmative answer for a suitable class of generalized fractal strings. Indeed, if η is an arbitrary languid[5] generalized fractal string of dimension $D \in (0,1)$, its oscillations in dimension D are described by its complex dimensions with real part D. We choose a screen S to the left of $\operatorname{Re} s = D$ so that only the complex dimensions with real part D are visible. Hence we need to assume that η allows such a screen (see Examples 5.32 and 5.33). Then, by Theorems 5.18 and 5.30,

$$N_\eta(x) = \sum_{\operatorname{Re}\omega = D} \operatorname{res}\left(\frac{\zeta_\eta(s)x^s}{s}; \omega\right) + o(x^D) \tag{9.9}$$

and

$$N_\nu(x) = \zeta_\eta(1)x + \sum_{\operatorname{Re}\omega = D} \operatorname{res}\left(\frac{\zeta_\eta(s)\zeta(s)x^s}{s}; \omega\right) + o(x^D), \tag{9.10}$$

as $x \to \infty$ (these formulas have to be interpreted distributionally). If ζ has no zeros on the line $\operatorname{Re} s = D$, all terms in the second series (which represents an almost periodic function, in an extended sense) remain, and we obtain a positive answer to the inverse spectral problem considered in [LapMa2].

Remark 9.4. In [LapMa1–2], the converse was established by using the Wiener–Ikehara(–Landau) Tauberian Theorem [Pos, Section 27, pp. 109–112] rather than an explicit formula. Accordingly, the argument is based on the assumption of a suitable error term beyond the asymptotic second term in the spectral counting function $N_\nu(x)$, rather than on the existence of a suitable screen. See [LapMa2, Theorem 2.3 and 3.2, pp. 20 and 23] and the discussion in Section 9.1.

We summarize the above discussion in the following theorem:

Theorem 9.5. *For a given D in the critical interval $(0,1)$, the inverse spectral problem* (S) *in dimension D—suitably interpreted as above—has a positive answer if and only if $\zeta(s)$ does not have any zero on the vertical line* $\operatorname{Re} s = D$.

Since $\zeta(s)$ has zeros on the critical line $\operatorname{Re} s = 1/2$, we also obtain the following corollary.

[5] That is, η satisfies **L1** and **L2**.

Corollary 9.6. *The inverse spectral problem* (S) *is not true in the midfractal case (i.e., when $D = 1/2$, see Figure 9.1).*

On the other hand, it is true for every $D \in (0,1)$, $D \neq 1/2$, if and only if the Riemann hypothesis holds. In the terminology of Section 6.3.2, the spectral operator is invertible for all fractal strings of dimension $D \neq 1/2$ if and only if the Riemann hypothesis holds. In that case, the inverse spectral operator is given as in Section 6.3.2, Remark 6.9.

Remark 9.7. The first part of the above corollary, stating that (S) is not true in the midfractal case, does not necessarily have a counterpart for the general zeta functions $\zeta_B(s)$ considered in Section 9.3 below.

Remark 9.8. The reformulation of the Riemann hypothesis in Theorem 9.5 raises the question of determining what is different about fractal strings of dimension 1/2. The authors have introduced the notion of the *dual of a fractal string* to be able to take into account the functional equation of the Riemann zeta function. The only strings that can be self-dual are those of dimension 1/2.

One could also ask to characterize the basic shapes B for which the associated zeta function has an Euler product and satisfies a functional equation. The authors do not know such a characterization other than in terms of ζ_B.

9.3 Fractal Sprays and the Generalized Riemann Hypothesis

Instead of considering fractal strings, we consider other fractal sprays on the basic shape B, as defined in Section 1.4. The spectral zeta function of the fractal spray of $\mathcal{L}$ on B is

$$\zeta_\nu(s) = \zeta_\mathcal{L}(s) \cdot \zeta_B(s), \tag{9.11}$$

where

$$\zeta_B(s) = \sum_\lambda \lambda^{-s/2}. \tag{9.12}$$

Here, λ runs over the normalized eigenvalues of the positive Laplacian on B, with Dirichlet boundary conditions (and hence, the frequencies of B are the numbers $f = \sqrt{\lambda}$).

Example 9.9. The frequencies of the fundamental domain of the lattice $\mathbb{Z}^m$ in $\mathbb{R}^m$ ($m \in \mathbb{N}^*$), with identification of opposite sides, are described by the classical Epstein zeta functions

$$\zeta_m(s) = \sum_{v \in \mathbb{Z}^m \setminus \{0\}} \|v\|^{-s},$$

where $\|v\|$ is the Euclidean norm. These functions satisfy a functional equation, relating $\zeta_m(s)$ and $\zeta_m(m-s)$; see [Ter, Theorem 1, p. 59] or Section A.4 of Appendix A. Note that $\zeta_1(s) = 2\zeta(s)$.

In general, we do not need to assume the existence of B as a subset of $\mathbb{R}^m$. In other words, we consider virtual basic shapes and the associated generalized fractal sprays. Indeed, for our theory to apply, we only need to assume that $\zeta_B(s) = \sum_f w_f f^{-s}$ is a zeta function that satisfies **L1** and **L2** for some suitable screen S, as in Section 9.2. If the coefficients w_f—to be thought of as the (complex) multiplicities of f—are real, we use the above generalized fractal string μ (with complex dimensions D, $D + i\gamma$ and $D - i\gamma$) to test whether ζ_B has zeros at $D \pm i\gamma$, by applying the analogue of formula (9.5). If the coefficients w_f are not real—such as for a Dirichlet L-series associated with a complex, nonreal character (see, e.g., [Lan] or Appendix A.2 below)—we cannot assume that the zeros come in complex conjugate pairs. However, if the window of ζ_B is symmetric with respect to the real axis (and ω is not a zero when $\overline{\omega}$ is a pole of ζ_B), we replace $\zeta_B(s)$ by the symmetrized Dirichlet series $\zeta_B(s)\overline{\zeta_B(\overline{s})}$, the zeros of which come in complex conjugate pairs. We can then go through exactly the same reasoning as above to characterize the absence of zeros of $\zeta_B(s)$ on the line $\operatorname{Re} s = D$ in terms of an inverse spectral problem for fractal sprays.

We summarize the above discussion in the following theorem:

Theorem 9.10. *Let $\zeta_B(s)$ be a languid*[6] *Dirichlet series (or Dirichlet integral) with associated window W. Let $D \in W \cap \mathbb{R}$. Then, if $\zeta_B(s)$ has real coefficients (or is associated with a real measure), the exact counterpart of the inverse spectral problem* (S) *for the corresponding generalized fractal spray is true for this value of D if and only if $\zeta_B(s)$ does not have any zero on the vertical line* $\operatorname{Re} s = D$.

Furthermore, in the general case when $\zeta_B(s)$ is not real-valued on the real axis, then, provided that W is symmetric with respect to the real axis, and ω is not a zero when $\overline{\omega}$ is a pole of $\zeta_B(s)$, the same criterion as above holds.

Remark 9.11. We note that, in the above theorem, $\zeta_B(s)$ need not satisfy a functional equation or an Euler product.

We thus obtain a criterion for the absence of zeros on vertical lines within the critical strip for all Epstein zeta functions [Ter], and for all the number-theoretic zeta functions for which the generalized Riemann hypothesis is expected to hold: for example, for the Dedekind zeta functions and, more generally, for the Hecke L-series associated with an algebraic number field [Lan, ParSh1–2], and also for the zeta functions of algebraic varieties over a finite field [ParSh1, Chapter 4, §1]. In all these cases, we

[6] That is, $\zeta_B(s)$ satisfies hypotheses **L1** and **L2** for a suitable screen S.

may choose the window W to be all of $\mathbb{C}$. (We refer to Appendix A for a brief introduction to the aforementioned zeta functions.)

Remark 9.12. In the case of the zeta function of a variety over a finite field, the Weyl term takes a special form. Consequently, the inverse spectral problem must be suitably interpreted. We refer to Section 11.5 below for a detailed discussion of the one-dimensional case of a curve over a finite field.

9.4 Notes

Figure 9.1: for related comments, see the figures in [Lap2, Figure 3.1, p. 165] and [Lap3, Figure 2.1, p. 143], along with the text surrounding them.

On the basis of [LapPo2, Theorem 2.1] and its proof, the authors of that paper have raised the question of finding a suitable notion of complex dimension that would extend, in particular, the usual notion of real fractal (i.e., Minkowski) dimension and would provide (when $d = 1$) a new interpretation for the critical strip $0 \leq \operatorname{Re} s \leq 1$ of the Riemann zeta function. See [LapPo1, p. 347] and [LapPo2, Section 4.4.b, p. 67], along with, for example, [Lap2, Remark 2.2 and Figure 3.1, pp. 142–143] and Figure 9.1. The later work in [LapMa1–2] used this intuition and also began to corroborate it (see, e.g., [LapMa2, Section 3.3 and Remark 3.21(c), (d), p. 32]), while [Lap-vF5] and the present work provides a rigorous theory for the notion of complex dimension.

A natural question raised in [LapPo2] (and further motivated by the results of [LapMa2]) consists in interpreting in terms of fractal strings and their (then hypothetical) complex dimensions the key symmetry of the Riemann zeta function with respect to the critical line $\operatorname{Re} s = 1/2$, which comes from the functional equation for ζ. (See also Remark 9.8 above.) Various aspects of this intriguing problem are discussed at some length in [Lap10], both from a physical and a mathematical point of view, for fractal strings and their quantization (fractal membranes, see Section 12.4.2 below). It is fair to say, however, that on a rigorous level, the question remains largely open at this point.

Remark 9.8: the authors' initial hope and expectation is that the coupling of a suitable (noncritical, i.e., of dimension $D \neq 1/2$) fractal string and of its dual string would help make the spectral oscillations of the resulting string manifest, by a phenomenon akin to the resonance phenomenon in the standard theory of vibrations, thereby showing (in view of Theorem 9.5 or 9.10 and the construction of Section 9.2, modified as suggested here) that $\zeta(s)$ cannot have any critical zeros such that $\operatorname{Re} s \neq 1/2$. Of course, if a fractal string is self-dual (i.e., critical or, equivalently, of dimension 1/2), then such a resonance phenomenon cannot occur, so that this idea would not preclude the existence of zeros on the line $\operatorname{Re} s = 1/2$. Although, intuitively, these expectations are very natural, they have proved so far to be very difficult to establish rigorously. A deeper understanding of the open problem mentioned in the previous paragraph seems to be necessary before any significant progress can be made in this direction. We note that in view of the methods and results of Chapter 11 (especially, in Section 11.1), a closely related comment can be made about a possible way to approach the main conjectures discussed in Section 12.1.

Section 9.3: we refer to [Ter, §1.4] for information about Epstein zeta functions, and to [ParSh1] for Hecke L-series associated with an algebraic number field. Information on zeta functions associated with algebraic varieties over a finite field can be found in [ParSh1, Chapter 4, §1].

10

Generalized Cantor Strings and their Oscillations

In this chapter, we analyze the oscillations in the geometry and the spectrum of the simplest type of generalized self-similar fractal strings. The complex dimensions of these generalized Cantor strings form a vertical arithmetic sequence $D + in\mathbf{p}$ $(n \in \mathbb{Z})$. We construct such a generalized Cantor string for any real-valued choice of D and positive $\mathbf{p}$. We also construct for each positive integer Λ the so-called truncated Cantor strings, which have a finite arithmetic progression $D + in\mathbf{p}$ of complex dimensions, where n is restricted by $-\Lambda < n < \Lambda$.

In Chapter 11, we shall apply the results of the present chapter and our explicit formulas from Chapter 5 to prove that suitable Dirichlet series with positive coefficients have no infinite vertical sequence of critical zeros in arithmetic progression. Using the truncated Cantor strings, we shall also obtain an upper bound for the possible length of such a progression of critical zeros.

10.1 The Geometry of a Generalized Cantor String

We begin by recalling the definition of the generalized Cantor string introduced in Section 8.4.1:

Definition 10.1. For real numbers $a > 1$ and $b > 0$, the *generalized Cantor string* (with parameters a and b) is the string

$$\mathcal{L}_{D,\mathbf{p}} = \sum_{n=0}^{\infty} b^n \, \delta_{\{a^{-n}\}}. \tag{10.1}$$

That is, it is the string with lengths $1, a^{-1}, a^{-2}, \ldots$, repeated with (possibly noninteger) multiplicity $1, b, b^2, \ldots$.

In Section 8.4.1, we allowed b to be an arbitrary complex number, but in this chapter, we assume that b is real. This corresponds to the requirement that the Minkowski dimension D of $\mathcal{L}_{D,\mathbf{p}}$ is real. Moreover, $0 < D < 1$ corresponds to $1 < b < a$.

The geometric zeta function of this string is

$$\zeta_{D,\mathbf{p}}(s) = \frac{1}{1 - b \cdot a^{-s}}. \tag{10.2}$$

Its dimension is $D = \log_a b$ and $\mathbf{p} = 2\pi / \log a$ is its oscillatory period.

One deduces from (10.2) that the complex dimensions are located at the points $\omega = D + in\mathbf{p}$, with $n \in \mathbb{Z}$. Hence, they lie on a single vertical line, $\operatorname{Re} s = D$, and their imaginary parts form a doubly infinite arithmetic progression (see Figure 10.1). Further, the poles are simple, with residue $1/\log a$.

Remark also that the pairs $(D, \mathbf{p})$ and (a, b) determine each other: D and $\mathbf{p}$ are given above in terms of a and b, and conversely, $a = e^{2\pi/\mathbf{p}}$ and $b = a^D$. (See also Section 8.4.1, Equation (8.28).) Note that if $a > b > 1$, then $D \in (0,1)$ as in our usual framework. However, we need to consider here for later use in Chapter 11 the more general case when D is an arbitrary positive or negative real number.

The geometric counting function is easy to compute. There are

$$1 + b + b^2 + \cdots + b^n$$

lengths less than x, with $n = [\log_a x]$. Thus

$$N_{D,\mathbf{p}}(x) = \frac{b^{n+1} - 1}{b - 1} = \frac{b}{b-1} b^{[\log_a x]} - \frac{1}{b-1}. \tag{10.3}$$

Much as in Section 6.4.1, and using the Fourier series of the periodic function $u \mapsto b^{-\{u\}}$ (see formula (1.13)), we obtain

$$N_{D,\mathbf{p}}(x) = \frac{x^D}{\log a} \sum_{n \in \mathbb{Z}} \frac{x^{in\mathbf{p}}}{D + in\mathbf{p}} - \frac{1}{b-1}. \tag{10.4}$$

Indeed, this coincides with the expression given by the explicit formula of Theorem 5.14.

More generally, for the k-th integrated counting function, the pointwise explicit formula without error term, Theorem 5.14, yields for every $k \geq 1$

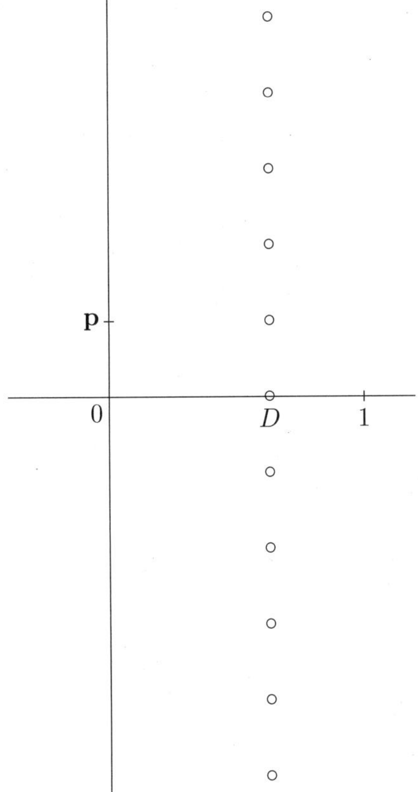

Figure 10.1: The complex dimensions of the generalized Cantor string with parameters a and b. Here, $D = \log_a b$ and $\mathbf{p} = 2\pi/\log a$.

(with $(s)_k$ given by Equation (5.12)):

$$N_{D,\mathbf{p}}^{[k]}(x) = \frac{x^{D+k-1}}{\log a} \sum_{n\in\mathbb{Z}} \frac{x^{in\mathbf{p}}}{(D+in\mathbf{p})_k} + \frac{1}{(k-1)!} \sum_{j=0}^{k-1} \binom{k-1}{j} (-1)^j \frac{x^{k-1-j}}{1-b\cdot a^j}. \quad (10.5)$$

This formula could also be derived from (10.4) by repeated integration, keeping in mind that the constants of integration are fixed by the condition that $N_{D,\mathbf{p}}^{[k]}(0) = 0$ for $k \geq 1$. It is not an easy matter, however, to compute the second sum over j in this way.

In view of formula (10.3), and since $a^D = b$, we clearly have the following result:

Theorem 10.2. *The counting function of the lengths of a generalized Cantor string is monotonic, locally constant, with jumps of* $b^n = x^D$ *at the points* $x = a^n$ $(n \in \mathbb{N})$.

10.2 The Spectrum of a Generalized Cantor String

Theorem 5.10, the pointwise explicit formula with error term, gives us the following expansion for the spectral counting function:

$$N_{\nu,D,\mathbf{p}}^{[k]}(x) = \frac{a}{(a-b)}\frac{x^k}{k!} + \frac{x^{D+k-1}}{\log a}\sum_{n\in\mathbb{Z}}\frac{\zeta(D+in\mathbf{p})}{(D+in\mathbf{p})_k}x^{in\mathbf{p}}$$
$$+ \frac{1}{(k-1)!}\sum_{j=0}^{k-1}\binom{k-1}{j}(-1)^j\zeta(-j)\frac{x^{k-1-j}}{1-b\cdot a^j} + R_\nu(x), \quad (10.6)$$

for $k \geq 2$. For $k = 1$, $\mathcal{P}_\nu^{[1]}$ is given by the same formula, as a distribution. This follows from Theorem 5.18, the distributional formula with error term. (See Remark 3.27, in conjunction with Section 6.4, which explains why our explicit formulas can be applied in this situation to obtain (10.6).)

In case $k = 1$, using the explicit formula (10.4) for $N_{D,\mathbf{p}}(x)$, we can even derive a closed formula for the error term $R_\nu(x)$, by a device similar to the one used in the papers [LapPo2, LapMa2]. For reasons that will become clear below, we write the explicit formula for $N_{D,\mathbf{p}}$, formula (10.4), as a limit,

$$N_{D,\mathbf{p}}(x) = \frac{b}{b-1}b^{[\log_a x]} - \frac{1}{b-1} = \lim_{N\to\infty}\frac{1}{\log a}\sum_{|n|\leq N}\frac{x^{D+in\mathbf{p}}}{D+in\mathbf{p}} - \frac{1}{b-1}.$$

At $x = 1$, the value of the sum is the average of the limits at 1_+ and 1_-, hence we obtain

$$\frac{1}{\log a}\sum_{n=-\infty}^{\infty}\frac{1}{D+in\mathbf{p}} = \frac{1}{b-1} + \frac{1}{2}.$$

Similarly, the total length of the generalized Cantor string is obtained as

$$\frac{1}{1-b/a} = \frac{1}{\log a}\sum_{n=-\infty}^{\infty}\frac{1}{1-D-in\mathbf{p}} + \frac{1}{2}.$$

Since $N_{D,\mathbf{p}}(x)$ vanishes for $0 \leq x < 1$, we have

$$N_{\nu,D,\mathbf{p}}(x) = \sum_{k=1}^{[x]} N_{D,\mathbf{p}}(x/k).$$

We substitute the explicit formula for $N_{D,\mathbf{p}}(x/k)$ to obtain

$$N_{\nu,D,\mathbf{p}}(x) = -\frac{[x]}{b-1} + \lim_{N\to\infty} \frac{1}{\log a} \sum_{|n|\le N} \sum_{k=1}^{[x]} \frac{(x/k)^{D+in\mathbf{p}}}{D+in\mathbf{p}}. \tag{10.7}$$

We need the following expression for the truncated sum $\frac{x^s}{s}\sum_{k=1}^{[x]} k^{-s}$,

$$\frac{1}{s}\sum_{k=1}^{[x]}\left(\frac{x}{k}\right)^s = \frac{x}{1-s} + \frac{[x]}{s} + \frac{x^s}{s}\zeta(s) + \log a \int_0^\infty \{xa^t\}\, a^{-st}\,dt, \tag{10.8}$$

where, as before, $\{x\}$ denotes the fractional part of x. Collecting the terms of order x, we deduce from (10.7) and (10.8) that

$$\begin{aligned} N_{\nu,D,\mathbf{p}}(x) = \frac{x}{1-b/a} &+ \lim_{N\to\infty}\frac{1}{\log a}\sum_{|n|\le N}\frac{x^{D+in\mathbf{p}}}{D+in\mathbf{p}}\zeta(D+in\mathbf{p}) - \frac{\{x\}}{2} \\ &+ \lim_{N\to\infty}\int_0^\infty \{xa^t\}\, b^{-t}\sum_{|n|\le N} e^{-2\pi int}\,dt. \end{aligned}$$

Both limits exist, and it is well known that the second limit, $\sum_{n\in\mathbb{Z}} e^{-2\pi int}$, is a sum of unit point masses at each integer. Since the integral starts at 0, this point is counted with mass 1/2, which cancels the term $-\frac{1}{2}\{x\}$. Thus we obtain

$$N_{\nu,D,\mathbf{p}}(x) = \frac{x}{1-b/a} + \frac{1}{\log a}\sum_{n=-\infty}^{\infty}\frac{x^{D+in\mathbf{p}}}{D+in\mathbf{p}}\zeta(D+in\mathbf{p}) + \sum_{n=1}^{\infty}\{xa^n\}\, b^{-n}. \tag{10.9}$$

In particular, the error term $R_\nu(x)$ in Equation (10.6) with $k=1$ satisfies

$$\frac{\zeta(0)}{1-b} + R_\nu(x) = \sum_{n=1}^{\infty}\{xa^n\}\, b^{-n}.$$

Using the fact that $\zeta(0) = -1/2$, we deduce that

$$-\frac{1}{2(b-1)} \le R_\nu(x) \le \frac{1}{2(b-1)}. \tag{10.10}$$

Remark 10.3. There is an interesting alternative way to derive the above explicit formula for the spectral counting function. Note that in general

$$N_\nu(x) = \sum_{\mu=1}^{\infty}\sum_{l_j\le x/\mu} 1 = \sum_{l_j\le x}[l_j x].$$

Since $[x] = x - \{x\}$, and $[l_j x] = 0$ for $l_j < 1/x$, we obtain, as in the proof of Theorem 1.20, that

$$N_\nu(x) = \mathrm{vol}_1(\mathcal{L})x - \sum_{j=1}^{\infty} \{l_j x\}.$$

We apply this to the Cantor string, with lengths a^n of multiplicity b^n, to obtain

$$N_{\nu,D,\mathbf{p}}(x) = \frac{x}{1 - b/a} - \sum_{n=0}^{\infty} b^n \left\{ x a^{-n} \right\}.$$

The function $f(t) = \sum_{n=-\infty}^{\infty} b^{n-t} \left\{ a^{t-n} \right\}$ is clearly periodic with period 1. Hence it has a Fourier series $f(x) = \sum_{m=-\infty}^{\infty} c_m e^{2\pi imt}$, where

$$c_m = \int_0^1 f(t) e^{-2\pi imt} dt = \int_{-\infty}^{\infty} \{a^t\} b^{-t} e^{-2\pi imt} dt.$$

Using Equation (10.8) for $x = 1$, we can compute this integral to obtain

$$c_m = -\frac{\zeta(D + im\mathbf{p})}{D + im\mathbf{p}}, \qquad \text{for all } m \in \mathbb{Z}.$$

Thus we deduce formula (10.9). See also the proof of Theorem 10.16, where this method is applied.

10.2.1 Integral Cantor Strings: a-adic Analysis of the Geometric and Spectral Oscillations

There is an important difference in the spectrum between integral and nonintegral values of a. In this section, we study the first case, when a is integral, and in Section 10.2.2, we study the case when a is nonintegral.

Definition 10.4. The generalized Cantor string constructed with the parameters a and b, is called *integral* if $a \in \mathbb{N}$, $a \geq 2$, and otherwise it is said to be *nonintegral*.

Remark 10.5. When b is integral (i.e., when the multiplicities are positive integers, $b \in \mathbb{N}^*$), and $2 \leq b < a$ (so that $D = \log_a b$ lies in $(0, 1)$), then the Cantor string is an ordinary fractal string, and we call it geometrically realizable, or simply geometric, if no confusion can arise with our usage of "geometric" in opposition to "spectral".

For example, the (ternary) Cantor string is a generalized Cantor string with parameters $a = 3$ and $b = 2$. Hence, it is both an integral Cantor string (since $a = 3$ is an integer) and a geometric fractal string (since $b = 2$ is an integer).

Remark 10.6. It may be helpful to the reader to see the connection between our present setting and that of Chapters 2 and 3. First, in Chapter 3, we considered Dirichlet polynomials, and generalized Cantor strings correspond exactly to the linear case of polynomials $1 - m_1 r_1^s$.

In the language of Chapter 2, where we considered arbitrary (geometric) self-similar strings, assume that b is an integer greater than 1, and $a > b$ is real. Then $\eta = \sum_{n=0}^{\infty} b^n \, \delta_{\{a^{-n}\}}$ is geometrically realizable, and it is an (ordinary, lattice) self-similar string with b scaling ratios that are all equal to a^{-1}: $r_1 = r_2 = \cdots = r_b = a^{-1}$. Thus, in particular, its multiplicative generator equals $r = a^{-1}$ and $k_1 = \cdots = k_b = 1$, in the notation of Definition 2.14. The associated polynomial equation (2.38) is of degree 1, and takes the form $bz = 1$, $z = a^{-\omega}$. Therefore, we recover the fact that the complex dimensions of η lie on a single vertical line. Indeed, they are the points $\omega = D + in\mathbf{p}$, with $n \in \mathbb{Z}$, $D = \log_a b$ and $\mathbf{p} = 2\pi/\log a$.

The following result will enable us to determine very precisely the nature of the jumps of the spectral counting function of an integral Cantor string (see Corollary 10.8 below, along with its proof).

Recall from Theorem 1.19 that

$$N_{\nu,D,\mathbf{p}}(x) = N_{D,\mathbf{p}}(x) + N_{D,\mathbf{p}}(x/2) + N_{D,\mathbf{p}}(x/3) + \dots . \tag{10.11}$$

Thus, by a direct computation, we derive the spectral counterpart of formula (10.3) for $N_{D,\mathbf{p}}(x)$.

Theorem 10.7. *Let x be a positive real number and let*

$$x = \sum_{k \in \mathbb{Z}} x_k a^k$$

be the expansion of x in base a. Then the spectral counting function, $N_\nu(x)$, of the integral Cantor string with parameters a and b is given by the following formula:

$$N_\nu(x) = \frac{a}{a-b}\, x - \frac{1}{a-b}\left(a \sum_{k=-\infty}^{-1} x_k a^k + b \sum_{k=0}^{\infty} x_k b^k \right). \tag{10.12}$$

Proof. Observe that the digits of x form a finite sequence to the left in the sense that $x_k = 0$ for $k \ll 0$. The expansion of x in base a allows us to obtain an expression for the integer part of $a^{-n}x$:

$$\left[a^{-n}x\right] = \left[\sum_{k\in\mathbb{Z}} x_k a^{k-n}\right] = \sum_{k=n}^{\infty} x_k a^{k-n}.$$

Thus we can compute the counting function of the frequencies as follows:

$$N_\nu(x) = \sum_{n=0}^{\infty} \left[a^{-n}x\right] b^n = \sum_{n=0}^{\infty} \sum_{k=n}^{\infty} x_k a^{k-n} b^n.$$

Now we interchange the order of summation. The inner sum becomes the finite sum $\sum_{n=0}^{k} a^{k-n}b^n = (a^{k+1} - b^{k+1})/(a-b)$. Thus

$$N_\nu(x) = \sum_{k=0}^{\infty} x_k \sum_{n=0}^{k} a^{k-n}b^n = \sum_{k=0}^{\infty} x_k \frac{a^{k+1} - b^{k+1}}{a-b}.$$

Next observe that $\sum_{k\geq 0} x_k a^k = x - \sum_{k<0} x_k a^k$. Thus we obtain

$$N_\nu(x) = \frac{a}{a-b}x - \frac{1}{a-b}\left(a \sum_{k=-\infty}^{-1} x_k a^k + b \sum_{k=0}^{\infty} x_k b^k \right).$$

This is the desired formula. □

Corollary 10.8. *The counting function of the frequencies, $N_\nu(x)$, of an integral Cantor string jumps by*

$$\frac{b^{n+1}-1}{b-1}$$

at integral values of x that are divisible by a^n and not by a^{n+1} $(n \in \mathbb{N})$.

Proof. Let x be exactly divisible by a^n. Then, there are two ways to represent x in base a:

$$x_+ = \sum_{k=n}^{\infty} x_k a^k \quad \text{and} \quad x_- = \sum_{k=n+1}^{\infty} x_k a^k + (x_n - 1)a^n + \sum_{k=-\infty}^{n-1} (a-1)a^k.$$

Note that the n-th digit, x_n, does not vanish. By the above formula (10.11) for N_ν, we find the two values

$$N_\nu(x_+) = \frac{a}{a-b}x - \frac{1}{a-b}\left(b x_n b^n + b \sum_{k=n+1}^{\infty} x_k b^k \right)$$

and

$$N_\nu(x_-) = \frac{a}{a-b}x - \frac{1}{a-b}\left(a \sum_{k=-\infty}^{-1} (a-1)a^k + b \sum_{k=0}^{n-1} (a-1)b^k \right.$$
$$\left. + b(x_n - 1)b^n + b \sum_{k=n+1}^{\infty} x_k b^k \right).$$

These values are the limit of $N_\nu(t)$ when t approaches x from above and from below, respectively. Hence we conclude that, at the point x, N_ν jumps by

$$\frac{a}{a-b} \sum_{k=-\infty}^{-1} (a-1)a^k + \frac{b}{a-b} \sum_{k=0}^{n-1} (a-1)b^k - \frac{b}{a-b}b^n = \frac{b^{n+1}-1}{b-1},$$

as was to be proved. □

10.2.2 Nonintegral Cantor Strings: Analysis of the Jumps in the Spectral Counting Function

In this section, we study the oscillations in the spectrum of nonintegral Cantor strings. This is the most important case for us, as will be seen in Chapter 11. In applications, we need to choose the oscillatory period $\mathbf{p}$ to be large. Since $a = e^{2\pi/\mathbf{p}}$, this means that a is a nonintegral real number, slightly greater than 1.

Theorem 10.9. *At $x = a^n$ $(n \in \mathbb{N})$, the spectral counting function $N_\nu(x)$ jumps by at least $b^n = x^D$. Hence $N_\nu(x)$ has jumps of order D at such points.*

More precisely, let m be the smallest positive integer such that a^m is an integer, or let $m = \infty$ if a^m is never an integer for $m \geq 1$. Let $q = [n/m]$, the integer part of n/m, or $q = 0$ if $m = \infty$. Then N_ν jumps by

$$b^n + b^{n-m} + b^{n-2m} + \cdots + b^{n-qm}, \tag{10.13}$$

at $x = a^n$. In other words, the value of the limit

$$\lim_{h \to 0^+} N_\nu(a^n + h) - N_\nu(a^n - h)$$

is given by (10.13).

Remark 10.10. The case $m = 1$ corresponds to integral Cantor strings and has been treated in more detail in the previous section. When $m = \infty$, the sum (10.13) only contains the term b^n.

Proof of Theorem 10.9. The frequencies are the numbers $k \cdot a^j$, with multiplicity b^j (with $k \in \mathbb{N}^*$, $j \in \mathbb{N}$). Thus the frequency a^n is found for all possible choices of k and j such that $ka^j = a^n$. We find $j = n, n-m, \ldots, n-qm$ as possible choices, and the corresponding multiplicities add up to a positive jump of

$$b^n + b^{n-m} + \cdots + b^{n-qm}.$$

Since for $x = a^n$, we have $b^n = x^D$, the jump is of order x^D. □

Remark 10.11. We see that when passing from the geometry to the spectrum of a generalized Cantor string, the oscillations remain of the same order. Thus, in a sense, the spectral operator (defined in Section 6.3.2) is invertible when restricted to the class of generalized Cantor strings. This observation will play an important role in Chapter 11.

10.3 The Truncated Cantor String

In this section, we study the geometry and the spectrum of truncated Cantor strings. This will enable us in Section 11.1.1 to obtain an upper bound for the length of finite arithmetic progressions of critical zeros of $\zeta(s)$.

First, we begin with some technical preliminaries. As before, let δ_n denote the distribution

$$\delta_n(f) = \int_{-\infty}^{\infty} \delta_n(x) f(x)\, dx = f(n).$$

The following function, called the *Fejer kernel,* approximates the sum

$$\sum_{n \in \mathbb{Z}} \delta_n$$

of delta functions at the integers.

Definition 10.12. For an integer $\Lambda > 0$ and $x \in \mathbb{R}$,

$$K_\Lambda(x) = \sum_{|n|<\Lambda} \left(1 - \frac{|n|}{\Lambda}\right) e^{2\pi i n x} = \sum_{|n|<\Lambda} c_n e^{2\pi i n x}.$$

We write $c_n = 1 - |n|/\Lambda$ for the coefficients of K_Λ, Λ being fixed.

Proposition 10.13. *The function K_Λ has most of its mass concentrated around the integers:*

(i) $K_\Lambda(x) \geq 0$ *for all* $x \in \mathbb{R}$,

(ii) *the total mass* $\int_0^1 K_\Lambda(x)\, dx$ *equals* 1,

(iii) $\int_0^{1/\Lambda} K_\Lambda(x)\, dx > \int_0^1 \left(\frac{\sin \pi t}{\pi t}\right)^2 dt =: C_1 > 9/20$.

Proof. It is well known that

$$K_\Lambda(x) = \frac{1}{\Lambda} \left(\frac{\sin \pi \Lambda x}{\sin \pi x}\right)^2;$$

see, e.g., [Fol, Section 8.5 and Exercise 33, p. 269]. This proves (i).

Property (ii) is clear since the constant coefficient of the Fourier series of K_Λ equals 1.

For (iii), we use $\sin^2(\pi x) < (\pi x)^2$ to obtain

$$\int_0^{1/\Lambda} K_\Lambda(x)\, dx > \frac{1}{\Lambda} \int_0^{1/\Lambda} \left(\frac{\sin \pi \Lambda x}{\pi x}\right)^2 dx = \int_0^1 \left(\frac{\sin \pi t}{\pi t}\right)^2 dt$$

by the substitution $\Lambda x = t$. Using Maple, we find that the last integral is slightly larger than 0.45. □

Let, as usual, $a = e^{2\pi/\mathbf{p}}$, so that $\log a = 2\pi/\mathbf{p}$. We define the *truncated generalized Cantor string* T via the explicit formula for its geometric counting function,

$$N_T(x) = \frac{1}{\log a} \sum_{|n|<\Lambda} \left(1 - \frac{|n|}{\Lambda}\right) \frac{x^{D+in\mathbf{p}} - 1}{D + in\mathbf{p}}, \tag{10.14}$$

for $x \geq 1$, and $N_T(x) = 0$ for $x \leq 1$.

The function $N_T(x)$ is differentiable with nonnegative derivative given by $N_T'(x) = \frac{1}{\log a}\, x^{D-1} K_\Lambda(\log_a x)$. Thus the corresponding measure vanishes on $x \leq 1$ and is positive for $x > 1$. We obtain the direct formula for the geometric counting function as the integral of this function,

$$N_T(x) = \frac{1}{\log a} \int_1^x t^D K_\Lambda(\log_a t)\, \frac{dt}{t} = \int_0^{\log_a x} a^{Dt} K_\Lambda(t)\, dt. \tag{10.15}$$

Thus $N_T(x)$ is increasing, with slope $a^{m(D-1)}\Lambda / \log a$ at integral powers $x = a^m$ of a.

The 'volume' of the truncated Cantor string is given by (see Remark 8.2)

$$\mathrm{vol}_1(T) = \frac{1}{\log a} \sum_{|n|<\Lambda} \left(1 - \frac{|n|}{\Lambda}\right) \frac{1}{1 - D - in\mathbf{p}} = \int_0^\infty a^{(D-1)t} K_\Lambda(t)\, dt. \tag{10.16}$$

Using the following lemma, we can estimate $\mathrm{vol}_1(T)$.

Lemma 10.14. *For $d \in (0,1)$ we have the estimate*

$$\sum_{|n|<\Lambda} \left(1 - \frac{|n|}{\Lambda}\right) \frac{1}{d + in\mathbf{p}} < \frac{1}{d} e^{2\zeta(2)/\mathbf{p}^2}.$$

Proof. We combine the terms for positive and negative n to estimate

$$\left(1 - \frac{|n|}{\Lambda}\right) \left(\frac{1}{d - in\mathbf{p}} + \frac{1}{d + in\mathbf{p}}\right) < \frac{2d}{d^2 + n^2\mathbf{p}^2} < \frac{2}{n^2\mathbf{p}^2}.$$

Hence their sum is bounded by

$$\frac{1}{d} + \sum_{n=1}^{\infty} \frac{2}{n^2\mathbf{p}^2} < \frac{1}{d}\left(1 + \frac{2\zeta(2)}{\mathbf{p}^2}\right) < \frac{1}{d} e^{2\zeta(2)/\mathbf{p}^2},$$

as claimed. □

Let $C_2 := e^{2\zeta(2)}$. We then deduce from Lemma 10.14 applied to $d = 1 - D$ that

$$\frac{1}{1-D} < \mathrm{vol}_1(T) \log a < C_2^{1/\mathbf{p}^2} \frac{1}{1-D}. \tag{10.17}$$

Remark 10.15. Note that $C_2 \approx 26.84$. In the applications in Chapter 11, we will usually assume that $\mathbf{p}$ is large, so that $C_2^{1/\mathbf{p}^2}$ is only slightly larger than 1. Since $\log a = 2\pi/\mathbf{p}$, this implies that

$$\mathrm{vol}_1(T) \approx \frac{\mathbf{p}}{2\pi(1-D)}$$

for large values of the oscillatory period.

10.3.1 The Spectrum of the Truncated Cantor String

The counting function of the spectrum of T is given by

$$N_\nu(x) = \sum_{\mu=1}^{[x]} N_T(x/\mu).$$

Using the direct formula (10.15), we obtain

$$N_\nu(x) = \int_0^\infty \left[xa^{-t}\right] a^{Dt} K_\Lambda(t)\, dt = \mathrm{vol}_1(T)x - \int_0^\infty \left\{xa^{-t}\right\} a^{Dt} K_\Lambda(t)\, dt. \tag{10.18}$$

Recall the notation $c_n = 1 - |n|/\Lambda$ and formula (10.16) for $\mathrm{vol}_1(T)$.

Theorem 10.16. *The function $N_\nu(x) = N_{\nu,T}(x)$ is given by the following explicit formula:*

$$N_\nu(x) = \mathrm{vol}_1(T)x + \frac{1}{\log a} \sum_{|n|<\Lambda} c_n \frac{x^{D+in\mathbf{p}}}{D+in\mathbf{p}} \zeta(D+in\mathbf{p}) + \int_0^\infty \left\{xa^t\right\} a^{-Dt} K_\Lambda(t)\, dt. \tag{10.19}$$

Proof. By Equation (10.18) above,

$$N_\nu(x) = \mathrm{vol}_1(T)x - f(x) + \int_0^\infty \left\{xa^t\right\} a^{-Dt} K_\Lambda(t)\, dt,$$

where $f(x) = \int_{-\infty}^\infty \{xa^t\} a^{-Dt} K_\Lambda(t)\, dt$. Observe that $f(ax) = a^D f(x)$ so that the function $x^{-D} f(x)$ is multiplicatively periodic. Its Fourier series is of the form

$$x^{-D} f(x) = \sum_{m=-\infty}^{\infty} a_m x^{im\mathbf{p}},$$

where $\mathbf{p} = 2\pi/\log a$. The coefficients are given by

$$a_m = \int_0^1 e^{-2\pi imx} a^{-Dx} f(a^x)\, dx.$$

Thus

$$\begin{aligned} a_m &= \int_0^1 e^{-2\pi imx} a^{-Dx} \int_{-\infty}^\infty \left\{a^{x+t}\right\} a^{-Dt} K_\Lambda(t)\, dt\, dx \\ &= \sum_{|n|<\Lambda} c_n \int_0^1 e^{-2\pi imx} a^{-Dx} \int_{-\infty}^\infty \left\{a^{x+t}\right\} a^{-Dt} e^{2\pi int}\, dt\, dx. \end{aligned}$$

We substitute $t = u - x$ to obtain

$$a_m = \sum_{|n|<\Lambda} c_n \int_0^1 e^{-2\pi i(m+n)x} \int_{-\infty}^{\infty} \{a^u\} a^{-Du} e^{2\pi inu}\, du\, dx.$$

Since the inner integral does not depend on x, there is only a contribution for $n = -m$, and we obtain

$$a_m = c_m \int_{-\infty}^{\infty} \{a^t\} a^{-Dt} e^{-2\pi imt}\, dt = c_m \int_{-\infty}^{\infty} \{a^t\} a^{-(D+im\mathbf{p})t}\, dt.$$

By formula (10.8) above for $x = 1$ we obtain that

$$f(x) = -\frac{1}{\log a} \sum_{|n|<\Lambda} c_n \frac{x^{D+in\mathbf{p}}}{D+in\mathbf{p}} \zeta(D+in\mathbf{p}),$$

thus proving (10.19). □

Remark 10.17. As in Remark 10.3, we could also derive this explicit formula in a way analogous to Section 10.2. That approach was taken in the paper [vF3].

10.4 Notes

Section 10.2.1 greatly improves some of the results of [LapPo2, Example 4.5, pp. 65–67], dealing with the (ternary) Cantor string. The Cantor string was studied, in particular, in the papers [LapPo1; LapPo2, Example 4.5; Lap2–3], as well as in Chapter 1 and Sections 2.3.1, 6.4.1 and 8.4.1 of this book.

11

The Critical Zeros of Zeta Functions

As we saw in Chapter 10, the complex dimensions of a generalized Cantor string form an arithmetic progression $\{D + in\mathbf{p}\}_{n \in \mathbb{Z}}$, with $0 < D < 1$ and $\mathbf{p} > 0$. In this chapter, we use this fact to study arithmetic progressions of critical zeros of zeta functions.

By combining our explicit formulas with the analysis of the oscillations in the geometry and spectrum of generalized Cantor strings carried out in Chapter 10, we show, in Theorems 11.1, 11.12 and 11.16, that the Riemann zeta function—and other zeta functions from a large class of Dirichlet series not necessarily satisfying a functional equation or an Euler product—does not have an infinite sequence of critical zeros forming an (almost) arithmetic progression. C. R. Putnam [Pu1–2] was the first to obtain such a result in the special case of the Riemann zeta function. We have found this result independently, first for the Riemann and Epstein zeta functions [Lap-vF2], and then in the generality presented here (see [Lap-vF4] and [Lap-vF5, Chapter 9]), which is natural in our framework and for which Putnam's method does not work.

In Section 11.1, we first present our proof for $\zeta(s)$, the Riemann zeta function, and then indicate in Sections 11.2 and 11.3 the changes necessary to obtain the general theorem, as well as to extend it to almost arithmetic progressions. In Section 11.4, we combine a special case of Theorem 11.16 with results from algebraic number theory to extend some of our results to Hecke L-series. Finally, in Section 11.5, we discuss a situation where our argument does not apply, and where the conclusion of Theorem 11.16 does not hold, the case of the zeta function of a curve over a finite field.

These results give rise to the problem of determining how long an arithmetic progression of zeros can be. This question is partially answered in Sections 11.1.1 and 11.4.1, where we derive an upper bound for the length of such a progression of zeros, for the Riemann zeta function (Theorem 11.5) and for Dirichlet L-series (Theorem 11.23). These latter results were first obtained in [vF3] and [Watk, vFWatk], respectively.

As will be clear to the reader, the tools developed in Chapter 5—namely, our distributional explicit formula (Theorem 5.18) along with the corresponding distributional error estimate (Theorem 5.30)—once again play an essential role in our proof of the main results of this chapter, Theorems 11.1, 11.12 and 11.16.

11.1 The Riemann Zeta Function: No Critical Zeros in Arithmetic Progression

Let $\mathcal{L}_{D,\mathbf{p}}$ be the generalized Cantor string of Sections 8.4.1 and 10.1, with lengths a^{-n} repeated with multiplicity b^n ($1 < b < a$, $n = 0, 1, 2, \ldots$). As before, we denote the geometric and spectral counting function by $N_{D,\mathbf{p}}(x)$ and $N_\nu(x) = N_{\nu,D,\mathbf{p}}(x)$, respectively. In the previous chapter, we computed $N_{D,\mathbf{p}}(x)$ and $N_\nu(x)$ by a direct calculation. In this chapter, we compute them again, but this time by using our explicit formulas in order to obtain the desired connection with the zeros of the Riemann zeta function.

Theorem 5.14, applied to $N_{D,\mathbf{p}}(x)$, yields

$$N_{D,\mathbf{p}}(x) = \frac{1}{\log a} \sum_{n=-\infty}^{\infty} \frac{x^{D+in\mathbf{p}}}{D+in\mathbf{p}} - \frac{1}{b-1}, \tag{11.1}$$

where $\mathbf{p} = \frac{2\pi}{\log a}$ and $D = \log_a b$ is determined by $a^D = b$.

We have to interpret $N_\nu(x)$ as a distribution.[1] We choose a screen to the left of $\operatorname{Re} s = 0$. Theorem 5.18 in conjunction with Theorem 5.30 yields

$$N_\nu(x) = \frac{a}{a-b} x + \frac{1}{\log a} \sum_{n=-\infty}^{\infty} \zeta(D+in\mathbf{p}) \frac{x^{D+in\mathbf{p}}}{D+in\mathbf{p}} + O(1), \tag{11.2}$$

as $x \to \infty$ (see Remark 3.27 and Section 6.4.3).

Theorem 11.1. *Let $0 < D < 1$ and $\mathbf{p} > 0$ be given. Then there exists an integer $n \neq 0$ such that $\zeta(D+in\mathbf{p}) \neq 0$. That is, the Riemann zeta function does not have an infinite sequence of critical zeros forming an arithmetic progression.*

---

[1] Throughout this section, for simplicity, we use the symbol $N_\nu(x)$, instead of the more precise notation $N_{\nu,D,\mathbf{p}}(x)$ used in Chapter 10, to refer to the spectral counting function of $\mathcal{L}_{D,\mathbf{p}}$.

Remark 11.2. Note that the theorem implies that $\zeta(D+in\mathbf{p}) \neq 0$ for infinitely many integers n. Indeed, if $\zeta(D+in\mathbf{p})$ were to be nonzero for only finitely many $n \in \mathbb{Z}$, say for $|n| < M$, then we would obtain a contradiction by applying the theorem to D and $M\mathbf{p}$ instead of to D and $\mathbf{p}$. In Section 11.3, we also obtain information about the density of the set $\{n \in \mathbb{Z} \colon \zeta(D+in\mathbf{p}) \neq 0\}$ in the set of integers. The same remark applies to Theorem 11.12 below.

Remark 11.3. A priori, our result implies that ζ does not have a vertical doubly infinite sequence of critical zeros in arithmetic progression. However, since the Riemann zeta function satisfies a functional equation relating $\zeta(s)$ and $\zeta(1-s)$, and $\zeta(s)$ does not vanish for $\operatorname{Re} s \geq 1$ by the Euler product, it follows immediately that every doubly infinite sequence of zeros on one line must be vertical, say on the line $\operatorname{Re} s = D$, with $D \in (0,1)$. On the other hand, since $\zeta(s)$ vanishes at $s = -2, -4, -6, \ldots$, the Riemann zeta function does have a horizontal arithmetic sequence of zeros that is infinite to the left.

The analogue of this remark applies to all the natural arithmetic zeta functions since they also satisfy a functional equation and have a convergent Euler product to the right of the critical strip. This is the case, for example, of the Hecke L-series of number fields, as well as of the more general L-series, to which the results of Section 11.2 or 11.4 can be applied.

Proof of Theorem 11.1. Assume that $\zeta(D+in\mathbf{p}) = 0$ for all $n \neq 0$. Let $a = e^{2\pi/\mathbf{p}}$ and $b = a^D$. The generalized Cantor string $\mathcal{L}_{D,\mathbf{p}}$ with these parameters has complex dimensions at all the points $D+in\mathbf{p}$ $(n \in \mathbb{Z})$; see Section 10.1 above. Equation (11.2), the explicit formula for the frequencies, then becomes very simple. Indeed, all the terms corresponding to $x^{D+in\mathbf{p}}$, with $n \neq 0$, disappear since $\zeta(D+in\mathbf{p}) = 0$. The resulting formula therefore reads

$$N_\nu(x) = \frac{a}{a-b}x + \zeta(D)\frac{x^D}{D\log a} + o\big(x^D\big), \qquad \text{as } x \to \infty, \tag{11.3}$$

as a distribution on $(0,\infty)$. We see that the frequencies of this Cantor string do not have oscillations of order D. On the other hand, we have seen in Theorem 10.9 that $N_\nu(x)$ jumps by at least b^q at a^q, for each $q \in \mathbb{N}$. Since $b = a^D$, we see that $N_\nu(x)$ jumps by at least x^D at $x = a^q$. But this means that N_ν has oscillations of order D. Since this is a contradiction, we conclude that $\zeta(D+in\mathbf{p}) \neq 0$ for some integer $n \neq 0$. □

Second proof of Theorem 11.1. It suffices to prove the theorem for $D \geq \frac{1}{2}$ because of the functional equation satisfied by ζ (see, for example, [Ti, Chapter II]). Now, let $D > 1/4$ and $\mathbf{p} > 0$ be such that $\zeta(D+in\mathbf{p}) \neq 0$ for at most finitely many values of n. Put $a = e^{2\pi/\mathbf{p}}$ and $b = a^D$ and consider the same generalized Cantor string $\mathcal{L}_{D,\mathbf{p}}$ as above. Instead of applying the distributional explicit formula at level 1 to obtain a formula for $N_\nu(x)$, we can apply the pointwise explicit formula at level 2, as we now explain.

We choose some value σ_0 strictly between $-\frac{1}{2}$ and 0. Then hypotheses **L1** and **L2** are satisfied—meaning that $\mathcal{L}_{D,\mathbf{p}}$ is languid—with $\kappa = \frac{1}{2} - \sigma_0$ for the screen $S\colon \sigma_0 + it$, $t \in \mathbb{R}$ (see Section 6.2.2). By Theorem 5.10, applied for $k = 2$, the integrated counting function of the frequencies of $\mathcal{L}_{D,\mathbf{p}}$ is given by

$$N_\nu^{[2]}(x) = \frac{a}{(a-b)}\frac{x^2}{2} + \frac{x^{D+1}}{\log a}\sum_{n\in\mathbb{Z}} \frac{\zeta(D+in\mathbf{p})}{(D+in\mathbf{p})_2} x^{in\mathbf{p}} + \frac{x}{1-b}\zeta(0) + O\left(x^{\sigma_0+1}\right), \quad (11.4)$$

as $x \to \infty$. Here, $(\omega)_2 = \omega(\omega+1)$, as in Equation (5.12). Note that by assumption, the sum over n in formula (11.4) has finitely many nonzero terms.

Much as in [In, Theorem C, p. 35], we deduce from Equation (11.4) by a Tauberian argument that

$$N_\nu(x) = N_\nu^{[1]}(x) = \frac{a}{a-b}x + \sum_{n\in\mathbb{Z}} \frac{x^{D+in\mathbf{p}}}{\log a}\frac{\zeta(D+in\mathbf{p})}{D+in\mathbf{p}} + O\left(x^{\frac{\sigma_0+1}{2}}\right), \quad (11.5)$$

as $x \to \infty$. Again, the sum over n in formula (11.5) has finitely many nonzero terms, and hence defines a continuous function of x. Since we assumed that $D > 1/4$, we can choose $\sigma_0 > -\frac{1}{2}$ such that $(\sigma_0+1)/2 < D$. We deduce that the counting function of the frequencies of this Cantor string does not have jumps of order D. But this contradicts Theorem 10.9. □

One remaining problem is to estimate the number of zeros of the Riemann zeta function that possibly lie in a vertical arithmetic progression. This question is the subject of the next section, where we consider the spectrum of truncated generalized Cantor strings. The argument in Section 11.1.1 depends on a closed expression for the error term in the explicit formula for $N_\nu(x)$. It was first published in [vF3]. We present yet another proof of Theorem 11.1 to illustrate the advantage of having such a closed expression for the error term.

Third proof of Theorem 11.1. Using formula (10.9), we can even determine an explicit value of x for which we obtain a contradiction. If $\zeta(D+in\mathbf{p}) = 0$ for all integers $n \neq 0$, it then follows from (10.9) that

$$N_\nu(x) = \frac{x}{1-b/a} - \frac{1}{\log a}\frac{x^D}{D}(-\zeta(D)) + \sum_{n=1}^{\infty} \{xa^n\}\, b^{-n}.$$

The last sum is estimated by

$$0 \leq \sum_{n=1}^{\infty} \{xa^n\} b^{-n} \leq \sum_{n=1}^{\infty} b^{-n} = \frac{1}{b-1}.$$

Thus the jump $N_\nu(x_+) - N_\nu(x_-)$ is at most $\frac{1}{b-1}$. Since $b = e^{2\pi D/\mathbf{p}}$, we have that $\frac{1}{b-1} < \frac{\mathbf{p}}{2\pi D}$. On the other hand, we have seen that the jump at a^n is at least b^n. Hence, we obtain a contradiction when $b^n \geq \frac{\mathbf{p}}{2\pi D}$. For

$$n = \left[\frac{\mathbf{p}}{2\pi D} \log \frac{\mathbf{p}}{2\pi D}\right] + 1$$

we have $b^n > \frac{\mathbf{p}}{2\pi D}$. For this integer, $x = a^n$ is slightly larger than $\left(\frac{\mathbf{p}}{2\pi D}\right)^{1/D}$. □

In order to obtain the analogue of this result for a broader class of zeta functions, we have to use our distributional explicit formula (Theorems 5.18 and 5.30) in Section 11.2 below, as in the first proof of Theorem 11.1 presented above, rather than our pointwise formula (Theorem 5.10) as in the second proof of this theorem. This was one of our main original motivations for developing the distributional theory of explicit formulas, as completed in Section 5.4.2 and Theorem 5.30.

As was alluded to in Remark 10.11, the following corollary captures the essence of our method of proof in this context (compare Corollary 9.6 above and review Section 6.3.2, where we have defined the spectral and inverse spectral operators for fractal strings):

Corollary 11.4. *The spectral operator is invertible when restricted to the class of generalized Cantor strings.*

11.1.1 Finite Arithmetic Progressions of Zeros

One generally conjectures that the Riemann zeta function has no zeros off the line $\operatorname{Re} s = 1/2$ inside the critical strip, and that the zeros on this line are never in an arithmetic progression (i.e., $1/2 + i\mathbf{p}$ and $1/2 + 2i\mathbf{p}$ are never both zeros of $\zeta(s)$).[2] This can be verified numerically using a table of zeros from Odlyzko [Od3], which lists the first three million zeros, the last one of which has an imaginary part of $1.13 \cdot 10^6$. Thus Theorem 11.5 below has been verified numerically for all $\mathbf{p} < 5.6 \cdot 10^5$.

For the 924,280th zero in this table,

$$\tfrac{1}{2} + it = \tfrac{1}{2} + 558{,}652.035\,125\,523\,i,$$

[2]A much more general conjecture will be discussed in Section 12.1; see Conjecture 12.1.

the number $\frac{1}{2} + 2it$ is very close to the 1,971,817th zero,

$$\tfrac{1}{2} + 1{,}117{,}304.070\,251\,415\, i,$$

but there are still three significant digits to distinguish these points from each other. This is the closest approximation to an arithmetic progression of zeros found in this table.

We present here the argument of [vF3] giving an upper bound for the length of an arithmetic progression of zeros of $\zeta(s)$.

Theorem 11.5. *Let $\mathbf{p} > 0$ and $D \in (0,1)$ be real numbers and let $\Lambda \geq 2$. Suppose that $\zeta(D + in\mathbf{p}) = 0$ for all integers n such that $0 < |n| < \Lambda$. Then*

$$\Lambda < 60 \log \mathbf{p} \left(\frac{\mathbf{p}}{2\pi} \right)^{\frac{1}{D} - 1}.$$

Moreover, $\Lambda < 13\,\mathbf{p}$ if $D = 1/2$ and $\Lambda < 80(\mathbf{p}/2\pi)^{1/D-1}$ if $D < 0.96$.

In light of the functional equation satisfied by $\zeta(s)$, we can assume without loss of generality that $D \geq 1/2$. Thus the length of an arithmetic progression is bounded by $O(\mathbf{p})$ for $D = 1/2$, and by $o(\mathbf{p})$ for $D > 1/2$.

To establish the theorem, we use the truncated Cantor string (of dimension D) defined by the geometric counting function (10.14) in Section 10.3, and with spectral counting function $N_\nu(x)$ (as studied in Section 10.3.1). Note that like the generalized Cantor strings used earlier in this section (i.e., in Section 11.1), truncated Cantor strings are "virtual fractal strings" in the sense that they do not have a geometric realization as an open subset of the real line.

We obtain a bound for Λ by showing in the next lemma using a direct computation that the function $N_\nu(x)$ increases by a large amount as x increases from a^m to $a^{m+\varepsilon}$, and then, using the explicit formula, that $N_\nu(x)$ does not increase by much if $\zeta(s)$ vanishes for many values of s in an arithmetic progession (Lemma 11.7).

Lemma 11.6. *Let $m \in \mathbb{N}$ and $\varepsilon = 1/\Lambda$. Then*

$$N_\nu\left(a^{m+\varepsilon}\right) - N_\nu\left(a^m\right) > C_1 a^{mD},$$

where $C_1 = \int_0^1 \left(\frac{\sin \pi t}{\pi t}\right)^2 dt > 9/20$ is the absolute constant of Proposition 10.13(iii).

Proof. As in Section 10.3, let $N_T(x)$ denote the geometric counting function of the truncated Cantor string, given by formula (10.14), or equivalently, by (10.15). Since $N_\nu(x) = \sum_{\mu=1}^\infty N_T(x/\mu)$, and $N_T(x)$ is nondecreasing for $x \geq 1$, we have for $y > x$ that

$$N_\nu(y) - N_\nu(x) \geq N_T(y) - N_T(x) = \int_{\log_a x}^{\log_a y} a^{tD} K_\Lambda(t)\, dt,$$

by formula (10.15). For $x = a^m$, $y = a^{m+\varepsilon}$ and $\varepsilon = 1/\Lambda$, we obtain an increase of at least $a^{mD} \int_0^\varepsilon K_\Lambda(t)\,dt > C_1 a^{mD}$, by Proposition 10.13(iii). □

Next we assume that $\zeta(D+in\mathbf{p}) = 0$ for all n, $0 < |n| \leq \Lambda - 1$, and show that $N_\nu(x)$ does not increase by much.

Lemma 11.7. *Assume that* $\zeta(D+in\mathbf{p}) = 0$ *for* $0 < |n| \leq \Lambda - 1$. *Then, with* $\varepsilon = 1/\Lambda$ *and for all integers* $m \geq 0$, *we have*

$$\frac{N_\nu\left(a^{m+\varepsilon}\right) - N_\nu\left(a^m\right)}{\varepsilon} < C_2^{1/\mathbf{p}^2} \frac{a^{m+\varepsilon}}{1-D} + a^{mD}\zeta(D) + C_2^{1/\mathbf{p}^2} \frac{\Lambda}{D\log a},$$

where $C_2 = e^{2\zeta(2)}$ *is the absolute constant occurring in Equation* (10.17).

Proof. By Theorem 10.16 we have

$$N_\nu\left(a^{m+\varepsilon}\right) - N_\nu\left(a^m\right) = a^m\left(a^\varepsilon - 1\right)\mathrm{vol}_1(T) + \frac{a^{mD}}{\log a}\frac{a^{\varepsilon D} - 1}{D}\zeta(D)$$
$$+ \int_0^\infty \left(\left\{a^{m+\varepsilon+t}\right\} - \left\{a^{m+t}\right\}\right) a^{-tD} K_\Lambda(t)\,dt, \quad (11.6)$$

where $\mathrm{vol}_1(T)$ denotes the 'volume' of the truncated Cantor string, as given in formula (10.16).

Next, we use that $\{x\} \leq 1$ and $K_\Lambda(t) \geq 0$ (by Proposition 10.13(i)) to estimate the integral on the right-hand side of (11.6),

$$\left|\int_0^\infty \left(\left\{a^{m+\varepsilon+t}\right\} - \left\{a^{m+t}\right\}\right) a^{-tD} K_\Lambda(t)\,dt\right| \leq \int_0^\infty a^{-tD} K_\Lambda(t)\,dt.$$

This last integral evaluates to

$$\frac{1}{\log a} \sum_{|n|<\Lambda} \frac{c_n}{D + in\mathbf{p}} < C_2^{1/\mathbf{p}^2} \frac{1}{D\log a},$$

by Lemma 10.14 (recall that $c_n = 1 - |n|/\Lambda$ and C_2 is given in Lemma 11.7).

We also estimate $\mathrm{vol}_1(T)$ by (10.17). Then we multiply by $\Lambda = 1/\varepsilon$ and estimate $(a^\varepsilon - 1)/\varepsilon < a^\varepsilon \log a$ and $(a^{\varepsilon D} - 1)/\varepsilon D > \log a$ by using the Mean Value Theorem and the fact that $\zeta(D) < 0$. This completes the proof of Lemma 11.7. □

To complete the proof of Theorem 11.5, we combine the lower bound of Lemma 11.6 for the jump of N_ν at a^m with the upper bound of Lemma 11.7. Write

$$A = C_1\Lambda(1-D) - \zeta(D)(1-D) \quad (11.7)$$

and

$$B = a^{\varepsilon} C_2^{1/\mathbf{p}^2}.$$

Combining Lemmas 11.6 and 11.7, we obtain, since $\zeta(D) < 0$ and using that $\log a = 2\pi/\mathbf{p}$, that

$$Aa^{mD} - Ba^m < C_2^{1/\mathbf{p}^2} \frac{\Lambda(1-D)}{D \log a} < C_2^{1/\mathbf{p}^2} \frac{\mathbf{p}A}{2\pi C_1 D}. \tag{11.8}$$

The function $Ax^D - Bx$ attains its maximum at the value $x = a^t$ such that $DAx^D = Bx$, i.e.,

$$a^t = x = D^{1/(1-D)} A^{1/(1-D)} B^{-1/(1-D)}, \tag{11.9}$$

or equivalently,

$$t = \frac{1}{1-D} \log_a(DA/B).$$

Note that $a^{\varepsilon} \leq e^{\pi/\mathbf{p}}$ since $\Lambda \geq 2$. Therefore, $B < e^{4/\mathbf{p}}$. If $DA \leq B$, we obtain

$$C_1 \Lambda < \frac{e^{4/\mathbf{p}}}{1-D} + \zeta(D) = \frac{1}{D} + \left(\zeta(D) - \frac{1}{D-1}\right) + \frac{e^{4/\mathbf{p}} - 1}{D(1-D)}.$$

We now use the following zero-free region of the Riemann zeta function (see [vF3, Theorem 2.4]):

Lemma 11.8. *If $\zeta(D + i\mathbf{p}) = 0$, then $\frac{1}{1-D} < 18 \log \mathbf{p}$.*

We obtain that

$$C_1 \Lambda < \frac{1}{D} + \left(\zeta(D) - \frac{1}{D-1}\right) + \frac{18(e^{4/\mathbf{p}} - 1) \log \mathbf{p}}{D}.$$

Now $\zeta(D) - \frac{1}{D-1}$ is a bounded function for $1/2 \leq D < 1$. In fact, the function on the right is decreasing for $1/2 \leq D < 1$. Since the theorem has been verified for all $\mathbf{p} < 5.6 \cdot 10^5$, we obtain $\Lambda \leq 5$.

If, on the other hand, $DA > B$, then $t > 0$, and we choose $m = \lceil t \rceil$. Then

$$Aa^{mD} - Ba^m > Aa^{tD} - Ba \cdot a^t = Aa^{tD}(1 - aD). \tag{11.10}$$

Lemma 11.9. *For $\mathbf{p} > 10{,}000$, we have $1 - aD > (1-D)e^{-120(\log \mathbf{p})/\mathbf{p}}$.*

Proof. Use $a = e^{2\pi/\mathbf{p}}$ and $e^x - 1 < \frac{x}{1-x}$ to estimate

$$\frac{1-aD}{1-D} = 1 - (a-1)\frac{D}{1-D} > 1 - \frac{2\pi D}{(\mathbf{p} - 2\pi)(1-D)}.$$

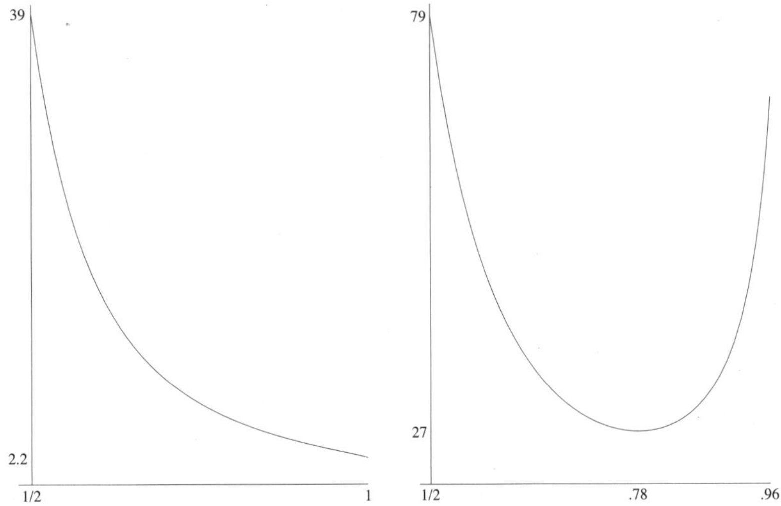

Figure 11.1: (a) The function $f(D) = (1-D)^{1-1/D}(C_1 D)^{-1/D}$.
(b) The function $\frac{1}{1-D}\, f(D)$.

Estimating D by 1 and $1/(1-D)$ by Lemma 11.8, we obtain

$$\frac{1-aD}{1-D} > 1 - \frac{36\pi \log \mathbf{p}}{\mathbf{p}-2\pi}.$$

Using Maple, one can verify that for $\mathbf{p} > 10,000$, the function on the right is bounded from below by $e^{-120\log \mathbf{p}/\mathbf{p}}$. □

Combining (11.8), (11.10) and Lemma 11.9, we obtain a bound for a^{tD}. Using (11.9), this yields the following bound for A, and hence for Λ,

$$\Lambda < \frac{A}{C_1(1-D)} < \frac{1}{1-D}\left(\frac{\mathbf{p}}{2\pi}\right)^{\frac{1}{D}-1}(1-D)^{1-1/D}(C_1 D)^{-1/D} e^{160\log \mathbf{p}/\mathbf{p}}. \tag{11.11}$$

For $D = 1/2$ and $\mathbf{p} > 44{,}000$, we thus obtain

$$\Lambda < 13\,\mathbf{p}.$$

The function $(1-D)^{1-1/D}(C_1 D)^{-1/D}$ is bounded away from 0 and decreasing (see Figure 11.1(a)), whereas $(1-D)^{-1}$ increases without bound as $D \uparrow 1$. The product of these two functions (see Figure 11.1(b)) decreases for $1/2 \le D < 0.78$, and then increases. For $D < 0.96$, the value of this product is less than the value at $D = 1/2$. Thus for $1/2 \le D < 0.96$ and large enough $\mathbf{p}$, we obtain

$$\Lambda < 80\left(\frac{\mathbf{p}}{2\pi}\right)^{\frac{1}{D}-1}.$$

Close to the minimum at $D = 0.78$, for $0.69 < D < 0.86$, we even obtain

$$\Lambda < 30\left(\frac{\mathbf{p}}{2\pi}\right)^{\frac{1}{D}-1}.$$

And for $D \geq 0.96$, we use again the zero-free region of Lemma 11.8 to obtain

$$\Lambda < 60 \log \mathbf{p}\left(\frac{\mathbf{p}}{2\pi}\right)^{\frac{1}{D}-1}.$$

This completes the proof of Theorem 11.5.

Remark 11.10. Lemma 11.7 is quite weak. The integrand in (11.6) is highly oscillatory, but we only estimate the difference of the fractional parts by 1, and therefore the integral by $O(\mathbf{p})$. It may be the case, however, that this integral has a fixed bound, and this would imply a uniform bound for Λ, independent of $\mathbf{p}$. This would be a highly significant result, as we now explain.

In (11.11), we estimated $C_1\Lambda(1-D) < A$ by formula (11.7), ignoring the term $\zeta(D)(1-D)$. If Λ could be uniformly bounded, then taking this term into account would reduce the bound for Λ even more, especially for D close to 1. This could yield a zero-free region for the Riemann zeta function of the form "if $\zeta(D + i\mathbf{p}) = 0$ then $D \leq \sigma$" for some fixed $\sigma < 1$.

Note that a bound $\Lambda \leq 2$ would exclude any arithmetic progression, and $\Lambda \leq 1$ (which means a contradiction) for $D \neq 1/2$ would imply the Riemann hypothesis. This last bound is unattainable by these methods for D close to $1/2$, because it cannot be attained at $D = 1/2$. However, if Lemma 11.7 can be improved to yield a uniform bound for Λ, then the present methods may be applied to the *doubled truncated Cantor string*, defined by

$$N_{TT}(x) = \alpha N_{\Lambda,D,\mathbf{p}}(x) + N_{\Lambda,1-D,\mathbf{p}}(x),$$

where $N_{\Lambda,D,\mathbf{p}}(x) = N_T(x)$ is as in (10.14), and α is a positive parameter. Since the nature of this string is different for $D = 1/2$ than for $D \neq 1/2$, it might yield a result valid for all $D > 1/2$.[3]

Clearly, for $D = 1/2$, the string TT is a trivial multiple of T, whereas for $D > 1/2$, the increase of N_T at $x = a^m$ is reinforced in N_{TT} by the extra complex dimensions at $1 - D + in\mathbf{p}$. Thus the counterpart of Lemma 11.6 would be a stronger result. On the other hand, since $\zeta(D + in\mathbf{p}) = 0$ implies $\zeta(1 - D + in\mathbf{p}) = 0$, the explicit formula for the spectral counting function $N_\nu(x)$ for the generalized fractal string TT still simplifies, and the counterpart of Lemma 11.7 remains essentially the same, with an extra

[3] For closely related reasons, such a "doubling" of suitable fractal strings was used by the authors in an unpublished work aimed at improving the results and methods of [LapMa1, 2] and of [Lap-vF5, Chapter 7] (corresponding to Chapter 9 in the present book).

erm containing $\zeta(1-D)$. This term then even improves the subsequent counterpart of inequality (11.8).

Remark 11.11. We refer to Section 12.1 for a discussion of the irrationality conjecture for the critical zeros of the Riemann and other zeta functions, as well as for conjectures regarding the relationship between the number of poles of a Dirichlet series and the maximal possible length of an arithmetic progression of zeros.

11.2 Extension to Other Zeta Functions

The first (distributional) proof of Theorem 11.1 generalizes naturally to a large subclass of the class of zeta functions introduced at the end of Chapter 9. As was mentioned in the introduction, this subclass includes all Epstein zeta functions and all Dedekind zeta functions of algebraic number fields (see [Ter] and [Lan], along with Appendix A). By a refinement of the argument, we also obtain an estimate for the density of nonzeros in arithmetic progressions, in Section 11.3 below.

We now precisely state the assumptions on the zeta functions ζ_B for which our results apply:

(*P*) Given a sequence w_f of positive coefficients associated with a sequence of positive real numbers f, let $\zeta_B(s) = \sum_f w_f f^{-s}$ be the corresponding (generalized) Dirichlet series. Assume that for some screen S, this function is languid and has only finitely many poles contained in the associated window W.

Recall that ζ_B is said to be languid if it satisfies the growth hypotheses **L1** and **L2** of Definition 5.2. Loosely speaking, this means that the meromorphic continuation of ζ_B grows polynomially along horizontal lines and along the vertical direction of the screen. Under these assumptions, we prove

Theorem 11.12. *Let ζ_B be a zeta function satisfying hypothesis (P) above. Let $\mathbf{p} > 0$ be arbitrary and let $D \in W \cap \mathbb{R}$ be such that the vertical line $\operatorname{Re} s = D$ lies entirely within W. Then there exists an integer $n \neq 0$ such that $\zeta_B(D + in\mathbf{p}) \neq 0$. That is, $\zeta_B(s)$ does not have an infinite vertical sequence of zeros forming an arithmetic progression within the window W.*

Before proving Theorem 11.12, we indicate how to associate a generalized fractal spray to ζ_B. First, we view the Dirichlet series ζ_B as the spectral zeta function of a virtual basic shape B. By definition, the 'frequency' f has multiplicity w_f; see Section 4.3. Then, given D and $\mathbf{p}$ as in Theorem 11.12, let $\mathcal{L} = \mathcal{L}_{D,\mathbf{p}}$ be the generalized Cantor string of dimension D and oscillatory period $\mathbf{p}$, as defined in Sections 8.4.1 and 10.1. Note that here, in contrast to Section 11.1, we do not restrict D to be between 0 and 1. Next, consider the generalized Cantor spray of $\mathcal{L}$ with virtual basic shape B. Recall from Section 4.3, Equation (4.35), that the spectral zeta function of this spray is given by

$$\zeta_\nu(s) = \zeta_B(s) \cdot \zeta_{\mathcal{L}}(s).$$

We have to interpret the corresponding spectral counting function $N_\nu(x)$ as a distribution. Theorem 5.18 in conjunction with Theorem 5.30 yields the following explicit formula for the frequencies of this spray:

$$N_\nu(x) = W_{B,\mathcal{L}}^{[1]}(x) + \frac{1}{\log a} \sum_{n=-\infty}^{\infty} \zeta_B(D + in\mathbf{p}) \frac{x^{D+in\mathbf{p}}}{D + in\mathbf{p}} + O\left(x^{\sup S}\right), \tag{11.12}$$

as $x \to \infty$, where $\sup S$ is given by (5.17b) and

$$W_{B,\mathcal{L}}^{[1]}(x) = \sum_{u\colon \text{pole of } \zeta_B} \operatorname{res}\left(\frac{\zeta_B(s)\zeta_{\mathcal{L}}(s)}{s} x^s; u\right) \tag{11.13}$$

is the Weyl term (see Section 6.6, formula (6.76), along with Remark 3.27 and Section 6.4). By the assumption in hypothesis (P), the sum over the poles of ζ_B in (11.13) has only finitely many nonzero terms.

Proof of Theorem 11.12. Assume that $\zeta_B(D + in\mathbf{p}) = 0$ for all $n \neq 0$. As above, let $\mathcal{L} = \mathcal{L}_{D,\mathbf{p}}$ be the generalized Cantor string with $a = e^{2\pi/\mathbf{p}}$ and $b = a^D$. Then the generalized Cantor spray of $\mathcal{L}$ on B has complex dimensions at all the points $D + in\mathbf{p}$ ($n \in \mathbb{Z}$), but since $\zeta_B(D + in\mathbf{p}) = 0$ for all $n \in \mathbb{Z}\backslash\{0\}$, Equation (11.12) simplifies, just as in the proof of Theorem 11.1:

$$N_\nu(x) = W_{B,\mathcal{L}}^{[1]}(x) + \zeta_B(D) \frac{x^D}{D \log a} + o\left(x^D\right), \quad \text{as } x \to \infty, \tag{11.14}$$

as a distribution on $(0, \infty)$. It follows that the frequencies of this Cantor spray have no oscillations of order D.

On the other hand, the counting function of the frequencies of the spray is related to the geometric counting function $N_{\mathcal{L}}(x) = N_{D,\mathbf{p}}(x)$ of the generalized Cantor string by

$$N_\nu(x) = \sum_f w_f N_{\mathcal{L}}\left(\frac{x}{f}\right). \tag{11.15}$$

Since $N_{\mathcal{L}}$ has jumps (by Theorem 10.2), the positivity of w_f guarantees that N_ν has jumps as well, at the points $x = f \cdot a^n$, where f runs through the frequencies of B and $n = 0, 1, \ldots$. Thus we obtain the analogue of Theorem 10.9: N_ν has jumps of order D. But this means that N_ν has oscillations of order D. Since this is in contradiction with formula (11.14), we conclude that $\zeta_B(D + in\mathbf{p}) \neq 0$ for some integer $n \neq 0$. $\square$

Note that the counterpart of Remark 11.2 applies, so that we deduce from Theorem 11.12 that there are infinitely many integers n such that $\zeta_B(D + in\mathbf{p}) \neq 0$.

The following corollary is the exact analogue of Corollary 11.4 in the present more general setting. See Definition 6.22 in Section 6.6 for the definition of the spectral operator for fractal sprays.

Corollary 11.13. *The spectral operator is invertible when restricted to the class of all generalized Cantor sprays, with virtual basic shape B defined by a zeta function ζ_B satisfying hypothesis (P).*

11.3 Density of Nonzeros on Vertical Lines

Using a refinement of the argument of Section 11.2, we can obtain a lower bound for the density of points where $\zeta_B(D + in\mathbf{p})$ is nonzero.

Let ζ_B be a zeta function satisfying the above hypothesis (P). Let $\rho \geq 0$ be such that

$$\zeta_B(D + it) = O\big(t^\rho\big), \quad \text{as } |t| \to \infty. \tag{11.16}$$

Theorem 11.14. *Let $\delta > 0$ and assume that $\rho < 1$, where ρ is the exponent of Equation* (11.16)*. Then, for infinitely many values of T, tending to infinity, the set*

$$\{n \in \mathbb{Z}\colon |n| \leq T,\ |\zeta_B(D + in\mathbf{p})| \neq 0\} \tag{11.17}$$

contains more than $T^{1-\rho-\delta}$ elements.

Proof. Suppose that the set defined by (11.17) contains fewer than $T^{1-\rho-\delta}$ elements, for all sufficiently large T. Let $0 < n_1 < n_2 < n_3 < \dots$ be the sequence of positive elements of the set (11.17) for $T = \infty$. Then we have that $n_j \geq j^{1/(1-\rho-\delta)}$, except possibly for the first few integers n_j. Moreover, if n is not in the sequence, then $\zeta_B(D + in\mathbf{p}) = 0$. Thus, by (11.12),

$$\begin{aligned} N_\nu(x) = W^{[1]}_{B,\mathcal{L}}(x) &+ \frac{x^D}{D \log a}\zeta_B(D) \\ &+ \frac{1}{\log a}\sum_{j=1}^{\infty} \frac{x^{D \pm in_j\mathbf{p}}}{D \pm in_j\mathbf{p}}\zeta_B(D \pm in_j\mathbf{p}) + o\big(x^D\big), \end{aligned} \tag{11.18}$$

as $x \to \infty$. The j-th term in this series is bounded by a constant times $n_j^{\rho-1}$, hence the series is absolutely convergent. Therefore its sum is a continuous function of x. By (11.18), so is $N_\nu(x)$. But this is a contradiction, since, by the analogue of Theorem 10.9 established in the proof of Theorem 11.12, the function N_ν has jumps of order D, and hence is discontinuous. $\square$

For the Riemann zeta function $\zeta(s)$, we thus obtain, choosing $D = 1/2$ and $\rho = 1/6$ (see [Ti, Theorems 5.5 and 5.12]),[4] the following corollary.

Corollary 11.15. *For every* $\mathbf{p} > 0$ *and* $\delta > 0$,

$$\#\left\{n \in \mathbb{Z}\colon |n| \leq T,\ \zeta\left(\tfrac{1}{2} + in\mathbf{p}\right) \neq 0\right\} \geq T^{5/6-\delta},$$

for infinitely many values of T, *tending to infinity.*

11.3.1 Almost Arithmetic Progressions of Zeros

By a classical result from Fourier theory, a Fourier series $\sum_{n\in\mathbb{Z}} a_n e^{inx}$ does not have jump discontinuities if $a_n = o(n^{-1})$ as $|n| \to \infty$, see [Zyg, Theorem 9.6, p. 108] or [Ru1, §5.6.9, p. 118]. Since this result is formulated in terms of the derivative of this Fourier series (i.e., for measures), we formulate the following argument on level $k = 0$.

Theorem 11.16. *Let* ζ_B *satisfy hypothesis* (P) *above. Then there do not exist* $D \in W \cap \mathbb{R}$ *and* $\mathbf{p} > 0$ *such that* $\zeta_B(D + in\mathbf{p}) \to 0$ *as* $|n| \to \infty$.[5]

Proof. By Theorems 5.18 and 5.30 applied at level $k = 0$, we obtain the analogue of formula (11.12):

$$\nu = W^{[0]}_{B,\mathcal{L}}(x)\,dx + \frac{1}{\log a}\sum_{n\in\mathbb{Z}} x^{D-1+in\mathbf{p}}\zeta_B(D + in\mathbf{p})\,dx + o\big(x^{D-1}\,dx\big), \tag{11.19}$$

as $x \to \infty$. Here, $W^{[0]}_{B,\mathcal{L}}(x)$ is the distributional derivative of the Weyl term, given by

$$W^{[0]}_{B,\mathcal{L}}(x) = \sum_{u\colon \text{pole of } \zeta_B} \operatorname{res}\left(\zeta_B(s)\zeta_\eta(s)x^{s-1}; u\right). \tag{11.20}$$

If $\zeta_B(D + in\mathbf{p}) \to 0$ as $|n| \to \infty$, then, by the classical result from Fourier analysis cited just above, this measure is continuous. That is, it does not have any atoms.

On the other hand, by Equation (11.15) and the argument following it, N_ν has jumps of order D. Since $N_\nu(x) = \int_0^x \nu(dx)$, this shows that ν has atoms. This contradiction shows that $\zeta_B(D + in\mathbf{p})$ does not converge to 0 as $|n| \to \infty$. □

---

[4] By [Ti, Theorem 5.18, p. 99], one could even take $\rho = 27/164$. Further, if the Lindelöf hypothesis holds, then one can take $\rho = 0$.

[5] We assume as in Theorem 11.12 that the line $\operatorname{Re} s = D$ lies within the window W.

11.4 Extension to L-Series

Using some well-known results from algebraic number theory [Lan, ParSh1–2], we can also obtain information about the zeros of certain Dirichlet series with complex (rather than positive) coefficients. For example, we can deduce that given any Hecke L-series (see Appendix A, Section A.2) associated with an algebraic number field (and with a complex-valued character), its critical zeros do not form an arithmetic progression. In particular, this is true of any Dirichlet L-series.[6] This statement is a simple consequence of the special case of our results for Dedekind zeta functions (which, by definition, have positive coefficients) and of class field theory.

We now precisely state the resulting theorem. Further extensions of Theorem 11.1 are possible using higher-dimensional representations (as, for example, in [Lan, ParSh1–2, RudSar]) rather than one-dimensional representations (i.e., characters), but we omit this discussion here for simplicity of exposition.

Theorem 11.17. *Let K be a number field and χ_0 a character of a generalized ideal class group of K. Then the associated Hecke L-series, $L(s, \chi_0)$, has no infinite sequence of critical zeros forming an arithmetic progression.*

Proof. Let L be the class field associated with this ideal class group, and let ζ_L be the Dedekind zeta function of L. This is the Hecke L-series associated with the trivial character. Let χ run over the characters of the ideal class group. According to a well-known result from class field theory (see Equation (A.10) in Appendix A or, for example, [Lan, Chapter XII, Theorem 1, p. 230] and [ParSh2, Chapter 2, Theorem 2.24, p. 106]), we have

$$\zeta_L(s) = \prod_{\chi} L(s, \chi). \tag{11.21}$$

We deduce from (11.21) that if the factor $L(s, \chi_0)$ has an infinite sequence of zeros in an arithmetic progression, then so does $\zeta_L(s)$. But this Dirichlet series has positive coefficients. Hence, by Theorem 11.12, it does not have such a sequence of zeros. It follows that $L(s, \chi_0)$ does not have such a sequence of zeros either. □

Remark 11.18. The reader unfamiliar with the terminology used in the proof of Theorem 11.17 may assume that $K = \mathbb{Q}$, the field of rational numbers. Then $L(s, \chi_0)$ is an ordinary Dirichlet L-series (see [Da, Lan; Ser, §VI.3] or Example A.2 in Appendix A). We note that the corresponding special case of formula (11.21) is established in [Da, Chapter 6] or [Ser, §VI.3.4].

---

[6] We are grateful to Ofer Gabber for suggesting to us this extension of our results [Gab].

11.4.1 Finite Arithmetic Progressions of Zeros of L-Series

After learning about our work in [Lap-vF1–4], Mark Watkins obtained an extension of these results to Dirichlet L-series for arithmetic progressions of zeros of *finite* length that *do not necessarily start on the real line* (see [Watk, vFWatk]).

In this section, we present Watkins' proof of the following theorem:[7]

Theorem 11.19. *Let*

$$L(s,\chi)=\sum_{\mu=1}^{\infty}\chi(\mu)\mu^{-s}$$

be a Dirichlet L-series with character χ. *Let* ω *be a complex number with real part* $d \geq 1/2$, *and let* $p>0$. *Then*[8] $L(\omega+2\pi ikp,\chi)\neq 0$ *for some integer* k *such that*

$$0<|k|\leq \exp\Big(\tfrac{2+o(1)}{d^2(d+1)}\,p\log^2 p\Big).$$

The argument is based on the idea of studying the frequencies of the "shifted" generalized Cantor spray with lengths a^{-n} of (possibly nonreal) multiplicity $b^n=a^{\omega n}$, on the virtual basic shape $\mathcal{B}$ with frequencies $\chi(k)^{-1}k$ $(k=1,2,3,\dots)$, and hence with spectral zeta function $L(s,\chi)$. Using the pointwise explicit formula, we estimate the jump of the counting function of the frequencies. Since for finite progressions, the jumps are not localized, we consider the jump around a^n over an interval of width about $a^{n/3}$.

The argument consists of two steps. First, we show by a direct computation that the frequency counting function has large jumps. Then we show, using the explicit formula of Theorem 5.10, that this function does not have large jumps if $L(s,\chi)$ has an arithmetic progression of zeros of sufficient length. From this we derive the contradiction.

Lemmas

We use $\log x$ for the natural logarithm, and $\log_a x$ for the logarithm with base a. Also, $[x]$ denotes the integer part of x, $\{x\}$ denotes its fractional part, and $\|x\|=\min(\{x\},1-\{x\})$ is the distance to the nearest integer.

Lemma 11.20. *Let* $N,M\geq 0$, $0<r<1$ *and* $a>1$. *Then there exists an integer* $\tilde{n}$ *satisfying*

$$N<\tilde{n}<N+(1-\log_a r)M+1, \tag{11.22}$$

such that if $\|a^{\mu}\|\leq a^{\mu-\tilde{n}}$ *then* $\|a^{\mu}\|\leq ra^{\mu-\tilde{n}}$ *for every* μ, $0\leq\mu\leq M$.

[7] This result will be restated in more detail in Theorem 11.23 below.

[8] We write $\mathbf{p}=2\pi p$ for the oscillatory period of the corresponding generalized Cantor string.

Proof. Order the numbers $\mu - \log_a \|a^\mu\|$ that are greater than N in increasing order, together with N, to obtain a sequence of values

$$N = v_0 < v_1 < v_2 < \cdots < \infty.$$

For $\mu = 0$, and in general if a^μ is an integer, we set $\mu - \log_a \|a^\mu\| = \infty$. Let v_j be the smallest value such that $v_{j+1} \geq v_j - \log_a r + 1$ (since ∞ is the greatest value, such a value exists). Put

$$\tilde{n} = [v_j] + 1.$$

Then $\mu - \log_a \|a^\mu\| \geq \tilde{n}$ implies $\mu - \log_a \|a^\mu\| \geq v_{j+1} \geq \tilde{n} - \log_a r$. In other words, if $\|a^\mu\| \leq a^{\mu-\tilde{n}}$ then $\|a^\mu\| \leq r a^{\mu-\tilde{n}}$. Moreover, $j \leq M$, hence $v_j < v_{j-1} - \log_a r + 1 < \cdots < v_0 + j(1 - \log_a r)$. It follows that $\tilde{n}$ satisfies (11.22). □

Lemma 11.21. *Let the numbers $N, M \geq 0$, $0 < r < 1$ and $a > 1$ be such that*

$$a\left(1 - ra^{-N}\right) \geq 1$$

and

$$a^{M^2}\left(\left(1 + ra^{-N}\right)^M - \left(1 - ra^{-N}\right)^M\right) \leq 1.$$

Let $\tilde{n}$ be as in Lemma 11.20. If there do not exist integers μ, $1 \leq \mu \leq M$, such that $\|a^\mu\| \leq a^{\mu-\tilde{n}}$, then we put $m = \infty$ and $\alpha = \infty$. Otherwise, we let m be the smallest such μ,

$$m = \min\left\{\mu\colon 1 \leq \mu \leq M\colon \|a^\mu\| \leq a^{\mu-\tilde{n}}\right\},$$

and let α be the integer nearest to a^m. Then

$$\alpha \geq 2, \tag{11.23}$$

and for every $\mu \leq M$ such that $\|a^\mu\| \leq a^{\mu-\tilde{n}}$, we have that μ is an integer multiple of m, and $\alpha^{\mu/m}$ is the integer closest to a^μ.

Proof. By Lemma 11.20, $\alpha = a^m(1 + \theta a^{-\tilde{n}})$ for some real number θ such that $|\theta| \leq r$. If $\alpha = 1$, then $a^m(1 - ra^{-\tilde{n}}) \leq 1$ and hence $a(1 - ra^{-N}) < 1$. But this contradicts the first assumption, hence $\alpha \geq 2$. Further, for $\mu \leq M$ such that $\|a^\mu\| \leq a^{\mu-\tilde{n}}$, we have, again by Lemma 11.20,

$$\beta = a^\mu(1 + \theta' a^{-\tilde{n}}),$$

where β is the integer closest to a^μ and $|\theta'| \leq r$. Then

$$|\alpha^\mu - \beta^m| = a^{\mu m}\left|\left(1 + \theta a^{-\tilde{n}}\right)^\mu - \left(1 + \theta' a^{-\tilde{n}}\right)^m\right|.$$

Since $\mu, m \leq M$ and $\tilde{n} > N$, the second assumption implies $\alpha^\mu = \beta^m$. It follows that $\alpha^{\mu/m}$ is the integer closest to a^μ.

Let $g = \gcd(\mu, m)$ be the greatest common divisor of μ and m, and choose integers λ and l such that $g = \lambda\mu + lm$. Then $\alpha^{g/m} = \alpha^l \beta^\lambda$ is an integer, and

$$\|a^g\| \leq |a^g - \alpha^{g/m}| = a^g \left|1 - (1 + \theta a^{-\tilde{n}})^{g/m}\right|.$$

Since the graph $y = (1+x)^{g/m}$ is concave downward, we obtain

$$\|a^g\| \leq a^g \left(1 - (1 - ra^{-\tilde{n}})^{g/m}\right) \leq ra^{g-\tilde{n}}.$$

By minimality of m, we conclude that $g = m$ and $m|\mu$. □

By the Mean Value Theorem,

$$a^{M^2}\left((1+ra^{-N})^M - (1-ra^{-N})^M\right) \leq a^{M^2-N} 2rM(1+ra^{-N})^{M-1}. \tag{11.24}$$

The right-hand side of (11.24) is bounded by 1 if

$$(M^2 - N) + \log_a(2rM) + (M-1)\log_a(1+ra^{-N}) \leq 0.$$

Moreover, $\log(1+x) \leq x$, hence $(M-1)\log(1+ra^{-N}) < rMa^{-N}$. We will choose

$$N \geq M^2 + \log_a(4rM). \tag{11.25}$$

Then we have, since $a > 1$,

$$a^N \geq 4rMa^{M^2} > 4rM, \tag{11.26}$$

so that $(M-1)\log(1+ra^{-N}) < 1/4 < \log 2$ and

$$\log(2rM) + (M-1)\log(1+ra^{-N}) < \log(4rM).$$

Together with (11.25) and (11.24), this yields the second assumption of Lemma 11.21. Further, $a(1 - ra^{-N}) \geq 1$ is equivalent to $a^N \geq \frac{a}{a-1}r$. Since $\log a < a - 1$, this is implied by (11.26) provided that $4M \geq 1/\log a$ and $M \geq 1$. Hence we will choose

$$M \geq \max\left\{1, \frac{1}{4\log a}\right\}. \tag{11.27}$$

Then (11.25) and (11.27) imply the two conditions of Lemma 11.21.

Lemma 11.22. *Let the numbers a, N, M, r and the integers $\tilde{n}, m, \alpha$ be as in Lemmas 11.20 and 11.21. Assume that*

$$a^N \log a \geq 1, \tag{11.28}$$

$$N - M \geq \log_a(3/2), \tag{11.29}$$

$$r \leq 1/2. \tag{11.30}$$

Let $w \in \mathbb{N}$ *and put* $n = \tilde{n} + w$.

If the integers $l \geq n - M$ *and* $k \geq 1$ *are such that* $|ka^l - a^n| \leq a^w$, *then* $= n - \lambda m$ *for some integer* λ *with* $0 \leq \lambda \leq M/m$, *and the integer nearest o* a^{n-l} *equals* $k = \alpha^\lambda$. *It then follows further that* $|ka^l - a^n| \leq \frac{1}{2}a^w$.

This conclusion remains valid if there are no integers μ, $1 \leq \mu \leq M$, *such that* $\|a^\mu\| \leq a^{\mu - \tilde{n}}$ *(the case* $m = \infty$ *in Lemma 11.21). Then* $l = n$, $k = 1$ *is the only pair of values for which* $|ka^l - a^n| \leq a^w$.

Proof. If $l \geq n + 1$, then

$$ka^l - a^n \geq a^{\tilde{n}+w}(a-1) > a^w a^N (a-1) > a^w a^N \log a.$$

Hence by (11.28), if $|ka^l - a^n| \leq a^w$, then $l \leq n$. Write $\mu = n - l$, so that $0 \leq \mu \leq M$. Then $\|a^\mu\| \leq |k - a^{n-l}| \leq a^{w-l} = a^{\mu - \tilde{n}}$. By Lemma 11.21, we have $\|a^\mu\| \leq ra^{\mu - \tilde{n}}$ and $\mu = \lambda m$. Further,

$$|k - \alpha^\lambda| \leq |k - a^\mu| + \|a^\mu\| \leq a^{\mu - \tilde{n}}(1 + r) < \frac{3}{2}a^{M-N}$$

by (11.30). Hence by (11.29), $k = \alpha^\lambda$. Finally,

$$|ka^l - a^n| = \|a^\mu\| a^l \leq ra^{\mu - \tilde{n}}a^l = ra^w \leq \frac{1}{2}a^w.$$

This completes the proof of Lemma 11.22. □

The Direct Computation

Let χ be a Dirichlet character, $1/2 \leq d < 1$ and ω be a complex number with real part d. Put $b = a^\omega$, so that $|b| = a^d$. Assume that

$$a \leq 4. \tag{11.31}$$

Define $N_\nu(x) = \sum_{ka^l \leq x} \chi(k) b^l$. Fix an integer $w \geq 0$, which will parametrize the width of the jump. Then for $\frac{1}{2}a^w \leq \eta \leq a^w$ and $n > w$, the antiderivative $N_\nu^{[2]}(x)$ of $N_\nu(x)$ satisfies

$$\begin{aligned} &N_\nu^{[2]}(a^n + \eta) - N_\nu^{[2]}(a^n - a^w) - (a^w + \eta) N_\nu(a^n - a^w) \\ &= \sum\nolimits_{-a^w < ka^l - a^n < \eta,\, l \geq n - M} (a^n - ka^l + \eta) \chi(k) b^l + (a^w + \eta) A, \end{aligned} \tag{11.32a}$$

where A is estimated by

$$|A| \leq \frac{4}{(1-d)\log a} a^w + \frac{2}{\log a} a^{-dM} a^{dn}. \tag{11.32b}$$

Proof of (11.32). The left-hand side of (11.32a) is equal to the integral

$$\int_{a^n - a^w}^{a^n + \eta} (N_\nu(t) - N_\nu(a^n - a^w))\, dt = \sum_{-a^w < ka^l - a^n < \eta} (a^n - ka^l + \eta)\, \chi(k) b^l.$$

Since k lies between $(-a^w + a^n)a^{-l}$ and $(a^w + a^n)a^{-l}$, there are at most $2a^w a^{-l} + 1$ possible values for k for each given value of l. Hence the sum for $l < n - M$ is bounded by

$$(a^w + \eta) \sum_{l=0}^{n-M-1} a^{dl}(2a^w a^{-l} + 1)$$
$$= (a^w + \eta)\left(2a^w \frac{1 - a^{(d-1)(n-M)}}{1 - a^{d-1}} + \frac{a^{d(n-M)} - 1}{a^d - 1}\right).$$

Using $\log a^y < a^y - 1$ for $y = d$ and $1 - d$, we obtain a bound of

$$(a^w + \eta)\left(2a^w \frac{a^{1-d}}{(1-d)\log a} + \frac{a^{d(n-M)}}{d \log a}\right).$$

Now $d \geq \frac{1}{2}$ implies that $a^{1-d} \leq 2$, by (11.31). Hence we find the desired bound (11.32b) for $|A|$. □

The Explicit Formula

The following information about L-series follows from the functional equation (see [Lan, Corollary 3, p. 300]). In the left half-plane $\operatorname{Re} s < -\sigma$, we have

$$L(-\sigma + it, \chi) = O\big(t^{1/2+\sigma}\big) \quad \text{for } \sigma > 0, \tag{11.33}$$

and in the critical strip $0 < \operatorname{Re} s < 1$, we have for every $\varepsilon > 0$ that

$$L(\sigma + it, \chi) = O\big(t^{(1-\sigma)/2+\varepsilon}\big) \quad \text{for } 0 < \sigma < 1. \tag{11.34}$$

Let $p > 0$ and assume that $L(\omega + 2\pi i k p, \chi) = 0$ for all $0 < |k| \leq K$. Let $a = e^{1/p}$ and let $0 < \sigma < \frac{1}{2}$ be such that $w = (1-\sigma)n/2$ is an integer. Then

$$N_\nu^{[2]}(a^n + \eta) - N_\nu^{[2]}(a^n - a^w)$$
$$= (a^w + \eta)\frac{\operatorname{res}(L(s,\chi);1)}{1 - b/a} a^n + (a^w + \eta)\frac{L(\omega,\chi)}{\omega \log a} a^{n\omega} + B, \tag{11.35a}$$

where B is bounded by

$$|B| \leq C_3 p a^{2w} + C_4 a^{dn+n} (Kp)^{-(d+1)/2+\varepsilon}. \tag{11.35b}$$

Both constants C_3 and C_4 are positive and depend on χ. Further, C_3 also depends on σ, and may become unbounded as $\sigma \to 1/2$.

Proof of (11.35). We apply the pointwise explicit formula with error term, Theorem 5.10, to $N_\nu^{[2]}(x)$, with screen the vertical line $\operatorname{Re} s = -\sigma$. Let

$b = a^{\omega}$. The spectral zeta function of the shifted Cantor spray on the basic shape $\mathcal{B}$ (see the introduction of this section, along with Equations (4.35) and (10.2)) is given by

$$\frac{L(s,\chi)}{1-ba^{-s}}. \tag{11.36}$$

The pole at $s = 1$ (if χ is the trivial character) gives the Weyl term. There are also poles at ω and at $\omega + 2\pi ik$, for $|k| > K$, but the poles for $0 < |k| \leq K$ are canceled by the zeros of $L(s,\chi)$. The contribution of $\omega + 2\pi ikp$ to the explicit formula for $N_{\nu}^{[2]}(x)$ equals

$$\frac{x^{\omega+1+2\pi ikp} L(\omega + 2\pi ikp, \chi)}{(\omega + 2\pi ikp)(\omega + 1 + 2\pi ikp)\log a} = O\big(x^{d+1}(kp)^{(1-d)/2+\varepsilon}k^{-2}p^{-1}\big).$$

The implied constant depends on the values of $L(s,\chi)$ on the line $\operatorname{Re} s = d$. Summing over $|k| > K$, these contributions add up to

$$O\big(x^{d+1}(Kp)^{-(1+d)/2+\varepsilon}\big).$$

Further, hypothesis **L2** (see (5.20)) is satisfied for the spectral zeta function (11.36) with $\kappa = 1/2 + \sigma < 1$ and the constant C of **L2** estimated by a constant multiple of $(1 - |b|a^{\sigma})^{-1} = O(p)$, by (11.33) and using $a^{y} - 1 > y/p$ for $y = \sigma + d$ to estimate $1 - |b|a^{\sigma}$. By inequality (5.36), applied with $k = 2$, the error term in Theorem 5.10, the pointwise explicit formula with error term, is estimated by $O_{\chi}\left(x^{1-\sigma}p\right)$. By this explicit formula, we find that[9]

$$N_{\nu}^{[2]}(x) = \frac{\operatorname{res}(L(s,\chi);1)}{1-b/a}\frac{x^2}{2} + \frac{x^{\omega+1}L(\omega,\chi)}{\omega(\omega+1)\log a} + \frac{xL(0,\chi)}{1-b} + B',$$

where

$$B' = O\big(x^{1-\sigma}p + x^{d+1}(Kp)^{-(1+d)/2+\varepsilon}\big). \tag{11.37}$$

We substitute $x = a^n + \eta$ and $x = a^n - a^w$ and substract. Using Taylor's Theorem (or the binomial formula if e is an integer), we obtain

$$(y+\xi)^e = y^e + e\xi y^{e-1} + \xi^2 y^{e-2} \sum_{m=2}^{\infty} \binom{e}{m} y^{2-m}\xi^{m-2}. \tag{11.38}$$

In particular, applying (11.38) for $y = a^n$, $\xi = \eta$ and $\xi = -a^w$, we obtain the following identity for $e = 2$:

$$(a^n + \eta)^2 - (a^n - a^w)^2 = 2(\eta + a^w)a^n + (\eta^2 - a^{2w}).$$

[9]Here, $\operatorname{res}(L(s,\chi);1) = 0$ if χ is not the trivial character, since in that case $L(s,\chi)$ is an entire function.

The second term on the right-hand side is of order $a^{2w} = a^{n(1-\sigma)} = x^{1-\sigma}$, which gives a term of the same order as the first error term in Equation (11.37). For $e = \omega + 1$, we obtain

$$(a^n + \eta)^{\omega+1} - (a^n - a^w)^{\omega+1} = (\omega+1)(\eta + a^w)a^{\omega n} + O\big(a^{n(d-1)+2w}\big),$$

since the sum for $m \geq 2$ in the Taylor expansion (11.38) is bounded. Since $d - 1 < 0$, the error term is of lesser order than the first term of B' above. Also for $e = 1$, we obtain a contribution of lesser order than the first term in B'. Thus we obtain (11.35a). Further, B' is bounded as in (11.37), with $x = a^n$. Since $n(1-\sigma) = 2w$, the resulting upper bound for $|B|$ is as stated in (11.35b). □

Zeros in Arithmetic Progression

By the functional equation, we may restrict our attention to zeros with real part at least 1/2. Moreover, it is well known that there are no zeros on the line $\operatorname{Re} s = 1$ (see, e.g., [Ti, Chapter III]).

We now state Watkin's theorem [Watk, vFWatk]:

Theorem 11.23. *Let $L(s, \chi)$ be a Dirichlet L-series, let*

$$p \geq \frac{1}{\log 4},$$

and let ω have real part $d \geq 1/2$. Then there exists an integer $k \neq 0$ such that

$$\log|k| \leq \left(\frac{2}{(d+1)d^2} + o(1)\right) p \log^2 p$$

and $L(\omega + 2\pi ikp, \chi) \neq 0$. Here, $o(1)$ denotes a function that tends to 0 as $p \to \infty$.

Proof. Let K be such that $L(\omega+2\pi ikp, \chi) = 0$ for every k, $0 < |k| \leq K$. Let $N, M, \tilde{n}, r, m, \alpha$ be as in Lemmas 11.20 and 11.21. Choose w to be between $\tilde{n}/3$ and $\tilde{n}$ and put $n = \tilde{n} + w$, so that $0 < \sigma = 1 - 2w/n < 1/2$. By (11.32a) and (11.35a), dividing by $a^w + \eta$, we find that

$$\frac{\operatorname{res}(L(s,\chi);1)}{1-b/a} a^n + \frac{L(\omega,\chi)}{\omega \log a} a^{n\omega} - N_\nu\,(a^n - a^w)$$
$$= \frac{1}{a^w + \eta} \sum_{\substack{-a^w < ka^l - a^n < \eta \\ l \geq n-M}} (a^n - ka^l + \eta)\chi(k)b^l + A - \frac{B}{a^w + \eta}.$$

By Lemma 11.22, in this sum, $l = n - \lambda m$, $0 \leq \lambda \leq M/m$, $k = \alpha^\lambda$ and the range of the sum does not depend on η. The left-hand side does not depend on η either. We take $\eta = a^w$ and $\eta = a^w/2$ and substract, to find, after multiplying by $6a^w$,

$$\left|\sum_{\lambda=0}^{[M/m]} (a^w - a^n + \alpha^\lambda a^{n-\lambda m})\chi(\alpha)^\lambda b^{n-\lambda m}\right| \leq 12|A|a^w + 7|B|$$
$$\leq \left(\tfrac{48}{1-d} + 7C_3\right)pa^{2w} + \left(24pa^{-dM} + 7C_4 a^{\tilde{n}}(Kp)^{-(d+1)/2+\varepsilon}\right) a^{dn+w}. \tag{11.39}$$

We split the sum into the difference of two sums

$$b^n \left(a^w \sum_{\lambda=0}^{[M/m]} \chi(\alpha)^\lambda b^{-\lambda m} - \sum_{\lambda=0}^{[M/m]} (a^n - \alpha^\lambda a^{n-\lambda m})\chi(\alpha)^\lambda b^{-\lambda m} \right).$$

By (11.23), (11.29) and (11.30),

$$a^m \geq \alpha - ra^{m-n} \geq 2 - \tfrac{2}{3}r \geq \tfrac{5}{3} > \left(\tfrac{9}{7}\right)^2,$$

and hence we have $a^{dm} > \frac{9}{7}$. The first sum is smallest if it is alternating (i.e., if $\chi(\alpha)b^{-m} < 0$), so its value is at least

$$\left|\sum_{\lambda=0}^{[M/m]} \chi(\alpha)^\lambda b^{-\lambda m}\right| \geq \frac{1 - a^{-dM}}{1 + a^{-dm}} > \frac{9}{16}\left(1 - a^{-dM}\right).$$

The first term (for $\lambda = 0$) in the second sum vanishes. Using

$$|a^n - \alpha^\lambda a^{n-\lambda m}| = a^{n-\lambda m}\|a^{\lambda m}\| \leq ra^w$$

by Lemma 11.20, the second sum is bounded by

$$ra^w \sum_{\lambda=1}^{[M/m]} a^{-\lambda dm} < \frac{ra^w}{a^{dm} - 1} < \frac{7}{2}ra^w.$$

Hence we obtain

$$\left|\sum_{\lambda=0}^{[M/m]} (a^w - a^n + \alpha^\lambda a^{n-\lambda m})\chi(\alpha)^\lambda b^{n-\lambda m}\right|$$
$$> \left(\frac{9}{16} - \frac{9}{16}a^{-dM} - \frac{7}{2}r\right) a^{dn+w}. \tag{11.40}$$

We choose $r = 1/56$. Combining (11.39) and (11.40), we obtain after dividing by a^{dn+w} that

$$\tfrac{4}{8} - \tfrac{9}{16}a^{-dM} < \left(\tfrac{48}{1-d} + 7C_3\right)pa^{w-dn} + 24pa^{-dM} + 7C_4 a^{\tilde{n}}(Kp)^{-(d+1-2\varepsilon)/2}. \tag{11.41}$$

Since $p \geq 1/\log 4 \geq \frac{9}{16}$ by hypothesis, we have $24p + \frac{9}{16} \leq 25p$. Choose

$$M = \frac{p}{d}\log(8 \cdot 25p), \tag{11.42}$$

so that

$$\left(24p + \tfrac{9}{16}\right) a^{-dM} \leq \tfrac{1}{8}.$$

Also, we estimate the exponent $w - dn < -(2d - 1 + \sigma)N/(1+\sigma)$ in the first term on the right of (11.41), using $w = \tilde{n}\frac{1-\sigma}{1+\sigma}$, $n = \tilde{n}\frac{2}{1+\sigma}$ and $\tilde{n} > N$. We then choose

$$N \geq \frac{p(1+\sigma)}{2d - 1 + \sigma}\log\left(8p\left(\tfrac{48}{1-d} + 7C_3\right)\right), \tag{11.43}$$

so this term is $< 1/8$. We now obtain a contradiction for

$$Kp = (28C_4)^{2/(d+1-2\varepsilon)} a^{2(N+(1-\log_a r)M+1)/(d+1-2\varepsilon)}.$$

Note that by our choice of M and r, (11.27) and (11.30) are satisfied. Further, by (11.25) and (11.42),

$$N \geq \frac{p^2}{d^2}\log^2(200p) + p\log\left(\frac{p\log(200p)}{14d}\right), \tag{11.44}$$

and this implies (11.28) and (11.29). Unless d is close to 1 and p is small,[10] this is more restrictive than the bound (11.43). Hence the choice of w is immaterial, and we choose w to be about $n/3$, so that $\sigma \approx \frac{1}{3}$. We thus obtain a contradiction for

$$\log K = \frac{2 + o(1)}{d^2(d+1)}\, p\log^2 p$$

as $p \to \infty$. This completes the proof of Theorems 11.19 and 11.23. □

11.5 Zeta Functions of Curves Over Finite Fields

The zeta function of a curve over a finite field is periodic with a purely imaginary period. If the curve is not the projective line, then the zeta function has zeros, and by periodicity, each zero gives rise to a shifted arithmetic progression of zeros. Moreover, depending on the curve, the zeta function can have a real zero, and then it has a vertical arithmetic progression of zeros starting on the real axis. So the conclusion of Theorem 11.12 does

---

[10] Roughly, when $1 - a^{-M^2} < d < 1$.

not hold in this case. However, this situation is very similar to the one considered in Section 11.2. Indeed, the associated measure is positive, and the zeta function has an Euler product and satisfies a functional equation. The main goal of this section is to explain why our proof does not go through in this situation. The key reason for this is that the Weyl term itself has large jumps, see the comment following Equation (11.60).

The interested reader can find additional examples and further information about the theory of curves over a finite field in [vF1, vL-vdG, Wei2], as well as, for instance, in [ParSh1, Chapter 4, §1]. In Appendix A, we define a two-variable zeta function of which the zeta functions in this section are a special case.

Let $\mathbb{F}_q$ be the finite field with q elements, where q is a power of a prime number. Let C be a curve defined over $\mathbb{F}_q$. We denote the function field of C by $\mathbb{F}_q(C)$. Thus, $\mathbb{F}_q(C)$ is the field of algebraic functions from C to $\mathbb{P}^1\left(\overline{\mathbb{F}}_q\right)$, the projective line over the algebraic closure of $\mathbb{F}_q$. We view C as embedded in projective space, with homogeneous coordinates $(x_0 : \ldots : x_n)$. The affine part of C is then given as the subset where $x_0 \neq 0$, and this gives rise to the affine coordinate ring $R = \mathbb{F}_q[C]$, the ring of functions that have no poles in the affine part of C. Note that the field of fractions of R is $\mathbb{F}_q(C)$.

The ring R has ideals. For an ideal $\mathfrak{a}$ of R, the ring $R/\mathfrak{a}$ is a finite-dimensional vector space over $\mathbb{F}_q$, and we denote its dimension by $\deg \mathfrak{a}$, the *degree of* $\mathfrak{a}$. Ideals have a unique factorization into prime ideals, and this fact is expressed by the equality

$$\sum_{\mathfrak{a}} q^{-s \deg \mathfrak{a}} = \prod_{\mathfrak{p}} \frac{1}{1 - q^{-s \deg \mathfrak{p}}}, \tag{11.45}$$

where $\mathfrak{a}$ runs over all ideals and $\mathfrak{p}$ over all prime ideals of R. This function of the complex variable s is the incomplete zeta function associated with C. We complete it with the factors corresponding to the points of C with $x_0 = 0$. Also, C may have singularities, and the factors corresponding to the singular points may have to be modified.

Example 11.24 (The projective line). If $C = \mathbb{P}^1$, that is, if the curve has genus $g = 0$, then $R = \mathbb{F}_q[X]$, the ring of polynomials over $\mathbb{F}_q$. In this case, every ideal of R is generated by a single polynomial. The number of ideals of degree n is equal to the number of monic polynomials of degree n, which is q^n. Hence

$$\sum_{\mathfrak{a}} q^{-s \deg \mathfrak{a}} = \sum_{n=0}^{\infty} q^n q^{-ns} = \frac{1}{1 - q^{1-s}}.$$

There is only one point at infinity, and the corresponding factor in the Euler product is

$$\frac{1}{1 - q^{-s}}.$$

Thus, the zeta function of $\mathbb{P}^1$ over $\mathbb{F}_q$ is

$$\zeta_{\mathbb{P}^1/\mathbb{F}_q}(s) = q^{-s}\frac{1}{(1-q^{-s})(1-q^{1-s})}, \tag{11.46}$$

where the factor q^{-s} has been inserted so that this function satisfies the functional equation

$$\zeta_{\mathbb{P}^1/\mathbb{F}_q}(1-s) = \zeta_{\mathbb{P}^1/\mathbb{F}_q}(s). \tag{11.47}$$

Finally, we note that the logarithmic derivative of this zeta function is the generating function of the number of points of $\mathbb{P}^1$ with values in algebraic extensions $\mathbb{F}_{q^n}$ of $\mathbb{F}_q$, for $n = 1, 2, \ldots$. Indeed, $\mathbb{P}^1(\mathbb{F}_{q^n})$ contains $q^n + 1$ points, and

$$\begin{aligned} -\frac{\zeta'_{\mathbb{P}^1/\mathbb{F}_q}}{\zeta_{\mathbb{P}^1/\mathbb{F}_q}}(s) &= \log q\Big(1 + \frac{q^{-s}}{1-q^{-s}} + \frac{q^{1-s}}{1-q^{1-s}}\Big) \\ &= \log q\Big(1 + \sum_{n=1}^{\infty}(q^n+1)q^{-ns}\Big). \end{aligned} \tag{11.48}$$

This result holds for the logarithmic derivative of the zeta function of an arbitrary curve over a finite field, as the reader may check after the subsequent discussion. We do not need this fact here, but it was the original motivation for introducing these functions. (See [Wei1–3].)

Now let C be a curve over $\mathbb{F}_q$ of arbitrary genus g. We describe a way to obtain the completed zeta function of C without first having to choose an affine part of C, as we did in Example 11.24 for genus 0. The field $\mathbb{F}_q(C)$ has valuations. Associated with a valuation v, we have a residue class field $\mathbb{F}_v$, which is a finite extension of $\mathbb{F}_q$. The degree of this extension is denoted $\deg v$, the *degree* of v. A formal sum of valuations

$$\mathfrak{D} = \sum_v m_v v, \quad \text{with } m_v \in \mathbb{Z}, \tag{11.49}$$

with only finitely many nonzero coefficients m_v, is called a *divisor* of C. A valuation will also be called a *prime divisor*. Thus, divisors form a group, the free group generated by the prime divisors. The *degree* of a divisor is $\deg\mathfrak{D} = \sum_v m_v \deg v$. A divisor is *positive*, $\mathfrak{D} \geq 0$, if $m_v \geq 0$ for all v.

Recall that g is the genus of C. The zeta function of C is defined by

$$\zeta_C(s) = q^{-s(1-g)} \sum_{\mathfrak{D}\geq 0} q^{-s\deg\mathfrak{D}}. \tag{11.50}$$

Since, by definition, the factorization of a divisor into prime divisors is unique, we have the Euler product

$$\zeta_C(s) = q^{-s(1-g)} \prod_v \frac{1}{1-q^{-s\deg v}}. \tag{11.51}$$

It is known that $\zeta_C(s)$ is a rational function of q^{-s}, of the form

$$\zeta_C(s) = x^{1-g}\frac{P(x)}{(1-x)(1-qx)}, \qquad x = q^{-s}, \tag{11.52}$$

where

$$P(x) = p_0 + p_1x + p_2x^2 + \cdots + p_{2g}x^{2g}$$

has degree $2g$, integer coefficients, and $p_0 = 1$. It can be shown that

$$P(1) = p_0 + p_1 + p_2 + \cdots + p_{2g}$$

is the *class number* of C; i.e., the number of divisor classes up to linear equivalence. Note that

$$P(1) = -\operatorname{res}(\zeta_C; 0)(q-1)\log q,$$

and that $\operatorname{res}(\zeta_C; 0) = -\operatorname{res}(\zeta_C; 1)$.

The function $\zeta_C(s)$ satisfies the functional equation

$$\zeta_C(s) = \zeta_C(1-s), \tag{11.53}$$

which, in terms of P, means that

$$P\left(\frac{1}{x}\right) = q^g x^{-2g} P\left(\frac{x}{q}\right).$$

In other words, $p_{2g-j} = q^{g-j}p_j$ for $0 \leq j \leq 2g$.

By (11.52), ζ_C is periodic with period $2\pi i/\log q$. The poles of ζ_C are simple and located at $1+2k\pi i/\log q$ and $2k\pi i/\log q$ ($k \in \mathbb{Z}$). It is also known that the zeros of ζ_C have real part $1/2$. In other words, the zeros of $P(x)$ have absolute value $q^{-1/2}$. This is the Riemann hypothesis for curves over a finite field, established by André Weil in [Weil–3]. See Remark A.6 and also, for example, [Roq, Step, Bom] or [ParSh1, Chapter 4, §1] and the relevant references therein. We do not use the Riemann hypothesis in the sequel.

Using the Dirichlet series for $\zeta_{\mathbb{P}^1}$ over $\mathbb{F}_q$,

$$\frac{x}{(1-x)(1-qx)} = \sum_{k=0}^{\infty} \frac{q^k-1}{q-1}x^k,$$

we find that

$$\zeta_C(s) = \sum_{j=-g}^{g} p_{j+g}x^j \sum_{k=0}^{\infty} \frac{q^k-1}{q-1}x^k,$$

where $x = q^{-s}$, and

$$
\begin{aligned}
\zeta_C(s) &= \sum_{n=1-g}^{\infty} x^n \sum_{\substack{j=-g \\ j\le n}}^{g} p_{j+g} \frac{q^{n-j}-1}{q-1} \\
&= \sum_{n=1-g}^{g-1} x^n \sum_{j=-g}^{n} p_{j+g} \frac{q^{n-j}-1}{q-1} + \sum_{n=g}^{\infty} x^n \sum_{j=-g}^{g} p_{j+g} \frac{q^{n-j}-1}{q-1} \\
&= \sum_{n=1-g}^{g-1} x^n \sum_{j=-g}^{n} p_{j+g} \frac{q^{n-j}-1}{q-1} + P(1) \sum_{n=g}^{\infty} x^n \frac{q^n-1}{q-1}, \qquad (11.54)
\end{aligned}
$$

where in the last equality we have used that $\sum_{j=-g}^{g} p_{j+g} = P(1)$ and $\sum_{j=-g}^{g} p_{j+g} q^{-j} = q^g P(1/q) = P(1)$. Note that for $g = 0$, the first sum is to be interpreted as 0.

We interpret this zeta function as the spectral zeta function of a (virtual) basic shape B. Thus, the frequencies of B are

$$
q^n, \text{ with multiplicity } P(1)\,\frac{q^n-1}{q-1}, \text{ when } n \ge g, \qquad (11.55)
$$

and the small frequencies are

$$
q^n, \text{ with multiplicity } \sum_{j=-g}^{n} p_{j+g} \frac{q^{n-j}-1}{q-1}, \text{ when } 1-g \le n \le g-1. \qquad (11.56)
$$

Now, let η be a (generalized) fractal string, and consider the fractal spray of η on B. (See Sections 1.4, 4.3, and 6.6.) Its frequencies are counted by

$$
N_\nu(x) = \sum_{n=1-g}^{g-1} N_\eta\left(\frac{x}{q^n}\right) \sum_{j=-g}^{n} p_{j+g} \frac{q^{n-j}-1}{q-1} + P(1) \sum_{n=g}^{\infty} N_\eta\left(\frac{x}{q^n}\right) \frac{q^n-1}{q-1}. \qquad (11.57)
$$

The first term in the asymptotic expansion of $N_\nu(x)$ is the 'Weyl term' (in the sense of Section 6.6). If $\eta = \mathcal{L}$ is an ordinary fractal string of dimension D with lengths l_j, we can find the corresponding Weyl term as follows: First, we compute

$$
\sum_{n=0}^{\infty} N_{\mathcal{L}}\left(\frac{x}{q^n}\right) \frac{q^n-1}{q-1} = \sum_{n=0}^{\infty} \sum_{j:\, l_j^{-1} \le x q^{-n}} \frac{q^n-1}{q-1} = \sum_{j:\, l_j^{-1} \le x} \sum_{n=0}^{[\log_q x l_j]} \frac{q^n-1}{q-1},
$$

from which we deduce

$$\begin{aligned}\sum_{n=0}^{\infty} N_{\mathcal{L}}\left(\frac{x}{q^n}\right)\frac{q^n-1}{q-1} &= \sum_{j:\, l_j^{-1}\leq x} \frac{q^{[\log_q x l_j]+1}-1}{(q-1)^2} - \frac{[\log_q x l_j]+1}{q-1} \\ &= x\sum_{j=1}^{\infty} l_j \frac{q^{1-\{\log_q x l_j\}}-1}{(q-1)^2} + O\big(x^D \log x\big),\end{aligned}$$

as $x \to \infty$. The first sum is multiplicatively periodic with period q. The other terms in formula (11.57), namely

$$\sum_{n=1-g}^{g-1} N_{\mathcal{L}}\left(\frac{x}{q^n}\right)\sum_{j=-g}^{n} p_{j+g}\frac{q^{n-j}-1}{q-1} - P(1)\sum_{n=0}^{g-1} N_\eta\left(\frac{x}{q^n}\right)\frac{q^n-1}{q-1},$$

add up to $O\big(x^D \log x\big)$. Thus, the Weyl term is given by

$$W(x) = x\cdot P(1)\sum_{j=1}^{\infty}\frac{q^{1-\{\log_q x l_j\}}}{(q-1)^2}\, l_j = x\cdot G\left(\log_q x\right), \tag{11.58}$$

where G is the periodic function (of period 1) defined by

$$G(u) = P(1)\sum_{j=1}^{\infty}\frac{q^{1-\{u+\log_q l_j\}}}{(q-1)^2}\, l_j. \tag{11.59}$$

Since $1 \leq q^{1-\{t\}} \leq q$ for any value of t, we deduce from (11.59) that G is bounded in terms of the total length L of $\mathcal{L}$:

$$L\frac{P(1)}{(q-1)^2} \leq G(u) \leq L\frac{qP(1)}{(q-1)^2}.$$

Moreover, we also see that G is discontinuous, so that the Weyl term has jumps of order x at the points $x = q^m l_1^{-1}$, for $m = 0, 1, 2, \ldots$. By our explicit formulas, we also immediately find the Fourier expansion of the Weyl term, by applying formula (6.77) for $k = 1$:

$$W(x) = \operatorname{res}\left(\zeta_C; 1\right)\sum_{n\in\mathbb{Z}}\zeta_{\mathcal{L}}\left(1+n\frac{2\pi i}{\log q}\right)\frac{x^{1+2n\pi i/\log q}}{1+2n\pi i/\log q}. \tag{11.60}$$

We apply this to the generalized Cantor string $\eta = \sum_{m=0}^{\infty} b^m \delta_{\{q^m\}}$ with parameters $a = q$ and $b > 0$, so that the oscillatory period is $\mathbf{p} = 2\pi/\log q$. (See Figures 11.2 and 11.3 for a graph of the Weyl term associated with the spray of the ordinary Cantor string on the basic shape $\mathbb{P}^1(\mathbb{F}_3)$.) We deduce that at $x = q^m$, $m = 0, 1, 2, \ldots$, the counting function $N_\nu(x)$ jumps by

$$\sum_{n=0}^{m} b^{m-n}\frac{q^n-1}{q-1} = \frac{1}{q-1}\left(\frac{q^{m+1}-b^{m+1}}{q-b} - \frac{b^{m+1}-1}{b-1}\right),$$

up to jumps of order b^m, caused by the terms with $n \leq g-1$. These latter jumps are smaller since $b < q$. We see that the jump is of order x, and

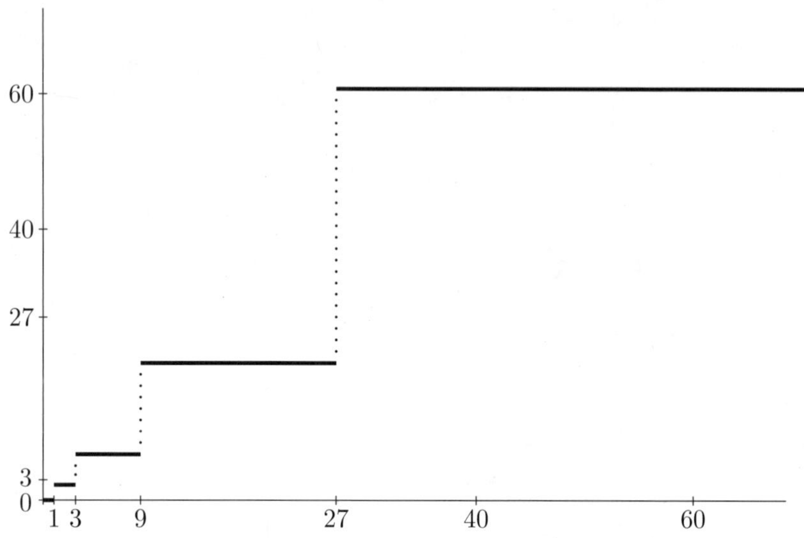

Figure 11.2: The Weyl term of the ordinary Cantor spray on $\mathbb{P}^1(\mathbb{F}_3)$, viewed additively.

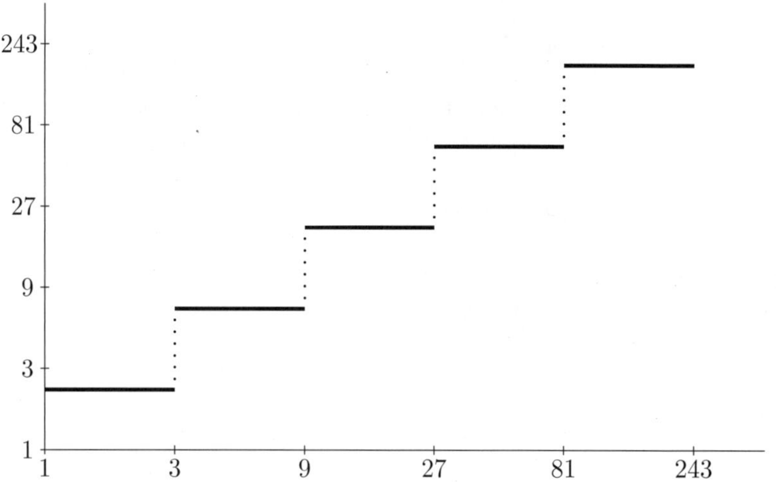

Figure 11.3: The Weyl term of the ordinary Cantor spray on $\mathbb{P}^1(\mathbb{F}_3)$, viewed multiplicatively.

not of order x^D. Since these jumps could be caused by the Weyl term and not by the second asymptotic term, we cannot conclude that there must be oscillatory terms of order D in the explicit formula for $N_\nu(x)$. Moreover, in contrast to the situation in Sections 11.1 and 11.2, we see that the jump of $N_\nu(x)$ is caused by the terms for large values of n, and not by the first terms in the sum for $N_\nu(x)$.

We illustrate this discussion in two examples (see [vF1]).

Example 11.25. The curve given by $y^2 = x^3 - x$ over $\mathbb{F}_3$ has genus 1. Its zeta function is

$$\frac{1+3^{1-2s}}{(1-3^{-s})(1-3^{1-s})} = 1 + 4\sum_{n=1}^{\infty} 3^{-ns}\frac{3^n-1}{2},$$

and the Weyl term associated with a string $\mathcal{L} = \{l_j\}_{j=1}^{\infty}$ is

$$W(x) = x\sum_{j=1}^{\infty} 3^{1-\{\log_3 xl_j\}}\, l_j.$$

Example 11.26. The *Klein curve* is given by the equation

$$x^3y + y^3z + z^3x = 0.$$

Over $\mathbb{F}_2$ it has genus 3, with zeta function given by

$$2^{2s}\frac{1+5\cdot 2^{-3s}+8\cdot 2^{-6s}}{(1-2^{-s})(1-2^{1-s})}$$
$$= 2^{2s} + 3\cdot 2^s + 7 + 20\cdot 2^{-s} + 46\cdot 2^{-2s} + 14\sum_{n=3}^{\infty} 2^{-ns}\left(2^n-1\right),$$

and the Weyl term associated with a string $\mathcal{L} = \{l_j\}_{j=1}^{\infty}$ is

$$W(x) = 14\,x\sum_{j=1}^{\infty} 2^{1-\{\log_2 xl_j\}}\, l_j.$$

We close this section with an example of a Dirichlet series that is a rational function of q^{-s}, has a functional equation and an Euler product, but does not satisfy the analogue of the Riemann hypothesis.

Example 11.27. Consider the zeta function

$$\zeta_X(s) = \frac{1-5x+5x^2}{(1-x)(1-5x)}, \qquad x = 5^{-s}.$$

As a Dirichlet series,

$$\zeta_X(s) = 1 + \sum_{n=1}^{\infty} \frac{5^n-1}{4}\, x^n, \qquad x = 5^{-s}.$$

One checks that this function satisfies the functional equation (11.53). Moreover, the numerator factors as $1-5x+5x^2 = (1-(3-\phi)x)(1-(2+\phi)x)$, where ϕ is the golden ratio. Since $3-\phi < \sqrt{5} < 2+\phi$, the zeros of $\zeta_X(s)$ do not have real part 1/2. The logarithmic derivative of $\zeta_X(s)$ is given by

$$-\frac{\zeta_X'(s)}{\zeta_X(s)} = (\log 5)\sum_{n=1}^{\infty} x^n \left(1 + 5^n - (3-\phi)^n - (2+\phi)^n\right).$$

Let $a_n = (3-\phi)^n + (2+\phi)^n$, for $n = 0, 1, 2, \ldots$. One checks that these numbers satisfy the relation $a_{n+2} = 5\,(a_{n+1} - a_n)$ and $a_0 = 2$, $a_1 = 5$. Thus $a_2 = 15$, $a_3 = 50$, $a_4 = 175, \ldots$. If we want to write $\zeta_X(s)$ as an Euler product of the form

$$\zeta_X(s) = \prod_{n=1}^{\infty} (1 - x^n)^{-v_n},$$

where v_n stands for the number of valuations of degree n, then we find the following formula for v_n:

$$\sum_{d\,|\,n} dv_d = 1 + 5^n - (3-\phi)^n - (2+\phi)^n = 1 + 5^n - a_n.$$

From this, v_n can be successively computed: $v_1 = 1$, $v_2 = 5$, $v_3 = 25$, $v_4 = 110$, $v_5 = 500, \ldots$. In particular, $v_n > 0$ for all $n \geq 1$.

The Weyl term associated with an arbitrary string $\mathcal{L} = \{l_j\}_{j=1}^{\infty}$ is

$$W(x) = x \sum_{j=1}^{\infty} \frac{5^{1-\{\log_5 xl_j\}}}{16}\, l_j,$$

just as in the case of a curve C. However, ζ_X is not the zeta function of a curve, since otherwise it would satisfy the counterpart of the Riemann hypothesis.

12 Concluding Comments, Open Problems, and Perspectives

In this chapter, we make several suggestions for the direction of future research related to, and naturally extending in various ways, the theory developed in this book. In several places, we also provide some additional background material that may be helpful to the reader.

In Section 12.1, we formulate general conjectures about the zeros of Dirichlet series, going beyond our results in Chapter 11, concerning zeros in infinite arithmetic progressions. We also give examples showing the necessity of the assumptions which we are led to make in the resulting "irrationality conjectures".

In Section 12.2, we propose a new definition of fractality, involving the notion of complex dimension. Namely, we propose to call a given set *fractal* if it has at least one nonreal complex dimension (with positive real part). We illustrate this definition by means of two examples, and compare it with other definitions of fractality that have previously been suggested in the literature. This enables us, in particular, to resolve several paradoxes concerning the fractality of certain geometric objects, such as the a-string and the Devil's staircase (i.e., the graph of the Cantor function).

In Section 12.3, we explore some of the relationships between fractality (as defined in Section 12.2) and self-similarity. In particular, we discuss the geometric aspects of this relationship for self-similar fractal drums, and propose a natural interpretation (expressed in terms of complex dimensions) of the lattice–nonlattice dichotomy for higher-dimensional self-similar sets. We illustrate our conjecture by discussing the example of the (von Koch) snowflake curve, a lattice fractal of which we determine the complex dimensions. (See Section 12.3.1.) We had already done so in [Lap-vF5], where we had given an approximate tube formula for the Koch snowflake (see [Lap-vF5, Section 10.3, esp. pp. 209–211]). Here, however, we discuss the joint work of the first author with E. Pearse [LapPe1], who obtain a formula for the area of the inner tubular neighborhoods of the Koch snowflake curve, from which the possible complex dimensions of the snowflake are deduced. One important problem left open by that work is to define a suitable geometric zeta function whose poles coincide with the underlying complex dimensions (see [LapPe2–4]).

We also point out in Section 12.3, as well as in Section 12.5, some of the challenges associated with the extension of our theory of complex dimensions to higher-dimensional objects, such as self-similar fractal drums. In Section 12.3.2, we briefly comment on the recent work ([Pe] and [LapPe2–4]) in which, in the geometric setting, a higher-dimensional theory of the complex dimensions of self-similar fractals is developed via associated self-similar tilings and corresponding tube formulas.

In Section 12.4, we discuss two different types of extensions of the theory of fractal strings. In Section 12.4.1, we provide an overview of the recent joint work of B. Hambly and the first author [HamLap] on random fractal strings and their associated random zeta functions and complex dimensions. Two main classes of random strings are considered: random self-similar strings and their natural generalizations, and random stable strings, such as, for example, one defined in terms of the zero set of Brownian motion. This leads us to refine the definition of fractality proposed in Section 12.2, by allowing the possibility of a natural boundary for the underlying geometric zeta function. In Section 12.4.2, we briefly discuss the new notion of fractal membrane (or quantized fractal string), introduced by the first author in the forthcoming book [Lap10] and further studied by the first author and R. Nest in the papers in preparation [LapNes1–3]. Two related aspects of fractal membranes are directly relevant to potential extensions of the theory in this book: First, their zeta function (or partition function) has an Euler product representation (in which the reciprocal lengths of the underlying fractal string play the role of the primes). And second, because of this Euler product, the corresponding zeros and poles are on an equal footing, and can be viewed as natural invariants of the fractal membrane.

In Section 12.5.1, we discuss the problem of describing the spectrum of a fractal drum, as initially formulated in the Weyl–Berry Conjecture. In Section 12.5.2, we consider this problem for the important special case of self-similar drums, including the snowflake drum and its natural generalizations. In Section 12.5.3, we briefly examine the question of understanding the spectrum of a fractal drum in terms of the periodic orbits of a dynamical system naturally associated to it. In general, and for example, for the snowflake drum, this is a very difficult problem, the formulation of which is only in a preliminary phase. We focus here on the case of a self-similar string for which more can be said at this point.

We close this chapter with a very speculative section, Section 12.7, in which we raise the question of whether a suitable homological interpretation can be found for the complex dimensions of a fractal. In Section 12.6, we make a preliminary suggestion for a possible way to construct such a theory of complex homology (also called fractal cohomology).

12.1 Conjectures about Zeros of Dirichlet Series

We state and briefly discuss several conjectures that could be tackled, or at least receive geometric meaning, within our framework.

We begin by formulating a natural conjecture regarding the vertical distribution of the zeros of the Riemann zeta function $\zeta = \zeta(s)$:

Conjecture 12.1 (Irrationality Conjecture). *The imaginary parts of the critical zeros of ζ on one vertical line in the upper half-plane are rationally independent. In particular, the critical zeros of ζ are all simple.*

We refer, for example, to the paper by Odlyzko and te Riele [Od-tR] and the relevant references therein for numerical and theoretical evidence in support of this conjecture. (Concerning the numerical evidence in support of the simplicity of the zeros, see also, e.g., [Od1–2] and [vL-tR-W], where additional relevant information can be found.) To the knowledge of the authors, since then there have not been significant new developments in this direction.

Remark 12.2. (i) Since $\zeta(s) = \sum_{n=1}^{\infty} n^{-s}$ is defined by a Dirichlet series with real coefficients, its zeros come in complex conjugate pairs. Thus, in the statement of Conjecture 12.1, only the zeros of $\zeta(s)$ with positive imaginary part are considered, say, those on a given vertical half-line $\operatorname{Re} s = D$, $\operatorname{Im} s > 0$. A similar convention is assumed, most often implicitly, in the statement of all the conjectures discussed in the present section. Indeed, the more general zeta functions considered below also have complex conjugate pairs of zeros because they are defined by Dirichlet series with real coefficients.

(ii) We have formulated Conjecture 12.1 in such a way that it is independent of the truth of the Riemann hypothesis. This will enable us, in particular, to extend this conjecture to more general zeta functions, without undue restrictions from the perspective of our present theory. On the other hand, in the literature on this subject, the Riemann hypothesis is often assumed. That is, the critical zeros of $\zeta(s)$ are assumed to be of the form $1/2 \pm i\gamma_n$, $n = 1, 2, \ldots$, with γ_n real and positive. Conjecture 12.1 then asserts that the sequence $\{\gamma_n\}_{n=1}^{\infty}$ is linearly independent over the rationals, or, equivalently, over the ring of rational integers.

(iii) In the statement of Conjecture 12.1, one could even replace "rational independence" by "algebraic independence". We prefer not to do so here, however, because it is less clear how to interpret the resulting statement in the framework of fractal strings. We also point out that Conjecture 12.1 could be recast in the language of fractal strings as the question of invertibility of the spectral operator when restricted to a certain class of fractal strings (much as was done in Corollary 11.4 in a related context).

(iv) Conjecture 12.1 implies, in particular, that $\zeta(s)$ does not have an infinite vertical sequence of critical zeros in arithmetic progression, as was first shown by Putnam in [Pu1–2] and was reproved by a different method in Theorem 11.1 above. It also implies that (nontrivial) finite vertical arithmetic progressions of zeros of $\zeta(s)$ do not exist. We note that even though it is much less general than Conjecture 12.1, the latter statement would provide a significant step towards that conjecture. In Sections 11.1.1 and 11.4.1 we presented partial solutions to this problem. However, to obtain a bound on the length of an arithmetic progression of zeros that is independent of the period of the sequence, significant additional technical difficulties still need to be overcome in order to be able to analyze the nature of the oscillations in the frequency spectrum of the resulting fractal strings.

We have used above the important example of the Riemann zeta function as a motivation for stating and exploring further conjectures regarding a broader class of Dirichlet series. The reader will remember that in Chapter 11 we were able to show that many zeta functions do not have an infinite vertical arithmetic progression of critical zeros by considering generalized Cantor sprays instead of Cantor strings. However, both the zeros and the poles of the zeta function of a curve over a finite field are periodically distributed along vertical lines. (See Section 11.5 above.) Hence, it seems that some restrictions on the multiplicity or the number of the poles or zeros are necessary in order to avoid obvious counterexamples, such as Examples 12.7 and 12.9 below. We therefore make the following assumptions. (In the following, all the poles and zeros are counted according to their multiplicity.)

We will formulate our next conjectures for a general Dirichlet series ζ_B associated with a measure. But first we show by means of an example that we need to assume positivity and discreteness of the underlying measure.

Example 12.3. Let ζ be the Riemann zeta function, and let $L_1, \dots, L_{2n}$ be L-series associated with $2n$ independent nontrivial real characters (see Appendix A, Section A.2).[1] Then, for any choice of real numbers $c_1, \dots, c_{2n}$, the function

$$\zeta(s) + c_1 L_1(s) + \cdots + c_{2n} L_{2n}(s)$$

has a simple pole at $s = 1$. By suitably choosing the coefficients, one can arrange for this function to have a sequence of zeros in an arithmetic pro-

[1] Note that by assumption, only $\zeta(s)$ has a pole (at $s = 1$).

gression of length n. Indeed, choosing any sequence $D + ik\mathbf{p}$, $k = 1, \ldots, n$, with $\mathbf{p} > 0$, this amounts to solving n linear equations with complex coefficients in $2n$ real variables. Note that $D - ik\mathbf{p}$, $k = 1, \ldots, n$, will automatically be an arithmetic progression of zeros as well. However, according to the following conjectures, the resulting Dirichlet series will not have real positive coefficients.

If one chooses $\zeta, L_1, \ldots, L_{2n}$ to be completed zeta functions (see Section A.3 of Appendix A), associated with real-valued characters, we can choose the coefficients $c_1, \ldots, c_{2n}$ to be real and small, so that the resulting Dirichlet integral is associated with a positive measure and has a sequence of zeros in an arithmetic progression of length n. However, in that case, it is no longer a Dirichlet series, associated with a discrete measure.

Let $\zeta_B(s)$ be a Dirichlet series with positive coefficients satisfying hypothesis (P), formulated on page 303 of Section 11.2. At this point, we invite the reader to review the statement of this hypothesis, which involves the existence of a window W for $\zeta_B(s)$. Note that according to hypothesis (P), $\zeta_B(s)$ is required neither to satisfy a functional equation nor to have an Euler product. Moreover, we recall that we use throughout this section the convention described in comment (i) above. Thus, according to this convention, only the zeros of $\zeta_B(s)$ with positive imaginary part should be taken into account in the statement of Conjectures 12.4–12.6 and 12.10 below.

By analogy with the Riemann zeta function, we formulate the following conjectures for $\zeta_B(s)$:

Conjecture 12.4. *If ζ_B has at most one pole in a window, then it does not have a vertical arithmetic progression of zeros of length two in this window.*

More generally, if ζ_B has one pole, then it does not have two zeros with positive imaginary part on one vertical line in this window, the imaginary parts of which are rationally dependent.

Conjecture 12.5. *If ζ_B has n poles in a window, then it can have a vertical arithmetic progression of zeros of length at most n in this window.*

More generally, if ζ_B has n poles in a window, then it can have at most n zeros with positive imaginary part on one vertical line in this window, the imaginary parts of which are rationally dependent.

The next conjecture is the exact counterpart in this context of Conjecture 12.1 concerning the Riemann zeta function.

Conjecture 12.6 (General Irrationality Conjecture). *If ζ_B has one pole in a window, then the imaginary parts of the zeros of ζ_B in the upper half-plane on a given vertical line in this window are rationally independent. In particular, the zeros of ζ_B are all simple in this window.*

Note that since $\zeta(s)$ is a particular instance of the zeta functions which we consider, this conjecture implies Conjecture 12.1.

Before stating our last conjecture, we provide three examples showing that there are some further restrictions on what may be expected to hold.

Example 12.7. The function

$$\prod_{k=1}^{n} \zeta\left(\frac{1}{2}+\frac{s}{k}\right)$$

shows that a function with n poles can have n zeros in arithmetic progression. Indeed, let $1/2+i\gamma$ be a critical zero of $\zeta(s)$, with $\gamma>0$. Then the values $s=ki\gamma$, for $k=1,\dots,n$, are zeros of the above function in the upper half-plane.

The following example shows that it is necessary to assume hypothesis (P) in our conjectures.

Example 12.8. In the previous example, if instead of the Riemann zeta function, we use a Dirichlet L-series $L(s,\chi)$ associated with a character χ, we obtain a function without poles,

$$\prod_{k=1}^{n} L\left(\frac{1}{2}+\frac{s}{k},\chi\right),$$

with a vertical arithmetic progression of zeros of length n, since it is known that at least some of the zeros of $L(s,\chi)$ lie on the critical line $\operatorname{Re} s=1/2$. Indeed, as for $\zeta(s)$, it is conjectured that all nontrivial zeros lie on this line.

Example 12.9. Let $\zeta_K(s)$ be the zeta function of an algebraic number field K. As is well known, this function has zeros with real part $1/2$. Then the function

$$\zeta\left(\frac{1}{2}+s\right)+c_1\zeta_K\left(\frac{1}{2}+c_2 s\right),$$

with a suitable choice of the real constant c_1 and the positive constant c_2, shows that a Dirichlet series with two poles can have a double zero. Indeed, if ζ_K does not have a double zero itself, then one can first choose c_2 so that both functions have a common zero at a point $s=i\gamma$ (here, γ is as in the previous example). One can then adjust c_1 so that the derivative also vanishes at that point.

Conjecture 12.10. *If ζ_B has at most n poles in a window, then its zeros have multiplicity at most n in this window.*

12.2 A New Definition of Fractality

Our work shows that the complex dimensions of a fractal string contain important geometric information; see especially Section 8.2 and Theorems 2.17, 8.1 and 8.15, along with Section 8.4 and Chapters 6 and 7.

Motivated by the fact that the complex dimensions of a fractal string describe very precisely the oscillations of the fractal, and also, for example, by the fact that self-similar strings always have nonreal complex dimensions, whereas the dimensions of the a-string are all real,

> we propose to define "fractality" as *the presence of at least one nonreal complex dimension with positive real part.*[2]

Loosely speaking, we propose to define a geometric object as being "fractal" if its geometry has oscillations. We could refine this by defining a set to be *fractal in dimension* d if there exists a nonreal complex dimension ω of the set such that $\mathrm{Re}\,\omega = d$. (See Sections 12.3 and 3.7.)

Note that nonreal complex dimensions come in conjugate pairs ω, $\bar{\omega}$, since $\zeta_{\mathcal{L}}(s)$ is real-valued for $s \in \mathbb{R}$. Hence a fractal set, in this definition, has at least two nonreal complex dimensions.

12.2.1 Fractal Geometers' Intuition of Fractality

We now explain how our proposed definition agrees with the intuition of fractal geometers, even in the cases where other definitions of fractality disagree with generally accepted intuition.

We first consider the a-string,[3] viewed as the complement in $(0, 1)$ of the sequence j^{-a} $(j = 1, 2, 3, \dots)$. Thus, the boundary of the a-string is

$$F = \{1, 2^{-a}, 3^{-a}, \dots, 0\}.$$

We quote from [Fa3, p. 45],

> *No one would regard this set, with all but one of its points isolated, as a fractal, yet it has fractional box dimension.*[4]

We agree with Falconer that the a-string is not fractal. However, as explained in Remark 12.12, we disagree with his stated reason.

Secondly, all fractal geometers agree that self-similar (or, more generally, self-affine or self-alike) objects are fractal, because they have fine structure at all scales. For example, Mandelbrot writes in [Man1, p. 82] of the Devil's staircase (i.e., the graph of the Cantor function, see Figure 12.1 or [Man1, Plate 83, p. 83], along with Remark 12.11 below):

> *One would love to call the present curve a fractal, but to achieve this goal we would have to define* fractals *less stringently, on the basis of notions other than D alone.*[5]

We agree that the Devil's staircase, which is self-affine, is a fractal. In Section 12.2.2, we explain how this intuition fits exactly with our proposed definition.

---

[2] In view of recent results on random fractal strings obtained in [HamLap], this definition will be completed at the end of Section 12.4.1.

[3] Already known to Bouligand [Bou] in a different terminology.

[4] That is, noninteger Minkowski dimension.

[5] In the notation of Mandelbrot, D denotes the Hausdorff dimension.

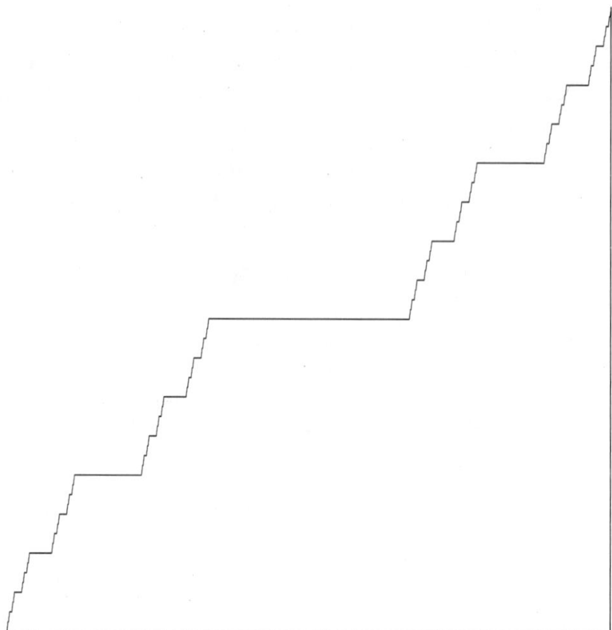

Figure 12.1: The Devil's staircase.

Remark 12.11. Recall that the Devil's staircase is defined as the graph of the Cantor function $\mathcal{C}$. The latter is a nondecreasing continuous function on $[0, 1]$ such that $\mathcal{C}(0) = 0$ and $\mathcal{C}(1) = 1$. Further, $\mathcal{C}$ is constant on each interval in the complement of the ternary Cantor set in $[0, 1]$. In fact, $\mathcal{C}$ is nothing but the primitive of the natural (Θ-dimensional) Hausdorff measure on the Cantor set, where $\Theta = \log_3 2$. (See, for example, [Coh, p. 55 and pp. 22–24] or [ReSi1, pp. 20–23].)

Remark 12.12. In our theory, at least in the one-dimensional case, fractality is independent of the geometric realization of a fractal. For example, one could realize the Cantor string as a sequence of intervals of lengths 1, 1/3, 1/3, 1/9, ..., in such a way that the endpoints of these intervals form an infinite, decreasing sequence with a single limit point at the origin. According to Falconer's quote above, this set would not be fractal since it has only one nonisolated point. In our setting, however, the fractality of the Cantor string is independent of its geometric realization. Nevertheless, see also Remark 12.15 at the end of Section 12.2.2.

The intuition of fractal geometers has been guided by the following defining properties of fractality: self-similarity, or, more generally, approximate self-alikeness; usually fractional box (i.e., Minkowski), packing, or Hausdorff dimension; fine geometric structure at all scales; a simple recursive construction; singularity of an associated measure (e.g., Hausdorff measure) with respect to Lebesgue measure; and other properties. However, a clear definition of fractality is still missing. Indeed, we quote from the introduction of Falconer's book [Fa3, p. xx]:

> *My personal feeling is that the definition of a 'fractal' should be regarded in the same way as the biologist regards the definition of 'life'. There is no hard and fast definition, but just a list of properties characteristic of a living thing, such as the ability to reproduce or to move or to exist to some extent independently of the environment. Most living things have most of the characteristics on the list, though there are living objects that are exceptions to each of them.*

The classical definition of fractality, as stated by Mandelbrot in [Man1, p. 15], is the following:

> *A fractal is by definition a set for which the Hausdorff–Besicovitch dimension strictly exceeds the topological dimension.*[6]

In other words, if H denotes the Hausdorff dimension of a set $F \subset \mathbb{R}^d$ and T denotes its topological dimension, then F is fractal, according to Mandelbrot, if and only if $H > T$.[7] Mandelbrot discusses in detail in [Man1, Chapter 3] and elsewhere in his book the implications as well as the limitations of his definition. (See also the introduction of [Man3].)

In the case of the Devil's staircase $\mathcal{C}$, one has $H(= D) = T = 1$ because $\mathcal{C}$ is rectifiable; see, for example, [Man1, p. 82] and [Fed2, Theorem 3.2.39, p. 275]. Thus the Devil's staircase is not fractal according to this definition.

For certain applications, one could prefer to use the Minkowski dimension over the Hausdorff dimension in order to define fractality. But a definition such as "the Minkowski dimension strictly exceeds the topological dimension" would classify the a-string as fractal. And also, if the lengths of the a-string are rearranged so that the Hausdorff and the Minkowski dimensions of this set coincide, as is clearly possible, then the above definition, unmodified, already classifies the a-string as fractal.

Other possible definitions of fractality have been introduced in the literature, involving various notions of (real) fractal dimensions, such as the packing dimension P [Tr2] (which is in some sense dual to the Hausdorff dimension) and, as discussed just above, the Minkowski(–Bouligand) (or box) dimension D. All these definitions run into similar problems. For example,

---

[6]A dichotomy fractal vs. nonfractal—based on D rather than H—was used in [Lap1] (and in later papers, such as [Lap2–3, LapPo2, LapMa2]) for pragmatic reasons in the context of drums with fractal boundary; it was not intended to provide a definition of fractality.

[7]Recall that T is a nonnegative integer and that we always have $H \geq T$; see, e.g., [HurWa] and [Rog].

the packing dimension fails to classify the Devil's staircase as fractal, since we also have $P = 1$ (because $H \le P \le D$, by Remark 1.5).

12.2.2 Our Definition of Fractality

We already mentioned that the a-string is not fractal in our proposed new sense, since $\zeta_{\mathcal{L}}$ has a meromorphic continuation to $\operatorname{Re} s \ge 0$, with only one pole at $D = 1/(1+a)$. (See Theorem 6.20 in Section 6.5.1 for a complete analysis of the complex dimensions of this string and compare this with [Lap1, Example 5.1 and Appendix C].) By definition, the same statement applies to the boundary of $\mathcal{L}$, which is equal to the compact set $\{1, 2^{-a}, 3^{-a}, \ldots, 0\}$.

On the other hand, by Corollary 2.21, a self-similar string (and thus a self-similar set in $\mathbb{R}$ other than a single interval) is always fractal since it has infinitely many complex dimensions with positive real part. According to Conjecture 12.18 in Section 12.3 below, any (nontrivial) self-similar set in $\mathbb{R}^d$ would be fractal in our new sense. We point out that when $d = 1$, Conjecture 12.18 has been proved in Theorems 2.17 and 3.6 above, and for $d \ge 2$, it has been proved in [LapPe2, 3] for a suitable class of self-similar sets (or tilings).[8] (See Section 12.3.2.)

We now analyze our proposed definition for the Devil's staircase. This will clearly expose the problem of defining and determining the complex dimensions of a set. We will regard the exponents of ε in the asymptotic expansion of the volume of the (inner) tubular neighborhoods as the complex codimensions. This is motivated by our tube formula for fractal strings, as stated in the introduction (Equation (**)) or, e.g., in Theorem 8.1 (Equation (8.4)). (See also Sections 12.3.1 and 12.3.2, along with [LapPe1–3].)

The volume $V(\varepsilon)$ of the inner tubular neighborhoods of the Devil's staircase (Figure 12.2) is approximated by

$$2\varepsilon - \varepsilon^{2-\Theta}\left(1 - \frac{\pi}{4}\right)\left(\left(\frac{1}{2}\right)^{\{x\}} + \left(\frac{3}{2}\right)^{\{x\}}\right), \tag{12.1}$$

where $x = \log_3 \varepsilon^{-1}$ and $\Theta = \log_3 2$. By formula (1.13) on page 15, writing $\mathbf{p} = 2\pi/\log 3$, we obtain the expression

$$2\varepsilon^{2-1} + \frac{4-\pi}{8\log 3}\sum_{n=-\infty}^{\infty} \frac{\varepsilon^{2-\Theta-in\mathbf{p}}}{(\Theta + in\mathbf{p})(\Theta - 1 + in\mathbf{p})} \tag{12.2}$$

for this approximate volume (see also Remark 12.14 below). Thus the complex dimensions of the Devil's staircase are 1 and $\Theta + in\mathbf{p}$, $n \in \mathbb{Z}$, and they are all simple. (See Figure 12.3.) Hence the Devil's staircase is fractal in our new sense, and more precisely, it is fractal in dimension Θ.

---

[8] When $d = 1$, this class coincides with that of all self-similar sets in $\mathbb{R}$ considered in this book; see Section 2.1.1.

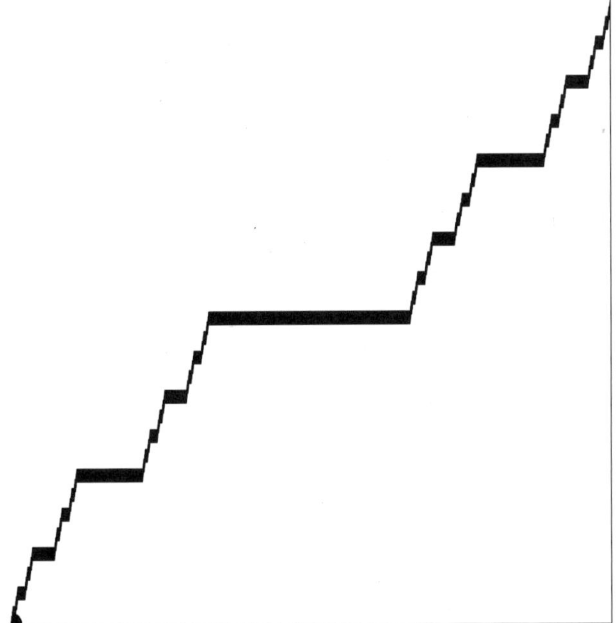

Figure 12.2: The inner ε-neighborhood of the Devil's staircase.

Remark 12.13. Observe that $\Theta = \log_3 2$ is the dimension of the Cantor set, which is the set where $\mathcal{C}$ is increasing (by infinitesimal amounts), while 1 is the dimension of the complement of the Cantor set in $[0, 1]$, the open set which is composed of the open intervals on which $\mathcal{C}$ is constant. Further note that in formula (12.1) (or (12.2)), the leading term 2ε can be interpreted as follows: The Devil's staircase is rectifiable with finite length equal to 2. This well-known fact (see, e.g., [Man1, p. 82]) can be understood intuitively by adding up the infinitesimal steps of the Devil's staircase. This amounts to projecting the Devil's staircase onto the horizontal and vertical axes in Figure 12.1 and adding up the resulting lengths.

Remark 12.14. We caution the reader that even though the complex dimensions with positive real part of the Devil's staircase are determined by the approximate formula (12.2), which was obtained in [Lap-vF5, Section 10.2, p. 202], the residues corresponding to the poles other than 1 in (12.2) are not accurate. (Compare this comment with [LapPe1] and Section 12.3.1, especially Theorem 12.21 and Remark 12.23.)

Thus our definition using the complex dimensions of a fractal classifies each of the above examples correctly, including the "borderline examples". Indeed, we are not aware of a single example where our definition classifies

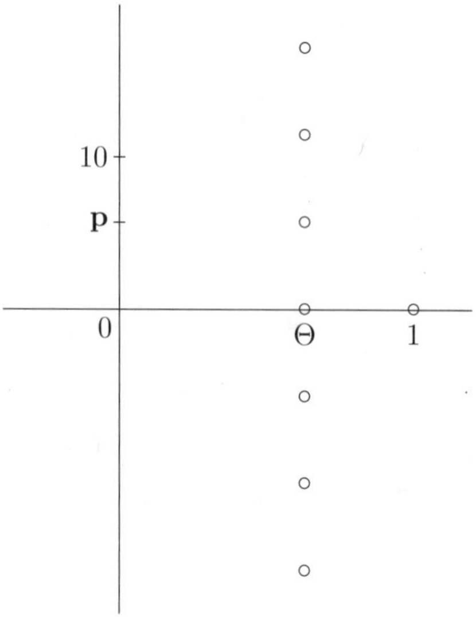

Figure 12.3: The complex dimensions of the Devil's staircase. $\Theta = \log_3 2$ and $\mathbf{p} = 2\pi / \log 3$.

a set incorrectly. Thus with our definition, there would be no borderline cases where the definition would disagree with intuition. An exception could be the zero set of Brownian motion and related stable random strings, for which the determination of the complex dimensions (or the natural boundary of analyticity) is an open problem; see Problem 12.35 and the discussion surrounding it at the end of Section 12.4.1. Thus our present knowledge is still rather limited, since it is clearly a difficult task, both conceptually and practically to determine the complex dimensions of a higher-dimensional geometric object. Indeed, the definition that we adopted for the Devil's staircase, via the asymptotic expansion of $V(\varepsilon)$, is only provisional. We plan to address the question of defining the complex dimensions of higher-dimensional fractal sets in a subsequent work.

In most cases of interest in the applications—such as multifractals, Julia sets, the Mandelbrot set, and strange attractors in the theory of dynamical systems—this definition of fractality remains to be tested and the theory to be further developed along the following lines. (i) First, define a suitable geometric, spectral, or dynamical zeta function associated with the object. (ii) Second, investigate the resulting complex dimensions, defined as the poles of the meromorphic continuation of the appropriate zeta function. In particular, establish the existence of nonreal complex dimensions with positive real part (or, in view of the discussion surrounding Problem 12.35 at the end of Section 12.4.1, the existence of a natural boundary along a suitable screen, part of which is contained in the half-plane $\operatorname{Re} s > 0$).

(iii) Finally, show the relevance of these definitions and results (in conjunction with the theory developed in this monograph) to the study of the geometric, spectral, or dynamical object under investigation. This can be achieved, for example, by using the explicit formulas of Chapter 5 and considerations specific to the problem at hand in order to obtain appropriate tube formulas, spectral asymptotics, or analogues of the Prime Orbit Theorem and the associated explicit formula.

In practice, steps (i) and (iii) will often have to be interchanged. Hence, for example, one might first obtain a suitable tube formula, and then deduce from it the possible complex dimensions of the object, or even better, a suitable definition of the (geometric) zeta function of which these are the poles. We refer to Section 12.2 above and Section 12.3 below, based in part on [Lap-vF5], [LapPe1–3] and [Pe], for various examples illustrating this approach, from the Cantor staircase to the Koch snowflake curve (Sections 12.2 and 12.3.1, respectively) and self-similar sets (or tilings) in higher-dimensional Euclidean spaces (Section 12.3.2). Furthermore, we mention that in several works in progress, a related method is used to study the complex dimensions of certain multifractals [JafLap, LapLevRo] and quasicrystals.

We note that more refined information about fractal geometries can be obtained by taking into account the structure of the complex dimensions (and the associated residues). In particular, we believe that the (conjectured) quasiperiodicity of the complex dimensions—along with the presence of geometric oscillations of order D or of order arbitrarily close to but less than D—is an important feature of self-similar geometries. This quasiperiodicity was established in the case of self-similar strings in Chapter 3. In view of Remark 12.20 below, an analogous statement should hold for approximately self-similar fractals as well. In the next subsection, we discuss such an application of complex dimensions, namely to understanding the notion of lacunarity.

Remark 12.15. In higher dimensions, one needs significantly more than just a counterpart of the lengths of a string in order to obtain a suitable notion of geometric zeta function and, therefore, of the complex dimensions. See [LapPe1–4] along with Section 12.3.2 below. Thus, in general, our proposed notion of fractality will truly depend on the geometry of the object.

Furthermore, even in the one-dimensional case (i.e., for fractal strings), it is possible to extend some aspects of the theory developed in this book in order to distinguish, for example, a Cantor string Ω_1 realized as having for boundary the classic ternary Cantor set from another such string Ω_2 having for boundary a sequence of points with a single limit point, say at the origin. This is done in [LapLevRo] where a one-parameter family of multifractal zeta functions $\zeta_\Omega(\alpha, s)$ (with $-\infty < \alpha \leq +\infty$) is introduced for a given fractal string Ω (viewed as a bounded open subset of $\mathbb{R}$, or more generally, for a suitable Borel measure on $\mathbb{R}$). For $\alpha = 1$, under mild assumptions, $\zeta_\Omega(1, s)$ coincides with $\zeta_\Omega(s)$, the standard geometric zeta function of the fractal string Ω. Hence, it does not depend on the realization of the length sequence $\{l_j\}_{j=1}^\infty$ of Ω. For other values of α, however, this is no longer the case in general, and $\zeta_\Omega(\alpha, \cdot)$ usually depends on finer topological and geometric information than just the lengths of the string. This is the case, in particular, for the aforementioned two realizations Ω_1 and Ω_2 of the Cantor string.

It follows that our proposed definition of fractality extended in the obvious way to each value of α (fractality for the regularity parameter α) can be used to provide finer information than is done in this book.[9] In particular, there are many interesting examples which have a nontrivial multifractal spectrum of zeta functions and hence, in general, of complex dimensions. On the other hand, even in this more flexible setting, the a-string is still nonfractal for every value of α.

We refer to [LapLevRo] for further information about this intriguing subject, in which many important problems remain wide open, especially in the standard multifractal case, for more general singular measures, and in higher dimension.

12.2.3 Possible Connections with the Notion of Lacunarity

In his book [Man1, Chapter X], Mandelbrot suggests complementing the notion of (real) fractal dimension with that of *lacunarity,* which is aimed at better taking into account the texture of a fractal. (See [Man1, Section X.34] and, for a more quantitative approach, [Man2], along with the relevant references therein, including [BedFi].)

We quote from the introduction of [Man2, pp. 16–17]:

> *Fractal lacunarity is an aspect of "texture" that is dominated by the sizes of the largest open components of the complement, which are perceived as "holes" or "lacunas". ...*
>
> *Lacunarity is very small when a fractal is nearly translation invariant, being made of "diffuse" clumps separated by "very small" empty* lacunas, *and lacunarity is high when this set is made of "tight" clumps separated by "large" empty* gaps *or* lacunas.

Our present theory of complex dimensions may help shed some new light on this somewhat elusive notion of lacunarity, especially in the one-dimensional situation. In particular, we hope to explain in more detail elsewhere how one of the main examples discussed in [Man1, Section X.34] and [Man2]—namely, in our notation, the sequence of Cantor sets (or "Can-

[9] Note that the explicit formulas and other results obtained in our theory can be applied to each multifractal zeta function $\zeta(\alpha, \cdot)$.

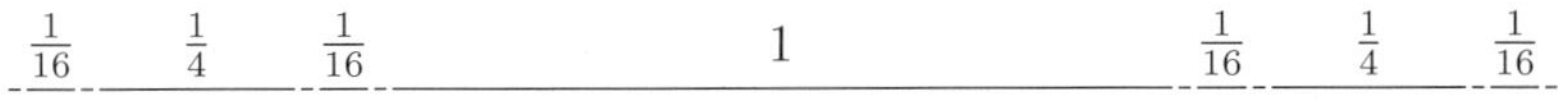

$\frac{1}{16}$ $\frac{1}{4}$ $\frac{1}{16}$ 1 $\frac{1}{16}$ $\frac{1}{4}$ $\frac{1}{16}$

Figure 12.4: A lacunary Cantor string, with $a = 4$, $b = 2$ ($D = 1/2$, $\mathbf{p} = \pi/\log 2$ and $k = 1$).

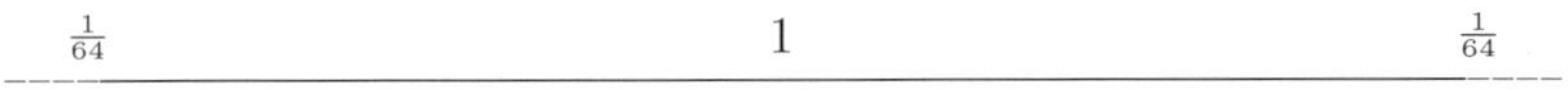

$\frac{1}{64}$ 1 $\frac{1}{64}$

Figure 12.5: A less lacunary Cantor string, with $a = 64$, $b = 8$ ($D = 1/2$, $\mathbf{p} = \pi/(3\log 2)$ and $k = 3$).

tor dusts") $\eta_k = \sum_{j=0}^{\infty} 2^{jk}\delta_{\{4^{jk}\}}$ $(k \geq 1)$[10] represented in [Man2, Figure 1, pp. 18–19] and discussed in [Man2, Section 2.2, p. 21]—can be understood in terms of the complex dimensions of the associated Cantor strings. More specifically, for each fixed $k \geq 1$, the complex dimensions of the Cantor set η_k are $1/2 + in\pi/(k\log 2)$ $(n \in \mathbb{Z})$ and the corresponding residues are independent of n and equal to $1/(2k\log 2)$. (See Figure 10.1 in Chapter 10 above, with the choice of parameters $a = 4^k$ and $b = 2^k = \sqrt{a}$.) Therefore, even though these Cantor strings have the same real dimension $D = 1/2$, they have very different sets of complex dimensions (and residues), which accounts for their different lacunarities (in the language of [Man1–2]).

In particular, as $k \uparrow \infty$, the oscillatory period $\mathbf{p}_k = \pi/(k\log 2)$ decreases to 0, so the complex dimensions become denser and denser on the vertical line $\operatorname{Re} s = 1/2$, with smaller and smaller residues. Since the imaginary parts of the complex dimensions vanish in the limit, the nonreal complex dimensions eventually disappear, as might be expected intuitively for the homogeneous limiting set, which has zero lacunarity according to [Man2]; see Figure 12.5. Note that as $k \uparrow \infty$, the gaps (also called holes or lacunas in [Man1–2]) in the complement of the Cantor set defined by η_k become smaller and smaller, while the Cantor sets themselves appear to be more and more translation invariant or homogeneous and therefore "less and less fractal" or "lacunary". (See the discussion in Section 2.1, page 21, of [Man2].)

In contrast, as $k \downarrow 0$ (a limiting case that is not considered in [Man1–2] because it does not correspond to a geometric Cantor set but that is compatible with our notion of generalized Cantor strings, see Chapter 10), $\mathbf{p}_k = \pi/(k\log 2)$ increases to infinity and so do the associated residues $1/(2k\log 2)$. Therefore, the complex dimensions become sparser and sparser on the vertical line $\operatorname{Re} s = 1/2$, but with larger and larger

[10] In ordinary language, in the j-th stage of its construction, the complement in $(0,1)$ of the Cantor set η_k consists of b^j open intervals of length a^{-j}, with $a = 4^k$ and $b = 2^k = \sqrt{a}$.

residues. This is in agreement with the intuition that the (generalized) Cantor string η_k is more and more lacunary (i.e., has larger and larger gaps or holes, see Figure 12.4) as the real number k decreases to 0.

Remark 12.16. Recall that in Chapter 8, we have obtained precise tube formulas for the volume of the ε-neighborhoods of a fractal string $\mathcal{L}$, expressed in terms of the complex codimensions of $\mathcal{L}$ and of the associated residues (or principal parts) of $\zeta_{\mathcal{L}}$. This is of interest in the present context because in [Man1] and especially in [Man2], lacunarity has also been linked heuristically to the (possibly oscillatory) prefactor occurring in the definition of the upper or lower Minkowski content (prior to taking the corresponding limit as $\varepsilon \to 0^+$). See formula (1.6) on page 11, Section 8.4.3 and the notes to Chapter 8.

We leave it as an exercise for the interested reader to specialize to the above example the tube formulas obtained for generalized Cantor strings in Section 8.4.1, and to interpret the resulting expressions as $k \uparrow \infty$ and as $k \downarrow 0$.

Remark 12.17. We note that the notion of fractal lacunarity may also help connect aspects of our work with earlier physical work of which we have recently become aware and that was aimed in part at understanding the relationship between lacunarity and turbulence, crack propagation or fractal growth, among other physical applications. (See, for example, [BadPo, BallBlu1–3, BessGM, FouTuVa, ShlW, SmiFoSp] along with the semi-expository articles [SalS, Sor] and the relevant references therein.)

12.3 Fractality and Self-Similarity

We refine our definition of fractality as follows. The roughness of a fractal set is first of all characterized by its Minkowski dimension D. Then, either it has nonreal complex dimensions with real part D, or it only has nonreal complex dimensions with smaller real part. These two cases correspond to, respectively, a fractal set that is not Minkowski measurable, and one that is. (See Chapter 8, especially Sections 8.3, 8.4.2 and 8.4.4.) In the first case, we say that the set is maximally fractal, whereas in the second case, the set is fractal only in its less-than-D-dimensional features.

From our analysis of self-similar strings in Chapter 2 (Theorem 2.17), it follows that lattice strings are maximally fractal in this sense, whereas nonlattice strings are not fractal in dimension D, but are fractal in infinitely many dimensions less than and arbitrarily close to D.

Conjecturally, the dichotomy lattice vs. nonlattice and the corresponding characterization of the nature of fractality applies to a wide variety of situations, including drums with self-similar fractal boundary, drums with self-similar fractal membrane, random fractals as well as "approximately self-similar sets", such as limit sets of Fuchsian and Kleinian groups or hyperbolic Julia sets [Su, BedKS, Lal2]. This conjecture was first formulated in [Lap3, Conjectures 2–6, pp. 159, 163, 169, 175, 190, 198], to which we refer for a more detailed description of these situations. One obtains, conjecturally, a corresponding lattice–nonlattice dichotomy in the description of the shape and the sound of a self-similar drum, in the nature of random walks and Brownian motion on a self-similar fractal, and in the nature of several function spaces associated with a self-similar fractal.

The following conjecture is very natural in our framework and significantly supplements the geometric aspects of [Lap3, Conjecture 3, p. 163] for self-similar drums. It is the higher-dimensional analogue of Theorems 2.17 and 3.6, which corresponds to self-similar sets in $\mathbb{R}$.

Conjecture 12.18. *The exact analogues of Theorems 2.17 and 3.6 hold for every (strictly) self-similar set*[11] *in* $\mathbb{R}^d$ *(and thus for any drum with self-similar fractal boundary), satisfying the open set condition.*[12]

In particular, the geometric zeta function has a meromorphic continuation to all of $\mathbb{C}$*, with a single (simple) pole at* $s = D$*, where* $D \in (d-1, d)$ *is the Minkowski dimension of the set.*[13] *All the other complex dimensions lie in the closed half-plane* $\operatorname{Re} s \leq D$*. Furthermore, the complex dimensions of a lattice self-similar set are periodically distributed on finitely many vertical lines, with period equal to the oscillatory period of the set, defined exactly as in Equation* (2.35) *and Definition 2.14. On the other hand, a nonlattice self-similar set does not have any nonreal complex dimension on the line* $\operatorname{Re} s = D$*, but it has infinitely many complex dimensions arbitrary close (from the left) to this line. Moreover, its complex dimensions exhibit a quasiperiodic structure.*

Remark 12.19 (Tube formulas for self-similar sets). We further conjecture that a tube formula holds for self-similar fractal sets in $\mathbb{R}^d$, with $d \geq 1$, naturally extending the tube formula for self-similar fractal strings in $\mathbb{R}$ obtained in Section 8.4.2, Theorem 8.23 and Section 8.4.4, Theorem 8.36. (In particular, for self-similar fractal sprays—studied in Section 6.6.2 and in Section 6.6.1 for the example of the Sierpinski drum—this clearly follows from our explicit formulas exactly as in the case of self-similar strings.) Much as in the proof of Theorems 8.36 and 8.23, it would then follow that a nonlattice self-similar set in $\mathbb{R}^d$ is Minkowski measurable, whereas a lattice self-similar set, such as the Sierpinski gasket or the von Koch snowflake curve (viewed as self-similar fractal boundaries), is not. This statement is

---

[11] As in [Lap3, Conjecture 3], here and in Section 12.5, we also allow for the set (viewed as the boundary of a self-similar drum) to be composed of a finite union of congruent copies of such a self-similar set, just as for the snowflake drum.

[12] See Section 2.1.1 and Remark 2.22.

[13] The Devil's staircase is self-affine but not self-similar. Thus Figure 12.3 does not contradict this conjecture.

exactly the geometric content of [Lap3, Conjecture 3, p. 163] (and, in particular, of [Lap3, Conjecture 2, p. 159] for the special case of snowflake-type curves); see Remark 8.39 above. After [Lap-vF4] and [Lap-vF5] were completed, the statement concerning the Minkowski measurability of nonlattice self-similar sets was proved by Gatzouras [Gat] by means of the Renewal Theorem (much as was done in [Lap3, Section 4.4.1b] and in [Fa4] for the $d = 1$ case, and earlier in [Lal1] in a related situation, see Remark 8.40 above). Because of the use of the Renewal Theorem, however, the statement concerning the non-Minkowski measurability of lattice self-similar sets does not follow from the result of [Gat], although our methods from Section 8.4.2 yield it directly.

Recently, a tube formula for the example of the von Koch snowflake curve was obtained in [LapPe1], as will be briefly discussed in Section 12.3.1 below. This tube formula is of a form compatible with that expected for drums with self-similar boundaries and also yields a set of possible complex dimensions compatible with Conjecture 12.18. See also Section 12.3.2 for related but different work on tube formulas for self-similar tilings and systems, rather than sets.

Remark 12.20 (Approximately self-similar sets). In light of [Lap3, Conjecture 4, p. 175] (motivated in part by work of Lalley in [Lal1–3]), Conjecture 12.18 has a natural counterpart for approximately self-similar fractals (in the sense of [Lap3, §4.5] alluded to above). In that case, the dichotomy lattice vs. nonlattice must be defined by means of Lalley's nonlinear analogue of the Renewal Theorem [Lal2–3].

12.3.1 Complex Dimensions and Tube Formula for the Koch Snowflake Curve

Consider the snowflake drum Ω, whose boundary $\partial\Omega$, the snowflake curve, consists of three congruent von Koch curves fitted together (see Figures 12.6 and 12.7, along with [Fa3, Figure 0.2, p. xv]). The construction of the von Koch curve is illustrated in Figure 12.7. Note that this curve is an example of a lattice self-similar fractal in $\mathbb{R}^2$ and a nowhere differentiable plane curve (with infinite length).

By decomposing an ε-neighborhood of the von Koch curve K as in Figure 12.8, Erin Pearse and the first author derived in [LapPe1] the following formula for the volume (i.e., area) of the ε-neighborhood:

Theorem 12.21 (Tube Formula). *The volume of the inner ε-neighborhood of the von Koch snowflake curve $\partial\Omega$ is given by the following pointwise formula:*

$$V(\varepsilon) = G_1(\varepsilon)\varepsilon^{2-D} + G_2(\varepsilon)\varepsilon^2, \tag{12.3}$$

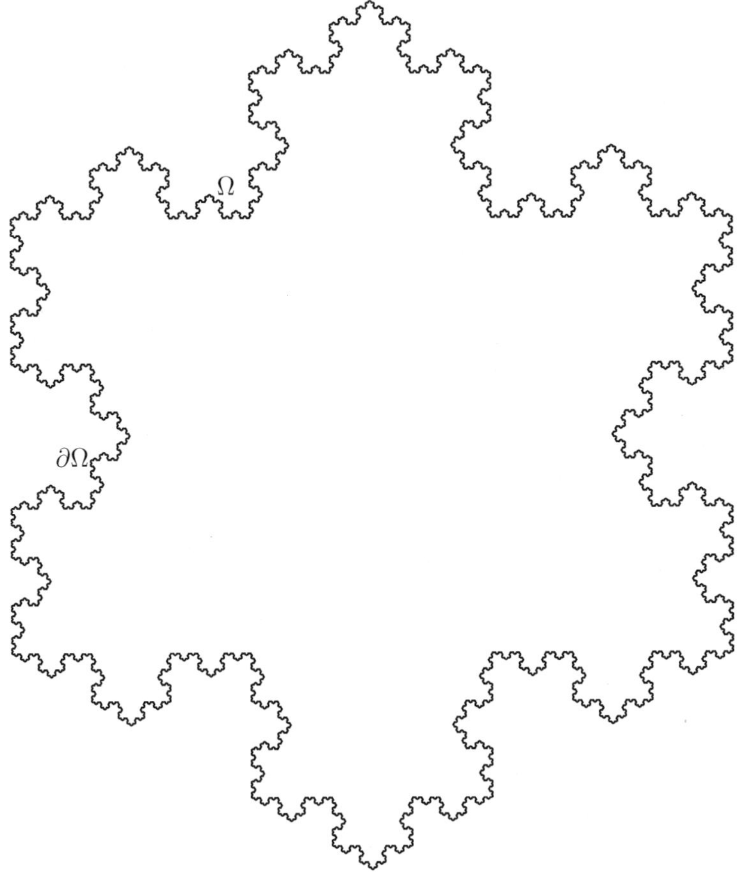

Figure 12.6: The snowflake drum Ω. Its boundary is the von Koch snowflake curve ∂Ω.

where G_1 and G_2 are periodic functions (of multiplicative period 3) given by

$$G_1(\varepsilon) = \frac{1}{\log 3} \sum_{n=-\infty}^{\infty} \left(a_n + \sum_{\nu=-\infty}^{\infty} f_{n-\nu} b_\nu \right) (-1)^n \varepsilon^{-in\mathbf{p}} \tag{12.4}$$

and

$$G_2(\varepsilon) = \frac{1}{\log 3} \sum_{n=-\infty}^{\infty} \left(\sigma_n + \sum_{\nu=-\infty}^{\infty} f_{n-\nu} \tau_\nu \right) (-1)^n \varepsilon^{-in\mathbf{p}}, \tag{12.5}$$

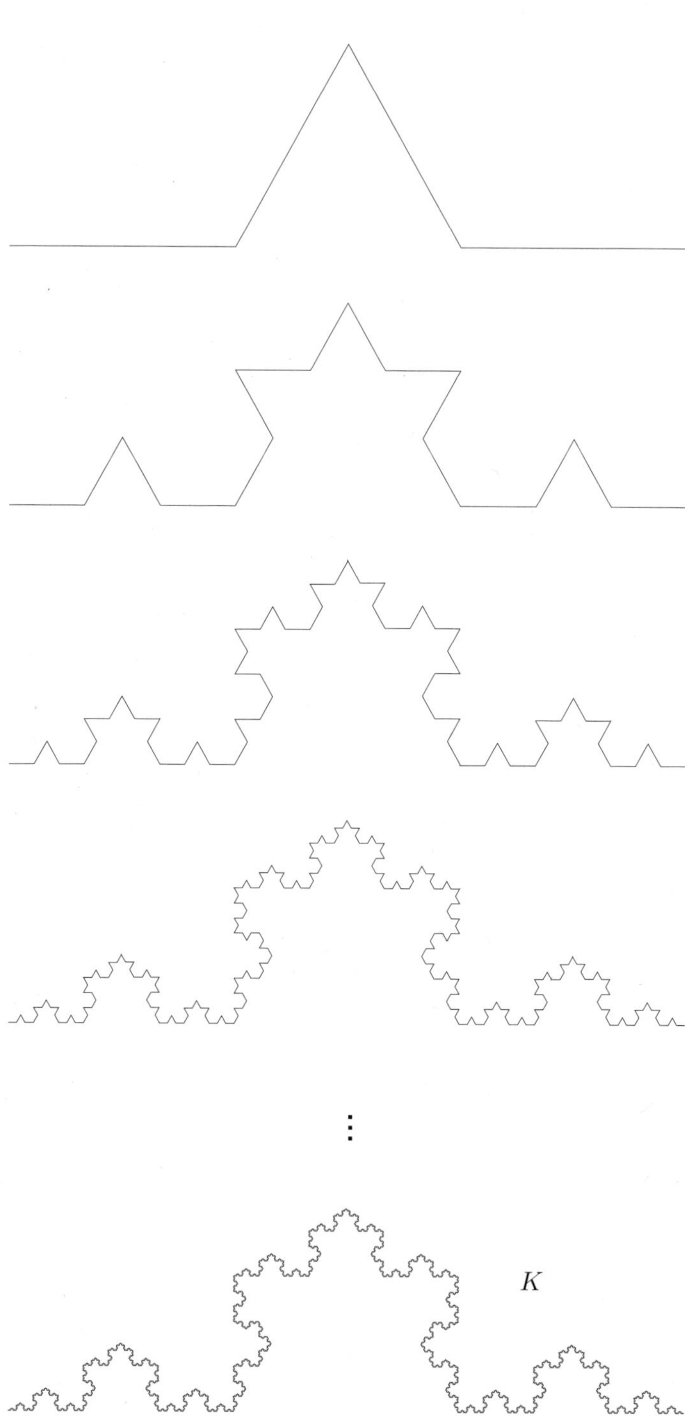

Figure 12.7: The von Koch curve K.

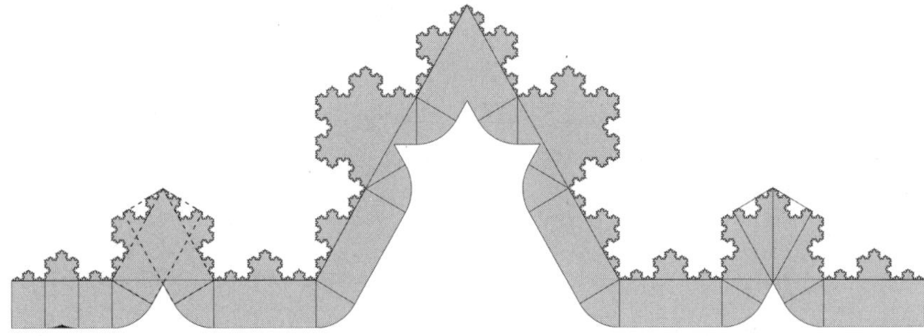

Figure 12.8: The inner ε-neighborhood of the Koch curve.

where the coefficients $f_{n-\nu}$ *are explained below* (*see* (12.8)), *and, for* $n \in \mathbb{Z}$,

$$a_n = -\frac{2^{-5}3^{5/2}}{D-2+in\mathbf{p}} + \frac{2^{-3}3^{3/2}}{D-1+in\mathbf{p}} + \frac{2^{-3}(\pi - 3^{3/2})}{D+in\mathbf{p}} + \frac{b_n}{2}, \tag{12.6a}$$

$$b_n = \sum_{m=1}^{\infty} \frac{3^{2m+1}-4}{3^{2m+1}-2}\binom{2m}{m}\frac{2^{-2-4m}}{(4m^2-1)(D-2m-1+in\mathbf{p})}, \tag{12.6b}$$

$$\sigma_n = -\log 3\left(\frac{\pi}{3}+2\sqrt{3}\right)\delta_{n0} - \tau_n, \tag{12.6c}$$

$$\tau_n = \sum_{m=1}^{\infty} \frac{3^{2m+1}-1}{3^{2m+1}-2}\binom{2m}{m}\frac{2^{2-4m}}{(4m^2-1)(-2m-1+in\mathbf{p})}, \tag{12.6d}$$

where the Kronecker-delta $\delta_{n0} = 1$ *for* $n = 0$ *and* 0 *for* $n \neq 0$.[14] *Here,* $D = \log_3 4$ *is the Minkowski dimension of the von Koch snowflake curve* $\partial\Omega$ *and* $\mathbf{p} = 2\pi/\log 3$ *is its oscillatory period* (*following the terminology of Section 2.4*). *The numbers* f_ν *are the Fourier coefficients of the periodic function* $h(\varepsilon)$, *a suitable nonlinear analogue of the Cantor–Lebesgue function, which is discussed in Remark 12.22 below and reflects the self-similarity of the von Koch curve.*

Now, by analogy with the tube formula (∗∗) in the introduction (or more precisely, by analogy with (8.14) in Corollary 8.3), we interpret the exponents of ε (in the expansion (12.3) of $V(\varepsilon)$) as the complex codimensions of $\partial\Omega$. Hence, we can simply read off the possible complex dimensions from (12.3). Since some of the Fourier coefficients of G_1 and G_2 in (12.4) and (12.5) may vanish, the corresponding exponent of ε may not be a complex codimension. As depicted in Figure 12.9, upon collecting these exponents from each series we obtain the following set of possible complex

[14] Since $\binom{2m}{m} < 2^{2m}$, the series for b_n and τ_n converge as fast as the geometric series $\sum_{m=1}^{\infty} 4^{-m}$.

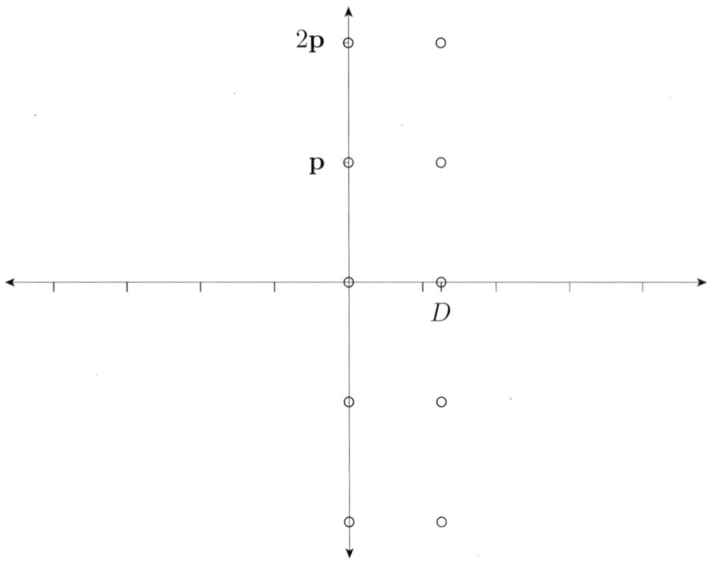

Figure 12.9: The possible complex dimensions of the snowflake curve $\partial\Omega$. $D = \log_3 4$ and $\mathbf{p} = 2\pi/\log 3$.

dimensions of the von Koch snowflake curve:

$$\mathcal{D}_{\partial\Omega} = \{D + in\mathbf{p} \colon n \in \mathbb{Z}\} \cup \{in\mathbf{p} \colon n \in \mathbb{Z}\}. \tag{12.7}$$

These dimensions are simple. We note that it follows from [LapPe1] that the set $\mathcal{D}_K$ of possible complex dimensions of the Koch curve is also given by (12.7).

The following comment (in conjunction with Figures 12.8 and 12.10) may help the reader gain an intuitive idea of why the periodic function $h(\varepsilon)$ enters into the statement of Theorem 12.21 and its geometric interpretation.

Remark 12.22. The factors f_ν appearing in (12.4) and (12.5) are the coefficients of a function which may be written variously as

$$h(\varepsilon) = \sum_{\nu\in\mathbb{Z}} f_\nu (-1)^\nu \varepsilon^{-i\nu\mathbf{p}} = \sum_{\nu\in\mathbb{Z}} f_\nu e^{2\pi i\nu x} = f(x), \tag{12.8}$$

with $x = -\log_3(\varepsilon\sqrt{3})$ and $\mathbf{p} = 2\pi/\log 3$ as in Theorem 12.21. While we do not know $h(\varepsilon)$ analytically, or its Fourier coefficients f_ν explicitly, we know that it is a nonlinear and multiplicatively periodic counterpart of Cantor's classical staircase singular function.[15] Specifically, $h = h(\varepsilon)$ has multiplicative period 3; i.e., $h(\varepsilon) = h(\varepsilon/3)$. Alternatively, it can be thought of as an additively periodic function $f = f(x)$ with period 1 and with Fourier expansion given by (12.8) (here, $x = -\log_3(\varepsilon\sqrt{3})$ as above). These

[15] Recall from Section 12.2 above that the graph of the classical Cantor–Lebesgue function is also coined the Devil's staircase in [Man1, Plate 83].

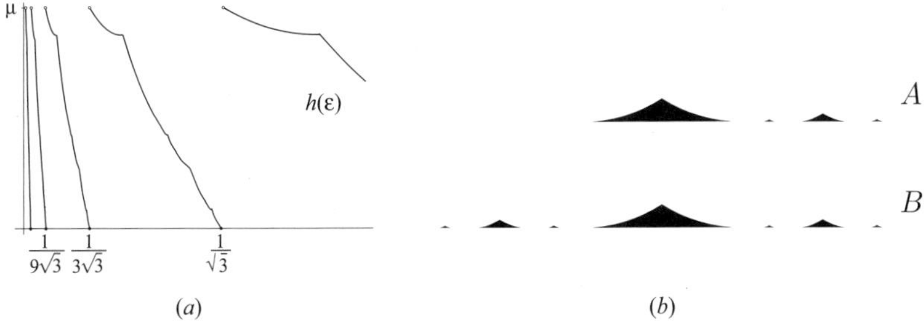

Figure 12.10: (a) The function h. (b) The number $\mu = \mathrm{vol}_2(A)/\mathrm{vol}_2(B)$, given as a ratio of areas.

properties, among others, are directly induced by the (lattice) self-similarity of the von Koch curve K.

The function $h = h(\varepsilon)$ is continuous and even monotonic decreasing when restricted to one of its periods $\left(3^{-(n+1)-1/2}, 3^{-n-1/2}\right]$. Further, it has a finite jump discontinuity at the left-hand point of this interval, and its jump at that point is shown in [LapPe1, Section 3.4] to be equal to μ (as given geometrically in Figure 12.10(b)). (The function h itself is defined geometrically in [LapPe1, Section 3.2] as the proportion of certain "error blocks" formed as ε crosses one of its period intervals. In particular, we have $0 \leq h(\varepsilon) \leq \mu < 1$, for all $\varepsilon > 0$.) Since f is monotonic, and hence of bounded variation, it follows from [Zyg, Theorem II.8.1][16] that the Fourier series expansion (12.8) of f (or of $h = h(\varepsilon)$) converges pointwise (clearly, the convergence is conditional). Moreover, by [Zyg, Theorem II.4.12], the Fourier coefficients of f (and h) satisfy $f_\nu = O(1/\nu)$ as $|\nu| \to \infty$.

A sketch of the graph of h is given in Figure 12.10(a). To see the ternary nature of this function, consider that each connected component of the graph is similar to a skewed version of the black figure depicted at top right in Figure 12.10(b). We refer the interested reader to [LapPe1] for further information.

In the next remark, we briefly compare the main result of [LapPe1] discussed in this section with the earlier result obtained in [Lap-vF5, Section 10.3].

Remark 12.23 (Exact versus approximate tube formula). In [Lap-vF5, formulas (10.3) and (10.4)], we had provided without proof a provisional approximate formula for the volume of the tubular neighborhood of the

[16]Namely, the Fourier series of a periodic function of bounded variation converges pointwise at every point x to the mean value $(f(x+) + f(x-))/2$, where

$$f(x+) = \lim_{t \to x^+} f(t) \quad \text{and} \quad f(x-) = \lim_{t \to x^-} f(t).$$

In particular, it converges to $f(x)$ wherever f is continuous. (See, e.g., [Zyg, Theorem II.8.1] or [Fol, Theorem 8.43, p. 266] for this extension of Dirichlet's Theorem, due to Camille Jordan.)

von Koch snowflake curve, of the form

$$V(\varepsilon) \approx \sum_{n=-\infty}^{\infty} e_n \varepsilon^{2-D-in\mathbf{p}}, \tag{12.9}$$

where $e_n \neq 0$ for each $n \in \mathbb{Z}$; specifically,

$$e_n = -\frac{3\sqrt{3}\,\log 3\,(4\log 4 + \log 3)}{8(-\log(9/4) + 2\pi in)(\log(4/3) + 2\pi in)(\log 4 + 2\pi in)}. \tag{12.10}$$

(See [Lap-vF5, Eq. (10.4), p. 210].) We then stated in [Lap-vF5, Section 10.3] that the set of complex dimensions of the von Koch snowflake curve should be $\{D + in\mathbf{p} \colon n \in \mathbb{Z}\}$ with D and $\mathbf{p}$ as in Theorem 12.21. By contrast, $V(\varepsilon)$ is precisely equal to the right-hand side of (12.3), except for the fact that the Fourier coefficients of the function h are not explicitly known. It is compatible with the approximate formula (12.9), although it also points out the possibility of having one line of complex dimensions above zero.

We note that both the exact formula (12.3) and the earlier approximate formula (12.9) are in agreement with Conjecture 12.18 above,[17] because the von Koch curve is a lattice self-similar set with Minkowski dimension D and oscillatory period $\mathbf{p} = 2\pi/\log 3$ (since, in the notation of Theorem 2.17, we have $N = 4$, $K = 1$, and $r_1 = r_2 = r_3 = r_4 = 1/3$).

Remark 12.24 (Reality principle). As is the case for the complex dimensions of self-similar strings (see Chapter 2), the *possible complex dimensions of $\partial\Omega$ come in complex conjugate pairs, with attached complex conjugate coefficients.* Indeed, $\bar{f}_n = f_{-n}$ (because f is real-valued), and a simple inspection of the formulas in (12.6) shows that for every $n \in \mathbb{Z}$,

$$\bar{a}_n = a_{-n},\ \bar{b}_n = b_{-n},\ \bar{\sigma}_n = \sigma_{-n} \text{ and } \bar{\tau}_n = \tau_{-n}. \tag{12.11}$$

It follows that a_0, b_0, σ_0 and τ_0 are real numbers and that each infinite sum on the right-hand side of (12.3) is real, in agreement with the fact that $V(\varepsilon)$ represents an area. For the same reason, the periodic functions G_1 and G_2 in (12.4) and (12.5) of Theorem 12.21 are real-valued.

The reader can easily check that the tube formula (12.3) can be rewritten in the following form:

$$V(\varepsilon) = \sum_{n\in\mathbb{Z}} c_n \varepsilon^{2-D-in\mathbf{p}} + \sum_{n\in\mathbb{Z}} d_n \varepsilon^{2-in\mathbf{p}}, \tag{12.12}$$

for suitable complex numbers c_n and d_n such that $\bar{c}_n = c_{-n}$ and $\bar{d}_n = d_{-n}$ for all $n \in \mathbb{Z}$. One caveat should be mentioned here: of course, it is always possible that one of the coefficients c_n (or d_n) vanishes, in which case $D + in\mathbf{p}$ (or $in\mathbf{p}$) is not a complex dimension of Ω.

[17] Possibly with appropriate adjustments above 0, in view of (12.3).

12.3.2 Towards a Higher-Dimensional Theory of Complex Dimensions

Theorem 12.21 may be viewed as providing a first step towards the beginning of a higher-dimensional theory of complex dimensions of self-similar fractals. Therefore, in the long term, it may help to extend to two or more dimensions some of the results obtained in this book, especially in parts of Chapters 2, 3 and 8, as well as of Section 6.4. A main challenge associated with Theorem 12.21 (apart from extending it to other interesting higher-dimensional examples), however, consists in defining in this situation an appropriate analogue ζ_Ω of the geometric zeta function, and showing that it admits a meromorphic continuation to some half-plane $\operatorname{Re} s > \delta$, with $-\infty \leq \delta < D$. In light of (12.3) (and by analogy with Corollary 8.3), one would then expect all the poles of ζ_Ω to be simple and given by (12.7) or a suitable subset thereof.

In [LapPe2–4], the beginning of a higher-dimensional theory of complex dimensions is developed for a suitable class of self-similar systems in $\mathbb{R}^d$, via tube formulas and geometric zeta functions for certain tilings naturally associated with these systems.

More precisely, the self-similar tilings introduced in [Pe] are used in the papers [LapPe2–3] to define an appropriate geometric (or scaling) zeta function, whose poles yield the complex dimensions of the self-similar system (i.e., the system of similarity transformations $\{\Phi_j : j = 1, \ldots, N\}$ defining the given self-similar set, see Section 2.1.1).[18] The resulting zeta function is then used in conjunction with our extended distributional explicit formula of Chapter 5 (see Theorem 5.26 in Section 5.4.1) to obtain a (distributional) explicit inner tube formula for the self-similar system or tiling in $\mathbb{R}^d$. In the one-dimensional case (i.e., for self-similar fractal strings in $\mathbb{R}$), this tube formula coincides with the one obtained in Chapter 8 (Theorem 8.1 and Section 8.4). In higher dimension, however, the geometry of the underlying basic shapes—called *generators* in [Pe] and [LapPe2–4] because they generate the self-similar tiling via the action of the semigroup defined by the iterates of the maps Φ_j (for $j = 1, \ldots, N$)—plays a key role in the expression of the tube formula. (For fractal strings, there is essentially only one kind of generator, the unit interval.)

In fact, in the papers [LapPe2–4], several new connections between aspects of geometric measure theory [Fed1–2, Schn] and fractal geometry are also developed in the process. These connections involve the notion of mixed volumes (of the generators) encountered in Steiner's classic formula for the volume of tubes of convex bodies [Stein, Fed2, Schn] and that of curvature measure (à la Weyl [Wey3, Gra] and Federer [Fed1], but in a more

[18] The canonical self-similar tiling in question is a tiling of the complement of the self-similar set in its convex hull. For example, for the Koch tiling, depicted in Figure 12.11, the convex hull of the Koch curve is an isosceles triangle.

general context; see [LapPe2–4] along with Section 8.2). In the long term, one expects to be able to interpret the coefficients of the tube formulas obtained in [LapPe2–3] in terms of suitably defined fractal curvature measures associated with the complex dimensions of the underlying self-similar system (and its corresponding tiling). A step in this direction is carried out in the work [LapPe4], building upon [LapPe2–3] and partially motivated by some of the questions raised in Section 8.2 above and in Section 12.7 below (or in [Lap-vF5, Sections 6.1.1 and 10.5]).

Remark 12.25. We point out that although the methods and main result (Theorem 12.21 above) of [LapPe1] were useful to develop the theory of [Pe] and [LapPe2–4], they are different and do not follow from the latter more recent work. Indeed, the Koch tube formula (12.3) from [LapPe1] is associated with the self-similar set K whereas the tube formulas in [LapPe2–4] are expressed in terms of the tiles of the self-similar tiling, or equivalently, in terms of the cells of the underlying self-similar system corresponding to the self-similar set.

We close this subsection by giving a more precise and quantitative discussion of the main results of [LapPe2], which extend both the classic Steiner tube formula for convex bodies [Stein] and the tube formula for fractal strings (Theorem 8.1 above).

For a fractal spray in $\mathbb{R}^d$ $(d \geq 1)$, viewed as a tiling with possibly multiple generators (or, in our present terminology, basic shapes), it is shown in [LapPe2] that, under mild hypotheses,[19] the following (distributional) inner tube formula holds:

$$V_{\mathrm{til}}(\varepsilon) = \sum_{\omega \in \mathcal{D}_{\mathfrak{s}}(W) \cup \{0,\dots,d-1\}} c_{\o mega} \varepsilon^{d-\omega} + \mathcal{R}(\varepsilon), \tag{12.13}$$

where ω ranges not only over the integral dimensions $\{0,\dots,d-1\}$, as would be the case in the classic Steiner–Weyl–Federer tube formula for compact convex sets and smooth submanifolds of $\mathbb{R}^d$, but also over the (typically infinite) set of complex dimensions $\mathcal{D}_{\mathfrak{s}}(W)$ of the spray (i.e., the poles of the associated scaling zeta function), and the coefficients $c_{\omega,j}$ are expressed as the residues of an inner product involving a vector-valued zeta function (defined in terms of the inner radii of the scaled generators) and the curvature matrix of the generators. Further, the error term can be estimated as follows:

$$\mathcal{R}(\varepsilon) = O\big(\varepsilon^{d-\sup S}\big), \quad \text{as } \varepsilon \to 0^+. \tag{12.14}$$

---

[19] Namely, the generators of the spray are assumed to be "Steiner-like" (i.e., the corresponding inner tube formula is a suitable polynomial in ε, as defined in [LapPe2] and further investigated in [LapPe3]) and the underlying scaling measure (or generalized fractal string) is assumed to be languid along the screen. Further, the screen may not pass through $0, 1, \dots, d$.

Moreover, when the fractal string associated with the spray is strongly languid (which is the case for self-similar tilings), we may take $W = \mathbb{C}$ and $\mathcal{R}(\varepsilon) \equiv 0$ in the tube formula (12.13).

Remark 12.26. The usual Steiner formula refers to the full (not just inner) ε-neighborhoods of the set in question. Under somewhat different assumptions, there is a corresponding version of the tube formula (12.13) for the volume of such ε-neighborhoods. We prefer, however, to focus on the aforementioned inner tube formula because in the key special case of self-similar tilings (as discussed in (ii) below), this has an intrinsic geometric meaning. This also fits precisely the type of tube formula for fractal strings obtained in this book (see (i) just below).

The following two special cases of the tube formula (12.13) are of particular interest:

(i) When $d = 1$, that is, for a (possibly generalized) fractal string, one recovers our (distributional) tube formula for fractal strings, Equation (8.4), under exactly the same hypotheses as in Theorem 8.1.

(ii) For a self-similar tiling (as defined in [Pe]), associated with a self-similar system of N contractive similarity transformations of $\mathbb{R}^d$ $(d \geq 1)$ and corresponding attractor the self-similar set F, the inner tube formula (12.13) holds with $W = \mathbb{C}$ and $\mathcal{R}(\varepsilon) \equiv 0$ (i.e., without error term).[20] In that case, the scaling zeta function of the self-similar tiling is strongly languid and equals the geometric zeta function of a (normalized) self-similar string with a single gap and with the same scaling ratios $\{r_j\}_{j=1}^N$. Namely,

$$\zeta_{\mathfrak{s}}(s) = \frac{1}{1 - \sum_{j=1}^N r_j^s}, \qquad \text{for } s \in \mathbb{C}, \tag{12.15}$$

as in Equation (2.15) following Theorem 2.9. Hence, the set of complex dimensions $\mathcal{D}_{\mathfrak{s}}(\mathbb{C})$ of the tiling consists of the complex solutions of the equation

$$\sum_{j=1}^N r_j^s = 1, \tag{12.16}$$

exactly as in Section 2.2.1. These can also be viewed as the complex dimensions of the self-similar set F, or, more accurately, of the self-similar system defining it.

We note that in higher dimensions, the true complexity of the geometry of the self-similar tiling is hidden in the curvature coefficients c_ω of the tube

[20] This is under the hypothesis that the generators of the tiling are Steiner-like, which is satisfied, for example, for the self-similar tiling naturally associated with the Koch curve, the Sierpinski gasket, the Sierpinski carpet and its higher-dimensional analogue, the Menger sponge, as well as with a broader class of self-similar systems [LapPe3].

formula (12.13). In particular, the *geometric zeta function* of the self-similar tiling (or *tiling zeta function*) is defined in [LapPe2] to be the meromorphic distribution-valued function

$$\zeta_{\mathrm{til}}(\varepsilon, s) := \zeta_{\mathfrak{s}}(s)\big\langle \mathbf{g}(s), \mathcal{E}(\varepsilon, s)\big\rangle_{\boldsymbol{\kappa}(\varepsilon)}. \tag{12.17}$$

In (12.17), $\zeta_{\mathfrak{s}}$ is the complex valued scaling zeta function given in (12.15) and the bilinear map $\langle\cdot,\cdot\rangle_{\boldsymbol{\kappa}}$ induced by $\boldsymbol{\kappa}$ is defined as the matrix product

$$\big\langle \mathbf{g}, \mathcal{E}\big\rangle_{\boldsymbol{\kappa}} := \mathbf{g}^T \boldsymbol{\kappa}\, \mathcal{E}, \tag{12.18}$$

where $\mathbf{g}(s)^T$ denotes the transpose of the vector-valued function

$$\mathbf{g}(s) := \big[g_1^s, \dots, g_K^s\big]. \tag{12.19}$$

In (12.19), $g_1, \dots, g_K$ are the inner radii of the K generators of the tiling.[21]
Further, in (12.17), $\mathcal{E}(\varepsilon, s)$ is the vector-valued function

$$\mathcal{E}(\varepsilon, s) = \left[\frac{1}{s}, \frac{1}{s-1}, \dots, \frac{1}{s-d}\right]\varepsilon^{d-s}. \tag{12.20}$$

Several comments are required before we can explain the curvature matrix $\boldsymbol{\kappa}$. We assume that each of the finitely many generators is Steiner-like, that is, for each $k = 1, \dots, K$, the generator G_k satisfies a generator tube formula of the form (with $0 < \varepsilon < g_k$)[22]

$$V_{G_k}(\varepsilon) = \sum_{j=0}^{d-1} \kappa_j(G_k; \varepsilon)\varepsilon^{d-j}, \tag{12.21}$$

for certain real-valued curvature coefficients $\kappa_j(G_k; \varepsilon)$ for $j = 0, \dots, d-1$, which we assume to be measurable and bounded on $(0, \infty)$ as functions of ε. These coefficients are rigid motion invariant and homogenous of degree j, in the sense that for all $x > 0$, $\kappa_j(xG_k; \varepsilon) = x^j \kappa_j(G_k; \varepsilon)$, where xG_k is the homothetic image of G_k. We define the *curvature matrix* $\boldsymbol{\kappa}$ of the generators as the $K \times (d+1)$ matrix

$$\boldsymbol{\kappa} = \{\kappa_j(G_k; \varepsilon)\}_{1 \le k \le K, 0 \le j \le d}. \tag{12.22}$$

Here, for $k = 1, \dots, K$, we define $\kappa_d(G_k; \varepsilon)$ by $\kappa_d(G_k; \varepsilon) = -\mathrm{vol}_d(G_k)$ for $\varepsilon \ge g_k$. In all known examples, the coefficients $\kappa_j(G_k; \varepsilon)$, and hence the entries of the matrix $\boldsymbol{\kappa}$, depend on ε in a piecewise constant manner.

We can now state the main result of [LapPe2] in the special case of self-similar tilings.

---

[21]The inner radius of $G \subset \mathbb{R}^d$ is the radius of the largest ball contained in G.

[22]Clearly, $V_{G_k}(\varepsilon) = \mathrm{vol}_d(G_k)$ for all $\varepsilon \ge g_k$.

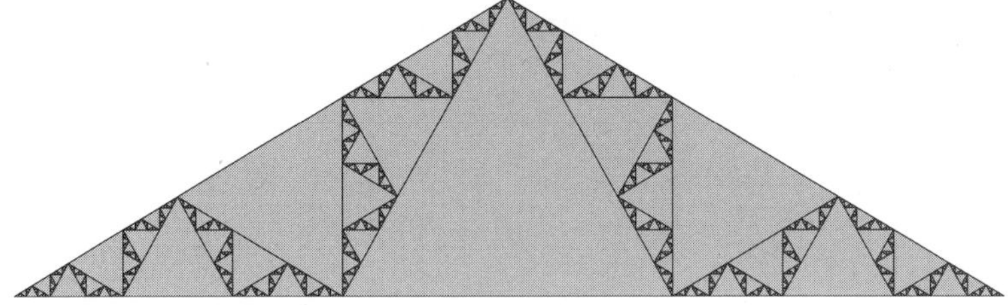

Figure 12.11: The Koch tiling.

Theorem 12.27. *The tube formula of a self-similar tiling is given by*

$$V_{\mathrm{til}}(\varepsilon) = \sum_{\omega \in \mathcal{D}_{\mathrm{til}}} \operatorname{res}\big(\zeta_{\mathrm{til}}(\varepsilon, s); \omega\big), \tag{12.23}$$

where $\mathcal{D}_{\mathrm{til}} := \mathcal{D}_{\mathfrak{s}}(\mathbb{C}) \cup \{0, \dots, d-1\}$ *is the set of poles of* ζ_{til}*, and* $\mathcal{D}_{\mathfrak{s}}$ *is the set of solutions of* (12.16).

In the special case when all the poles of ζ_{til} are simple and the curvature matrix $\boldsymbol{\kappa}$ is constant for $\varepsilon < \min\{g_1, \dots, g_K\}$, the tube formula is of the form (12.13), with $W = \mathbb{C}$ and $\mathcal{R}(\varepsilon) \equiv 0$:

$$V_{\mathrm{til}}(\varepsilon) = \sum_{\omega \in \mathcal{D}_{\mathrm{til}}} c_\omega \varepsilon^{d-\omega}, \tag{12.24}$$

where for $\omega \in \mathcal{D}_{\mathfrak{s}}$, the coefficients c_ω are given by

$$c_\omega = \operatorname{res}(\zeta_{\mathfrak{s}}(s); \omega) \sum_{j=0}^{d} \sum_{k=1}^{K} \frac{g_k^\omega}{\omega - j} \kappa_j(G_k), \tag{12.25}$$

and for $\omega \in \{0, \dots, d-1\}$, the coefficients c_ω are given by

$$c_\omega = \zeta_{\mathfrak{s}}(\omega) \sum_{k=1}^{K} g_k^\omega \kappa_\omega(G_k). \tag{12.26}$$

In light of (12.15), (12.17), (12.20) and (12.22), the function defined by $s \mapsto \zeta_{\mathfrak{s}}(s)\langle \mathbf{g}(s), \mathcal{E}(\cdot, s)\rangle_{\boldsymbol{\kappa}}$ is distribution-valued and meromorphic in all of $\mathbb{C}$. It follows that on the right-hand side of (12.23), each summand is well defined as the residue of ζ_{til} at $\omega \in \mathcal{D}_{\mathrm{til}}$ and is itself a distribution on the open interval $(0, \infty)$. These facts are verified in [LapPe2, Appendix B].

The tube formula (12.24) is illustrated in [LapPe2] in a variety of examples, including the Cantor, Koch, Sierpinski, and pentagasket tilings, which are all lattice self-similar tilings with a single line of simple complex dimensions. All of those tilings have a single generator, except the pentagasket tiling, which has six generators (one regular pentagon and five isosceles triangles).

We now conclude this subsection by the following example of the Koch tiling, depicted in Figure 12.11 and discussed in more detail in [LapPe2].

Example 12.28 (The Koch Tiling). The tube formula for the Koch tiling KT is of the form

$$V_{\mathrm{KT}}(\varepsilon) = \frac{g}{\log 3} \sum_{n \in \mathbb{Z}} c_n \left(\frac{\varepsilon}{g}\right)^{2-D-in\mathbf{p}} + 3^{3/2}\varepsilon^2 + \frac{1}{1-2\cdot 3^{-1/2}}\varepsilon, \tag{12.27}$$

where $g = \frac{\sqrt{3}}{18}$ is the inner radius of the single generator of the tiling (an equilateral triangle), $D := \log_3 4$ and $\mathbf{p} := 4\pi/\log 3$, and for $n \in \mathbb{Z}$,

$$c_n := -\frac{1}{D+in\mathbf{p}} + \frac{2}{D-1+in\mathbf{p}} - \frac{1}{D-2+in\mathbf{p}}.$$

From (12.27), we clearly read off that

$$\mathcal{D}_{\mathfrak{S}} = \{D + in\mathbf{p} \colon n \in \mathbb{Z}\} \quad \text{and} \quad \mathcal{D}_{\mathrm{KT}} = \mathcal{D}_{\mathfrak{S}} \cup \{0, 1\}. \tag{12.28}$$

Indeed, the elements of $\mathcal{D}_{\mathfrak{S}}$ are precisely the poles of the scaling zeta function $\zeta_{\mathfrak{S}}(s) = \frac{1}{1-2\cdot 3^{-s/2}}$ (since $d = 2$, $N = 2$ and $r_1 = r_2 = 3^{-1/2}$ in this case). Hence, $\mathcal{D}_{\mathfrak{S}}$ is the set of complex dimensions of the Koch tiling.

12.4 Random and Quantized Fractal Strings

In this section, we present some extensions of the framework of this book, that of fractal strings. In Section 12.4.1, we discuss random fractal strings, as studied in [HamLap], and in Section 12.4.2, we briefly discuss the notion of fractal membrane (or quantized fractal string) introduced in the forthcoming book [Lap10] and further studied in the work in preparation [LapNes1–3].

12.4.1 Random Fractal Strings and their Zeta Functions

In this section, we discuss recent work of Ben Hambly and the first author in the paper [HamLap]. In that paper, the authors develop a random counterpart of the theory of fractal strings and their associated complex dimensions. Typical examples of random fractal strings studied in [HamLap] include random Cantor-type sets, random self-similar sets or random recursive constructions (defined via random trees and branching processes), as well as the zero set of Brownian motion (and its analogue for a one-parameter family of stable strings). An interesting aspect of this work is that random fractal strings and their associated complex dimensions and zeta functions exhibit a broader variety of behaviors than their deterministic counterparts.

More specifically, the authors obtain tube formulas, as well as explicit formulas for the geometric and spectral counting functions of random fractal strings, expressed (as in the deterministic case) in terms of the underlying complex dimensions. The latter complex dimensions are defined as the poles of a suitable *random zeta function* (defined almost surely and denoted henceforth $\zeta_{\mathcal{L}}(s)$ or $\zeta_\eta(s)$) associated with the random fractal string.

It is useful in this context to consider the *random measure* (or random generalized fractal string) η representing the random string, much as in the deterministic theory developed in this book (see especially Chapters 4 and 5) and in [Lap-vF5].[23] The *random geometric zeta function* of $\mathcal{L}$ (or of η) is defined as the Mellin transform of the random measure η,

$$\zeta_\eta(s) = \zeta_{\mathcal{L}}(s) = \int_0^\infty x^{-s}\eta(dx), \qquad \text{for } \operatorname{Re} s \gg 1, \tag{12.29}$$

and then meromorphically continued (wherever possible). As such, ζ_η is a random zeta function (i.e., a zeta function-valued random variable) and is therefore only defined almost surely (i.e., for almost every realization of η). Then, as could be expected in a probabilistic setting, most of the statements in [HamLap] hold almost surely; that is, for almost every realization of the random fractal string. These statements include, for example, the existence of the analytic continuation of the random zeta function, the tube formulas, and the spectral and geometric asymptotic expansions, but exclude those concerning the mean zeta function to be discussed below.

A main new difficulty in this probabilistic context is to show that under suitable conditions the (pointwise) random zeta function $\zeta_{\mathcal{L}}(s)$ admits almost surely a meromorphic continuation to a nontrivial region (i.e., beyond the abscissa of convergence, which coincides with the dimension of $\mathcal{L}$) and is languid there. (The latter is significantly easier to verify than the former for the type of random strings studied in that work.) It follows that the explicit formulas of Chapter 5 (or of [Lap-vF5, Chapter 4]) can be applied to almost every sample path (i.e., to almost every realization of the random string) to yield the desired tube formula or eigenvalue asymptotics.

Typically, in [HamLap], one first studies the *mean zeta function*, $\mathbb{E}(\zeta_{\mathcal{L}})$, defined as the average (or expected value) of the (pointwise) random zeta function $\zeta_{\mathcal{L}}$ over all possible realizations of the random string $\mathcal{L}$. Then, after having established the existence of the meromorphic continuation of $\mathbb{E}(\zeta_{\mathcal{L}})$ to a suitable domain, one uses an argument based on a variant of the Central Limit Theorem to control the fluctuations of the random zeta function around its mean and deduce the existence almost surely of the desired analytic continuation of $\zeta_{\mathcal{L}}$ to some half-plane of the form $\operatorname{Re} s > \alpha$, with $-\infty \leq \alpha < D$ and where D is almost surely the dimension of $\mathcal{L}$.

---

[23] As usual, we have $\eta = \sum_{\ell \in \mathcal{L}} \delta_{\ell^{-1}}$, where the sum is extended over the lengths of the random fractal string $\mathcal{L}$. However, the lengths ℓ of a random fractal string are themselves random variables and hence η is a random measure (i.e., a measure-valued random variable).

This method works well for random self-similar strings (obtained via random recursive constructions), as well as for stable strings (including the random fractal string having for boundary the zero set of Brownian motion), as will be further discussed below. On the other hand, in [HamLap, Section 6], one exhibits a spatially homogeneous random string arising from a random Cantor set C for which these techniques break down. Indeed, the random Cantor set C is chosen in such a way as to have violent random irregularities at all scales. (Random fractals of this type were also studied earlier from a different point of view in [BarHam].) This has the effect of introducing much stronger fluctuations in the scale geometry of the string and prevents one from controlling the error term (beyond the mean) in the Euler–Maclaurin approach to analytic continuation. It is possible (and is raised as an open problem in [HamLap]) that for such random strings, almost surely, the random zeta function $\zeta_{\mathcal{L}}(s)$ admits the vertical line $\operatorname{Re} s = D$ as a natural boundary (i.e., cannot be meromorphically continued beyond the region where it is holomorphic); see the end of Problem 12.35.

We now briefly comment on one of the main classes of random strings studied in [HamLap]. One type of randomness considered in [HamLap] is expressed in terms of the theory of branching processes [Harr], with (roughly speaking) each sample path corresponding to a suitable choice of random tree and associated functions (such as the reproduction process describing the offspring being produced, and the life-span function, a random characteristic that assigns a score or weight to each individual). We refer, for example, to Section 3 of [HamLap] for a description of the general notion of branching process and of the hypotheses made in that work, as well as to Section 4 of [HamLap] for the way in which this general construction is applied to random recursive strings (also called random self-similar strings, but allowing for a random and possibly infinite number of scaling ratios).

Remark 12.29. In short, branching processes—also called Galton–Watson processes—arose from Galton's 19th century study of the extinction of family names and provide useful mathematical models for understanding the time evolution of populations whose members reproduce and die according to suitable probabilistic laws. In their simplest form, they can be described as follows (see, e.g., [Harr]). The initial ancestor (represented by the root of the associated random tree) has a random number of children (or offspring); in turn, these children have offspring (or families), etc. Each successive generation corresponds to a different level of the tree. It is important to note that all the offspring are assumed to reproduce independently of one another. Furthermore, each additional offspring is an independent copy of the initial one. In the case of random self-similar strings modeled by random trees, this property enables one to obtain a suitable functional equation satisfied by the mean zeta function $\mathbb{E}(\zeta(s))$, and to deduce from it an explicit expression for this function. (See [HamLap, Section 4].)

Random Self-Similar Strings

We now discuss in more detail some of the results obtained in [HamLap], first in the case of random self-similar strings, and then for random stable strings. The random recursive constructions considered in [HamLap] enable one to have a different number of scaling ratios $\{r_j\}_{j=1}^N$ and gaps $\{g_k\}_{k=1}^K$ for each generation (i.e., at each level of the random tree). Moreover, these numbers can be infinite; i.e., $N = \infty$ or $K = \infty$ is allowed. In the simplest cases, the potential complex dimensions of the resulting random self-similar string are contained among the complex solutions of an expectation Moran equation

$$\mathbb{E}\left(\sum_{j=1}^{N} r_j^s\right) = 1. \tag{12.30}$$

(Cancellations may occur with the solutions of $\mathbb{E}(\sum_{k=1}^K g_k^s) = 0$, much as was shown in the deterministic case in Section 2.3.3 above.) Indeed, it is shown in [HamLap, Lemma 4.4] that

$$\mathbb{E}(\zeta_{\mathcal{L}})(s) = \frac{\mathbb{E}\left(\sum_{k=1}^K g_k^s\right)}{1 - \mathbb{E}\left(\sum_{j=1}^N r_j^s\right)}. \tag{12.31}$$

In other situations when the underlying randomness is of a more intricate nature, one may obtain more complicated expressions for $\mathbb{E}(\zeta_{\mathcal{L}}(s))$ (and hence for the ensuing expectation equation), involving certain continuous integrals, for example. Note that the expectation symbol $\mathbb{E}$ (with respect to the underlying probability measure) is needed in Equations (12.30) and (12.31) because both the scaling ratios r_j and the gaps g_k are random variables. If the family size is finite (see Remark 12.29 above), then $\mathbb{E}(\zeta_{\mathcal{L}})$ can be meromorphically continued to all of $\mathbb{C}$. Otherwise, under suitable assumptions, it can be analytically continued up to the vertical line $\operatorname{Re} s = 0$. In either case, (12.31) (and hence (12.30)) holds for s in $\mathbb{C}$ or $\{s\colon \operatorname{Re} s > 0\}$, respectively.

The main result in [HamLap, Section 4] asserts the existence of a suitable analytic continuation for the pointwise or random (and not just the mean) zeta function, for almost every realization of the random fractal string $\mathcal{L}$.

Theorem 12.30 ([HamLap, Theorem 4.5]). *Almost surely, the random zeta function $\zeta_{\mathcal{L}}(s)$ of the random self-similar string admits a meromorphic continuation to a nontrivial open half-plane $\operatorname{Re} s > D - \tau$, where D is the Minkowski dimension of $\mathcal{L}$ and the positive constant τ is estimated in terms of the parameters specifying the underlying branching process. Further, $\mathcal{L}$ is languid in the corresponding window. It follows that the set $\mathcal{D}_{\mathcal{L}}$ of visible complex dimensions of $\mathcal{L}$ (i.e., the poles of the random zeta function) is contained in the set of complex solutions s to the expectation equation (12.30) such that $\operatorname{Re} s > D - \tau$.*

One deduces from the above result (and our earlier work on explicit formulas, see Chapters 5 and 8) an inner tube formula for random self-similar strings (see Theorem 7.6 in [HamLap]). In particular, for almost every realization of $\mathcal{L}$, the volume of the inner tubular neighborhoods is given, for $\varepsilon \to 0^+$, by

$$V(\varepsilon) = \sum_{\omega \in \mathcal{D}_{\mathcal{L}}} \operatorname{res}\left(\frac{\zeta_{\mathcal{L}}(s)(2\varepsilon)^{1-s}}{s(1-s)}; \omega \right) + \{2\varepsilon\zeta_{\mathcal{L}}(0)\} + o\left(\varepsilon^{1-D+\tau-\delta}\right) \quad (12.32)$$

for every fixed $\delta > 0$, where the term between braces is included only if 0 is not a complex dimension of $\mathcal{L}$. We refer to part (1) of Theorem 7.8 of [HamLap] for the resulting estimate in the nonlattice case.

Stable Random Strings

We now discuss the family of *stable random strings* $\{\mathcal{L}_\alpha\}_{\alpha \in (0,1)}$, induced by the one-parameter family $\{C_\alpha\}_{\alpha \in (0,1)}$ of random Cantor sets defined by the excursions of stable subordinators, as studied in [HamLap, Section 5] (see [PitY] for an extensive survey of such random sets).[24] Hence, for example for $\alpha = 1/2$, the boundary of $\mathcal{L}_{1/2}$ coincides with $C_{1/2}$, the set of zeros of one-dimensional Brownian motion. If, for each $0 < \alpha < 1$, we write the lengths of the α-stable string in the form $\mathcal{L}_\alpha = \{\ell_{j,\alpha}\}_{j=1}^\infty$, then it follows from a result obtained in [PitY] that almost surely

$$l_{j,\alpha} j^{1/\alpha} \longrightarrow M_\alpha \quad \text{as } j \to \infty, \quad (12.33)$$

for some positive and finite constant M_α. Hence, according to the Minkowski measurability criterion obtained in [LapPo2] (see Remark 8.19 above), almost surely, $\mathcal{L}_\alpha$ is Minkowski measurable with Minkowski dimension α and Minkowski content

$$\mathcal{M}_\alpha = \frac{2^{1-\alpha}}{1-\alpha}(M_\alpha)^\alpha. \quad (12.34)$$

In particular, almost surely, the zero set of a Brownian path is Minkowski measurable and has dimension 1/2.

Moreover, it is shown in [HamLap, Lemma 5.3] that the mean zeta function $\mathbb{E}(\zeta_{\mathcal{L}_\alpha})$ can be meromorphically continued to the whole complex plane and is given by

$$\mathbb{E}(\zeta_{\mathcal{L}_\alpha})(s) = \frac{\Gamma(s-\alpha)}{\Gamma(1-\alpha)\Gamma(s)}, \quad \text{for } s \in \mathbb{C}. \quad (12.35)$$

Recall that the gamma function $\Gamma(s)$ is meromorphic in all of $\mathbb{C}$. Further, it has no zeros and has only simple poles, which lie on the real axis

[24] We note that the focus in [PitY] was not on the fractality of the resulting random set.

at $s = 0, -1, -2, -3, \ldots$. Therefore, in view of (12.35), the poles of $\mathbb{E}(\zeta_{\mathcal{L}_\alpha})$ are all simple and lie on the real axis at the values $\alpha, \alpha - 1, \alpha - 2, \alpha - 3, \ldots$. Recall that $\alpha = D$ is the Minkowski dimension of almost every realization of the random fractal string $\mathcal{L}_\alpha$. It follows, in particular, that almost surely, the random (or pointwise) zeta function is holomorphic for $\operatorname{Re} s > \alpha$.

Remark 12.31. To avoid any possible confusion, we note that unlike random self-similar strings, random stable strings are not defined in terms of random trees and branching processes. Instead, they are derived from stable subordinators (increasing functions).

Remark 12.32. The aforementioned results concerning the Minkowski content and the poles of the mean zeta function of a random stable string (with parameter α) are reminiscent of those obtained earlier for the a-string defined in Section 6.5.1 and for which all the complex dimensions were also located on the real axis. Compare, for example, the present results and those stated in Theorem 6.20 and Section 8.1.2. (Note, however, that even if we set $\alpha = \frac{1}{a+1}$ or, equivalently, $a = \frac{1-\alpha}{\alpha}$, so that the Minkowski dimensions of the strings coincide, the other poles on the real axis do not coincide.)

The next result provides information about the analytic continuation of the random zeta function $\zeta_{\mathcal{L}_\alpha}$.

Theorem 12.33 ([HamLap, Theorem 5.7]). *Almost surely, the geometric zeta function $\zeta_{\mathcal{L}_\alpha}$ of the random stable string $\mathcal{L}_\alpha$ admits a meromorphic continuation to the open half-plane $\operatorname{Re} s > \alpha/2$ and has a simple pole at $s = \alpha$ but no other visible poles (in that window). Moreover, the value of the residue at $s = \alpha$ is $\alpha(M_\alpha)^\alpha$, where M_α is given by Equation (12.33).*

One deduces that for every fixed $\delta > 0$, the volume $V(\varepsilon)$ of the tubular neighborhoods of $\mathcal{L}_\alpha$ is almost surely given by

$$V(\varepsilon) = \frac{M_\alpha (2\varepsilon)^{1-\alpha}}{1-\alpha} + o\big(\varepsilon^{1-\alpha/2-\delta}\big), \quad \text{as } \varepsilon \to 0^+, \tag{12.36}$$

and that for every fixed $\delta > 0$, the spectral counting function $N_\nu(x)$ of $\mathcal{L}_\alpha$ is almost surely given by

$$N_\nu(x) = x + M_\alpha \zeta(\alpha) x^\alpha + o\big(x^{\alpha/2+\delta}\big), \quad \text{as } x \to \infty \tag{12.37}$$

(see [HamLap, Theorems 7.8 and 7.9]). For the expert reader, we mention that the α-th power of the (positive and finite) constant M_α occurring in Equations (12.33), (12.34), as well as in (12.36) and (12.37), has an interesting probabilistic interpretation. In particular, the square root of $M_{1/2}$ is equal to the (Lévy) local time at 0 of Brownian motion run for a unit length of time.

The following comment may help to clarify one aspect of Theorem 12.33 that is not immediately obvious. This will be helpful in the following discussion.

Remark 12.34. According to Theorem 12.33, $D = \alpha$ is the only pole in the window $W = \{s\colon \operatorname{Re} s > \alpha/2\}$, which is the window for which $\zeta_{\mathcal{L}}$ has been shown in [HamLap] to be both languid and meromorphic, almost surely. (Indeed, $\alpha - 1 < \alpha/2$, since $0 < \alpha < 1$.) However, by (12.35), even if $\zeta_{\mathcal{L}}$ were shown to have, almost surely, a meromorphic continuation that is languid in a region W', larger than W, the only poles of $\zeta_{\mathcal{L}}$ in W' would still be located on the real axis. Indeed, according to (12.35) and the discussion following it, the only poles of $\mathcal{L}$ in W' are the same as those of the mean zeta function (12.35) and hence are all simple and located at $\alpha - n$, for all n such that $\alpha - n \in W'$.

A Refinement of the Notion of Fractality

In light of the above discussion, it is natural to wonder whether the random zeta function $\zeta_{\mathcal{L}}$ of a random fractal string may have a natural boundary (i.e., cannot be meromorphically continued beyond a certain vertical line, or more generally curve, then called a natural boundary for $\zeta_{\mathcal{L}}$ or for $\mathcal{L}$, in short). The following open problem—stated (in somewhat more concise form) at the very end of [HamLap]—addresses this question.

Problem 12.35. Find natural classes of random fractal strings $\mathcal{L}$ for which, almost surely, the random zeta function $\zeta_{\mathcal{L}}$ admits a natural boundary. Consider, especially, the case of random self-similar strings and that of stable random strings (studied in [HamLap, Sections 4 and 5]).

In particular, because of their strong scale-irregularity, the random Cantor strings (or homogeneous random strings) considered in [HamLap, Section 6] are potential candidates for having the vertical line $\operatorname{Re} s = D$ itself as a natural boundary. Is that really the case?

We close this section by recalling that by Theorem 12.33 above, the visible complex dimensions of a stable random string $\mathcal{L}$ (i.e., the visible poles of $\zeta_{\mathcal{L}}$) must all lie on the real axis; see Remark 12.34 above. In light of the definition of fractality proposed at the beginning of Section 12.2, this would seem to go against our intuition that the zero set of Brownian motion, for example, is a (random) fractal.[25] However, if $\zeta_{\mathcal{L}}$ does have a natural boundary $\mathbb{L}$ (say, along some vertical line $\operatorname{Re} s = \sigma$, with $\sigma \leq D/2$)—as may be expected according to Problem 12.35 above—then, in some intuitive sense, it implies that $\zeta_{\mathcal{L}}$ has a dense subset of singularities accumulating along $\mathbb{L}$. Hence, almost surely, $\zeta_{\mathcal{L}}$ would have infinitely many *nonreal* singularities accumulating along $\mathbb{L}$. An analogous comment would apply to the strongly scale-irregular random Cantor sets considered in [HamLap, Section 6], except that the natural boundary might be along the line $\operatorname{Re} s = D$, in which

[25] Another possibility would be to draw on the analogy with the a-string (as discussed in Remark 12.32 above) and recall from Section 12.2.2 that the latter string is not fractal in our sense.

case the object would be strongly fractal. This leads us naturally to supplement the definition of fractality given at the very beginning of Section 12.2:

> A geometric object is fractal if it has at least one nonreal complex dimension (with positive real part) or if the associated zeta function has a natural boundary along a screen[26] (located in the open half-plane $\operatorname{Re} s > 0$).

Clearly, such a definition can be easily adapted at the spectral or dynamical level, as well as to the random case.

12.4.2 Fractal Membranes: Quantized Fractal Strings

In the book [Lap10] (announced in [Lap9]), the first author has proposed the notion of quantized fractal string, called *fractal membrane,* and has developed an analogy between self-similar geometries and arithmetic geometries. In particular, lattice strings (or membranes) correspond to varieties over finite fields whereas nonlattice strings (or membranes) are viewed as a counterpart of algebraic number fields.[27] Moreover, the scaling ratios of self-similar geometries (or of self-similar membranes [Lap10]) play the role of the generalized primes attached to fractal membranes, while expressions like (2.10) for the geometric zeta function of a self-similar string (with possibly infinitely many scaling ratios and gaps) correspond to the Euler product representation (of the same nature as that for the classical Riemann zeta function, see [In, Pat, Ti]) of the partition function of a fractal membrane (as obtained in [Lap10]). Finally, in joint work of the first author with Ryszard Nest [LapNes1], it was recently shown that the fractal membranes (respectively, self-similar membranes) introduced in [Lap10] can be rigorously constructed as the second quantization of fractal strings by using bosonic (respectively, Gibbs–Boltzmann or free) statistics, along with aspects of the theory of operator algebras and of Connes' noncommutative geometry [Con]. (See also [LapNes2, 3].)

We next provide some additional information about fractal membranes and their associated zeta functions (or partition functions).

Fractal Membranes as Noncommutative, Infinite Dimensional Tori

We first discuss the basic intuition behind the notion of fractal membrane. Heuristically, a fractal membrane (or quantized fractal string) can be thought of as a noncommutative, adelic, infinite dimensional torus. (See [Lap10, Chapters 3 and 4].) In particular, it can roughly be considered as an infinite dimensional restricted product of circles, $\coprod_{j=1}^{\infty} S_j$

[26] In the sense of this book, see Sections 1.2.1 and 5.3.

[27] Finite extensions of the field of rational numbers; see, e.g., [ParSh1, I & II].

(or a restricted Hilbert cube, with opposite faces identified). Hence, instead of vibrating independently of one another, as is the case for a standard fractal string, all the intervals I_j (or circles S_j, for $j = 1, 2, \dots$) of the quantized fractal string[28] are now vibrating in mutually perpendicular directions within a potentially infinite dimensional space. One adds the physically natural constraint that for each given mode of vibration of the membrane (i.e., for each eigenfunction of the underlying Dirac operator $\mathcal{D}$, to be discussed in Remark 12.36 below), only finitely many circles (or pairs of opposite faces of the corresponding Hilbert cube) are actually vibrating. Of course, different modes of vibration usually involve different finite subsets of circles. The above constraint explains the use of the term "adelic" as a qualification of the infinite dimensional torus associated with a fractal membrane.[29] Furthermore, its mathematical counterpart in this context is the notion of restricted tensor product of Hilbert spaces, as discussed in the following remark.

Remark 12.36. The adjective "noncommutative" also used above to qualify the infinite dimensional torus associated with a fractal membrane $\mathcal{T}$ indicates that as a mathematical object, $\mathcal{T}$ is not truly a set of points. Instead, it is viewed as a noncommutative geometric object (in the sense of Alain Connes [Con]) and is given by a spectral triple

$$\mathcal{T} = (\mathcal{A}, \mathcal{H}, \mathcal{D}), \tag{12.38}$$

where $\mathcal{A}$ is a suitable algebra of operators (the noncommutative algebra of coordinates on the membrane) represented on a separable, infinite dimensional Hilbert space $\mathcal{H}$, and $\mathcal{D}$ is an appropriate analogue in this context of the Dirac operator. Namely, $\mathcal{D}$ is a suitable unbounded self-adjoint operator acting on the Hilbert space $\mathcal{H}$, with compact resolvent (and hence, discrete spectrum), and such that the commutators $[\mathcal{D}, a]$ are compact operators for all a in some dense subalgebra $\mathcal{A}_0$ of $\mathcal{A}$.

According to a well-known analogy in the theory of operator algebras, the C*-algebra $\mathcal{A}$ can be thought of as the space of continuous functions on the underlying noncommutative space. Similarly, the dense subalgebra $\mathcal{A}_0$ plays the role of the space of Lipschitz continuous functions[30] on this same noncommutative space. Indeed, intuitively, $[\mathcal{D}, a]$ is the (quantized) differential of a. The boundedness of this differential means heuristically that the gradient of a is bounded and hence that a is 'Lipschitz' for $a \in \mathcal{A}_0$. We note, in addition, that the Dirac operator $\mathcal{D}$ (or rather, the inverse of its restriction to the orthogonal complement of the kernel of $\mathcal{D}$) enables one to define an analogue of the infinitesimal length element or, more precisely,

---

[28] that is, all the strings of the fractal harp, see Chapter 1, Figure 1.1.

[29] Another image used in [Lap10] is that of a noncommutative, adelic, Riemann surface of infinite genus.

[30] Sometimes simply referred to as the space of smooth functions.

the noncommutative counterpart of a Riemannian metric in this context. (See, e.g., [Con, Chapters IV–VI].)

A spectral triple like (12.38) provides a way to completely describe the fractal membrane $\mathcal{T}$. In the rigorous construction given in [LapNes1], $\mathcal{H}$ is obtained as a restricted tensor product of the Hilbert spaces $\mathcal{H}_j$ associated with each circle S_j (or interval I_j, see below) composing the membrane, relative to a suitable vacuum vector (here, the zero energy mode of the Dirac operator $\mathcal{D}$). Furthermore, $\mathcal{A}$, the algebra of quantum observables, is defined in terms of a tensor product of Toeplitz algebras, one for each circle. Each of these Toeplitz algebras (see, e.g., [BotSil]) is an algebra of bounded linear operators acting on a Hilbert space of holomorphic functions on the unit disc (one unit disc for each circle S_j, or one for each endpoint of the j-th interval I_j).

An outline of the formal construction will be provided in the second to next unnumbered subsection. In addition, we refer to [LapNes1] for the precise and complete construction of a fractal membrane and for the relevant definitions. See also [Lap10, Chapter 3] for a heuristic definition, and Chapter 4, in conjunction with Chapter 2 of [Lap10], for the physical motivations (coming from conformal field theory and string theory) leading to the representation of a fractal membrane as a noncommutative geometric object; namely, as the stringy spacetime describing the propagation of strings in an adelic, infinite dimensional torus.

Finally, we note that one of the new heuristic and mathematical insights provided by the construction given in [LapNes1] is that once fractal strings have been quantized, their endpoints are no longer fixed on the real axis but are allowed to move freely within a suitable copy of the unit disc in the complex plane. This seems to be analogous to the notion of D-brane in nonperturbative string theory, in the spirit of (but somewhat different from) [Lap10, Chapter 2]. Hence, a fractal membrane may be viewed as some kind of fractal D-brane.

Fractal Membranes and their Zeta Functions

In order to describe the zeta function (or partition function) of a fractal membrane, we need to introduce some additional notation. Let

$$\Omega = \bigcup_{j=1}^{\infty} \left(a_j^-, a_j^+\right) \tag{12.39}$$

be a fractal string, viewed as an open subset of $\mathbb{R}$, or equivalently, a disjoint union of bounded open intervals $I_j = \left(a_j^-, a_j^+\right)$, each of length $l_j := a_j^+ - a_j^-$ and thought of heuristically as being attached to a circle of radius

$$R_j := \frac{2\pi}{\log l_j^{-1}}, \tag{12.40}$$

for $j = 1, 2, \ldots$. (This value of R_j is dictated by spectral considerations.) Moreover, let $\mathcal{L} = \{l_j\}_{j=1}^{\infty}$ denote the associated sequence of lengths, assumed to be written in nonincreasing order according to their multiplicities as follows:

$$1 > l_1 \geq l_2 \geq \cdots \geq l_j \geq \ldots, \tag{12.41}$$

with $l_j \to 0$ as $j \to \infty$.

Then the resulting fractal membrane $\mathcal{T}$, obtained after second (or Dirac) quantization of the fractal string $\mathcal{L}$, has a discrete eigenvalue (or energy) spectrum $\sigma(\mathcal{T}) = \{\lambda_n\}_{n=1}^{\infty}$ and its spectral (or quantum) *partition function* $Z_{\mathcal{T}}(s)$, defined by

$$Z_{\mathcal{T}}(s) := \sum_{n=1}^{\infty} e^{-s\lambda_n}, \quad \text{for } \operatorname{Re} s \gg 1, \tag{12.42}$$

is shown in [Lap10, Chapter 3] to be given by the Euler product

$$Z_{\mathcal{T}}(s) = \prod_{j=1}^{\infty} \left(1 - l_j^s\right)^{-1}, \quad \text{for } \operatorname{Re} s > D, \tag{12.43}$$

where $D = D_{\mathcal{L}}$ is the Minkowski dimension of $\mathcal{L}$.[31] In other words, the reciprocal lengths $p_j := l_j^{-1}$ are the generalized primes (also called g-primes or Beurling primes) of the fractal membrane $\mathcal{T}$ and the (spectral) partition function $Z_{\mathcal{T}}(s)$ is the associated *Beurling zeta function.*

Equivalently, the Euler product representation given by Equation (12.43) means that if the spectrum $\sigma(\mathcal{T}) = \{\lambda_n\}_{n=1}^{\infty}$ is written in nondecreasing order according to multiplicity, with $\lambda_n \to \infty$ as $n \to \infty$, it then consists of the logarithms of the corresponding generalized integers (also called g-integers or Beurling integers in the literature), obtained by taking all the finite products of (nonnegative integer) powers of the generalized primes $p_j = l_j^{-1}$, and counted according to their natural multiplicity—exactly like the positive integers are obtained from the standard rational primes.

Remark 12.37. Such zeta functions were considered by Arne Beurling in [Beu] for purely analytical reasons, in order to extend the classic Prime Number Theorem to a suitable sequence $\{p_j\}_{j=1}^{\infty}$ of g-primes, with

$$1 < p_1 \leq p_2 \leq \cdots \leq p_j \leq \ldots \tag{12.44}$$

---

[31] In fact, it can be proved that D is also equal to the abscissa of convergence of the Dirichlet series in (12.42), as in Equation (1.20) of Definition 1.9. It follows that $Z_{\mathcal{T}}(s)$ is given by both (12.42) and (12.43) for $\operatorname{Re} s > D$, much like the Riemann zeta function is given by both the Dirichlet series $\sum_{n=1}^{\infty} n^{-s}$ and the Euler product $\prod_p \left(1 - p^{-s}\right)^{-1}$ for $\operatorname{Re} s > 1$.

and $p_j \to \infty$ as $j \to \infty$. See, for example, [Lap10, Appendix D] and [HilLap], along with the relevant references therein, for an exposition of some of their main properties.

Remark 12.38. In the rigorous construction of the fractal membrane given in [LapNes1], the spectrum $\sigma(\mathcal{T})$ of the fractal membrane $\mathcal{T}$ is nothing but the spectrum of the Dirac operator $\mathcal{D}$ discussed in Remark 12.36 above: $\sigma(\mathcal{T}) = \sigma(\mathcal{D})$. Note that $\sigma(\mathcal{D})$ is shown in [LapNes1] to be discrete and hence consists of the sequence of eigenvalues of $\mathcal{D}$.

Remark 12.39. As was mentioned earlier, in [Lap10, Section 3.3], the first author has also introduced the parallel notion of *self-similar membrane*. Formally, it corresponds to allowing the above g-integers (i.e., the logarithms of the eigenvalues of the given self-similar membrane) to appear with the exact same multiplicities as the lengths of a self-similar string expressed as monomials in the scaling ratios (see Section 2.1 above and [Lap10, Section 3.3]).[32] Note that in this analogy, the lengths l_j now play the role of the scaling ratios r_j, allowed here to be in infinite number. Furthermore, as was shown in [LapNes1], this new choice of multiplicities corresponds to a different type of quantum statistics in the mathematical construction of the membrane; namely, free (or Gibbs–Boltzmann) statistics for self-similar membranes instead of bosonic (or Bose–Einstein) statistics for ordinary fractal membranes.

Let $\mathcal{L} = \{l_j\}_{j=1}^{\infty}$ be a fractal string, constructed with the sequence of lengths l_j. The (spectral) partition function $Z_{\mathcal{S}}(s)$ of the self-similar membrane $\mathcal{S}$ is then given by

$$Z_{\mathcal{S}}(s) = \frac{1}{1 - \sum_{j=1}^{\infty} l_j^s}, \quad \text{for } \operatorname{Re} s > D, \tag{12.45}$$

where $D = D_{\mathcal{L}}$ is the Minkowski dimension of the fractal string $\mathcal{L}$.[33] Hence, formally, $Z_{\mathcal{S}}(s)$ coincides with the geometric zeta function of a normalized self-similar string with a single gap (as in Section 2.2.1) and with infinitely many scaling ratios $r_j = l_j$, for $j = 1, 2, \dots$. (Compare formula (12.45) above with Equation (2.15) following Theorem 2.9.) This is the starting point of the aforementioned analogy between self-similar and arithmetic geometries developed in [Lap10].

---

[32] For simplicity, we only consider here the case of a single gap, as in Section 2.2.1.

[33] *Caution*: As is explained in [Lap10, Section 3.3], a self-similar membrane is not a special case of fractal membrane, even though the constructions of both types of membranes turn out to be very analogous (see [LapNes1]). Hence (12.45) is not incompatible with (12.43), since each of these formulas applies to a different type of mathematical object. See also the end of the next unnumbered subsection.

Construction of Fractal Membranes

We next outline the mathematical construction of fractal membranes (and of their self-similar counterparts) given in [LapNes1]. As was mentioned earlier, it builds on the heuristic notion of fractal membrane introduced in [Lap10] but also brings new insights into its properties. The present overview complements Remark 12.36 in which we gave a nontechnical description of the spectral triple $\mathcal{T} = (\mathcal{A}, \mathcal{H}, \mathcal{D})$ associated with a fractal membrane, viewed as a noncommutative space. (See also the exposition in [Lap10, Section 4.2], on which our description is based. Further, see [LapNes1] for more details, references, and definitions.) This spectral triple can be described as follows. (Recall from Remark 12.36 that $\mathcal{A}$ is a C*-algebra represented on the complex Hilbert space $\mathcal{H}$, and that $\mathcal{D}$ is a Dirac-like operator, thought of as a first-order differential operator and acting on $\mathcal{H}$ as an unbounded, self-adjoint operator.)

Let

$$\mathbb{D} = \{z \in \mathbb{C} \colon |z| \leq 1\} \tag{12.46}$$

denote the closed unit disc of $\mathbb{C}$ and let $\Omega = \bigcup_{j=1}^{\infty} \left(a_j^-, a_j^+\right)$ denote the fractal string to be quantized (as in formula (12.39) above), with boundary $\partial\Omega$. Consider the space $\mathcal{F}$ of functions $f = f(x,z) \colon \partial\Omega \times \mathbb{D} \to \mathbb{C}$ such that $|f| = 1$ and the map $x \mapsto f(x, \cdot)$ is a continuous map from $\partial\Omega$ to $L^\infty(\mathbb{D})$. Denote by $\mathcal{F}_0$ the subset of $\mathcal{F}$ consisting of functions $f = f(x,z)$ such that both the map $x \mapsto f(x, \cdot)$ and its derivative, given by $x \mapsto g(x, \cdot)$ with $g(x,z) := z\partial f/\partial z(x,z)$, are Lipschitz functions from $\partial\Omega$ to $L^\infty(\mathbb{D})$.[34] Here, $L^\infty(\mathbb{D})$ denotes the space of essentially bounded functions on $\mathbb{D}$, equipped with the supremum norm. Clearly, $\mathcal{F}$ is a commutative algebra, for the standard algebraic operations, and $\mathcal{F}_0$ is a subalgebra of $\mathcal{F}$.

For each fixed $j \geq 1$, let $\mathcal{H}_j$ be a copy of the Hilbert space L^2-Hol$(\mathbb{D})$ of square-integrable holomorphic functions on $\mathbb{D}$. We then let

$$\mathcal{H} = \bigotimes_{j=1}^{\infty} \mathcal{H}_j \tag{12.47}$$

be the (restricted) tensor product of the Hilbert spaces $\mathcal{H}_j$, relative to the vacuum vector $V := \bigotimes_{j=1}^{\infty} \mathbf{1}_j$, where $\mathbf{1}_j$ is the constant function equal to 1 on $\mathbb{D}$ (viewed as an element of $\mathcal{H}_j$).

[34] Some mild modification is required in order to take into account fractal strings of infinite length (such as the fractal string of lengths $\{1/p \colon p \text{ a rational prime}\}$) or, more generally, arbitrary fractal strings $\mathcal{L} = \{l_j\}_{j=1}^{\infty}$ such that $\sum_{j=1}^{\infty} (l_j)^\alpha < \infty$, for some $\alpha > 0$. Indeed, we must then replace "Lipschitz" by "uniformly continuous", with respect to a suitable modulus of continuity. (The above Lipschitz condition corresponds to fractal strings of finite length.) The resulting fractal membranes—including the prime membranes [Lap10, Chapter 3]—play an important role throughout [Lap10].

Given $\varphi \in L^\infty(\mathbb{D})$, an essentially bounded function on $\mathbb{D}$, we denote by T_φ the Toeplitz operator with symbol φ, viewed as a bounded linear operator on L^2-Hol($\mathbb{D}$) (or, equivalently, on any of the Hilbert spaces $\mathcal{H}_j$). Thus T_φ is defined as the compression to L^2-Hol($\mathbb{D}$) of the multiplication operator by φ acting on $L^2(\mathbb{D})$. (See, e.g., [BotSil, Chapters 1 and 2].) Given $f \in \mathcal{F}$, set

$$Q(f) = \bigotimes_{j=1}^{\infty} T_j(f), \tag{12.48}$$

viewed as a bounded linear operator on the Hilbert space $\mathcal{H}$, where for each $j \geq 1$, $T_j(f) := T_{\varphi_j}$ is the Toeplitz operator (acting on $\mathcal{H}_j$) with symbol

$$\varphi_j := \frac{f(a_j^+, \cdot)}{f(a_j^-, \cdot)}. \tag{12.49}$$

(The operator $Q(f)$ can be thought of as the quantization of the classical observable f.) The C*-algebra $\mathcal{A}$ is then defined as the C*-algebra generated (in $\mathcal{B}(\mathcal{H})$, the space of bounded linear operators on $\mathcal{H}$) by the set $\{Q(f)\colon f \in \mathcal{F}\}$. Further, using the notation of Remark 12.36, the dense subalgebra $\mathcal{A}_0 \subset \mathcal{A}$ is defined as the algebra generated (in $\mathcal{B}(\mathcal{H})$) by the set $\{Q(f)\colon f \in \mathcal{F}_0\}$, where $\mathcal{F}_0 \subset \mathcal{F}$ is the space of smooth classical observables defined earlier. Intuitively, $\mathcal{A}$ is the space of quantum observables associated with the space of classical observables $\mathcal{F}$, while $\mathcal{A}_0$ is the space of admissible quantum observables associated with the space of smooth classical observables $\mathcal{F}_0$.[35]

Finally, for each $j \geq 1$, with $R_j = 2\pi/\log l_j^{-1}$ as in formula (12.40) above, let

$$\mathcal{D}_j = \frac{1}{iR_j}\frac{d}{d\theta} \tag{12.50}$$

denote the scaled Dirac operator on the unit circle $S = \partial\mathbb{D}$,[36] restricted to its positive energy subspace—so that its eigenvalues are simple, nonnegative and consist of the numbers $2\pi n/R_j = n\log l_j^{-1}$, for $n = 0, 1, 2, \ldots$. (Note that viewed as acting on the isomorphic copy L^2-Hol($\mathbb{D}$), $\frac{1}{i}\frac{d}{d\theta}$ becomes $z\frac{d}{dz}$, and that $z\frac{d}{dz}(z^n) = nz^n$ for every $n = 0, 1, 2, \ldots$.) It then suffices to let

$$\mathcal{D} = \bigotimes_{j=1}^{\infty} \mathcal{D}_j, \tag{12.51}$$

---

[35] It can be checked that $\mathcal{A} = \bigotimes_{j=1}^{\infty} \mathcal{A}_j$, where for each $j \geq 1$, $\mathcal{A}_j$ is the Toeplitz algebra generated by the operators $T_j(f)$, with $f \in \mathcal{F}$.

[36] Here, θ is the angular variable along the circle. Further, $\mathcal{D}_j$ acts as an unbounded, self-adjoint operator on $\mathcal{H}_j$.

viewed as an unbounded, self-adjoint operator on $\mathcal{H}$. It follows that the spectrum of $\mathcal{D}$ consists of the logarithms of m, where m ranges through all the generalized integers based on the generalized primes $p_j := l_j^{-1}$ $(j \geq 1)$ (see Remark 12.38 and [Lap10, Chapter 3 and Appendix D]), and hence that the spectral partition function of the fractal membrane,

$$Z_{\mathcal{T}}(s) := \text{Trace}\left(e^{-s\mathcal{D}}\right),$$

is given by the Euler product (12.43), for $\text{Re}\, s > D_{\mathcal{L}}$, as stated in [Lap10, Section 3.2].

We close this brief overview by noting that if in the above construction, we replace the bosonic Fock space $\mathcal{H}$ by its free counterpart (in the sense of noncommutative probability theory [Voi], and thus with a free tensor product instead of a bosonic one in formula (12.47) above), and make analogous changes elsewhere, we then obtain a spectral triple defining a self-similar membrane $\mathcal{S}$ (in the sense of [Lap10, Section 3.3]), viewed as a noncommutative space. As discussed in Remark 12.39 above, the corresponding spectral partition function $Z_{\mathcal{S}}(s)$ is then given by formula (12.45), and therefore coincides with the geometric zeta function of a self-similar string with infinitely many scaling ratios $r_j := l_j$, for $j \geq 1$. Indeed, the spectrum of the associated Dirac operator now consists of the logarithms of the self-similar integers m' based on the primes $p_j = l_j^{-1}$. We note that the only difference between those integers m' and the g-integers m discussed just above consists in their multiplicities, which are defined exactly as the multiplicities of the lengths of a self-similar string with a single gap; see [Lap10, Section 3.3].

Flows on the Moduli Space of Fractal Membranes

In [Lap10, Chapter 5], a moduli space of fractal strings is introduced, $\mathcal{M}_{\text{fs}}$, and also its quantum counterpart, the moduli space of fractal membranes, denoted by $\mathcal{M}_{\text{fm}}$. These moduli spaces can be thought of as moduli spaces of zeta functions. Further, they are noncommutative spaces, initially defined as singular quotient spaces (of equivalence classes of fractal strings or of fractal membranes, respectively). For example, $\mathcal{M}_{\text{fs}}$ is viewed in [Lap10, Section 5.1] as a generalization of the noncommutative space of Penrose tilings ([Con] or [Lap10, Appendix F]); namely, two fractal strings $\mathcal{L} = \{l_j\}_{j=1}^{\infty}$ and $\mathcal{L}' = \{l'_j\}_{j=1}^{\infty}$ (written in nonincreasing order) are considered to be equivalent if there exist $j_0, q_0 \in \mathbb{N}$ such that $l'_j = l_{q_0+j}$ for all $j \geq j_0$ (so that $\zeta_{\mathcal{L}'}(s) - \zeta_{\mathcal{L}}(s)$ is an analytic function without poles). The moduli space $\mathcal{M}_{\text{fm}}$ is defined similarly as a quotient space (but then, for $\mathcal{L}$ and $\mathcal{L}'$ in the same equivalence class, $Z_{\mathcal{L}'}(s)/Z_{\mathcal{L}}(s)$ is a holomorphic function without zeros).

The advantage of $\mathcal{M}_{\text{fm}}$ over $\mathcal{M}_{\text{fs}}$ is that each equivalence class in $\mathcal{M}_{\text{fm}}$ is associated to a multiplicative class of zeta functions (i.e., partition functions) having the same poles and zeros, whereas only the poles are invariants of a given equivalence class in $\mathcal{M}_{\text{fs}}$ (i.e., of the associated additive class of zeta functions). This key difference between $\mathcal{M}_{\text{fm}}$ and $\mathcal{M}_{\text{fs}}$ is due to the fact that according to formula (12.43) above, the zeta function (i.e., spectral partition function) of a fractal membrane has an Euler product, whereas by Equation (12.45) above, this is not the case in general for the (geometric) zeta function of a fractal string, at least not in the standard sense of number theory.

Still in [Lap10, Chapter 5], the first author then proposes to consider a suitable noncommutative one-parameter flow on $\mathcal{M}_{\text{fm}}$, the modular flow. The latter induces continuous flows of zeta functions and their zeros (or poles), as well as corresponding continuous deformations of the g-primes. By considering the fixed points of these flows (or rather, their noncommutative analogue), the first author then gives a new dynamical interpretation of the Riemann hypothesis and of the class of zeta functions (or L-series) expected to satisfy its natural generalization. In particular, it is conjectured in [Lap10, Section 5.5] that a suitable section of zeta functions obeying a generalized functional equation (GFE)[37] can be chosen over the effective part of the moduli space $\mathcal{M}_{\text{fm}}$, and hence that the fixed points (or ω-limit sets) of the flow correspond to the zeta functions satisfying a self-dual functional equation,[38] i.e., to all the arithmetic zeta functions (such as the Riemann zeta function and other L-series).

Needless to say, these flows of fractal membranes, as well as of zeta functions and their associated zeros, still need to be rigorously constructed and their conjectured properties to be verified in order for the research program outlined at the end of [Lap10] to be fully realized.

12.5 The Spectrum of a Fractal Drum

In this section, we consider a bounded open set $\Omega \subset \mathbb{R}^d$, with a fractal boundary $\partial\Omega$, and equipped with a Laplace operator Δ. The counting function of the frequencies of Δ is denoted by $N_\nu(x)$. We refer to Appendix B for information and references about the case when Ω is a manifold with smooth boundary.

12.5.1 The Weyl–Berry Conjecture

Hermann Weyl's conjecture (about manifolds with smooth boundary, see Section B.4.1 in Appendix B), has been extended by Michael V. Berry

---

[37] In short, a GFE for $Z(s)$ is (after suitable completion by appropriate gamma-like factors) of the form $Z_{\mathcal{T}}(s) = Z_{\mathcal{T}^*}(1-s)$, where $\mathcal{T}^*$ is interpreted as a dual fractal membrane. See [LapNes2, 3] and [HilLap], along with the relevant references therein, for results along these lines in this and a related context.

[38] In [Lap10], a GFE is said to be self-dual if $\mathcal{T} = \mathcal{T}^*$; i.e., $Z_{\mathcal{T}}(s) = Z_{\mathcal{T}}(1-s)$.

in [Berr1–2]. Berry's intriguing and stimulating conjecture was formulated in terms of the Hausdorff dimension H of $\partial\Omega$. In [BroCa], Brossard and Carmona have disproved this conjecture and suggested that the Hausdorff dimension H should be replaced by the Minkowski dimension D of $\partial\Omega$. Using probabilistic techniques, they have also obtained an analogue of estimate (B.2) (expressed in terms of D) for the spectral partition function $\theta_\nu(t)$ (rather than for $N_\nu(x)$) in the case of the Dirichlet Laplacian.

Later on, using analytical techniques (extending, in particular, those of [Weyl1–2], [CouHi] and [Met]), Lapidus [Lap1] has shown that the counterpart of estimate (B.2) holds for the spectral counting function $N_\nu(x)$ itself. More precisely, provided that $D \in (d-1, d)$ and $\mathcal{M}^*(D; \partial\Omega) < \infty$, the following Weyl asymptotic formula with sharp error term holds pointwise:

$$N_\nu(x) = c_d \operatorname{vol}(M) x^d + O\big(x^D\big), \tag{12.52}$$

as $x \to \infty$.[39] (Here, the positive constant c_d depends only on d and is given just after formula (B.2) in Appendix B; see also Remark 12.40 below.) Analogous results were obtained in [Lap1] for the Neumann Laplacian (under a suitable assumption on $\partial\Omega$) and, more generally, for positive elliptic differential operators (with Dirichlet, Neumann, or mixed boundary conditions); see [Lap1, Theorem 2.1 and Corollaries 2.1–2.2, pp. 479–480]. We mention that the snowflake drum is a natural example of a (self-similar) fractal drum for which estimate (12.52) holds pointwise, either for Dirichlet or Neumann boundary conditions; see [Lap1, Example 5.4, pp. 518–519] along with Section 12.5.2 below.

We note that the a-string—introduced in [Lap1] and studied in Section 6.5.1 above (see also [LapPo1–2])—provides a simple example showing that the Hausdorff dimension (or its companion, the packing dimension) cannot be used in this context to measure the roughness of the boundary $\partial\Omega$ of a fractal drum. Indeed, a simple variant of that example enables one to exhibit a one-parameter family of open sets $\{\Omega_a\}_{a>0}$ such that (with the notation of Remark 12.40) $D(\partial\Omega_a) = n - 1 + \frac{1}{a+1}$ and $H(\partial\Omega_a) = n-1$, $0 < \mathcal{M}^*(D; \partial\Omega_a) < \infty$,[40] while the estimate in (12.52) is sharp with $D = D(\partial\Omega_a)$ for every value of $a > 0$. Observe that D takes every possible value in the admissible interval $(n-1, n)$ as the parameter a varies in $(0, \infty)$. See [Lap1, Examples 5.1 and 5.1$'$, pp. 512–515], along with Remark 1.5 in Chapter 1.

---

[39] If $D = d - 1$ (the nonfractal case in the terminology of [Lap1], see also Figure 9.1 on page 269), the result of [Lap1] yields instead the error term $O\big(x^D \log x\big)$ obtained previously by G. Métivier in [Met] in that situation and, in the special case of piecewise smooth boundaries, in [CouHi].

[40] Actually, $\partial\Omega_a$ is Minkowski measurable, as shown in [Lap1, Appendix C].

Remark 12.40. Note that since $\partial\Omega$ is the boundary of a bounded open set, we always have $n-1 \leq H \leq D \leq n$, where D denotes the Minkowski dimension of $\partial\Omega$ and H denotes its Hausdorff dimension. Further, if $\partial\Omega$ is piecewise smooth, or more generally, Lipschitz, we have $D = H = n-1$. (See, e.g., [Lap1, Section 3] and the references therein.)

12.5.2 The Spectrum of a Self-Similar Drum

It is conjectured in [Lap3] that the (frequency) spectrum of a drum with self-similar fractal boundary does not have oscillations of order D in the nonlattice case, whereas it has (multiplicatively) periodic oscillations of order D in the lattice case. Here, as before, D denotes the Minkowski dimension of the boundary of the drum. More specifically, as $x \to \infty$, it is conjectured in [Lap3] that in the nonlattice case (respectively, lattice case), the spectral counting function $N_\nu(x)$ has a monotonic (respectively, oscillatory) asymptotic second term of the form a constant times x^D (respectively, $f(\log x)x^D$, for a nonconstant periodic function f of period $2\pi/\mathbf{p}$, the additive generator of the boundary). By a standard Abelian argument ([Sim, Theorem 10.2, p. 107] or [Lap1, Appendix A, pp. 521–522]), the same statement would also hold for the spectral partition function $\theta_\nu(t)$ as $t \to 0^+$, with x replaced by t^{-1}. (For strictly self-similar drums, see [Lap3, Conjecture 3, p. 163] along with the important special case of snowflake-type drums discussed in [Lap3, Conjecture 2, p. 159], including the usual snowflake drum. See also [Lap3, Conjecture 4, p. 175] and Remark 12.20, for the broader class of approximately self-similar drums, in the sense of Section 12.3 above.)

In the special case of self-similar sprays (as defined in Section 6.6.2 and corresponding to the Dirichlet or Neumann Laplacian on a disconnected open set), the results of Section 6.6 above, Theorem 6.25 (for the Sierpinski drum), and especially Theorem 6.28, agree exactly with this conjecture. (See also the earlier references given in Section 6.6, including [FlVa, Ger, GerSc1–2, Lap2–3, LeVa, vB-Le].)

Moreover, in the case of the Dirichlet Laplacian on a connected domain, the interesting example of the snowflake drum (see Figure 12.6), a lattice self-similar drum, has been studied in [FlLeVa] from the point of view of the spectral partition function $\theta_\nu(t)$ (or of the closely related notion of heat content). See also the recent extension in [vB-Gi] to a one-parameter family of snowflake-type drums (much as in [Lap3, Conjecture 2, p. 159]). Again, these results agree with [Lap3, Conjectures 2 and 3]. However, we point out two open problems in this situation:

(i) As far as we know, in the lattice case, the counterpart for $\theta_\nu(t)$ of the periodic function f has not been proved to be nonconstant, even for the example of the snowflake drum. We note that for the latter example, the counterpart of f was verified numerically to be nontrivial in [FlLeVa].

(ii) To our knowledge, no such (pointwise) results have been obtained for the spectral counting function $N_\nu(x)$ rather than for $\theta_\nu(t)$. We note that technically, the conjectured pointwise result for $N_\nu(x)$ is significantly harder to establish than for $\theta_\nu(t)$.

Remark 12.41. We refer the interested reader to [LapNeuRnGri] and [GriLap] for a computer graphics-aided study of the frequency spectrum and of the normal modes of vibration of the snowflake drum (coined *snowflake harmonics*). Mathematically, the latter are the eigenfunctions of the Dirichlet Laplacian on the snowflake domain; that is, on the bounded domain having the snowflake curve for fractal boundary (see Figure 12.6). We also refer to the earlier paper [LapPan] for a mathematical study of the pointwise behavior of these eigenfunctions (and of their gradient) near the fractal boundary, for a class of simply connected domains including the snowflake domain. The work of [LapNeuRnGri] and [LapPan] was motivated in part by intriguing physical work and experiments in [SapGoM] on the vibrations of fractal drums, and was aimed in the long term at understanding how (and why) fractal shapes—such as coastlines, blood vessels and trees—arise in nature. We note that the numerical data in [LapNeuRnGri] (along with those in [SapGoM], [FlLeVa] and [GriLap]) should be useful to investigate the open problems and conjectures stated in the present subsection and in Section 12.5.3 below.

Given the theory developed in this book, in particular Theorems 2.17 and 3.6, it is natural to complement the spectral aspects of [Lap3, Conjecture 3, p. 163] by the following conjecture, which is a partial analogue of Conjecture 12.18 in Section 12.3 for the spectral[41] (rather than for the geometric) complex dimensions of a self-similar drum.

Conjecture 12.42. *Assume, for simplicity, that $d-1 < D < d$, where D is the Minkowski dimension of the boundary of a (strictly) self-similar drum $\Omega \subset \mathbb{R}^d$ (satisfying the open set condition, see Section 2.1.1 and Remark 2.22). Then D is the only real spectral complex dimension of the drum, other than d itself, and it is simple. (See also the end of Remark 12.43 below.)*

Moreover, on the vertical line $\operatorname{Re} s = D$, *a nonlattice drum does not have any nonreal spectral complex dimensions, whereas a lattice drum has an infinite sequence of spectral complex dimensions, contained in the arithmetic progression $\{D + in\mathbf{p} : n \in \mathbb{Z}\}$, where $\mathbf{p}$ is the (spectral) oscillatory period of the drum.*

[41] By definition, the (visible) spectral dimensions of a fractal drum are the poles (within a given window W) of a meromorphic extension of $\zeta_\nu(s)$, the spectral zeta function of this drum. Henceforth, we omit the adjective "visible" when referring to spectral complex dimensions and it is implicitly understood that W is a suitable neighborhood of $\{s : \operatorname{Re} s \geq D\}$.

In addition, a nonlattice drum has spectral complex dimensions arbitrarily close (from the left) to the line $\operatorname{Re} s = D$, *whereas this is not the case for a lattice drum.*

Finally, the spectral complex dimensions of a nonlattice drum are quasiperiodically distributed (in the sense of Theorems 2.17 and 3.6); that is, they can be approximated by those of a suitable sequence of lattice drums.

We point out that it follows from Conjecture 12.42 and from our explicit formulas (namely, from Theorem 5.26) that the conclusion of [Lap3, Conjectures 2 and 3] holds pointwise for the spectral partition function $\theta_\nu(t)$ and distributionally for the eigenvalue counting function $N_\nu(x)$.

Remark 12.43. As was noted in [Lap2–3], it follows easily from Weyl's asymptotic formula with error term ([Lap1, Theorem 2.1], estimate (12.52) above) that for any fractal drum such that $\mathcal{M}^*(D; \partial\Omega) < \infty$ and $D < d$, the spectral zeta function $\zeta_\nu(s)$ has a meromorphic extension to the open half-plane $\operatorname{Re} s > D$, with a single (simple) pole at $s = d$, either for Dirichlet or Neumann boundary conditions.[42] Therefore, all the spectral complex dimensions of the fractal drum other than d itself lie in the closed half-plane $\operatorname{Re} s \leq D$. This is the case, in particular, for the self-similar drums considered in Conjecture 12.42 (under the hypothesis of footnote 42).

Remark 12.44. In view of [Lap3, Conjecture 4, p. 175] and Remark 12.20 above, an entirely analogous conjecture can be made about the spectrum of approximately self-similar fractal drums. (For a discussion of the case of drums with self-similar fractal membrane, see [Lap6, Section 8] which would complete the rigorous results of [KiLap1] regarding an analogue of Weyl's formula for Laplacians on self-similar fractals.)

Remark 12.45. A conjecture more precise than Conjecture 12.42 could be made for the spectrum of self-similar drums. It would be much closer to the statement of Conjecture 12.18 for the geometry of self-similar drums. At this point, however, we prefer to refrain from formulating it until a better understanding of a suitable counterpart of the spectral operator has been obtained in the context of self-similar fractal drums. (Recall that the spectral operator discussed in this book relates the geometric and the spectral complex dimensions of a fractal string or, more generally, of a fractal spray; see Section 6.3.2 and Definition 6.22 in Section 6.6.) Further insight into this difficult problem may be gained by examining the question raised in the next remark and subsection, regarding the relationship between the spectrum and the dynamics of a fractal drum.

---

[42] Here and in Conjecture 12.42 above, only for the Neumann Laplacian, we assume as in [Lap1] or [Lap3, Conjecture 3] that Ω is a quasidisc, which is the case, for instance, for the snowflake domain or its natural generalizations; see [Lap1, Section 4.2.B and Example 5.4] along with [LapPan].

Remark 12.46. In a recent work [Tep1, 2]—motivated in part by some of the results in [RamTo, BessGM, Lap2–3, LapPo2, LapMa2, Ki1, FukSh, KiLap1] and in Chapter 2 of [Lap-vF5]—Alexander Teplyaev has obtained an interesting expression for the spectral zeta function[43] $\zeta_\nu(s)$ of certain drums with self-similar fractal membrane, that is, for Laplacians *on* a self-similar fractal as in, e.g., [Ki1–2, KiLap1–2]. It extends to the present situation the factorization formula (1.38) of Theorem 1.19.[44] Indeed, it is of the form

$$\zeta_\nu(s) = \zeta_{\mathcal{L}}(s)\zeta_{\mathcal{P}}(s), \tag{12.53}$$

where $\zeta_{\mathcal{L}}(s)$ is the geometric zeta function of a (virtual) self-similar string $\mathcal{L}$ (as in Chapters 2 and 3, see, e.g., Equation (2.10) of Theorem 2.4) and $\zeta_{\mathcal{P}}(s)$ is defined in terms of an underlying complex dynamical system (i.e., of the inverse iterates of a suitable polynomial of one complex variable).[45] Also compare (12.53) with the factorization formula for the spectral zeta function of a fractal spray (see Equation (1.46) in Section 1.4 above, along with [LapPo3] and [Lap-vF5, Equation (1.41), p. 22]). In particular, $\zeta_{\mathcal{P}}(s)$ reduces to the Riemann zeta function $\zeta(s)$ (and hence formula (12.53) to (1.38)) in the case corresponding to a self-similar string (with possibly multiple gaps), viewed as a unit interval with a (self-similar) fractal structure. So far, the type of (self-similar) drum with fractal membrane for which formula (12.53) is known to be true is rather limited (besides the one-dimensional self-similar case, it includes the interesting case of the Sierpinski gasket), but it is natural to conjecture [Lap7] that appropriate (and certainly significant) modifications should enable one to considerably broaden its domain of validity. In view of the recent work in [Sab1, 2] extending the decimation method [RamTo, FukSh], the underlying complex dynamics should involve, in general, rational functions of several complex variables. (See [Lap7].)

We note that for the Laplacian on the Sierpinski gasket F, an interesting feature of formula (12.53) is that the zeros of the numerator of $\zeta_{\mathcal{L}}(s)$ cancel the poles of $\zeta_{\mathcal{P}}(s)$, a phenomenon in some sense dual to that encountered in [LapMa1, 2], where, for a suitable fractal string $\mathcal{L}$, the poles of $\zeta_{\mathcal{L}}(s)$ (i.e., the geometric complex dimensions of $\mathcal{L}$) are canceled by the corresponding (critical) zeros of the Riemann zeta function $\zeta(s) = \zeta_{\mathcal{P}}(s)$, resulting in fewer poles of the spectral zeta function $\zeta_\nu(s) = \zeta_{\mathcal{L}}(s)\zeta(s)$, by formula (1.38).[46] It follows [Tep1, 2] that the spectral complex dimensions of F (i.e., the

[43] See Section B.3 of Appendix B for a definition of the spectral zeta function.

[44] Formula (1.38) was observed in [Lap2] and used extensively in later work on fractal strings, including the present book. (See, e.g., [Lap3, LapPo2–3, LapMa2, HeLap2, Lap-vF4–5]).

[45] More precisely, $\zeta_\nu(s)$ may be a suitable finite linear combination of expressions as in (12.53). For simplicity, this fact will be ignored in the rest of this discussion.

[46] Recall that this was the main idea in Chapters 9 and 11.

poles of $\zeta_\nu(s)$) lie on a single vertical line and are of the form $D_\nu + in\mathbf{p}$ (for $n \in \mathbb{Z}$), where $D_\nu = \log 9/\log 5$ is the spectral dimension of F (as defined in [KiLap1] and the relevant references therein, including [FukSh]) and with oscillatory period $\mathbf{p} = 2\pi/\log 5$, as conjectured in [Lap6, §8] (see especially [Lap6, Conjecture 8.2, p. 236, and Example 8.9, p. 239]).

In closing this remark, we mention the recent paper [DerGrVo] on the spectrum of the Laplacian for a class of decimable self-similar fractals including the Sierpinski gasket.[47] In that work, the factorization formula (12.53) is not investigated but results supplementing those of [FukSh, KiLap1, Lap-vF5, Tep1–2] (among other references) are obtained. In particular, the spectral zeta function of the Laplacian is shown to have a meromorphic continuation to all of $\mathbb{C}$ (rather than in a half-plane, as in [Tep1–2]) and the eigenvalue counting function is shown to have oscillations (of leading order), as conjectured in [KiLap1, p. 105] and further discussed in [Lap5] and [Lap6, Section 8].

12.5.3 Spectrum and Periodic Orbits

In [Gut2, Section 16.5 and Chapter 17], it is explained how the spectrum of the Laplacian on a Riemannian manifold is related to the periodic orbits of a particle moving in the manifold. By a trace formula, reminiscent of the Selberg Trace Formula, the spectral partition function—or rather, its quantum-mechanical analogue, a suitable distributional trace of the unitary group $\{e^{it\Delta}\}_{t\in\mathbb{R}}$, where Δ is the Laplacian on the manifold—is expressed as a sum over the periodic orbits. This relationship—which is rather surprising at first and is usually known in the physics literature as the Gutzwiller Trace Formula [Gut1–2]—follows from a heuristic application of the method of stationary phase to a Feynman path integral.[48]

As an example, we consider fractal strings. In this case, all the quantities involved can be worked out explicitly. As in Gutzwiller [Gut2, Section 16.5, formula (16.13)], we show how counting frequencies can be transformed into counting periodic orbits,[49] using the Poisson Summation Formula (see, e.g., [Ser, Section 6.1] or [Lap10, Appendix C]).

Given $t > 0$, let

$$z_{\nu,\mathcal{B}}(t) = \sum_{k\in\mathbb{Z}\setminus\{0\}} e^{-\pi k^2 t} = 2\sum_{k=1}^{\infty} e^{-\pi k^2 t}$$

---

[47] These are necessarily lattice self-similar fractals, in the spectral sense of [KiLap1].

[48] We wish to thank Michael V. Berry for a conversation about this subject. Also see [Berr4].

[49] Note that here we view the Bernoulli string as the unit circle, instead of as the unit interval $(0, 1)$. The periodic orbits of a particle are the orbits of this particle around the circle. Hence we have the basic periodic orbit corresponding to going around once, and repetitions thereof.

be the spectral partition function for the (normalized) squared frequencies (i.e., eigenvalues) of the Bernoulli string. (See footnote 4 on page 401 of Appendix B.) The squared frequencies of the fractal string $\mathcal{L} = \{l_j\}_{j=1}^{\infty}$ are then counted by[50]

$$z_{\nu,\mathcal{L}}(t) = 2\sum_{j=1}^{\infty}\sum_{k=1}^{\infty} e^{-\pi k^2 l_j^{-2} t}.$$

Since, by formula (A.22), we have

$$z_{\nu,\mathcal{B}}(t) + 1 = \frac{1}{\sqrt{t}} \sum_{k\in\mathbb{Z}} e^{-\pi k^2 t^{-1}}, \tag{12.54}$$

we find that

$$\begin{aligned} z_{\nu,\mathcal{L}}(t) &= \sum_{j=1}^{\infty}\left(\frac{l_j}{\sqrt{t}} - 1 + \frac{l_j}{\sqrt{t}} \sum_{k\in\mathbb{Z};\, k\neq 0} e^{-\pi k^2 l_j^2 t^{-1}}\right) \\ &= \frac{L}{\sqrt{t}} + \sum_{j=1}^{\infty}\left(-1 + \frac{l_j}{\sqrt{t}} \sum_{k\in\mathbb{Z};\, k\neq 0} e^{-\pi k^2 l_j^2 t^{-1}}\right), \end{aligned} \tag{12.55}$$

where $L = \sum_{j=1}^{\infty} l_j$ is the total length of the string. We recognize the analogue of the Weyl term, $L/\sqrt{t}$, and the double sum in (12.55) extends over the periodic orbits (of lengths $k \cdot l_j$, $k = 1, 2, \ldots$) of a particle on the string. The convergence of this series is subtle, and can be checked by noting that

$$\frac{l_j}{\sqrt{t}} \sum_{k\in\mathbb{Z};\, k\neq 0} e^{-\pi k^2 l_j^2 t^{-1}} = \int_{-\infty}^{\infty} e^{-\pi x^2}\, dx + O(l_j) = 1 + O(l_j),$$

as $j \to \infty$.

For the frequency counting function itself, we obtain a similar result. Note that the Fourier transform of the characteristic function of $(0, x)$ is

$$\frac{1 - e^{-2\pi i x y}}{2\pi i y}.$$

Thus we find, again by means of the Poisson Summation Formula,

$$\sum_{1\leq k\leq x} 1 = x - \frac{1}{2} + \sum_{k=1}^{\infty} \frac{\sin 2\pi x k}{\pi k}.$$

[50]Recall that the (normalized, nonzero) frequencies of the Laplacian on a circle of length l_j are $k \cdot l_j^{-1}$, with $k \in \mathbb{Z}\backslash\{0\}$. Here, we use a normalization such that the square of these frequencies is equal to $\pi k^2 l_j^{-2}$ (instead of $k^2 l_j^{-2}$), in order to obtain elegant formulas. Further, following the convention used, in particular, in Chapter 1 and in Appendix B, we exclude here the zero frequency.

Hence

$$N_\nu(x) = \sum_{j=1}^{\infty} \sum_{1 \le k \le x l_j} 1 = Lx - \sum_{j=1}^{\infty} \left(\frac{1}{2} - \sum_{k=1}^{\infty} \frac{\sin 2\pi x l_j k}{\pi k} \right). \tag{12.56}$$

We recover the Weyl term, Lx, and a sum over all the periodic orbits. That this series converges is again a subtle matter.

For higher-dimensional fractal sets, a similar method could possibly be applied, although this is still a remote prospect, as we briefly discuss below. Note that when the boundary of a bounded open set is fractal, as is the case for the snowflake drum, the determination of the periodic orbits becomes problematic, since a particle bounces off the boundary in unpredictable directions. The bouncing of a point-particle is not even well defined.

Remark 12.47 (Self-similar fractal drums and fractal billiards). Motivated in part by results or comments from [Berr1–3, Lal2–3, PaPol2; BedKS, esp. Sections 6 and 8] and the relevant references therein, including several of those mentioned at the beginning of this section, the first author has formulated in [Lap3, Conjecture 6, p. 198] a metaconjecture underlying many of the conjectures from Part II of [Lap3] briefly discussed in Sections 12.3 and 12.5.2, both for (approximately) self-similar drums with fractal membrane and with fractal boundary (as in the present situation). Roughly speaking, it suggests that there should exist a suitable dynamical system associated with a given self-similar drum (viewed, say, as a billiard table) such that an analogue of the Selberg (or, more generally, the Gutzwiller) Trace Formula holds in this context. Accordingly, the lattice–nonlattice dichotomy in the spectrum of self-similar drums could be understood as follows: In the nonlattice case, there would exist an invariant ergodic measure with respect to which the periodic orbits of the fractal billiard are equidistributed, whereas in the lattice case, the periodic orbits would be concentrated and their oscillatory behavior (of order D) would be described by their distribution with respect to a suitable measure. Therefore, conjecturally, this would explain dynamically the absence or the presence of oscillations of order D in the spectrum of self-similar drums, in the nonlattice or lattice case, respectively.

As is clear from the discussion in this section, in the present case of drums with fractal boundary, we are still far from being able to formulate such a conjecture precisely, let alone to prove or disprove it, even for the snowflake drum. Nevertheless, this is a challenging problem that appears to be worth investigating further in the future.

12.6 The Complex Dimensions as the Spectrum of Shifts

In the simplest case of the Cantor string, discussed in Section 1.1.2, it is easy to see that a shifted copy, CS $+ x$, overlaps the Cantor string only for shifts over $2 \cdot 3^{-n}$, $n = 0, 1, \ldots$, or suitable combinations of such shifts: $x = \sum_{n=0}^{\infty} a_n 3^{-n}$, $a_n = \pm 2$. For all other values of the real number x, there is no overlap: $\mathrm{CS} \cap (\mathrm{CS} + x) = \emptyset$. Moreover, the Minkowski dimension (or Hausdorff dimension) of the intersection is always $D = \log_3 2$ and its Minkowski content (or Hausdorff measure) is 2^k times smaller than that of the Cantor set itself, where k is the number of nonzero digits of x in the above representation. This information suffices to reconstruct the Cantor set and its complex dimensions.

In general, let F be a self-similar fractal subset of the real line. Let I_ε denote the interval $(-\varepsilon, \varepsilon)$. To measure the overlap of F with a shifted copy of it, we consider the function, for $x \in \mathbb{R}$,

$$f(\varepsilon, x) = \mathrm{vol}_1(F + I_\varepsilon) \cap (F + I_\varepsilon + x),$$

where for two subsets of $\mathbb{R}$, $A + B$ denotes the set $\{a + b \colon a \in A, b \in B\}$, so that $A + I_\varepsilon$ is the open ε-neighborhood of A. We then construct the dimension-like function

$$D(x) = \inf\{d \geq 0 \colon f(\varepsilon, x) = O\big(\varepsilon^{1-d}\big) \text{ as } \varepsilon \to 0^+\}$$

and the upper Minkowski content-like function

$$\mathcal{M}^*(x) = \limsup_{\varepsilon \to 0^+} f(\varepsilon, x) \varepsilon^{D(x)-1}.$$

We expect that these functions exhibit the following behavior. In the lattice case, $D(x) > 0$ for a bounded discrete set of values. The function $\mathcal{M}^*(x)$ is discontinuous at each of these values, and continuous and vanishing on the complement. In the nonlattice case, $D(x) > 0$ for every x in a countable dense subset of a compact connected interval and $\mathcal{M}^*(x)$ is continuous. The complex dimensions of F can be recovered from the functions $D(x)$ and $\mathcal{M}^*(x)$.

12.7 The Complex Dimensions as Geometric Invariants

We close this chapter with a section of a very speculative nature, regarding a possible cohomological interpretation of the complex dimensions of a fractal. In Section 8.2, we pointed out an analogy between the explicit formula for the volume of the tubular neighborhoods of a fractal string

Table 12.1: Homological properties compared with properties of the zeta function.

homology	*zeta function*
Poincaré duality	functional equation
grading by integers $0, 1, \dots, d$	lines of poles at $\operatorname{Re} s = 0, 1/2, \dots, d/2$
eigenvalues of Frobenius in dimension j on a variety over the finite field $\mathbb{F}_q$	q^ω, where ω is a complex dimension with $\operatorname{Re} \omega = j/2$
eigenvalues of the infinitesimal generator of a dynamical system on a manifold	ω, the complex dimensions
Lefschetz fixed point formula	coefficients of the zeta function
functoriality	the zeta function divides another one

and formula (8.23) for the volume of the tubular neighborhoods of a submanifold M of Euclidean space. Similarly, in Appendix B, Remark B.3 and Theorem B.4, it is explained that the principal parts of the spectral zeta function associated with the Laplacian on a Riemannian manifold M have a geometric interpretation and can be expressed in terms of invariants of the cotangent bundle on M; see the formulas (B.7) and (B.8).

Thus both the poles and the principal parts of the spectral zeta function of a manifold have a geometric interpretation in terms of the de Rham complex of the manifold. In part because of this analogy, we want to raise the question of whether a theory of complex cohomology of fractal strings, and, more generally, of fractal drums, could exist. This theory would be based on the geometric and spectral zeta functions of the fractal.

We propose the dictionary in Table 12.1 for translating homological properties in terms of zeta functions. This is motivated in part by analogy with the two situations mentioned above, i.e., the similarity in Chapter 8 between the explicit formula for the volume of the tubular neighborhoods of a fractal string and formula (8.23) for manifolds, as well as the formulas for the spectral zeta function and the spectral partition function of a manifold in Appendix B. It is also suggested by the situation of algebraic varieties over a finite field, along with the associated zeta functions (see Section 11.5) and étale cohomology theory (in particular, the Weil Conjectures [Wei3, 7], [ParSh1, Chapter 4, Sections 1.1–1.3]). This situation bears some similarity with Cantor strings and other lattice strings, especially in the case of curves over a finite field. (See also [Lap10, Chapters 3 and 4].)

Sometimes a stratification into different dimensions will be possible, such as for manifolds, and also for lattice strings. But for nonlattice strings, a stratification will be impossible. It is therefore a problem to know how such a theory would be set up algebraically, since there will not be a chain complex equipped with the usual boundary maps. Instead, the cycles in such a theory could be represented by Dirichlet series with analytic continuation up to a certain line $\operatorname{Re} s = \Theta$, whereas the boundaries would be represented by Dirichlet series that converge at $s = \Theta$ (i.e., with abscissa of convergence less than Θ). As a consequence, a homology class would be represented by a Dirichlet series with analytic continuation up to $\operatorname{Re} s = \Theta$, modulo Dirichlet series that converge there. Thus, only the information contained in the poles with their principal parts remains. In particular, only the asymptotic behavior of the coefficients of the Dirichlet series will be important in representing a complex cohomology class.

Remark 12.48. Recently, a different approach to fractal cohomology—as proposed conjecturally in our earlier book [Lap-vF5, Section 10.5] and further refined in parts of [Lap10] and [Lap-vF9]—was taken by R. Nest and the first author in the work in progress [LapNes4]. For now, it only applies to self-similar geometries. In addition to the theory of complex dimensions of fractal strings developed in [Lap-vF5] and in the present book, it makes use of and significantly extends, in particular, aspects of the higher-dimensional theory, developed in [Pe] and [LapPe1–4]. Perhaps not surprisingly, a large number of zeta functions plays a key role in [LapNes4].

12.7.1 Connection with Varieties over Finite Fields

The analogy between the geometric zeta function of self-similar fractal strings and the zeta function of a variety over a finite field becomes apparent in the following simplest example. The zeta function of the affine line[51] over $\mathbb{F}_p$ is defined as $\zeta_{\mathbb{A}^1/\mathbb{F}_p}(s) = (1 - p \cdot p^{-s})^{-1}$. We compare this with the geometric zeta function of the Cantor string of Example 1.1.2,

$$\zeta_{\mathrm{CS}}(s) = \frac{1}{1 - 2 \cdot 3^{-s}}.$$

Writing $D = \log_3 2 = \log 2 / \log 3$ for the Minkowski dimension of the Cantor string, and setting $p = 3$, we find

$$\zeta_{\mathbb{A}^1/\mathbb{F}_3}(s) = \zeta_{\mathrm{CS}}(s + D - 1).$$

Setting $p = 2$, we find

$$\zeta_{\mathbb{A}^1/\mathbb{F}_2}(s) = \zeta_{\mathrm{CS}}(sD).$$

The poles of these zeta functions form a vertical arithmetic progression on one line, namely, $\omega = 1 + ik\mathbf{p}$ with $k \in \mathbb{Z}$ and $\mathbf{p} = 2\pi / \log 3$ or $\mathbf{p} = 2\pi / \log 2$, respectively.

---

[51] Here, and henceforth, p is a prime number and $\mathbb{F}_p$ denotes the finite field with p elements.

In general, one defines the zeta function of a variety over a finite field as the generating function (or Mellin transform) of the counting measure of the positive divisors on the variety. As such, it is immediately clear that the zeta function can be obtained as an Euler product of factors that are defined in terms of the prime divisors of the variety. A variety over a finite field comes equipped with an action of the Frobenius endomorphism.[52] This defines a discrete-time flow on the variety, the orbits of which are conjugacy classes of points on the variety, which (for curves) are the prime divisors of the variety. The logarithmic derivative of the Euler product is the generating function of the counting measure of the orbits of Frobenius.

One of the most important developments in the theory of algebraic varieties was the definition and subsequent development of a cohomology theory. Indeed, the étale cohomology, which captures the combinatorics of families of étale covers of the variety, provides a theory that can be compared to the classical singular homology and cohomology theories of manifolds. Since the étale theory is defined purely algebraically, it allows application to varieties defined over a finite field. In particular, one recovers the zeta function of the variety as the alternating product of the characteristic polynomials of the induced action of Frobenius on the étale cohomology groups. The poles of this zeta function are located on integer vertical lines $\operatorname{Re} s = 0, 1, \dots, n$, where n is the dimension of the variety, and the zeros are located on half-integer vertical lines $\operatorname{Re} s = \frac{1}{2}, \frac{3}{2}, \dots, n - \frac{1}{2}$. (See, e.g., [Kat, FreKie, Den3, Wei1–2, ??, ParSh1] or [Lap10, Appendix B] for further information about this beautiful subject.)

This theory was first modeled on the theory of the Riemann zeta function, which is the first example of the zeta function of an arithmetic geometry, namely, the spectrum of $\mathbb{Z}$. There exist only one-dimensional[53] arithmetic geometries. The (completed) Riemann zeta function has simple poles at $s = 1$ and $s = 0$, hence one could say (by analogy with the case of a curve over a finite field) that the cohomology groups H^0 and H^2 are one-dimensional; i.e., the spectrum of $\mathbb{Z}$ (completed at the archimedean place) is connected and one-dimensional. On the other hand, if the Riemann hypothesis holds, then the middle cohomology group H^1 could possibly be defined, and should be infinite dimensional. Moreover, the logarithmic derivative of the Riemann zeta function is the generating function of the prime-power counting function. It is not known how the prime numbers p can be viewed as the (primitive) periodic orbits (of length $\log p$) of a flow, but there is some indication that such a flow, if it exists, should take into account the smooth (archimedean) structure [Den3, Haran2]. The simplest such flow is the shift on the real line.

---

[52]Which is induced on the variety by the automorphism $x \mapsto x^p$ of $\mathbb{F}_p$.

[53]Indeed, the product of the spectrum of $\mathbb{Z}$ with itself has coordinate ring $\mathbb{Z} \otimes \mathbb{Z} = \mathbb{Z}$, hence the product reduces to the diagonal [Haran1].

Table 12.2: Self-similar fractal geometries vs. varieties over finite fields: Analogies between the zeta functions.

self-similar geometries	*finite geometries*
lattice string	variety V over the field $\mathbb{F}_q$
nonlattice string $\mathcal{L}$	infinite dimensional variety
Cantor string	affine rational variety
geometric zeta function $\zeta_{\mathcal{L}}$ counts lengths	zeta function ζ_V counts divisors
lattice case is periodic	the zeta function is periodic
period $\mathbf{p}$ is typically large	period is typically small
nonlattice: $\mathbf{p} \to \infty$ (i.e., characteristic $\to 1$)	period of $\zeta_V \to 0$ if $\#\mathbb{F}_q \to \infty$
residue at D (Minkowski dimension) gives $\mathcal{M}(D; \mathcal{L})$	residue at n (dimension) gives the class number

12.7.2 Complex Cohomology of Self-Similar Strings

We propose constructing a cohomology theory, which to each (dynamical) complex dimension ω of such a geometry, would associate a nontrivial cohomology group H^ω (with coefficients in the field of complex numbers). In general, H^ω would be expected to be an infinite dimensional Hilbert space. This should be the case, for example, for the cohomology spaces associated with a nonlattice string (or, more generally, with a nonlattice self-similar geometry). Such a theory might be coined *complex cohomology* or *fractal cohomology*.

Fractal Geometries and Finite Geometries

The geometric zeta function of a fractal string can have complex dimensions of higher multiplicity, as in Example 2.3.4. Hence $\zeta_{\mathcal{L}}(s)$ should be compared to the zeta function of a variety over a finite field, and not for example to its logarithmic derivative. This is confirmed by the fact that the residue at a complex dimension is quite arbitrary, and not an integer in general. Another confirmation is that the logarithmic derivative of the geometric zeta function is the logarithmic derivative of the dynamical zeta function (see Section 7.1 and 7.2 in Chapter 7), which is the generating function of the counting function of the periodic orbits of a dynamical system. This corresponds to the logarithmic derivative of the zeta function of a variety, which is the generating function of the counting function of the periodic orbits of the Frobenius automorphism, as explained in Section 12.7.1. On the other hand, for the simplest self-similar strings (i.e., those with a single gap, as in Section 2.2.1), the geometric zeta function has only poles and no zeros. This would mean that there is only 'even dimensional cohomology' in that case, but the correct interpretation is as yet unclear.

This is explained in the following three tables. Table 12.2 summarizes the analogies between the corresponding zeta functions. We explain the sense in which $\mathbf{p} \to \infty$ for nonlattice strings, and the connection with the characteristic. Recall that the oscillatory period of a lattice string with scaling ratios $r_j = r^{k_j}$, $j = 1, \dots, N$ (for positive integers $k_1, \dots, k_N$ without common factor and $r \in (0,1)$) is $\mathbf{p} = 2\pi/\log r^{-1}$. It is the period of its geometric zeta function: $\zeta_{\mathcal{L}}(s) = \zeta_{\mathcal{L}}(s + i\mathbf{p})$, for $s \in \mathbb{C}$. On the other hand, if $\mathcal{L}$ is a nonlattice string, and $r_j \approx r^{k_j}$ gives a lattice approximation, then $\zeta_{\mathcal{L}}(s)$ and $\zeta_{\mathcal{L}}(s + in\mathbf{p})$ are close, for n not too large. The next (better) lattice approximation to $\mathcal{L}$ has larger values for the integers $k_1, \dots, k_N$ and a value of r that is closer to 1. Hence, its oscillatory period is larger (see Section 3.4 for more details). In that sense, $\mathbf{p} \to \infty$ for nonlattice strings, as $\mathbf{p}$ runs through the values of the oscillatory periods of lattice approximations. For a variety (of dimension n) over the finite field $\mathbb{F}_q$, on the other hand, the zeta function is periodic with period $\mathbf{p} = 2\pi/\log q$. Again, the zeta function of V is periodic: $\zeta_V(s) = \zeta_V(s + i\mathbf{p})$, for $s \in \mathbb{C}$. Here, q is a power of a prime number, the characteristic of the finite field. Hence the period is small typically. However, the limit as $q \downarrow 1$ (i.e., if the characteristic of $\mathbb{F}_q$ tended to 1, if this were possible), corresponds to the limit $r \uparrow 1$ that we observe for nonlattice strings.

Table 12.3: Self-similar fractal geometries vs. varieties over finite fields: Cohomological aspects.

self-similar geometries	*finite geometries*
poles from the scaling ratios	poles from even cohomology
zeros from the gaps	zeros from odd cohomology
lattice case is periodic	zeta function is periodic
nonlattice cohomology is infinite dimensional	the cohomology collapses to a finite-dimensional one

Table 12.3 summarizes the expected properties of a cohomology theory for self-similar fractal strings (and sets), and Table 12.4 presents the dynamical analogies between self-similar flows and the Frobenius flow of a variety.

The ideas in this last section are expanded from those in [Lap-vF5, Section 10.5]. They were first presented in the present form in [Lap-vF9]. We invite the reader to develop them further.

Table 12.4: Self-similar fractal geometries vs. varieties over finite fields: Dynamical aspects.

self-similar geometries	*finite geometries*
dynamical flow	Frobenius flow
dynamical zeta function $-\zeta'_{\mathcal{L}}/\zeta_{\mathcal{L}}$ counts the closed orbits	$-\zeta'_V/\zeta_V$ counts the Frobenius or Galois orbits of points
Euler-type product connects orbits with lengths	Euler product connects orbits with divisors
r^ω (poles) are solutions to (2.11)	q^ω (zeros and poles) are eigenvalues of Frobenius
number of lines is k_N = degree of (2.11)	number of lines is $2n+1$, $n = \dim V$

12.8 Notes

Various extensions of the present theory of complex dimensions are considered in [HamLap] and in several works in preparation. In short, they deal with the following situations: (i) Random (for example, statistically self-similar) fractals [HamLap]; see Section 12.4.1. (ii) Multifractals (nonhomogeneous fractals which typically have a continuum of real fractal dimensions [Fa3, Chapter 17; Man3; Lap-vF10, Part 2]); see [JafLap] and [LapLevRo]. (iii) Self-similar fractals (or systems) in higher dimension, along with related scaling fractals and tilings; see [Pe] and [LapPe2–4], building upon [LapPe1] and the present work, briefly discussed in Section 12.3.2. (iv) Also, along a somewhat different direction, fractal graphs; see [GuIsLap2].

Section 12.3.1: the snowflake curve was introduced by Helge von Koch in [vK1, 2] and has since been used as a prototypical example of self-similar fractal. We refer, for instance, to [Man1, Section II.6] and [Fa3, Introduction and Chapter 9] for a more detailed discussion of the von Koch and snowflake curves. See also, e.g., [Ed] and [Lap8] for additional historical background and references.

As is noted in [LapPe1], a result entirely analogous to Theorem 12.21 can be obtained by the same method for the square (instead of triangular) snowflake curve. Further, the same method as that leading to Theorem 12.21 is expected to eventually yield an inner tube formula for all lattice self-similar fractals and possibly, by the density arguments of Chapters 2 and 3 (see especially Section 3.4), to provide information about the complex dimensions of nonlattice fractals.

Section 12.4.1: random fractals of various types (and not necessarily in $\mathbb{R}$) have been studied earlier from several points of view in a number of papers, including [Fa2, MauWi, PitY, Ham1–2, Gat] and the relevant references therein. Among those papers, [MauWi] and [Ham1–2] also consider certain random recursive constructions. Moreover, we point out that as is well known (see, e.g., [Man1]), random fractals have been used in the applications to obtain more realistic models of natural phenomena (in physics, chemistry, geology and biology) and of certain problems motivated by computer science and electrical engineering. See also [Que] and the references therein for interesting earlier work on random Dirichlet series, with rather different motivations and goals.

A suitable form of the Renewal Theorem, adapted to branching processes, plays an important role in the theory of Section 12.4.1. See [HamLap, Section 3] for details and for the relevant terminology.

Section 12.4.2: in [LapNes2, 3], the authors introduce a model of (generalized) fractal membrane and of the associated moduli space—based on quasicrystals ([Sen, Moo; Lap10, Appendix F])—and a corresponding continuous flow of zeta functions having some of the properties expected in [Lap10].

Section 12.5.1: we note that the analogue of Weyl's formula (without error term) for nonsmooth domains was first obtained in [BiSo].

The one-dimensional case of the Weyl–Berry conjecture was studied in great detail in [LapPo1–2] and [LapMa1–2]. Most of the results in [Lap1, LapPo1–2, LapMa1–2] and [Ca1] are extended in [HeLap1–2] by using a notion of generalized Minkowski content which is defined by means of some suitable gauge functions other than the traditional power functions in measuring the irregularities of the fractal boundary. It may appear at first sight that, even in the one-dimensional case, the work in [HeLap1–2] does not lie within our framework of complex dimensions because it allows, for instance, for logarithmic singularities in the corresponding zeta functions. However, under mild hypotheses, an appropriate use of the functional calculus for self-adjoint operators enables one to reduce this more general situation to the present one and to apply the theory of complex dimensions developed in this book in order to deduce, after a suitable change of variables, the explicit formulas corresponding to the framework of [HeLap2]. This is done in a joint work in preparation by Daniele Guido, Tommaso Isola and the first author [GuIsLap1].

Various extensions of the results of Section 12.5.1 and partial results toward a suitable substitute for the Weyl–Berry Conjecture for drums with fractal boundary have now been obtained in a number of papers, including [Lap2–3, LapPo1–2, LapMa1–2, FlVa, Ger, GerSc1–2, Ca1–2, vB, vB-Le, HuaSl, FlLeVa, LeVa, MolVa, vB-Gi, HeLap1–2]. The interested reader will find in those papers, in particular, several examples of monotonic or oscillatory behavior in the asymptotics of $N_\nu(x)$ or of $\theta_\nu(t)$. The theory presented in this book—once suitably extended to higher dimensions (see, in particular, Sections 12.2, 12.3, 12.5.2 and 12.5.3)—should help shed new light on these examples. We note that many of these examples can be viewed as fractal sprays (as in [LapPo3]) and hence already lie within the scope of the present theory; see, for example, Section 6.6 above.

Section 12.5.3: a sample of physical and mathematical works related to the Gutzwiller Trace Formula includes [Gut1–2, BallanBlo, BalaVor, Berr3, Vor, BerrHow, BraBh] and [Col, Chaz, DuGu].

Appendix A

Zeta Functions in Number Theory

In this appendix we collect some basic facts about zeta functions in number theory to which our theory of explicit formulas can be applied. We refer to [Lan, Chapters VIII and XII–XV], [ParSh1, Chapter 4] and [ParSh2, Chapter 1, §6, Chapter 2, §1.13] for more complete information and proofs.

A.1 The Dedekind Zeta Function

Let K be an algebraic number field of degree d over $\mathbb{Q}$, and let $\mathcal{O}$ be the ring of integers of K. The *norm* of an ideal $\mathfrak{a}$ of $\mathcal{O}$ is defined as the number of elements of the ring $\mathcal{O}/\mathfrak{a}$:

$$N\mathfrak{a} = \#\mathcal{O}/\mathfrak{a}. \tag{A.1}$$

The norm is multiplicative: $N(\mathfrak{a}\mathfrak{b}) = N\mathfrak{a} \cdot N\mathfrak{b}$. Furthermore, an ideal has a unique factorization into prime ideals,

$$\mathfrak{a} = \mathfrak{p}_1^{a_1} \dots \mathfrak{p}_k^{a_k}, \tag{A.2}$$

where $\mathfrak{p}_1, \dots, \mathfrak{p}_k$ are prime ideals.

We denote by r_1 the number of real embeddings $K \to \mathbb{R}$ and by r_2 the number of pairs of complex conjugate embeddings $K \to \mathbb{C}$. Thus we have that $r_1 + 2r_2 = d$. Further, w stands for the number of roots of unity contained in K. Associated with K, we need the *discriminant* $\operatorname{disc}(K)$, the *class number* h, and the *regulator* $\mathfrak{R}$. We refer to [Lan, p. 64 and p. 109] for the definition of these notions.

The *Dedekind zeta function* of K is defined for $\operatorname{Re} s > 1$ by

$$\zeta_K(s) = \sum_{\mathfrak{a}} (N\mathfrak{a})^{-s} = \sum_{n=1}^{\infty} A_n n^{-s}. \tag{A.3}$$

Here, $\mathfrak{a}$ runs over the ideals of $\mathcal{O}$, and, in the second expression, A_n denotes the number of ideals of $\mathcal{O}$ of norm n. This function has a meromorphic continuation to the whole complex plane, with a unique (simple) pole at $s = 1$, with residue

$$\frac{2^{r_1}(2\pi)^{r_2} h\mathfrak{R}}{w\sqrt{\operatorname{disc}(K)}}. \tag{A.4}$$

(See [Lan, Theorem 5 and Corollary, p. 161].)

From unique factorization and the multiplicativity of the norm, we deduce the Euler product for $\zeta_K(s)$: for $\operatorname{Re} s > 1$,

$$\zeta_K(s) = \prod_{\mathfrak{p}} \frac{1}{1-(N\mathfrak{p})^{-s}}, \tag{A.5}$$

where $\mathfrak{p}$ runs over the prime ideals of $\mathcal{O}$.

Example A.1 (The case $K = \mathbb{Q}$). Ideals of $\mathbb{Z}$ are generated by positive integers, and the norm of the ideal $n\mathbb{Z}$ is n, for $n \geq 1$. Hence,

$$\zeta_{\mathbb{Q}}(s) = \sum_{n=1}^{\infty} n^{-s} = \zeta(s), \tag{A.6}$$

so that the Dedekind zeta function of $\mathbb{Q}$ is the Riemann zeta function. The Euler product for $\zeta_{\mathbb{Q}}(s)$ is given by

$$\zeta_{\mathbb{Q}}(s) = \prod_{p} \frac{1}{1-p^{-s}}, \tag{A.7}$$

where p runs over the rational prime numbers.

A.2 Characters and Hecke L-series

Let χ be an ideal-character of $\mathcal{O}$, belonging to the cycle $\mathfrak{c}$ (see [Lan, Chapter VIII, §3 and Chapter VI, §1] for a complete explanation of the terms). The *L-series* associated with χ is defined by

$$L_{\mathfrak{c}}(s,\chi) = \sum_{(\mathfrak{a},\mathfrak{c})=1} \chi(\mathfrak{a})\,(N\mathfrak{a})^{-s}. \tag{A.8}$$

By the multiplicativity of the norm, this function has an Euler product

$$L_{\mathfrak{c}}(s,\chi)=\prod_{\mathfrak{p}\nmid\mathfrak{c}}\frac{1}{1-\chi(\mathfrak{p})\,(N\mathfrak{p})^{-s}}. \tag{A.9}$$

This zeta function can be completed with factors corresponding to the divisors of $\mathfrak{c}$, to obtain a function $L(s,\chi)$, called the *Hecke L-series* associated with χ. It is related to the Dedekind zeta function as follows: Let L be the class field associated with an ideal class group of $\mathcal{O}$, and let ζ_L be the Dedekind zeta function of L. Let χ run over the characters of the ideal class group. Then,

$$\zeta_L(s)=\prod_{\chi}L(s,\chi). \tag{A.10}$$

Example A.2 (The case $K=\mathbb{Q}$). A multiplicative function χ on the positive integers gives rise to a *Dirichlet L-series*

$$L(s,\chi)=\sum_{n=1}^{\infty}\chi(n)n^{-s}. \tag{A.11}$$

The value $\chi(n)$ only depends on the class of n modulo a certain positive integer c. The minimal such c is called the *conductor* of χ.

A.3 Completion of L-Series, Functional Equation

The fundamental property of the Dedekind zeta function and of the L-series is that it can be completed to a function that is symmetric about $s=\frac{1}{2}$. Let $\Gamma(s)$ be the gamma function. Let

$$\zeta_{\mathbb{R}}(s)=\pi^{-s/2}\Gamma\left(s/2\right) \quad \text{and} \quad \zeta_{\mathbb{C}}(s)=(2\pi)^{-s}\Gamma\left(s\right).$$

Denote by $\operatorname{disc}(K)$ the discriminant of K. Then the function

$$\xi_K(s)=\operatorname{disc}(K)^{s/2}\left(\zeta_{\mathbb{R}}(s)\right)^{r_1}\left(\zeta_{\mathbb{C}}(s)\right)^{r_2}\zeta_K(s) \tag{A.12}$$

has a meromorphic continuation to the whole complex plane, with simple poles located only at $s=1$ and $s=0$, and it satisfies the functional equation

$$\xi_K(1-s)=\xi_K(s). \tag{A.13}$$

We deduce from this key fact that the function

$$\psi_K(\sigma):=\limsup_{t\to\infty}\frac{\log\left|\zeta_K\left(\sigma+it\right)\right|}{\log t}, \tag{A.14}$$

defined for $\sigma \in \mathbb{R}$, is given by the following simple formula for $\sigma \notin (0,1)$:

$$\psi_K(\sigma) = \begin{cases} 0, & \text{for } \sigma \geq 1, \\ \dfrac{d}{2}(1-2\sigma), & \text{for } \sigma \leq 0. \end{cases}$$

It is known that ψ_K is convex on the real line. (The Lindelöf hypothesis says that $\psi_K(1/2) = 0$.) We deduce the following property (see [Lan, Chapter XIII, §5]): For every real number σ and $\varepsilon > 0$, there exists a constant C, depending on σ and ε, such that for all real numbers t with $|t| > 1$,

$$|\zeta_K(\sigma + it)| \leq C|t|^{\psi_K(\sigma)+\varepsilon}. \tag{A.15}$$

Thus $\zeta_K(s)$ satisfies the hypotheses **L1** and **L2** of Chapter 5. (See also Remark A.5 below.)

The formalism required to prove for general L-series the functional equation (A.13) and the estimate (A.15) about the growth along vertical lines was developed in Tate's thesis [Ta]. We refer to [Lan, Chapter XIV] for the corresponding results.

Remark A.3. Our theory also applies to the more general L-series associated with nonabelian representations of $\mathbb{Q}$, such as those considered in [RudSar]. (The abelian case corresponds to the Hecke L-series discussed above.) These L-series can also be completed at infinity, and they satisfy a functional equation, relating the zeta function associated to a given representation with the zeta function associated to the contragredient representation. Moreover, they have an Euler product representation, much like that of $L(s,\chi)$, except that the p-th Euler factor may be a polynomial in p^{-s} of degree larger than one.

We note that these zeta functions are called primitive L-series in [RudSar]. According to the Langlands Conjectures, they are the building blocks of the most general L-series occurring in number theory. See, for example, [Gel, KatSar, RudSar].

A.4 Epstein Zeta Functions

A natural generalization of the Riemann zeta function is provided by the Epstein zeta functions [Ep].[1] Let $q = q(\mathbf{x})$ be a positive definite quadratic form of $\mathbf{x} \in \mathbb{R}^d$, with $d \geq 1$. Then the associated *Epstein zeta function* is

[1]See, for example, [Ter, Section 1.4] for detailed information about these functions. However, we use the convention of [Lap2, §4], which is different from the traditional one used in [Ter].

defined by

$$\zeta_q(s) = \sum_{\mathbf{n} \in \mathbb{Z}^d \setminus \{0\}} q(\mathbf{n})^{-s/2}. \tag{A.16}$$

It can be shown that $\zeta_q(s)$ has a meromorphic continuation to all of $\mathbb{C}$, with a simple pole at $s = d$ having residue $\pi^{d/2}(\det q)^{-1/2}\Gamma\left(d/2\right)^{-1}$. Further, $\zeta_q(s)$ satisfies a functional equation analogous to that satisfied by $\zeta(s)$; namely, for the completed zeta function

$$\xi_q(s) = \pi^{-s/2}\Gamma\left(s/2\right)\zeta_q(s),$$

we have

$$\xi_q(s) = (\det q)^{-1/2}\xi_{q^{-1}}(d-s), \tag{A.17}$$

where q^{-1} is the positive definite quadratic form associated with the inverse of the matrix of q. (See [Ter, Theorem 1, p. 59].)

In the important special case when $q(x) = x_1^2 + \cdots + x_d^2$, the corresponding Epstein zeta function is denoted $\zeta_d(s)$ in Section 1.4, Equation (1.47), and can be viewed as a natural higher-dimensional analogue of the Riemann zeta function $\zeta(s)$. Indeed, we have $\zeta_1(s) = 2\zeta(s)$. (See, for example, [Lap2, §4] and the end of Section 1.4 above.) Since in this case, q is associated with the d-dimensional identity matrix, we have $q^{-1} = q$ and $\det q = 1$, so that (A.17) takes the form of a true functional equation relating $\zeta_d(s)$ and $\zeta_d(d-s)$. Moreover, $\zeta_d(s)$ has an Euler product like that of $\zeta(s)$ if and only if $d = 1$, 2, 4 or 8, corresponding to the real numbers, the complex numbers, the quaternions and the octaves, respectively. (See, e.g., [Bae].)

Remark A.4. More generally, one can also consider the Epstein-like zeta functions considered in [Es1]. Such Dirichlet series are associated with suitable homogeneous polynomials of degree greater than or equal to one.

Remark A.5. For all the zeta functions in Sections A.1–A.4, hypotheses **L1** and **L2**$'$ (Equations (5.19) and (5.21), page 143) are satisfied with a window W equal to all of $\mathbb{C}$. Moreover, for the example from Section A.6 below, one can take W to be a right half-plane of the form $\operatorname{Re} s \geq \sigma_0$, for a suitable $\sigma_0 > 0$.

A.5 Two-Variable Zeta Functions

Let C be an algebraic curve over a finite field, as in Section 11.5. In that section, we introduced the zeta function of C, $\zeta_C(s)$. In the paper [Pel], Pellikaan introduced a zeta function $\zeta_C(s,t)$ that specializes to $\zeta_C(s)$ for $t = 1$. Later, Schoof and van der Geer introduced the analogue for the integers, inspired by their work on positivity for Arakelov divisors [SchoG]. We present here a brief summary of their results.

A.5.1 The Zeta Function of Pellikaan

We consider a complete nonsingular curve C over the finite field with q elements $\mathbb{F}_q$. A divisor of C is a formal sum of valuations of $k = \mathbb{F}_q(C)$, the field of functions on C. In particular, the divisor of a function f is

$$(f) = \sum_v \operatorname{ord}(f, v)v,$$

where the sum is over all valuations of k. The degree of a divisor

$$\mathfrak{D} = \sum_v D_v v$$

is given by

$$\deg \mathfrak{D} = \sum_v D_v \deg v,$$

where $\deg v$ is the dimension of the residue class field at v over $\mathbb{F}_q$. Two divisors are said to be (linearly) equivalent if their difference is the divisor of a function. Let $\mathrm{Cl} = \mathrm{Cl}(C)$ be the group of divisor classes of C. This group has a grading by the degree, since $\deg(f) = 0$ for any nonzero function f on C, and we write Cl_n for the subset[2] of classes of degree n. A divisor is said to be *positive* if $D_v \geq 0$ for all v. Let $l(\mathfrak{D})$ denote the dimension over $\mathbb{F}_q$ of the vector space of functions f such that $\mathfrak{D} + (f) \geq 0$. We will need the theorem of Riemann–Roch,

$$l(\mathfrak{D}) = \deg \mathfrak{D} + 1 - g + l(\mathfrak{K} - \mathfrak{D}),$$

where $\mathfrak{K}$ is the canonical divisor of C, and $g = l(\mathfrak{K})$ is the genus of C. It follows that $l(\mathfrak{D}) = 0$ for $\deg \mathfrak{D} < 0$, and $l(\mathfrak{D}) = \deg \mathfrak{D} + 1 - g$ for $\deg \mathfrak{D} > \deg \mathfrak{K} = 2g - 2$.

Define, for $\operatorname{Re} t < \operatorname{Re} s < 0$, the function of two complex variables

$$\zeta_C(s, t) = \frac{1}{q^t - 1} \sum_{\mathfrak{D} \in \mathrm{Cl}} q^{tl(\mathfrak{D})} q^{-s(\deg \mathfrak{D} + 1 - g)}. \tag{A.18}$$

We first derive an expression for this function that converges for all s and t. Note that for $g = 0$, that is, when $C = \mathbb{P}^1$ is the projective line, as discussed in Example 11.24 for the case of the one-variable zeta functions, we have $l(\mathfrak{D}) = 0$ for $\deg \mathfrak{D} < 0$ and $l(\mathfrak{D}) = \deg \mathfrak{D} + 1$ for $\deg \mathfrak{D} \geq 0$. Also, there is only one divisor class of degree 0; i.e., $h = 1$. Hence the sum over the divisor classes in (A.18) becomes, for $\operatorname{Re} t < \operatorname{Re} s < 0$,

$$\sum_{n=-\infty}^{-1} q^{-s(n+1)} + \sum_{n=0}^{\infty} q^{t(n+1)} q^{-s(n+1)}.$$

---

[2]Thus Cl is isomorphic to the product $\mathrm{Cl}_0 \times \mathbb{Z}$. It is known that Cl_0 is a finite group. We denote its order by h.

Thus we find

$$\zeta_{\mathbb{P}^1}(s,t) = \frac{1}{(1-q^{t-s})(q^s-1)}.$$

In general, we write $h = \#\mathrm{Cl}_0 = \#\mathrm{Cl}_n$ for the class number of C (see also Section 11.5). Then

$$\zeta_C(s,t) = \sum_{n=-\infty}^{\infty} \sum_{\mathfrak{D}\in\mathrm{Cl}_n} \frac{q^{tl(\mathfrak{D})} - q^{t\max\{0,n+1-g\}}}{q^t-1} q^{-s(n+1-g)} + \frac{h}{q^t-1} \sum_{n=-\infty}^{\infty} q^{t\max\{0,n+1-g\}} q^{-s(n+1-g)}.$$

The first sum is finite, and the second sum equals $h\zeta_{\mathbb{P}^1}(s,t)$. Hence we find

$$\zeta_C(s,t) = \sum_{n=-\infty}^{\infty} \sum_{\mathfrak{D}\in\mathrm{Cl}_n} \frac{q^{tl(\mathfrak{D})} - q^{t\max\{0,n+1-g\}}}{q^t-1} q^{-s(n+1-g)} + h\zeta_{\mathbb{P}^1}(s,t). \tag{A.19}$$

We see that $\zeta_C(s,t)$ has poles at $q^s = 1$ and at $q^s = q^t$. Using the Riemann–Roch formula, we can continue to verify that $\zeta_C(s,t) = \zeta_C(t-s,t)$. Moreover, we see from (A.19) that $q^{-sg}\zeta_C/\zeta_{\mathbb{P}^1}$ is a polynomial in q^{-s} and q^t of degree $2g$ in q^{-s}.

Remark A.6. In [Na], Naumann proves that this polynomial is irreducible. Thus the zeros of $\zeta_C(s,t)$ lie in an irreducible family. It is well known that for $t = 1$, $\zeta_C(s,t)$ satisfies the Riemann hypothesis; i.e., all zeros in s of $\zeta_C(s,1)$ have real part $1/2$.

To relate this function to the divisors of C (and not to the classes of divisors), and hence derive the Euler product for $\zeta_C(s,1)$, we use that

$$\frac{q^{l(\mathfrak{D})}-1}{q-1} \tag{A.20}$$

equals the number of positive divisors in the divisor class $\mathfrak{D}$. Note that

$$\sum_{n=0}^{\infty} \frac{q^{t\max\{0,n+1-g\}}-1}{q^t-1} q^{-s(n+1-g)} = \zeta_{\mathbb{P}^1}(s,t).$$

Hence we find for $\mathrm{Re}\, s > \max\{0, \mathrm{Re}\, t\}$ that

$$\zeta_C(s,t) = \sum_{n=0}^{\infty} \sum_{\mathfrak{D}\in\mathrm{Cl}_n} \frac{q^{tl(\mathfrak{D})}-1}{q^t-1} q^{-s(n+1-g)}. \tag{A.21}$$

Hence by (A.20), for $t = 1$, we find the Euler product

$$\zeta_C(s,1) = q^{s(g-1)} \prod_v \frac{1}{1 - q^{-s \deg v}}.$$

The value $t = 0$ is also interesting. Taking the limit as $t \to 0$ in Equation (A.21), we find that

$$\zeta_C(s,0) = \sum_{n=0}^{\infty} \sum_{\mathfrak{D} \in \mathrm{Cl}_n} l(\mathfrak{D}) q^{-s(n+1-g)}.$$

A.5.2 The Zeta Function of Schoof and van der Geer

An Arakelov divisor is a formal linear combination of valuations of $\mathbb{Q}$, where the archimedean valuation has a real coefficient and the p-adic valuations have an integer coefficient. For example, the divisor of a number f is

$$(f) = -(\log |f|) v_\infty + \sum_p \operatorname{ord}(f,p) v_p,$$

where the sum is over all prime numbers. The degree of a divisor

$$\mathfrak{D} = \sum_v D_v v$$

is given by

$$\deg \mathfrak{D} = D_\infty + \sum_p D_p \log p,$$

which is a real number, not necessarily an integer. Clearly, the group of divisor classes of $\mathbb{Q}$ is isomorphic to $\mathbb{R}$. Thus, there is only the grading by the degree, and $h = 1$. Let

$$\theta(u) = \sum_{n=-\infty}^{\infty} e^{-\pi n^2 u^2}.$$

We use this function to measure positivity of a divisor at the archimedean valuation. A divisor is said to be *positive* if $D_p \geq 0$ for all p-adic valuations, and $\log \theta\left(e^{-2D_\infty}\right)$ is large.[3] We have the theorem of Riemann–Roch,

$$\log \theta(1/u) = \log u + \log \theta(u), \tag{A.22}$$

for $u > 0$. This is proved using the Poisson Summation Formula; see [Ti, Section 2.3], [Pat, Theorem 2.2] or [Schw1, Eq. (VII.7.5)]. It follows that

$$\theta(u) = 1 + O\left(e^{-\pi u^2}\right) \quad \text{for } u \to \infty,$$

[3] Positivity is not a definite notion: the larger $\log \theta\left(e^{-2D_\infty}\right)$, the more positive the divisor is said to be.

and

$$\theta(u) = u^{-1} + O\Big(u^{-1}e^{-\pi/u^2}\Big) \quad \text{for } u \to 0.$$

Define, for $\operatorname{Re} t < \operatorname{Re} s < 0$,

$$\zeta_{\mathbb{Z}}(s,t) = \frac{1}{t}\int_0^\infty \theta(u)^t u^s \,\frac{du}{u}. \tag{A.23}$$

We first derive an expression that converges for all s and t:

$$\zeta_{\mathbb{Z}}(s,t) = \frac{1}{t}\int_0^\infty \left(\theta(u)^t - \max\{1/u,1\}^t\right) u^s \,\frac{du}{u} \\ + \frac{1}{t}\left(\int_0^1 u^{s-t}\,\frac{du}{u} + \int_1^\infty u^s\,\frac{du}{u}\right).$$

The first term converges for every s and t, and the second is defined for $\operatorname{Re} t < \operatorname{Re} s < 0$, but we can easily compute it to find

$$\zeta_{\mathbb{Z}}(s,t) = \frac{1}{t}\int_0^\infty \left(\theta(u)^t - \max\{1/u,1\}^t\right) u^s \frac{du}{u} + \frac{1}{s(s-t)}, \tag{A.24}$$

for every $s, t \in \mathbb{C}$. We see that $\zeta_{\mathbb{Z}}(s,t)$ has poles at $s = 0$ and at $s = t$. Using the Riemann–Roch formula, we can continue to verify the functional equation $\zeta_{\mathbb{Z}}(s,t) = \zeta_{\mathbb{Z}}(t-s,t)$.

Remark A.7. In analogy with the geometric case, we might conjecture that the zeros of the Riemann zeta function lie in an irreducible family.

To relate this function to the divisors of $\mathbb{Q}$, and hence derive the Euler product for $\zeta_{\mathbb{Z}}(s,1)$, we use that $\theta(u) - 1$ 'equals' the number of positive divisors in the same divisor class as u. Note that

$$\int_0^\infty \frac{\max\{1/u,1\}^t - 1}{t} u^s \,\frac{du}{u} = \frac{1}{s(s-t)}.$$

Hence we find for $\operatorname{Re} s > \max\{0, \operatorname{Re} t\}$ that

$$\zeta_{\mathbb{Z}}(s,t) = \int_0^\infty \frac{\theta(u)^t - 1}{t} u^s \,\frac{du}{u}.$$

For $t = 1$, we find the Euler product to be

$$\zeta_{\mathbb{Z}}(s,1) = \pi^{-s/2}\Gamma\left(\frac{s}{2}\right)\prod_p \frac{1}{1-p^{-s}}.$$

There is also an Euler product for $t = 2$, 4 and 8, corresponding to the Gaussian integers, the quaternions and the octonions, respectively. Surprisingly, there is also an Euler product for $t = 0$, as we verify directly. Taking the limit as $t \to 0$, we find that

$$\zeta_{\mathbb{Z}}(s,0) = \int_0^\infty (\log\theta(u))\, u^s \,\frac{du}{u},$$

for $\operatorname{Re} s > 0$. Using the Jacobi triple product identity [HardW, Theorem 352, p. 282],

$$\sum_{n=-\infty}^{\infty} q^{n^2} = \prod_{n=1}^{\infty} (1+q^{2n-1})^2(1-q^{2n}),$$

we can compute

$$\log\left(\sum_{n=-\infty}^{\infty} q^{n^2}\right) = 2\sum_{n=1}^{\infty}(-1)^{n-1}q^n\sigma_{-1}(n|n|_2),$$

where $|n|_2 = 2^{-\operatorname{ord}(n,2)}$ denotes the 2-adic valuation, so that $n|n|_2$ is the largest odd factor of n, and $\sigma_{-1}(n) = \sum_{d\,|\,n} \frac{1}{d}$. We thus obtain

$$\zeta_{\mathbb{Z}}(s,0) = \pi^{-s/2}\Gamma(s/2)\zeta(s/2)\zeta(1+s/2)(1-2^{1-s/2})(1-2^{-1-s/2}),$$

where $\zeta(s)$ is the Riemann zeta function. Substituting the Euler product for the Riemann zeta function, we obtain the Euler product for $\zeta_{\mathbb{Z}}(s,0)$, valid for $\operatorname{Re} s > 2$. Using the functional equation for the Riemann zeta function,

$$\Gamma(s)\Gamma(-s) = \frac{\pi}{-s\sin \pi s},$$

the doubling formula

$$2^{2s-1}\Gamma(s)\Gamma(s+1/2) = \Gamma(2s)\Gamma(1/2),$$

and the fact that $\Gamma(1/2) = \sqrt{\pi}$, we obtain the alternative expression

$$\zeta_{\mathbb{Z}}(s,0) = \zeta(s/2)\zeta(-s/2)(1-2^{-1-s/2})(1-2^{-1+s/2})\frac{4\pi}{s\sin(\pi s/4)},$$

which shows clearly that $\zeta_{\mathbb{Z}}(-s,0) = \zeta_{\mathbb{Z}}(s,0)$.

Remark A.8. It would be interesting to consider two-variable dynamical zeta functions in the context of Chapter 7, in the spirit of [Lag2]. We hope to do so in some future work.

A.6 Other Zeta Functions in Number Theory

The flexibility of our theory of explicit formulas with an error term allows us to apply it to other zeta functions that do not necessarily satisfy a functional equation.

As an example, we mention the zeta function

$$\mathcal{P}(s) = \sum_p (\log p)\, p^{-s},$$

which was studied by M. van Frankenhuijsen in [vF2, §3.9] in connection with the ABC conjecture. To obtain information about this function, we consider the logarithmic derivative of the Euler product of the Riemann zeta function,

$$-\frac{\zeta'(s)}{\zeta(s)} = \sum_{p,m} (\log p)\, p^{-ms},$$

where p ranges over the rational primes (as above), and m over the positive integers. This function has simple poles at $s = 1$ and at each zero of $\zeta(s)$. By Möbius inversion,

$$\begin{aligned} \mathcal{P}(s) &= -\frac{\zeta'(s)}{\zeta(s)} + \frac{\zeta'(2s)}{\zeta(2s)} + \frac{\zeta'(3s)}{\zeta(3s)} + \frac{\zeta'(5s)}{\zeta(5s)} - \frac{\zeta'(6s)}{\zeta(6s)} + \dots \\ &= \sum_{n=1}^{\infty} \mu(n) \left(-\frac{\zeta'(ns)}{\zeta(ns)} \right), \end{aligned}$$

where $\mu(n)$ is the Möbius function, defined on the positive integers by

$$\mu(n) = \begin{cases} 0, & \text{if } n \text{ is not square-free,} \\ (-1)^k, & \text{if } n = p_1 \dots p_k \text{ is square-free.} \end{cases}$$

Thus the poles of $\mathcal{P}(s)$ are contained in the set

$$\{s/n \colon s \text{ is a zero of } \zeta, n = 1, 2, \dots\}.$$

If the Riemann hypothesis holds, then each of these points is a pole of $\mathcal{P}$, and in general, the number of possible cancelations is small. It follows that the poles of $\mathcal{P}(s)$ accumulate on the line $\operatorname{Re} s = 0$. Hence this line is a natural boundary for the analytic continuation of $\mathcal{P}(s)$. But on a line $\operatorname{Re} s = \sigma > 0$ that does not meet any of the poles of $\mathcal{P}(s)$, this function is bounded by a constant times $\log |t|$, for $t = \operatorname{Im} s > 2$. Hence this function satisfies hypotheses **L1** and **L2** (Equations (5.19) and (5.20), page 143) with $W = \{s \colon \operatorname{Re} s \geq \sigma_0\}$, where $\sigma_0 > 0$ is suitably chosen. For example, if the Riemann hypothesis is true, one can take any positive value for σ_0 other than $1/n$, for $n = 1, 2, \dots$.

Appendix B
Zeta Functions of Laplacians and Spectral Asymptotics

In this appendix, we provide a brief overview of some of the results from spectral geometry that are relevant to the study of the spectral zeta function associated with a Laplacian Δ on a smooth compact Riemannian manifold M. For the simplicity of exposition, we focus on the case when M is a closed manifold (i.e., without boundary). However, as is briefly explained at the end of the appendix, all the results stated for closed manifolds are known to have a suitable counterpart for the case of a compact manifold with boundary. An important special case of the latter situation is that when M is a smooth bounded open set in Euclidean space $\mathbb{R}^d$ and $\Delta = -\sum_{k=1}^{d} \partial^2/\partial x_k^2$ is the associated Dirichlet or Neumann Laplacian.

By necessity of concision, our presentation is somewhat sketchy and imprecise. For a much more detailed treatment of these matters, we refer the interested reader to some of the articles and books cited below, including [Min1–2, MinPl, Kac, McKSin, Se1, BergGM, AtPSin, Gi, Hö1–3, Gru, AndLap1–2], along with the relevant references therein.

B.1 Weyl's Asymptotic Formula

Let M be a closed, d-dimensional, smooth, compact and connected Riemannian manifold. We assume throughout that the closed manifold M is equipped with a fixed Riemannian metric g. Let Δ be the (positive) Laplacian (or Laplace–Beltrami operator) on M associated with g.[1] It is well known that Δ has a discrete (frequency) spectrum, written in increasing order according to multiplicity:

$$0 < f_1 \leq f_2 \leq \cdots \leq f_j \leq \ldots,$$

where $f_j \to +\infty$ as $j \to \infty$. Here, by convention, the frequencies of Δ are defined as the square root of its eigenvalues λ_j.[2,3]

Next, let $N_\nu(x) = N_{\nu,M}(x)$ be the associated *spectral counting function* (or counting function of the frequencies):

$$N_\nu(x) = \#\{j \geq 1 \colon f_j \leq x\}, \quad \text{for } x > 0. \tag{B.1}$$

Then Weyl's classical asymptotic formula [Weyl1–2] states that

$$N_\nu(x) = c_d \operatorname{vol}(M) x^d + o\left(x^d\right), \tag{B.2}$$

as $x \to \infty$, where $c_d = (2\pi)^{-d}\mathcal{B}_d$ and $\mathcal{B}_d$ denotes the volume of the unit ball in $\mathbb{R}^d$. Recall that $\mathcal{B}_d = \pi^{d/2}/\Gamma(d/2+1)$, where $\Gamma = \Gamma(s)$ is the usual gamma function. Further, $\operatorname{vol}(M)$ denotes the Riemannian volume of M.

The leading term in (B.2),

$$W(x) = W_M(x) = (2\pi)^{-d}\mathcal{B}_d \operatorname{vol}(M) x^d, \tag{B.3}$$

is often referred to as the Weyl term in the literature.

Remark B.1. Weyl's original result has been improved in various ways. One extension consists in giving a (sharp) remainder estimate for Weyl's asymptotic law, of the form

$$N_\nu(x) = c_d \operatorname{vol}(M) x^d + O\left(x^{d-1}\right), \tag{B.4}$$

as $x \to \infty$. This result is due to Hörmander [Hö1] in the case of closed manifolds, and to Seeley [Se4–5] (for $d \leq 3$) or to Pham The Lai [Ph] (for $d \geq 4$) in the case of manifolds with boundary (for example, for the Dirichlet or Neumann Laplacian on a smooth bounded open set in $\mathbb{R}^d$). We refer the interested reader to [Hö2–3] for a detailed exposition of these results.

---

[1]Using Einstein's summation convention, Δ is given in local coordinates by

$$\Delta = -\frac{1}{\sqrt{\det g}}\frac{\partial}{\partial x_\alpha}\, g^{\alpha\beta}\sqrt{\det g}\,\frac{\partial}{\partial x_\beta},$$

where $g = \left(g_{\alpha\beta}\right)_{\alpha,\beta=1}^d$ and $g^{-1} = \left(g^{\alpha\beta}\right)_{\alpha,\beta=1}^d$.

[2]We have used a slightly different normalization in the rest of this book; see, for example, Section 1.3 and footnote 1 of the introduction.

[3]Throughout this discussion, we ignore the zero eigenvalue of the Neumann Laplacian. Alternatively, we can replace Δ by $\Delta + \alpha$, for some positive constant α.

B.2 Heat Asymptotic Expansion

We denote by $z_\nu(t) = z_{\nu,M}(t)$ the trace of the heat semigroup $\{e^{-t\Delta} : t \geq 0\}$ generated by Δ.[4] Thus, $z_\nu(t)$ is given by

$$z_\nu(t) = \operatorname{Trace}\left(e^{t\Delta}\right) = \sum_{j=1}^{\infty} e^{-t\lambda_j}, \tag{B.5}$$

for every $t > 0$.

A well-known Tauberian argument shows that Weyl's formula (B.2) is equivalent to the following asymptotic formula for $z_\nu(t)$ (see, for example, [Kac] or [Sim]):

$$z_\nu(t) = e_d \operatorname{vol}(M) t^{-d/2} + o\big(t^{-d/2}\big), \tag{B.6}$$

as $t \to 0^+$, where $e_d = \Gamma(d/2+1)c_d$. Using the above expression for c_d and $\mathcal{B}_d$, one finds $e_d = (4\pi)^{-d/2}$.

Remark B.2. The fact that (B.2) implies (B.6) is immediate and follows from a simple Abelian argument; see, e.g., [Sim, Theorem 10.2, p. 107] or [Lap1, Appendix A, pp. 521–522]. However, the converse relies on Karamata's Tauberian Theorem [Sim, Theorem 10.3, p. 108] (which is closely related to the Wiener–Ikehara Tauberian Theorem [Pos, Section 27, pp. 109–112; Shu, Theorem 14.1, p. 115]). In addition, even the existence of an error term in (B.6) does not imply the corresponding Weyl formula with error term in (B.2).[5]

More generally, a key result due in its original form to Minakshisundaram and Pleijel [MinPl] (building, in particular, on work of the first of these authors [Min1–2] in a closely related context) states that $z_\nu(t)$ has the following asymptotic expansion (in the sense of Poincaré):[6]

$$z_\nu(t) \sim \sum_{k\geq 0} \alpha_k t^{-(d-k)/2}, \tag{B.7}$$

---

[4] For convenience, we use here for $z_\nu(t)$ the standard convention encountered in the literature on spectral geometry; that is, we work with the trace of $e^{-t\Delta}$ rather than of $e^{-t\sqrt{\Delta}}$. The latter choice would correspond to the spectral partition function

$$\theta_\nu(t) = \operatorname{Trace}\left(e^{-t\sqrt{\Delta}}\right) = \sum_{j=1}^{\infty} e^{-tf_j} = \int_0^\infty e^{-tx}\, dN_\nu(x),$$

as we defined it in Section 6.2.3 above; see, for example, Equation (6.29).

[5] We stress that in contrast to much of the rest of this book, all the asymptotic formulas in this appendix are interpreted pointwise.

[6] This asymptotic formula can be interpreted as follows: For each fixed integer $k_0 \geq 0$,

$$z_\nu(t) = \sum_{k\leq k_0} \alpha_k t^{-(d-k)/2} + O\big(t^{(k_0+1-d)/2}\big),$$

as $t \to 0^+$.

as $t \to 0^+$, where the coefficients $\alpha_k = \alpha_k(M)$ are integrals with respect to the Riemannian volume measure of M of suitable local geometric invariants of M. Namely, for each $k = 0, 1, 2, \dots$,

$$\alpha_k(M) = \int_M \alpha_k(y, M)\, d\,\mathrm{vol}_M(y), \tag{B.8}$$

where the function $\alpha_k(\cdot, M)$ can be expressed as a (locally invariant) polynomial of (suitable contractions of) the Riemann curvature tensor of M and of its covariant derivatives. In this sense, it is a local invariant of M.

Moreover, in the present situation, one can show that

$$\alpha_k(M) = 0, \quad \text{if } k \text{ is odd.} \tag{B.9}$$

Remark B.3. For example, α_0 is equal to $e_d \operatorname{vol}(M)$, while α_2 is proportional to the integral over M of the scalar curvature of M.[7] In general, the explicit computation of the coefficients α_k is difficult but a large amount of information is now available, particularly in the present case of closed manifolds. We refer to Gilkey's book [Gi, Sections 1.7, 1.10, 4.8, and 4.9] for a detailed treatment of this matter.

B.3 The Spectral Zeta Function and its Poles

Let us next introduce the *spectral zeta function* of the Laplacian on M (or simply, the zeta function of Δ) $\zeta_\nu(s) = \zeta_{\nu,M}(s)$:[8]

$$\zeta_\nu(s) = \operatorname{Trace}\left(\Delta^{-s/2}\right) = \sum_{j=1}^{\infty} f_j^{-s}. \tag{B.10}$$

Note that in view of (B.5), we have the following relation between $\zeta_\nu(s)$ and $z_\nu(t)$:

$$\zeta_\nu(2s) = \frac{1}{\Gamma(s)} \int_0^\infty z_\nu(t) t^{s-1}\, dt. \tag{B.11}$$

Hence, by Weyl's asymptotic formula (B.2) (or equivalently, by (B.6)), $\zeta_\nu(s)$ extends holomorphically to the open right half-plane $\operatorname{Re} s > d$. Further,

---

[7] By application of the Gauss–Bonnet formula (as extended by S.-S. Chern [Chern1–2] to every dimension d), it follows from the latter statement that the Euler characteristic of M is audible (i.e., can be recovered from the spectrum of M); see, e.g., [McKSin, pp. 44–45]. (Recall that the Euler characteristic of M vanishes when d is odd.)

[8] In the usual terminology, $\zeta_\nu(s)$ is the zeta function of $\sqrt{\Delta}$, because, according to our present conventions, the frequencies f_j of Δ are given by $f_j = \sqrt{\lambda_j}$, where the λ_j's are the eigenvalues of Δ, written in nondecreasing order. The reader should keep this in mind when comparing our formulas with those in [Gi], for example.

according to (B.7), $\zeta_\nu(s)$ has a simple pole at $s = d$. It follows that the abscissa of convergence of the Dirichlet series $\zeta_\nu(s) = \sum_{j=1}^{\infty} f_j^s$ is equal to d, the dimension of the manifold M.

More generally, the asymptotic expansion (B.7) combined with relation (B.11) above yields the following key theorem (see [MinPl] and, for instance, [Gi, Section 1.10, especially Lemma 1.10.1, p. 79]):

Theorem B.4. *The spectral zeta function $\zeta_\nu(s)$ of a closed Riemannian manifold M has a meromorphic extension to the whole complex plane, with simple poles located at d and at a (subset of) the points $d-2, d-4, \ldots$. Further, for $k = 0, 1, 2, \ldots$, the residue at $s = d-k$ is equal to*

$$\frac{2\alpha_k(M)}{\Gamma((d-k)/2)},$$

where $\alpha_k = \alpha_k(M)$ is the k-th coefficient in the heat asymptotic expansion (B.7).

More precisely, $\zeta_\nu(s)$ is holomorphic except for simple poles located at

$$\begin{cases} s = d - 2q, \ q = 0, 1, 2, \ldots, & \text{if } d \text{ is odd}, \\ s = d, d-2, d-4, \ldots, 4, 2, & \text{if } d \text{ is even}. \end{cases}$$

Remark B.5. From our present point of view, the first part of Theorem B.4 is the most important one. It implies that all the poles of the spectral zeta function of a smooth manifold are located on the real axis, in contrast to what happens for fractal manifolds, as illustrated in the main body of this book.

Remark B.6. Strictly speaking, although $s = d$ is always a (simple) pole of $\zeta_\nu(s)$, as was explained above, the other points mentioned in Theorem B.4 may not be poles of $\zeta_\nu(s)$, because the associated residue may happen to vanish. For example, if $M = \mathbb{T}^d = \mathbb{R}^d/\mathbb{Z}^d$ is the standard flat d-dimensional torus (i.e., the unit cube $[0,1]^d$ with its faces identified, as at the end of Section 1.4), then $\zeta_\nu(s) = \zeta_{\nu,M}(s)$ is the normalized Epstein zeta function associated with the standard quadratic form $q_d(x) = x_1^2 + \cdots + x_d^2$ for $x = (x_1, \ldots, x_d) \in \mathbb{R}^d$; namely,

$$\zeta_\nu(s) = \zeta_d(s) = \sum_{(n_1,\ldots,n_d)\in\mathbb{Z}^d\setminus\{0\}} \left(n_1^2 + \cdots + n_d^2\right)^{-s/2}$$

as in Equation (1.47) above.[9] Therefore, for any $d \geq 1$, $s = d$ is the only pole of $\zeta_\nu(s) = \zeta_d(s)$, and it is simple. (See Appendix A, Section A.4, or [Ter, Section 1.4].) In order to reconcile this fact with the statement of

[9] For convenience, we are using here the normalized eigenvalues of Δ on $\mathbb{T}^d$.

Theorem B.4, it suffices to note that $M = \mathbb{T}^d$ has zero Euler characteristic and vanishing curvature. An entirely analogous comment can be made about a general Epstein zeta function $\zeta_q(s)$ considered in Section A.4 of Appendix A (or in [Ter, Section 1.4]), which can be viewed as the spectral zeta function of the Laplacian on a flat torus $M = \mathbb{R}^d/\Lambda$, where Λ is a lattice of $\mathbb{R}^d$ with associated positive definite quadratic form $q = q(x)$. (See Equation (A.16).)

B.4 Extensions

Various extensions of the above results are known in spectral geometry. We mention only a few, which are most relevant to our situation or that may help clarify certain issues:

(i) Formulas (B.2), (B.7) and Theorem B.4 apply to the more general situation of a (positive) elliptic differential operator $\mathcal{P}$ (instead of the Laplacian Δ). If $\mathcal{P}$ is of order $m > 0$, then we define the j-th frequency of $\mathcal{P}$ by $f_j = \lambda_j^{1/m}$, where λ_j is the j-th eigenvalue of $\mathcal{P}$, written in nondecreasing order according to multiplicity. With this convention, the exponent of x in (B.2) remains equal to d, while the exponent of t^{-1} in (B.6) and (B.7) is now equal to d/m and $(d-k)/m$, respectively. Further, the poles of $\zeta_\nu(s)$ also remain the same as in Theorem B.4. On the other hand, in (B.2) and in (B.3), the constant $(2\pi)^{-d}\mathcal{B}_d$ will be replaced by $(2\pi)^{-d}$ times a volume in phase space (i.e., in the cotangent bundle of M) determined by the principal symbol of $\mathcal{P}$. Moreover, with the obvious change in notation, in the analogue of (B.7) and (B.8), the local invariants $\alpha_k(\cdot, \mathcal{P})$ are now expressed as (locally invariant) polynomials of the total symbol of $\mathcal{P}$ and of its covariant derivatives.

(ii) Let us now assume that M is a (smooth, compact) manifold with boundary. For elliptic boundary value problems on M (and, in particular, for the prototypical cases of the Dirichlet and Neumann Laplacians on a smooth bounded open set of d-dimensional Euclidean space $\mathbb{R}^d$), the analogue of Weyl's asymptotic formula (B.2) and of the Minakshisundaram–Pleijel heat asymptotic expansion (B.7) still holds. It takes the same form as above, except that in the counterpart of (B.7), the coefficients α_k (or the corresponding local invariants) are more complicated to compute.[10] In addition, a suitable counterpart of Theorem B.4 also holds; see [MinPl] and [McKSin]. In particular, the poles of ζ_ν are all simple and located on the real axis. Perhaps the most complete treatment of these questions in the case of manifolds with boundary can be found in Grubb's book [Gru], which

[10] In fact, to our knowledge, no explicit algorithm is known to calculate every α_k in this case, although a great deal of information is available.

also deals with the more general case of elliptic pseudodifferential boundary value problems on M.[11] Besides the earlier papers [Min1–2, MinPl] (which study slightly different notions of spectral zeta functions of Laplacians, motivated by the work of Carleman [Car]), other useful references in this setting include the aforementioned paper by McKean and Singer [McKSin], along with the classical paper by Mark Kac [Kac] entitled *Can one hear the shape of a drum?*, which gives some related results on certain planar domains.

In a seminal paper, entitled *Complex powers of elliptic operators,* Seeley [Se1] has used modern analytical tools to study spectral zeta functions. In turn, Seeley's paper (along with its sequel for boundary value problems [Se2–3]) has stimulated a number of further developments related to the zeta functions of elliptic pseudodifferential operators. (See, for example, [Shu, Chapter II] and [Gru].)

B.4.1 Monotonic Second Term

Under the assumptions of Remark B.1 for a manifold with smooth boundary, it need not be the case that $N_\nu(x)$ admits (pointwise) an asymptotic second term as $x \to \infty$. (Contrast this statement with the fact that $z_\nu(t)$ has an asymptotic expansion of every order as $t \to 0^+$; see formula (B.7).) Knowing when $N_\nu(x)$ admits a monotonic asymptotic second term (i.e., of the form a nonzero constant times x^{d-1}) is the object of Hermann Weyl's conjecture [Wey1–2]. In a beautiful work, Ivrii [Ivr1–2] has partially solved this conjecture. More specifically, for example for the Dirichlet or Neumann Laplacian, respectively, he shows that on a manifold M with boundary ∂M, we have (with the obvious notation for the volume of M and ∂M),

$$N_\nu(x) = c_d \operatorname{vol}_d(M) x^d \mp g_{d-1} \operatorname{vol}_{d-1}(\partial M) x^{d-1} + o\big(x^{d-1}\big), \tag{B.12}$$

as $x \to \infty$, provided a suitable condition is satisfied.[12] (Here, the positive constant g_{d-1} is explicitly known in terms of $d-1$, the dimension of the smooth boundary ∂M.) Positive results toward Weyl's Conjecture were also obtained by Melrose [Mel1–2] for manifolds with concave boundary. We refer the interested reader to volumes III and IV of Hörmander's treatise [Hö3] as well as to Ivrii's recent book [Ivr3] for further information about this subject.

---

[11] For a general pseudodifferential operator $\mathcal{P}$ on M, the heat asymptotic expansion may contain logarithmic terms, corresponding to the singularities of the symbol of $\mathcal{P}$. This is not the case, however, for an elliptic differential operator, and hence for a Laplacian on M. (See Corollary 4.27, page 388 and the comment on page 390 in [Gru].)

[12] Roughly speaking, this condition says that the set of multiply reflected periodic geodesics of M forms a set of measure zero, with respect to Liouville measure in phase space (i.e., in the cotangent bundle of M). This condition (which is sufficient but not necessary) is known to be generic among smooth Euclidean domains, but is very difficult to verify in any concrete example.

Finally, we note that situations where $N_\nu(x)$ has an oscillatory behavior (beyond the Weyl term) have been analyzed, in particular, by Duistermaat and Guillemin [DuGu] in terms of the concentration of periodic geodesics (or, more generally, of bicharacteristics). See also the beginning of Section 12.5.3 for a sample of related mathematical and physical works, including the papers by Colin de Verdière [Col] and Chazarain [Chaz].

B.5 Notes

We note that Weyl's formula plays an important role in mathematical physics and can be given interesting physical interpretations; see, for example, [CouHi, Kac, ReSi3, Sim], along with [BaltHi].

Further information about heat asymptotic expansions and related issues can be found in the papers by McKean and Singer [McKSin] or Atiyah, Patodi and Singer [AtPSin], and in [AndLap1–2] or in the first unnumbered subsection of [JohLap, Section 20.2.B], along with the relevant references therein. See also [BergGM] for many interesting examples of spectra of Laplacians on Riemannian manifolds.

Additional information regarding spectral zeta functions and some of their connections with dynamical or with arithmetic zeta functions can be found in the book [Lap-vF8].

Appendix C
An Application of Nevanlinna Theory

In this appendix, we briefly discuss aspects of Nevanlinna theory and give in Section C.2, Theorem C.1, an application of that theory to the complex zeros of Dirichlet polynomials, as defined and studied in Chapter 3. This theorem is used in the proof of Equation (2.37) of Theorem 2.17 (see Sections 2.5 and 2.6). Note however, that in Chapter 3 we obtain a better asymptotic density estimate (with the $O(\sqrt{\varrho})$ of Equation (C.8) below replaced by $O(1)$ in Theorem 3.6, Equation (3.10)).

Nevanlinna theory was developed in the 1930s to study the value distribution of meromorphic functions; that is, the distribution of solutions in z of the equation $f(z) = a$ (so-called a-points). Recall that a meromorphic function is an analytic function

$$f\colon \mathbb{C} \to \mathbb{C} \cup \{\infty\},$$

where $f(x)$ is defined to be ∞ if f has a pole at x. In this case, the function $g(z) = 1/f(z)$ is defined in a neighborhood of x and g is holomorphic at $z = x$ if we set $g(x) = 0$. In other words, we can view f as a holomorphic function to the Riemann sphere

$$\mathbb{P}^1(\mathbb{C}) = \mathbb{C} \cup \{\infty\}.$$

The starting point of Nevanlinna theory is the Poisson–Jensen formula. It was Nevanlinna's insight that this formula can be interpreted as saying that the number of a-points of f in a disc plus the average closeness of f to a on the boundary of this disc (measured in a suitable way) equals the number of poles of f plus the average size of f on this boundary. We exploit this fact for a holomorphic function, by counting the number of a-points of f in a disc by computing the average size of f on the boundary of this disc, provided one can bound the average closeness of f to a on the boundary.

We refer the reader to the monographs [Hay] and [LanCh] for an exposition of Nevanlinna theory.

C.1 The Nevanlinna Height

Recall that $\mathbb{P}^1(\mathbb{C})$ denotes the complex projective line; i.e., the complex line $\mathbb{C}$, completed by a point at infinity, denoted ∞. Alternatively, $\mathbb{P}^1(\mathbb{C})$ could be realized as the Riemann sphere. The *distance* between two points a and a' in $\mathbb{P}^1(\mathbb{C})$ is defined as

$$\|a, a'\| = \frac{|a - a'|}{\sqrt{1 + |a|^2}\sqrt{1 + |a'|^2}}, \quad \text{if } a, a' \neq \infty, \tag{C.1a}$$

and the distance of a point to the point at infinity is given by

$$\|a, \infty\| = \frac{1}{\sqrt{1 + |a|^2}}, \quad \text{if } a \neq \infty. \tag{C.1b}$$

Here, $|z|$ denotes the ordinary absolute value of the complex number z. (See, e.g., [Bea, §2.1].) When one views $\mathbb{P}^1(\mathbb{C})$ as a sphere of diameter 1 in three-dimensional Euclidean space, the distance is simply the chordal distance between the inverse images of a and a' under stereographic projection.

Let f be a nonconstant meromorphic function and let $a \in \mathbb{P}^1(\mathbb{C})$. The *mean proximity* function of f is the function of the positive real variable ϱ given by

$$m_f(a, \varrho) = \int_{|z|=\varrho} -\log \|f(z), a\| \, \frac{dz}{2\pi i z}. \tag{C.2}$$

The *counting function* of f is defined as[1]

$$n_f(a, \varrho) = \#\{z \in \mathbb{C} \colon |z| \leq \varrho, \ f(z) = a\}, \tag{C.3}$$

and, for $a \neq f(0)$, we set

$$N_f(a, \varrho) = \int_0^\varrho n_f(a, t) \, \frac{dt}{t}. \tag{C.4}$$

[1] This function takes finite values since the zeros of a nonconstant meromorphic function form a discrete subset of $\mathbb{C}$.

Finally, the *Nevanlinna height* of f is defined, for $a \neq f(0)$, by

$$T_f(\varrho) = N_f(a,\varrho) + m_f(a,\varrho) + \log \|f(0), a\|, \tag{C.5}$$

which is independent of a (cf. [LanCh, Theorem 1.6, p. 19]). It is this independence that we will exploit, for $a = \infty$ and for $a = 0$.

C.2 Complex Zeros of Dirichlet Polynomials

We investigate the distribution of the roots of the equation

$$\sum_{j=0}^{M} m_j r_j^s = \sum_{j=0}^{M} m_j e^{-w_j s} = 0,$$

where $r_0 = 1 > r_1 > r_2 > \ldots > r_M > 0$. The weights w_j are defined by $r_j = e^{-w_j}$, so that $w_0 = 0 < w_1 < w_2 < \cdots < w_M$.

Our analysis is partly similar to that of Jorgenson and Lang [JorLan3]. In particular, the result that the zeros of a Dirichlet polynomial lie in a bounded strip can be found in [JorLan3, p. 58]. On the other hand, in the present situation, we obtain more precise results than in [JorLan3]. Similar results were also obtained by B. Jessen. (See [Bohr, Appendix II].)

Let σ_l and σ_r be defined by the equations

$$e^{(w_M - w_{M-1})\sigma_l} \sum_{j=0}^{M} |m_j| = \frac{1}{2} |m_M|, \tag{C.6a}$$

and

$$e^{-w_1\sigma_r} \sum_{j=0}^{M} |m_j| = \frac{1}{2} |m_0|. \tag{C.6b}$$

In other words, writing $\sum |m_j| = \sum_{j=0}^{M} |m_j|$, we have

$$\sigma_l = -\frac{\log\left(2\sum |m_j|/|m_M|\right)}{w_M - w_{M-1}} \quad \text{and} \quad \sigma_r = \frac{\log\left(2\sum |m_j|/|m_0|\right)}{w_1}. \tag{C.7}$$

Theorem C.1. *Let $w_0 = 0 < w_1 < \ldots < w_M$ and let $m_0, \ldots, m_M$ be arbitrary nonzero complex numbers. Define*

$$f(s) = \sum_{j=0}^{M} m_j e^{-w_j s}$$

and assume that $f(0) = \sum_{j=0}^{M} m_j \neq 0$. *Then the number of complex zeros of* $f(s)$, *counted according to multiplicity in the closed disc of radius* ϱ, *equals*[2]

$$\frac{\log r_M^{-1}}{\pi}\varrho + O(\sqrt{\varrho}), \quad \textit{as } \varrho \to \infty. \tag{C.8}$$

Moreover, the zeros lie in the horizontally bounded strip $\sigma_l \leq \operatorname{Re} s \leq \sigma_r$, *where* σ_l *and* σ_r *are given respectively by Equations* (C.6a) *and* (C.6b) *and formula* (C.7) *above.*

Proof. To obtain information about the zeros of $f(s)$, we estimate the counting function $n_f(0, \varrho)$. To accomplish this, we first compute the height

$$T_f(\varrho) = m_f(\infty, \varrho) + \log \|f(0), \infty\|$$

and then combine the relation

$$T_f(\varrho) = N_f(0, \varrho) + m_f(0, \varrho) + \log \|f(0), 0\|$$

with estimates for $m_f(0, \varrho)$ to obtain an estimate for $N_f(0, \varrho)$. Finally, we use Lemma C.2 below to deduce the estimate for $n_f(0, \varrho)$.

The height of f is

$$m_f(\infty, \varrho) = \int_{|s|=\varrho} \log \sqrt{1 + |f(s)|^2}\, \frac{ds}{2\pi i s} + \log \|f(0), \infty\|.$$

We need to estimate $|f(s)|$ in the above integral. For $\sigma = \operatorname{Re} s \geq \sigma_l$, the function $|f(s)|$ is bounded by a constant depending on w_j and m_j $(j = 0, \ldots, M)$. On the other hand, for $\sigma \leq \sigma_l$, we have

$$\begin{aligned} |f(s)| = \left| \sum_{j=0}^{M} m_j r_j^s \right| &\geq |m_M| r_M^\sigma - \sum_{j=0}^{M-1} |m_j| r_j^\sigma \\ &\geq r_M^\sigma \left(|m_M| - \sum_{j=0}^{M-1} |m_j| e^{(w_M - w_j)\sigma} \right) \\ &\geq r_M^\sigma \left(|m_M| - e^{(w_M - w_{M-1})\sigma_l} \sum_{j=0}^{M} |m_j| \right) \\ &= \frac{|m_M|}{2} r_M^\sigma. \end{aligned} \tag{C.9}$$

Putting these estimates together, we find that

$$\log \sqrt{1 + |f(s)|^2} = \begin{cases} w_M |\sigma| + O(1) & (\sigma \leq \sigma_l), \\ O(1) & (\sigma \geq \sigma_l), \end{cases}$$

---

[2]In Theorem 3.6, Equation (3.10), the density is given as $\frac{\log r_M^{-1}}{2\pi}\varrho + O(1)$, as $\varrho \to \infty$; i.e., half of the density as given here. The reason is that in Theorem 3.6, we count zeros in the upper half of a vertical strip, $\{s \colon 0 \leq \operatorname{Im} s \leq \varrho\}$.

as $|s| = \varrho \to \infty$. On the circle with radius ϱ, the real part of $s = \varrho e^{i\theta}$ equals $\varrho \cos\theta$. For the height, we thus find that

$$T_f(\varrho) = -w_M \int_{\pi/2}^{3\pi/2} \varrho \cos\theta \, \frac{d\theta}{2\pi} + O(1) = \frac{w_M}{\pi}\varrho + O(1),$$

as $|s| = \varrho \to \infty$, where the implied constant depends only on M and the numbers r_j and m_j.

Now, clearly, $m_f(0,\varrho)$ is positive. Hence $N_f(0,\varrho) \le \frac{w_M}{\pi}\varrho + O(1)$. To show that this is the correct asymptotic order for $N_f(0,\varrho)$, we have to bound the function $m_f(0,\varrho)$ from above. In view of (C.1), inequality (C.9) shows that

$$\|f(s), 0\|^{-1} = \frac{\sqrt{1+|f(s)|^2}}{|f(s)|} = \sqrt{1+|f(s)|^{-2}}$$

is uniformly bounded for $\sigma \le \sigma_l$. For $\sigma \ge \sigma_r$, on the other hand, we have

$$|f(s)| \ge |m_0| - \sum_{j=1}^{M} |m_j| r_j^\sigma \ge |m_0| - r_1^\sigma \sum_{j=0}^{M} |m_j| \ge \frac{|m_0|}{2}.$$

Thus $\log \|f(s), 0\|^{-1}$ is uniformly bounded for $\operatorname{Re} s \ge \sigma_r$ and $\operatorname{Re} s \le \sigma_l$. Observe that this shows that the complex zeros of f lie in the horizontally bounded strip $\sigma_l \le \operatorname{Re} s \le \sigma_r$. The integral for $m_f(0,\varrho)$ over the parts of the circle $|s| = \varrho$ between $\operatorname{Re} s = \sigma_l$ and $\operatorname{Re} s = \sigma_r$ is bounded since $x \mapsto \log|x|$ is an integrable function around $x = 0$.

This shows that

$$N_f(0,\varrho) = \frac{w_M}{\pi}\varrho + O(1),$$

as $\varrho \to \infty$. Finally, in view of (C.3) and (C.4), the statement for $n_f(0,\varrho)$ is a consequence of the following general calculus lemma, applied to the functions $n(t) := n_f(0,t)$ and $N(\varrho) := N_f(0,\varrho)$. □

Lemma C.2. *Let $n(t)$ be a nondecreasing, nonnegative function on $[0,\infty)$ for which there exists $t_0 > 0$ such that $n(t) = 0$ for $t \le t_0$. Let*

$$N(\varrho) = \int_0^\varrho n(t) \, \frac{dt}{t}$$

and suppose that there exist positive constants c and C such that

$$|N(\varrho) - c\varrho| \le C \qquad \text{for all } \varrho > 0.$$

Then

$$|n(\varrho) - c\varrho| \le \sqrt{8Cc\varrho}$$

for all sufficiently large positive values of ϱ.

Proof. Consider a value of ϱ for which $n(\varrho) > c\varrho$. For this value, we have

$$\begin{aligned} n(\varrho) + C &\geq N(n(\varrho)/c) \\ &= N(\varrho) + \int_{\varrho}^{n(\varrho)/c} n(t)\,\frac{dt}{t} \geq c\varrho - C + n(\varrho)\log\frac{n(\varrho)}{c\varrho}. \end{aligned}$$

Hence, writing $x = \frac{n(\varrho)}{c\varrho}$, we deduce that

$$2C \geq c\varrho - n(\varrho) + n(\varrho)\log\frac{n(\varrho)}{c\varrho} = c\varrho\,(1 - x + x\log x)\,.$$

Consider now a value of ϱ for which $n(\varrho) < c\varrho$. In the same way as above, we find

$$c\varrho - C \leq N(\varrho) = N(n(\varrho)/c) + \int_{n(\varrho)/c}^{\varrho} n(t)\,\frac{dt}{t} \leq n(\varrho) + C + n(\varrho)\log\frac{c\varrho}{n(\varrho)}.$$

Again, we deduce that $c\varrho\,(1 - x + x\log x) \leq 2C$, with $x = \frac{n(\varrho)}{c\varrho}$.

The function

$$1 - x + x\log x = \frac{(x-1)^2}{2} + O\left((x-1)^3\right) \qquad (\text{as } x \to 1)$$

is nonnegative and vanishes at $x = 1$. It follows that for ϱ sufficiently large, x is close to 1. Around $x = 1$, this function takes values larger than $(x-1)^2/4$. Hence for large positive ϱ, $(x-1)^2 \leq \frac{8C}{c\varrho}$. This is equivalent to $|n(\varrho) - c\varrho| \leq \sqrt{8Cc\varrho}$, as was to be proved. □

Remark C.3. In Chapter 3, we define numbers d_l and d_r such that the Dirichlet polynomial $f(s)$ does not vanish for $\operatorname{Re} s < d_l$ and $\operatorname{Re} s > d_r$ (see formula (3.17)). The numbers σ_l and σ_r defined above give weaker bounds, because they are defined to satisfy the stronger property that $|f(s)|$ (respectively, $|f(s)|r_M^{-s}$) is bounded away from 0 by a fixed distance for $\operatorname{Re} s \geq \sigma_r$ (respectively, $\operatorname{Re} s \leq \sigma_l$).

Bibliography

[Ahl] L. V. Ahlfors, *Complex Analysis*, 3d. ed., McGraw-Hill, London, 1985.

[AndLap1] S. I. Andersson and M. L. Lapidus (eds.), *Progress in Inverse Spectral Geometry*, Trends in Mathematics, vol. 1, Birkhäuser-Verlag, Basel and Boston, 1997.

[AndLap2] S. I. Andersson and M. L. Lapidus, Spectral Geometry: An introduction and background material for this volume, in [AndLap1, pp. 1–14].

[ArMazu] M. Artin and B. Mazur, On periodic points, *Annals of Math.* **81** (1965), 82–99.

[AtPSin] M. Atiyah, V. K. Patodi and I. M. Singer, Spectral asymmetry and Riemannian geometry, I, *Math. Proc. Cambridge Philos. Soc.* **77** (1975), 43–69; II, *ibid.* **78** (1975), 405–432; III, *ibid.* **79** (1976), 71–99.

[BadPo] R. Badii and A. Politi, Intrinsic oscillations in measuring the fractal dimension, *Phys. Lett. A* **104** (1984), 303–305.

[Bae] J. C. Baez, The octonions, *Bull. Amer. Math. Soc.* (N. S.) **39** (2002), 145–205, and Errata, *ibid* **42** (2005), 213.

[Báe] L. Báez-Duarte, A class of invariant unitary operators, *Adv. Math.* **144** (1999), 1–12.

[Bal1] V. Baladi, Dynamical zeta functions, in: *Real and Complex Dynamical Systems* (B. Branner and P. Hjorth, eds.), NATO

Adv. Sci. Inst. Ser. C Math. Phys. Sci., vol. 464, Kluwer, Dordrecht, 1995, pp. 1–26.

[Bal2] V. Baladi, Periodic orbits and dynamical spectra, *Ergodic Theory and Dynamical Systems* **18** (1998), 255–292.

[Bal3] V. Baladi, *Kneading operators and transfer operators in higher dimensions*, in [Lap-vF10, Part 1, pp. 407–416].

[BalKel] V. Baladi and G. Keller, Zeta functions and transfer operators for piecewise monotone transformations, *Commun. Math. Phys.* **127** (1990), 459–477.

[BalaVor] N. L. Balazs and A. Voros, Chaos on the pseudosphere, *Phys. Rep.*, No. 3, **143** (1986), 109–240.

[BallBlu1] R. C. Ball and R. Blumenfeld, Universal scaling of the stress field at the vicinity of a wedge crack in two dimensions and oscillatory self-similar corrections to scaling, *Phys. Rev. Lett.* **65** (1990), 1784–1787.

[BallBlu2] R. C. Ball and R. Blumenfeld, Sidebranch selection in fractal growth, *Europhys. Lett.* **16** (1991), 47–52.

[BallBlu3] R. C. Ball and R. Blumenfeld, Probe for morphology and hierarchical corrections in scale-invariant structures, *Phys. Rev. E* **47** (1993), 2298–3002.

[BallanBlo] R. Ballan and C. Bloch, Solution of the Schrödinger equation in terms of classical paths, *Ann. Physics* **85** (1974), 514–546.

[BaltHi] H. P. Baltes and E. R. Hilf, *Spectra of Finite Systems*, B. I.-Wissenschaftsverlag, Vienna, 1976.

[Ban] T. Banchoff, Critical points and curvature for embedded polyhedra, *J. Differential Geom.* **1** (1967), 245–256. II, in: *Differential Geometry* (Proc. Special Year, Maryland), Progress in Math., vol. 32, Birkhäuser, Boston, 1983, pp. 34–55.

[BarHam] M. T. Barlow and B. M. Hambly, Transition density estimates for Brownian motion on scale irregular Sierpinski gaskets, *Ann. Inst. H. Poincaré Probab. Statist.* **33** (1997), 531–557.

[Bar] K. Barner, On A. Weil's explicit formula, *J. Reine Angew. Math.* **323** (1981), 139–152.

[Bea] A. F. Beardon, *Iteration of Rational Functions*, Springer-Verlag, Berlin, 1991.

[BedFi] T. Bedford and A. M. Fisher, Analogues of the Lebesgue density theorem for fractal sets of reals and integers, *Proc. London Math. Soc.* (3) **64** (1992), 95–124.

[BedKS] T. Bedford, M. Keane and C. Series (eds.), *Ergodic Theory, Symbolic Dynamics and Hyperbolic Spaces*, Oxford Univ. Press, Oxford, 1991.

[Bér] P. Bérard, Spectres et groupes cristallographiques I: Domaines euclidiens, *Invent. Math.* **58** (1980), 179–199.

[BergGM] M. Berger, P. Gauduchon and E. Mazet, *Le Spectre d'une Variété Riemannienne*, Lecture Notes in Math., vol. 194, Springer-Verlag, New York, 1971.

[BergGo] M. Berger and B. Gostiaux, *Differential Geometry: Manifolds, Curves and Surfaces*, English translation, Springer-Verlag, Berlin, 1988.

[Berr1] M. V. Berry, Distribution of modes in fractal resonators, in: *Structural Stability in Physics* (W. Güttinger and H. Eikemeier, eds.), Springer-Verlag, Berlin, 1979, pp. 51–53.

[Berr2] M. V. Berry, Some geometric aspects of wave motion: Wavefront dislocations, diffraction catastrophes, diffractals, in: *Geometry of the Laplace Operator*, Proc. Sympos. Pure Math., vol. 36, Amer. Math. Soc., Providence, R. I., 1980, pp. 13–38.

[Berr3] M. V. Berry, The Bakerian lecture, 1987: Quantum chaology, *Proc. Roy. Soc. London Ser. A* **413** (1987), 183–198.

[Berr4] M. V. Berry, *Private communication*, January 1999.

[BerrHow] M. V. Berry and C. J. Howls, High orders of the Weyl expansion for quantum billiards: resurgence of periodic orbits, and the Stokes phenomenon, *Proc. Roy. Soc. London Ser. A* **447** (1994), 527–555.

[BesTa] A. S. Besicovitch and S. J. Taylor, On the complementary intervals of a linear closed set of zero Lebesgue measure, *J. London Math. Soc.* **29** (1954), 449–459.

[BessGM] D. Bessis, J. S. Geronimo and P. Moussa, Mellin transforms associated with Julia sets and physical applications, *J. Statist. Phys.* **34** (1984), 75–110.

[Beu] A. Beurling, Analyse de la loi asymptotique de la distribution des nombres premiers généralisés, I, *Acta Math.* **68** (1937), 255–291.

[BiSo] M. S. Birman and M. Z. Solomyak, Spectral asymptotics of nonsmooth elliptic operators, I, *Trans. Moscow Math. Soc.* **27** (1972), 3–52; II, *ibid.* **28** (1973), 3–34.

[Bl] W. Blaschke, *Integralgeometrie*, Chelsea, New York, 1949.

[BogKe] E. B. Bogomolny and J. P. Keating, Random matrix theory and the Riemann zeros I: Three-and-four-point correlations, *Nonlinearity* **8** (1995), 1115–1131.

[Bohr] H. Bohr, *Almost Periodic Functions*, Chelsea, New York, 1951.

[Bom] E. Bombieri, *Counting points on curves over finite fields (d'après S. A. Stepanov)*, Séminaire Bourbaki, 25ème année 1972/73, no. 430, Lecture Notes in Math., vol. 383, Springer-Verlag, New York, 1974, pp. 234–241.

[BorCP] J. M. Borwein, K.-K. S. Choi and W. Pigulla, Continued fractions of tails of hypergeometric series, *Amer. Math. Monthly* **112**, No. 6, (2005), 493–501.

[BotSil] A. Böttcher and B. Silbermann, *Analysis of Toeplitz Operators*, Springer-Verlag, Berlin, 1990.

[Bou] G. Bouligand, Ensembles impropres et nombre dimensionnel, *Bull. Sci. Math.* (2) **52** (1928), 320–344 and 361–376.

[Bow1] R. Bowen, Symbolic dynamics for hyperbolic flows, *Amer. J. Math.* **95** (1973), 429–460.

[Bow2] R. Bowen, *Equilibrium States and the Ergodic Theory of Anosov Diffeomorphisms*, Lecture Notes in Math., vol. 470, Springer-Verlag, New York, 1975.

[BraBh] M. Brack and R. K. Bhaduri, *Semiclassical Physics*, Frontiers in Physics, Addison-Wesley, Reading, 1997.

[Bré] H. Brézis, *Analyse Fonctionnelle: Théorie et Applications*, Masson, Paris, 1983.

[BroCa] J. Brossard and R. Carmona, Can one hear the dimension of a fractal?, *Commun. Math. Phys.* **104** (1986), 103–122.

[Bu1] J.-F. Burnol, The explicit formula in simple terms, e-print, `arXiv: math.NT/98100169`, 1998.

[Bu2] J.-F. Burnol, The explicit formula and the conductor operator, e-print, `arXiv:math.NT/9902080`, 1999.

[Bu3] J.-F. Burnol, Sur les formules explicites I: analyse invariante, *C. R. Acad. Sci. Paris Sér. I Math.* **331** (2000), 423–428.

[Ca1] A. M. Caetano, Some domains where the eigenvalues of the Dirichlet Laplacian have non-power second term asymptotic estimates, *J. London Math. Soc.* (2) **43** (1991), 431–450.

[Ca2] A. M. Caetano, On the search for the asymptotic behaviour of the eigenvalues of the Dirichlet Laplacian for bounded irregular domains, *Internat. J. Appl. Sci. Comput.* **2** (1995), 261–287.

[Car] T. Carleman, Propriétés asymptotiques des fonctions fondamentales des membranes vibrantes, *Scand. Math. Congress* (1934), 34–44.

[Chaz] J. Chazarain, Formule de Poisson pour les variétés riemanniennes, *Invent. Math.* **24** (1974), 65–82.

[CheeMüS1] J. Cheeger, W. Müller and R. Schrader, On the curvature of piecewise flat manifolds, *Commun. Math. Phys.* **92** (1984), 405–454.

[CheeMüS2] J. Cheeger, W. Müller and R. Schrader, Kinematic and tube formulas for piecewise linear spaces, *Indiana Univ. Math. J.* **35** (1986), 737–754.

[Chern1] S.-S. Chern, A simple intrinsic proof of the Gauss–Bonnet formula for closed Riemannian manifolds, *Ann. of Math.* **45** (1944), 747–752.

[Chern2] S.-S. Chern, On the curvature integrals in a Riemannian manifold, *Ann. of Math.* **46** (1945), 674–684.

[Chern3] S.-S. Chern, On the kinematic formula in integral geometry, *J. of Math. and Mech.* **16** (1966), 101–118.

[Coh] D. L. Cohn, *Measure Theory*, Birkhäuser, Boston, 1980.

[Col] Y. Colin de Verdière, Spectre du laplacien et longueur des géodésiques périodiques, I et II, *Compositio Math.* **27** (1973), 83–106 and 159–184.

[Con] A. Connes, *Noncommutative Geometry*, Academic Press, New York, 1994.

[CouHi] R. Courant and D. Hilbert, *Methods of Mathematical Physics*, vol. I, English translation, Interscience, New York, 1953.

[Cram] H. Cramér, Studien über die Nullstellen der Riemannschen Zetafunktion, *Math. Z.* **4** (1919), 104–130.

[CranMH] A. Crannell, S. May and L. Hilbert, Shifts of finite type and Fibonacci harps, preprint, 2005.

[Da] H. Davenport, *Multiplicative Number Theory*, 2nd ed., Springer-Verlag, New York, 1980.

[Del] J. Delsarte, Formules de Poisson avec reste, *J. Anal. Math.* **17** (1966), 419–431.

[dV1] C.-J. de la Vallée Poussin, Recherches analytiques sur la théorie des nombres; Première partie: La fonction $\zeta(s)$ de Riemann et les nombres premiers en général, *Ann. Soc. Sci. Bruxelles Sér. I* **20** (1896), 183–256.

[dV2] C.-J. de la Vallée Poussin, Sur la fonction $\zeta(s)$ de Riemann et le nombre des nombres premiers inférieurs à une limite donnée, *Mém. Couronnés et Autres Mém. Publ. Acad. Roy. Sci., des Lettres, Beaux-Arts Belg.* **59** (1899–1900).

[Den1] C. Deninger, Local L-factors of motives and regularized determinants, *Invent. Math.* **107** (1992), 135–150.

[Den2] C. Deninger, Lefschetz trace formulas and explicit formulas in analytic number theory, *J. Reine Angew. Math.* **441** (1993), 1–15.

[Den3] C. Deninger, Evidence for a cohomological approach to analytic number theory, in: *Proc. First European Congress of Mathematics* (A. Joseph *et al.*, eds.), vol. I, Paris, July 1992, Birkhäuser-Verlag, Basel, 1994, pp. 491–510.

[DenSchr] C. Deninger and M. Schröter, A distributional theoretic proof of Guinand's functional equation for Cramér's V-function and generalizations, *J. London Math. Soc.* **52** (1995), 48–60.

[DerGrVo] G. Derfel, P. Grabner and F. Vogl, The zeta function of the Laplacian on certain fractals, e-print, `arXiv:math.SP/0508315`, 2005.

[DodKr] M. M. Dodson and S. Kristensen, *Hausdorff dimension and Diophantine approximation*, in [Lap-vF10, Part 1, pp. 305–347].

[DolFr] J. D. Dollard and C. N. Friedman, Product Integration, with Application to Differential Equations, *Encyclopedia of Mathematics and Its Applications*, vol. 10, Addison-Wesley, Reading, 1979.

[DuGu] J. J. Duistermaat and V. Guillemin, The spectrum of positive elliptic operators and periodic bicharacteristics, *Invent. Math.* **29** (1975), 39–79.

[Ed] G. A. Edgar (ed.), *Classics on Fractals*, Addison-Wesley, Reading, 1993.

[EdmEv] D. E. Edmunds and W. D. Evans, *Spectral Theory of Differential Operators*, Oxford Mathematical Monographs, Oxford University Press, Oxford, 1987.

[Edw] H. M. Edwards, *Riemann's Zeta Function*, Academic Press, New York, 1974.

[El] N. D. Elkies, Rational points near curves and small nonzero $|x^3-y^2|$ via lattice reduction, in: Proc. of ANTS-4 (W. Bosma, ed.), *Lecture Notes in Computer Science,* vol. 1838, Springer-Verlag, New York, 2000, pp. 33–63.

[Ep] P. Epstein, Zur Theorie allgemeiner Zetafunktionen, I, *Math. Ann.* **56** (1903), 614–644; II, *ibid.* **63** (1907), 205–216.

[Es1] D. Essouabri, Singularités des séries de Dirichlet associées à des polynômes de plusieurs variables et applications en théorie analytique des nombres, *Ann. Inst. Fourier* (*Grenoble*) **47** (1996), 429–484.

[Es2] D. Essouabri, *Private communication*, June 1996.

[EvPeVo] C. J. G. Evertsz, H.-O. Peitgen and R. F. Voss (eds.), *Fractal Geometry and Analysis: The Mandelbrot Festschrift*, World Scientific, Singapore, 1996.

[Fa1] K. J. Falconer, *The Geometry of Fractal Sets*, Cambridge Univ. Press, Cambridge, 1985.

[Fa2] K. J. Falconer, Random fractals, *Math. Proc. Cambridge Philos. Soc.* **100** (1986), 559–582.

[Fa3] K. J. Falconer, *Fractal Geometry: Mathematical Foundations and Applications*, John Wiley & Sons, Chichester, 1990.

[Fa4] K. J. Falconer, On the Minkowski measurability of fractals, *Proc. Amer. Math. Soc.* **123** (1995), 1115–1124.

[Fed1] H. Federer, Curvature measures, *Trans. Amer. Math. Soc.* **93** (1959), 418–491.

[Fed2] H. Federer, *Geometric Measure Theory*, Springer-Verlag, New York, 1969.

[Fel] W. Feller, *An Introduction to Probability Theory and its Applications*, vol. II, John Wiley & Sons, New York, 1966.

[Fey] R. P. Feynman, *Statistical Mechanics: A Set of Lectures*, W. A. Benjamin, New York, 1962.

[FlLeVa] J. Fleckinger, M. Levitin and D. Vassiliev, Heat equation on the triadic von Koch snowflake, *Proc. London Math. Soc.* (3) **71** (1995), 372–396.

[FlVa] J. Fleckinger and D. Vassiliev, An example of a two-term asymptotics for the "counting function" of a fractal drum, *Trans. Amer. Math. Soc.* **337** (1993), 99–116.

[Fol] G. B. Folland, *Real Analysis: Modern Techniques and Their Applications*, 2nd. ed., John Wiley & Sons, Boston, 1999.

[FouTuVa] J.-D. Fournier, G. Turchetti and S. Vaienti, Singularity spectrum of generalized energy integrals, *Phys. Lett. A* **140** (1989), 331–335.

[Fra1] M. Frantz, Minkowski measurability and lacunarity of self-similar sets in ℝ, preprint, December 2001.

[Fra2] M. Frantz, Lacunarity, Minkowski content, and self-similar sets in ℝ, in [Lap-vF10, Part 1, pp. 77–91].

[FreKie] E. Freitag and R. Kiehl, *Etale Cohomology and the Weil Conjectures*, Springer-Verlag, Berlin, 1988.

[Fu1] J. H. G. Fu, Tubular neighborhoods in Euclidean spaces, *Duke Math. J.* **52** (1985), 1025–1046.

[Fu2] J. H. G. Fu, Curvature measures of subanalytic sets, *Amer. J. Math.* **116** (1994), 819–880.

[FukSh] M. Fukushima and T. Shima, On a spectral analysis for the Sierpinski gasket, *Potential Analysis* **1** (1992), 1–35.

[Gab] O. Gabber, *Private communication*, June 1997.

[Gat] D. Gatzouras, Lacunarity of self-similar and stochastically self-similar sets, *Trans. Amer. Math. Soc.* **352** (2000), 1953–1983.

[Gel] S. Gelbart, An elementary introduction to the Langlands program, *Bull. Amer. Math. Soc.* (N. S.) **10** (1984), 177–219.

[Ger] J. Gerling, *Untersuchungen zur Theorie von Weyl–Berry–Lapidus*, Graduate Thesis (Diplomarbeit), Dept. of Physics, Universität Osnabrück, Germany, May 1992.

[GerSc1] J. Gerling and H.-J. Schmidt, Self-similar drums and generalized Weierstrass functions, *Physica A* **191** (1992), 536–539.

[GerSc2] J. Gerling and H.-J. Schmidt, Three-term asymptotics of the spectrum of self-similar fractal drums, *J. Math. Sci. Univ. Tokyo* **6** (1999), 101–126.

[Gi] P. B. Gilkey, *Invariance Theory, the Heat Equation, and the Atiyah–Singer Index Theorem*, 2nd ed., Publish or Perish, Wilmington, 1984. (New rev. and enl. ed. in *Studies in Advanced Mathematics*, CRC Press, Boca Raton, 1995.)

[Gou] S. Gouëzel, Spectre de l'opérateur de transfert en dimension 1, *Manuscripta Math.* **106** (2001), 365–403.

[Gra] A. Gray, *Tubes*, 2nd ed. (of the 1990 ed.), Progress in Math., vol. 221, Birkhäuser, Boston, 2004.

[GriLap] C. A. Griffith and M. L. Lapidus, *Computer graphics and the eigenfunctions for the Koch snowflake drum*, in [AndLap1, pp. 95–109].

[GröLS] M. Grötschel, L. Lovász, A. Schrijver, *Geometric Algorithms and Combinatorial Optimization*, Springer-Verlag, Berlin, 1993.

[Gru] G. Grubb, *Functional Calculus of Pseudodifferential Boundary Problems*, 2nd ed. (of the 1986 ed.), Progress in Mathematics, vol. 65, Birkhäuser, Boston, 1996.

[GuIsLap1] D. Guido, T. Isola and M. L. Lapidus, Complex dimensions of fractal strings with non-power scaling laws (tentative title), work in progress, 2005.

[GuIsLap2] D. Guido, T. Isola and M. L. Lapidus, A trace on fractal graphs and the Ihara zeta function (tentative title), work in progress, 2005.

[Gui1] H. P. Guinand, A summation formula in the theory of prime numbers, *Proc. London Math. Soc.* (2) **50** (1948), 107–119.

[Gui2] H. P. Guinand, Fourier reciprocities and the Riemann zeta-function, *Proc. London Math. Soc.* (2) **51** (1950), 401–414.

[Gut1] M. C. Gutzwiller, Periodic orbits and classical quantization conditions, *J. Math. Phys.* **12** (1971), 343–358.

[Gut2] M. C. Gutzwiller, *Chaos in Classical and Quantum Mechanics*, Interdisciplinary Applied Mathematics, vol. 1, Springer-Verlag, Berlin, 1990.

[Had1] J. Hadamard, Étude sur les propriétés des fonctions entières et en particulier d'une fonction considérée par Riemann, *J. Math. Pures Appl.* (4) **9** (1893), 171–215. (Reprinted in [Had3, pp. 103–147].)

[Had2] J. Hadamard, Sur la distribution des zéros de la fonction $\zeta(s)$ et ses conséquences arithmétiques, *Bull. Soc. Math. France* **24** (1896), 199–220. (Reprinted in [Had3, pp. 189–210].)

[Had3] J. Hadamard, *Oeuvres de Jacques Hadamard*, Tome I, Editions du Centre National de la Recherche Scientifique, Paris, 1968.

[Ham1] B. M. Hambly, Brownian motion on a random recursive Sierpinski gasket, *Ann. Probab.* **25** (1997), 1059–1102.

[Ham2] B. M. Hambly, On the asymptotics of the eigenvalue counting function for random recursive Sierpinski gaskets, *Probab. Theory Related Fields* **117** (2000), 221–247.

[HamLap] B. M. Hambly and M. L. Lapidus, Random fractal strings: their zeta functions, complex dimensions and spectral asymptotics, *Trans. Amer. Math. Soc.*, No. 1, **358** (2006), 285–314.

[Haran1] S. Haran, Riesz potentials and explicit sums in arithmetic, *Invent. Math.* **101** (1990), 696–703.

[Haran2] S. Haran, *The Mysteries of the Real Prime*, London Mathematical Society Monographs, New Series, vol. 25, Clarendon Press, Oxford, 2001.

[HardW] G. H. Hardy and E. M. Wright, *An Introduction to the Theory of Numbers*, 4th ed., Oxford Univ. Press, Oxford, 1960.

[Harr] T. E. Harris, *The Theory of Branching Processes*, rev. ed. (of the 1963 ed.), Dover, New York, 2004.

[HasJLS] J. Hastad, B. Just, J. C. Lagarias, C. P. Schnorr, Polynomial time algorithms for finding integer relations among real numbers, *SIAM J. Comput.* **18** (1989), 859–881.

[Hay] W. K. Hayman, *Meromorphic Functions*, Oxford Mathematical Monographs, Oxford Univ. Press, 1975.

[HeLap1] C. Q. He and M. L. Lapidus, Generalized Minkowski content and the vibrations of fractal drums and strings, *Mathematical Research Letters* **3** (1996), 31–40.

[HeLap2] C. Q. He and M. L. Lapidus, Generalized Minkowski content, spectrum of fractal drums, fractal strings and the Riemann zeta-function, *Memoirs Amer. Math. Soc.*, No. 608, **127** (1997), 1–97.

[HilLap] T. W. Hilberdink and M. L. Lapidus, Beurling zeta functions, generalised primes, and fractal membranes, *Acta Applicandae Mathematicae*, in press. (Also: e-print, `arXiv:math.NT/0410270`, 2004.)

[Hö1] L. Hörmander, The spectral function of an elliptic operator, *Acta Math.* **121** (1968), 193–218.

[Hö2] L. Hörmander, *The Analysis of Linear Partial Differential Operators*, vol. I, *Distribution Theory and Fourier Analysis*, 2nd ed. (of the 1983 ed.), Springer-Verlag, Berlin, 1990.

[Hö3] L. Hörmander, *The Analysis of Linear Partial Differential Operators*, vols. II–IV, Springer-Verlag, Berlin, 1983 & 1985.

[HuaSl] C. Hua and B. D. Sleeman, Fractal drums and the n-dimensional modified Weyl–Berry conjecture, *Commun. Math. Phys.* **168** (1995), 581–607.

[Hub] H. Huber, Zur analytischen Theorie hyperbolischer Raumformen und Bewegungsgruppen, *Math. Ann.* **138** (1959), 1–26.

[HurWa] W. Hurewicz and H. Wallman, *Dimension Theory*, Princeton Univ. Press, Princeton, 1941.

[Hut] J. E. Hutchinson, Fractals and self-similarity, *Indiana Univ. Math. J.* **30** (1981), 713–747.

[In] A. E. Ingham, *The Distribution of Prime Numbers*, 2nd ed. (reprinted from the 1932 ed.), Cambridge Univ. Press, Cambridge, 1992.

[Ivi] A. Ivić, *The Riemann Zeta-Function: The Theory of the Riemann Zeta-Function with Applications*, John Wiley & Sons, New York, 1985.

[Ivr1] V. Ja. Ivrii, Second term of the spectral asymptotic expansion of the Laplace–Beltrami operator on manifolds with boundary, *Functional Anal. Appl.* **14** (1980), 98–106.

[Ivr2] V. Ja. Ivrii, *Precise Spectral Asymptotics for Elliptic Operators Acting in Fiberings over Manifolds with Boundary*, Lecture Notes in Math., vol. 1100, Springer-Verlag, New York, 1984.

[Ivr3] V. Ja. Ivrii, *Microlocal Analysis and Precise Spectral Asymptotics*, Springer-Verlag, Berlin, 1998.

[JafLap] S. Jaffard and M. L. Lapidus, Complex dimensions of multifractals (tentative title), work in progress, 2006.

[JafMey] S. Jaffard and Y. Meyer, Wavelet methods for pointwise regularity and local oscillations of functions, *Memoirs Amer. Math. Soc.*, No. 587, **123** (1996), 1–110.

[JohLap] G. W. Johnson and M. L. Lapidus, *The Feynman Integral and Feynman's Operational Calculus*, Oxford Mathematical Monographs, Oxford Univ. Press, Oxford, 2000. (Corrected reprinting and paperback edition, 2002.)

[JorLan1] J. Jorgenson and S. Lang, On Cramer's theorem for general Euler products with functional equation, *Math. Ann.* **297** (1993), 383–416.

[JorLan2] J. Jorgenson and S. Lang, *Basic Analysis of Regularized Series and Products*, Lecture Notes in Math., vol. 1564, Springer-Verlag, New York, 1993.

[JorLan3] J. Jorgenson and S. Lang, *Explicit Formulas for Regularized Products and Series*, Lecture Notes in Math., vol. 1593, Springer-Verlag, New York, 1994, pp. 1–134.

[Kac] M. Kac, Can one hear the shape of a drum?, *Amer. Math. Monthly* (Slaught Memorial Papers, No. 11) (4) **73** (1966), 1–23.

[KahSa] J.-P. Kahane and R. Salem, *Ensembles Parfaits et Séries Trigonométriques*, Hermann, Paris, 1963.

[Kat] N. M. Katz, An overview of Deligne's proof for varieties over finite fields, *Proc. Sympos. Pure Math.* **28** (1976), 275–305.

[KatSar] N. M. Katz and P. Sarnak, Zeroes of zeta functions and symmetry, *Bull. Amer. Math. Soc.* (N. S.) **36** (1999), 1–26.

[Ke] J. P. Keating, The Riemann zeta-function and quantum chaology, in: *Quantum Chaos* (G. Casati, I. Guarneri and U. Smilanski, eds.), North-Holland, Amsterdam, 1993, pp. 145–185.

[Ki1] J. Kigami, Harmonic calculus on p.c.f. self-similar sets, *Trans. Amer. Math. Soc.* **335** (1989), 721–755.

[Ki2] J. Kigami, *Analysis on Fractals*, Cambridge Univ. Press, Cambridge, 2001.

[KiLap1] J. Kigami and M. L. Lapidus, Weyl's problem for the spectral distribution of Laplacians on p.c.f. self-similar fractals, *Commun. Math. Phys.* **158** (1993), 93–125.

[KiLap2] J. Kigami and M. L. Lapidus, Self-similarity of volume measures for Laplacians on p.c.f. self-similar fractals, *Commun. Math. Phys.* **217** (2001), 165–180.

[Kor] J. Korevaar, A century of complex Tauberian theory, *Bull. Amer. Math. Soc.* (N. S.) **39** (2002), 475–531.

[Kow] O. Kowalski, Additive volume invariants of Riemannian manifolds, *Acta Math.* **145** (1980), 205–225.

[Lag1] J. C. Lagarias, The computational complexity of simultaneous diophantine approximation problems, *SIAM J. Comput.* **14** (1985), 196–209.

[Lag2] J. C. Lagarias, Number theory zeta functions and dynamical zeta functions, in: *Spectral Problems in Geometry and Arithmetic* (T. Branson, ed.), Contemporary Mathematics, vol. 237, Amer. Math. Soc., Providence, R. I., 1999, pp. 45–86.

[LagP] J. C. Lagarias and P. A. B. Pleasants, Repetitive Delone sets and quasicrystals, *Ergodic Theory and Dynamical Systems* **23** (2003), 831–867.

[LagR] J. C. Lagarias and E. Rains, On a two-variable zeta function for number fields, *Ann. Inst. Fourier* (*Grenoble*) **53** (2003), 237–258. (See also: e-print, `arXiv:math.NT/0104176`, 2001.)

[Lal1] S. P. Lalley, Packing and covering functions of some self-similar fractals, *Indiana Univ. Math. J.* **37** (1988), 699–709.

[Lal2] S. P. Lalley, Renewal theorems in symbolic dynamics, with applications to geodesic flows, noneuclidean tessellations and their fractal limits, *Acta Math.* **163** (1989), 1–55.

[Lal3] S. P. Lalley, *Probabilistic counting methods in certain counting problems of ergodic theory*, in [BedKS, pp. 223–258].

[Lan] S. Lang, *Algebraic Number Theory*, 3d ed. (of the 1970 ed.), Springer-Verlag, New York, 1994.

[LanCh] S. Lang and W. Cherry, *Topics in Nevanlinna Theory*, Lecture Notes in Math., vol. 1433, Springer-Verlag, New York, 1990.

[Lap1] M. L. Lapidus, Fractal drum, inverse spectral problems for elliptic operators and a partial resolution of the Weyl–Berry conjecture, *Trans. Amer. Math. Soc.* **325** (1991), 465–529.

[Lap2] M. L. Lapidus, Spectral and fractal geometry: From the Weyl–Berry conjecture for the vibrations of fractal drums to the Riemann zeta-function, in: *Differential Equations and Mathematical Physics* (C. Bennewitz, ed.), Proc. Fourth UAB Intern. Conf. (Birmingham, March 1990), Academic Press, New York, 1992, pp. 151–182.

[Lap3] M. L. Lapidus, Vibrations of fractal drums, the Riemann hypothesis, waves in fractal media, and the Weyl–Berry conjecture, in: *Ordinary and Partial Differential Equations* (B. D. Sleeman and R. J. Jarvis, eds.), vol. IV, Proc. Twelfth Internat. Conf. (Dundee, Scotland, UK, June 1992), Pitman Research Notes in Math. Series, vol. 289, Longman Scientific and Technical, London, 1993, pp. 126–209.

[Lap4] M. L. Lapidus, Fractals and vibrations: Can you hear the shape of a fractal drum?, *Fractals* **3**, No. 4 (1995), 725–736. (Special issue in honor of Benoît B. Mandelbrot's 70th birthday. Reprinted in [EvPeVo, pp. 321–332].)

[Lap5] M. L. Lapidus, Analysis on fractals, Laplacians on self-similar sets, noncommutative geometry and spectral dimensions, *Topological Methods in Nonlinear Analysis*, **4** (1994), 137–195.

[Lap6] M. L. Lapidus, Towards a noncommutative fractal geometry? Laplacians and volume measures on fractals, in: *Harmonic Analysis and Nonlinear Differential Equations*, Contemporary Mathematics, vol. 208, Amer. Math. Soc., Providence, R. I., 1997, pp. 211–252.

[Lap7] M. L. Lapidus, *The possible origins of fractality in Nature: The example of the Koch snowflake drum*, lecture delivered at the Internat. Sympos. on "New Geometric Methods in Modern Science—The Philosophies of Nature Today: The Contributions of Mathematics, Physics, and Biology", Paris, March 2003. (Also, plenary lecture at the "Fifth European Conf. on Elliptic and Parabolic Problems: A Special Tribute to the work of Haïm Brezis", Gaeta, Italy, May 2004.)

[Lap8] M. L. Lapidus, *Fractal Geometry and Applications—An Introduction to this Volume*, in [Lap-vF10, Part 1, pp. 1–25].

[Lap9] M. L. Lapidus, T-duality, functional equation, and noncommutative spacetime, in: *Geometries of Nature, Living Systems and Human Cognition: New interactions of Mathematics with Natural Sciences and Humanities* (L. Boi, ed.), World Scientific, Singapore, 2005, pp. 3–91.

[Lap10] M. L. Lapidus, *In Search of the Riemann Zeros: Strings, fractal membranes and noncommutative spacetimes*, Amer. Math. Soc., Providence, R. I., 2007, 486+(xv) pages, to appear.

[LapFl] M. L. Lapidus and J. Fleckinger-Pellé, Tambour fractal: vers une résolution de la conjecture de Weyl–Berry pour les valeurs propres du laplacien, *C. R. Acad. Sci. Paris Sér. I Math.* **306** (1988), 171–175.

[LapLevRo] M. L. Lapidus, J. Lévy Véhel and J. Rock, Fractal strings and multifractal zeta functions, work in progress, 2006.

[LapMa1] M. L. Lapidus and H. Maier, Hypothèse de Riemann, cordes fractales vibrantes et conjecture de Weyl–Berry modifiée, *C. R. Acad. Sci. Paris Sér. I Math.* **313** (1991), 19–24.

[LapMa2] M. L. Lapidus and H. Maier, The Riemann hypothesis and inverse spectral problems for fractal strings, *J. London Math. Soc.* (2) **52** (1995), 15–34.

[LapNes1] M. L. Lapidus and R. Nest, Fractal membranes as the second quantization of fractal strings, in preparation, 2006.

[LapNes2] M. L. Lapidus and R. Nest, Functional equations for zeta functions associated with quasicrystals and fractal membranes, in preparation, 2006.

[LapNes3] M. L. Lapidus and R. Nest, Quasicrystals, zeta functions, and noncommutative geometry, in preparation, 2006.

[LapNes4] M. L. Lapidus and R. Nest, Towards a fractal cohomology theory (tentative title), work in progress, 2006.

[LapNeuRnGri] M. L. Lapidus, J. W. Neuberger, R. J. Renka and C. A. Griffith, Snowflake harmonics and computer graphics: Numerical computation of spectra on fractal domains, *Internat. J. Bifurcation & Chaos* **6** (1996), 1185–1210.

[LapPan] M. L. Lapidus and M. M. H. Pang, Eigenfunctions of the Koch snowflake drum, *Commun. Math. Phys.* **172** (1995), 359–376.

[LapPe1] M. L. Lapidus and E. P. J. Pearse, A tube formula for the Koch snowflake curve, with applications to complex dimensions, *J. London Math. Soc.*, in press. (Also: e-print, `arXiv: math-ph/0412029`, 2005.)

[LapPe2] M. L. Lapidus and E. P. J. Pearse, Tube formulas and complex dimensions of self-similar tilings, e-print, `arXiv:math.DS/0605527`, 2006.

[LapPe3] M. L. Lapidus and E. P. J. Pearse, Tube formulas for the generators of a self-similar tiling, preprint, 2006.

[LapPe4] M. L. Lapidus and E. P. J. Pearse, Fractal curvature measures and local tube formulas (tentative title), in preparation.

[LapPo1] M. L. Lapidus and C. Pomerance, Fonction zêta de Riemann et conjecture de Weyl–Berry pour les tambours fractals, *C. R. Acad. Sci. Paris Sér. I Math.* **310** (1990), 343–348.

[LapPo2] M. L. Lapidus and C. Pomerance, The Riemann zeta-function and the one-dimensional Weyl–Berry conjecture for fractal drums, *Proc. London Math. Soc.* (3) **66** (1993), 41–69.

[LapPo3] M. L. Lapidus and C. Pomerance, Counterexamples to the modified Weyl–Berry conjecture on fractal drums, *Math. Proc. Cambridge Philos. Soc.* **119** (1996), 167–178.

[Lap-vF1] M. L. Lapidus and M. van Frankenhuijsen, *Complex dimensions of fractal strings and explicit formulas for geometric and spectral zeta-functions*, preprint, IHES/M/97/34, Institut des Hautes Études Scientifiques, Bures-sur-Yvette, France, April 1997.

[Lap-vF2] M. L. Lapidus and M. van Frankenhuijsen, *Complex dimensions and oscillatory phenomena, with applications to the geometry of fractal strings and to the critical zeros of zeta-functions*, preprint, IHES/M/97/38, Institut des Hautes Études Scientifiques, Bures-sur-Yvette, France, May 1997.

[Lap-vF3] M. L. Lapidus and M. van Frankenhuijsen, (i) *Complex dimensions of fractal strings and oscillatory phenomena*; (ii) *Zeta-functions and explicit formulas for the geometry and spectrum of fractal strings*, Abstracts #918-35-537 and 918-35-539, Abstracts Amer. Math. Soc. **18**, No. 1 (1997), pp. 82–83. (Presented by M. L. L. and M. v. F., respectively, at the Annual Meeting of the American Mathematical Society, San Diego, Calif., Jan. 11, 1997, Special Session on "Analysis, Diffusions and PDEs on Fractals"; AMS meeting 918, event code AMS SS M1.)

[Lap-vF4] M. L. Lapidus and M. van Frankenhuijsen, Complex dimensions of fractal strings and oscillatory phenomena in fractal geometry and arithmetic, in: *Spectral Problems in Geometry and Arithmetic* (T. Branson, ed.), Contemporary Mathematics, vol. 237, Amer. Math. Soc., Providence, R. I., 1999, pp. 87–105.

[Lap-vF5] M. L. Lapidus and M. van Frankenhuijsen, *Fractal Geometry and Number Theory* (*Complex dimensions of fractal strings and zeros of zeta functions*), Birkhäuser, Boston, 2000.

[Lap-vF6] M. L. Lapidus and M. van Frankenhuijsen, A prime orbit theorem for self-similar flows and Diophantine approximation, *Contemporary Mathematics* **290** (2001), 113–138. (Also: MSRI Preprint No. 2001–039, Mathematical Sciences Research Institute (MSRI), Berkeley, November 2001.)

[Lap-vF7] M. L. Lapidus and M. van Frankenhuijsen, Complex dimensions of self-similar fractal strings and Diophantine approximation, *J. Experimental Math.* **12** (2003), 41–69. (Also: MSRI Preprint No. 2001–040, Mathematical Sciences Research Institute (MSRI), Berkeley, November 2001.)

[Lap-vF8] M. L. Lapidus and M. van Frankenhuijsen (eds.), Dynamical, Spectral, and Arithmetic Zeta Functions, *Contemporary Mathematics* **290**, Amer. Math. Soc., Providence, R. I., 2001.

[Lap-vF9] M. L. Lapidus and M. van Frankenhuijsen, *Fractality, self-similarity and complex dimensions*, in [Lap-vF10, Part 1, pp. 349–372].

[Lap-vF10] M. L. Lapidus and M. van Frankenhuijsen (eds.), *Fractal Geometry and Applications: A Jubilee of Benoît Mandelbrot*, Proc. Symposia Pure Math. **72** (Part 1: *Analysis, Number Theory, and Dynamical Systems*; Part 2: *Multifractals, Probability and Statistical Mechanics, Applications*), Amer. Math. Soc., Providence, R. I., 2004.

[LeLeLo] A. K. Lenstra, H. W. Lenstra, Jr., L. Lovász, Factoring polynomials with rational coefficients, *Math. Ann.* **261** (1982), 515–534.

[LeVa] M. Levitin and D. Vassiliev, Spectral asymptotics, renewal theorem, and the Berry conjecture for a class of fractals, *Proc. London Math. Soc.* (3) **72** (1996), 188–214.

[Man1] B. B. Mandelbrot, *The Fractal Geometry of Nature*, rev. and enl. ed. (of the 1977 ed.), W. H. Freeman, New York, 1983.

[Man2] B. B. Mandelbrot, Measures of fractal lacunarity: Minkowski content and alternatives, in: *Fractal Geometry and Stochastics* (C. Bandt, S. Graf and M. Zähle, eds.), Progress in Probability, vol. 37, Birkhäuser-Verlag, Basel, 1995, pp. 15–42.

[Man3] B. B. Mandelbrot, *Multifractals and 1/f Noise (Wild Self-Affinity in Physics)*, Springer-Verlag, New York, 1998.

[Marg] G. Margulis, Certain applications of ergodic theory to the investigation of manifolds of negative curvature, *Functional Anal. Appl.* **3** (1969), 89–90.

[MartVu] O. Martio and M. Vuorinen, Whitney cubes, p-capacity, and Minkowski content, *Exposition Math.* **5** (1987), 17–40.

[Mat] P. Mattila, *Geometry of Sets and Measures in Euclidean Spaces (Fractals and Rectifiability)*, Cambridge Univ. Press, Cambridge, 1995.

[MauWi] R. D. Mauldin and S. G. Williams, Random recursive constructions: asymptotic geometric and topological properties, *Trans. Amer. Math. Soc.* **295** (1986), 325–346.

[Maz] V. G. Maz'ja, *Sobolev Spaces*, Springer-Verlag, Berlin, 1985.

[McKSin] H. P. McKean and I. M. Singer, Curvatures and the eigenvalues of the Laplacian, *J. Differential Geom.* **1** (1967), 43–69.

[Mel1] R. B. Melrose, Weyl's conjecture for manifolds with concave boundary, in: *Geometry of the Laplace Operator*, Proc. Sympos. Pure Math., vol. 36, Amer. Math. Soc., Providence, R. I., 1980, pp. 254–274.

[Mel2] R. B. Melrose, *The trace of the wave group*, Contemporary Mathematics, vol. 27, Amer. Math. Soc., Providence, R. I., 1984, pp. 127–167.

[Met] G. Métivier, Valeurs propres de problèmes aux limites elliptiques irréguliers, *Bull. Soc. Math. France, Mém.* **51–52** (1977), 125–219.

[Mey] Y. Meyer, *Wavelets, Vibrations and Scaling*, CRM Monographs Ser. (Centre de Recherches Mathématiques, Université de Montréal), vol. 9, Amer. Math. Soc., Providence, R. I., 1998.

[Mil] J. Milnor, Euler characteristic and finitely additive Steiner measure, in: *John Milnor: Collected Papers*, vol. 1, Geometry, Publish or Perish, Houston, 1994, pp. 213–234. (Previously unpublished.)

[Min1] S. Minakshisundaram, A generalization of Epstein zeta-functions, *Canad. J. Math.* **1** (1949), 320–329.

[Min2] S. Minakshisundaram, Eigenfunctions on Riemannian manifolds, *J. Indian Math. Soc.* **17** (1953), 158–165.

[MinPl] S. Minakshisundaram and Å. Pleijel, Some properties of the eigenfunctions of the Laplace-operator on Riemannian manifolds, *Canad. J. Math.* **1** (1949), 242–256.

[Mink] H. Minkowski, Theorie der konvexen Körper, insbesondere Begründung ihres Oberflächenbegriffs, in: *Gesammelte Abhandlungen von Hermann Minkowski* (part II, Chapter XXV), Chelsea, New York, 1967, pp. 131–229. (Originally reprinted in: Gesamm. Abh., vol. II, Leipzig, 1911.)

[MolVa] S. Molchanov and B. Vainberg, On spectral asymptotics for domains with fractal boundaries, *Commun. Math. Phys.* **183** (1997), 85–117.

[Moo] R. V. Moody, Model sets: A survey, in: *From Quasicrystals to More Complex Systems* (F. Axel and J.-P. Gazeau, eds.), Les Editions de Physique, Springer-Verlag, Berlin, 2000, pp. 145–166.

[Mor] P. A. P. Moran, Additive functions of intervals and Hausdorff measure, *Math. Proc. Cambridge Philos. Soc.* **42** (1946), 15–23.

[Na] N. Naumann, On the irreducibility of the two variable zeta-function for curves over finite fields, *C. R. Acad. Sci. Paris Sér.* I *Math.* **336** (2003), 289–292. (Also: e-print, `arXiv:math.AG/0209092`, 2002.)

[Od1] A. M. Odlyzko, On the distribution of spacings between zeros of the zeta-function, *Math. Comp.* **48** (1987), 273–308.

[Od2] A. M. Odlyzko, *The* 10^{20}*-th zero of the Riemann zeta-function and 175 millions of its neighbors*, preprint, AT&T Bell Labs, Murray Hill, 1991; available at ⟨http://www.dtc.umn.edu/~odlyzko/unpublished/index.html⟩.

[Od3] A. Odlyzko, http://www.dtc.umn.edu/~odlyzko/ .

[Od-tR] A. M. Odlyzko and H. J. J. te Riele, Disproof of the Mertens conjecture, *J. Reine Angew. Math.* **357** (1985), 138–160.

[Os] A. Ostrowski, Bemerkungen zur Theorie der Diophantischen Approximationen, *Abh. Math. Sem. Hamburg Univ.* **1** (1922), 77–98.

[PaPol1] W. Parry and M. Pollicott, An analogue of the prime number theorem and closed orbits of Axiom A flows, *Annals of Math.* **118** (1983), 573–591.

[PaPol2] W. Parry and M. Pollicott, Zeta Functions and the Periodic Orbit Structure of Hyperbolic Dynamics, *Astérisque,* vols. 187–188, Soc. Math. France, Paris, 1990.

[ParSh1] A. N. Parshin and I. R. Shafarevich (eds.), *Number Theory,* vol. I, *Introduction to Number Theory,* Encyclopedia of Mathematical Sciences, vol. 49, Springer-Verlag, Berlin, 1995. (Written by Yu. I. Manin and A. A. Panchishkin.)

[ParSh2] A. N. Parshin and I. R. Shafarevich (eds.), *Number Theory,* vol. II, *Algebraic Number Fields*, Encyclopedia of Mathematical Sciences, vol. 62, Springer-Verlag, Berlin, 1992. (Written by H. Koch.)

[Pat] S. J. Patterson, *An Introduction to the Theory of the Riemann Zeta-Function,* Cambridge Univ. Press, Cambridge, 1988.

[Pe] E. P. J. Pearse, Canonical self-similar tilings of IFS, preprint, 2006.

[Pel] R. Pellikaan, On special divisors and the two variable zeta function of algebraic curves over finite fields, in: *Arithmetic, Geometry and Coding Theory,* Proceedings of the International Conference held at CIRM, Luminy, France, 1993, pp. 175–184.

[Ph] Pham The Lai, Meilleures estimations asymptotiques des restes de la fonction spectrale et des valeurs propres relatifs au laplacien, *Math. Scand.* **48** (1981), 5–38.

[Pin] M. A. Pinsky, The eigenvalues of an equilateral triangle, *SIAM J. Math. Anal.* **11** (1980), 819–827.

[PitY] J. Pitman and M. Yor, The two parameter Poisson–Dirichlet distribution derived from a stable subordinator, *Ann. Probab.* **25** (1997), 855–900.

[Pos] A. G. Postnikov, *Tauberian theory and its applications*, Proc. Steklov Inst. of Math., vol. 144, No. 2, Amer. Math. Soc., Providence, R. I., 1980.

[Pu1] C. R. Putnam, On the non-periodicity of the zeros of the Riemann zeta-function, *Amer. J. Math.* **76** (1954), 97–99.

[Pu2] C. R. Putnam, Remarks on periodic sequences and the Riemann zeta-function, *Amer. J. Math.* **76** (1954), 828–830.

[Que] H. Queffélec, Propriétés presque sûres et quasi-sûres des séries de Dirichlet et des produits d'Euler, *Canad. J. Math.* **32** (1980), 531–558.

[RamTo] R. Rammal and G. Toulouse, Random walks on fractal structures and percolation cluster, *J. Physique Lettres* **44** (1983), L13–L22.

[ReSi1] M. Reed and B. Simon, *Methods of Modern Mathematical Physics*, vol. I, *Functional Analysis*, rev. and enl. ed. (of the 1975 ed.), Academic Press, New York, 1980.

[ReSi2] M. Reed and B. Simon, *Methods of Modern Mathematical Physics*, vol. II, *Fourier Analysis, Self-Adjointness*, Academic Press, New York, 1975.

[ReSi3] M. Reed and B. Simon, *Methods of Modern Mathematical Physics*, vol. IV, *Analysis of Operators*, Academic Press, New York, 1979.

[Rie1] B. Riemann, *Ueber die Anzahl der Primzahlen unter einer gegebenen Grösse*, Monatsb. der Berliner Akad., 1858/60, pp. 671–680. (Reprinted in [Rie2, pp. 145–155]; English translation in [Edw, Appendix, pp. 299–305].)

[Rie2] B. Riemann, *Gesammelte Mathematische Werke*, Teubner, Leipzig, 1892, No. VII. (Reprinted by Dover Books, New York, 1953.)

[Rog] C. A. Rogers, *Hausdorff Measures*, Cambridge Univ. Press, Cambridge, 1970.

[Roq] P. Roquette, Arithmetischer Beweis der Riemannschen Vermutung in Kongruenzfunktionenkörpern beliebigen Geslechts, *J. Reine Angew. Math.* **191** (1953), 199–252.

[RösS] C. Rössner, C. P. Schnorr, An optimal, stable continued fraction algorithm for arbitrary dimension, in: *Proc. of IPCO V* (Conference on Integer Programming and Combinatorial Optimization, Vancouver, June 3-5, 1996), Lecture Notes in Computer Science, vol. 1084, Springer-Verlag, New York, 1996, pp. 31-43.

[Ru1] W. Rudin, *Fourier Analysis on Groups*, Interscience Publishers, John Wiley & Sons, New York, 1962.

[Ru2] W. Rudin, *Real and Complex Analysis*, 3rd ed., McGraw-Hill, New York, 1987.

[Ru3] W. Rudin, *Functional Analysis*, 2nd ed. (of the 1973 ed.), McGraw-Hill, New York, 1991.

[RudSar] Z. Rudnick and P. Sarnak, Zeros of principal L-functions and random matrix theory, *Duke Math. J.* **81** (1996), 269–322.

[Rue1] D. Ruelle, Generalized zeta-functions for Axiom A basic sets, *Bull. Amer. Math. Soc.* **82** (1976), 153–156.

[Rue2] D. Ruelle, Zeta functions for expanding maps and Anosov flows, *Invent. Math.* **34** (1978), 231–242.

[Rue3] D. Ruelle, *Thermodynamic Formalism*, Addison-Wesley, Reading, 1978.

[Rue4] D. Ruelle, *Dynamical Zeta Functions for Piecewise Monotone Maps of the Interval*, CRM Monographs Ser. (Centre de Recherches Mathématiques, Université de Montréal), vol. 4, Amer. Math. Soc., Providence, R. I., 1994.

[Sab1] C. Sabot, Spectral properties of hierarchical lattices and iterations of rational maps, *Mémoirs Soc. Math. France* (N. S.), No. 92, 2003, 1–104.

[Sab2] C. Sabot, Electrical networks, symplectic reductions, and applications to the renormalization map of self-similar lattices, in [Lap-vF10, Part 1, pp. 155–205].

[SalS] H. Saleur and D. Sornette, Complex exponents and log-periodic correlations in frustrated systems, *J. Phys.* I *France* **6** (1996), 327–355.

[SapGoM] B. Sapoval, Th. Gobron and A. Margolina, Vibrations of fractal drums, *Phys. Rev. Lett.* **67** (1991), 2974–2977.

[Schm] W. M. Schmidt, *Diophantine Approximation*, Lecture Notes in Math., vol. 785, Springer-Verlag, New York, 1980.

[Schn] R. Schneider, *Convex Bodies: The Brunn–Minkowski Theory*, Encyclopedia of Mathematics and its Applications, vol. 44, Cambridge Univ. Press, 2003. (Reprinted from the 1993 edition.)

[SchoG] R. Schoof and G. van der Geer, Effectivity of Arakelov divisors and the theta divisor of a number field, *Selecta Math.* (N. S.) **6** (2000), 377–398. (Also: e-print, `arXiv:math.AG/9802121`, 1998.)

[SchrSo] M. Schröter and C. Soulé, On a result of Deninger concerning Riemann's zeta-function, in: *Motives, Proc. Sympos. Pure Math.*, vol. 55, Amer. Math. Soc., Providence, R. I., 1994, pp. 745–747.

[Schw1] L. Schwartz, *Théorie des Distributions*, rev. and enl. ed. (of the 1951 ed.), Hermann, Paris, 1966.

[Schw2] L. Schwartz, *Méthodes Mathématiques pour les Sciences Physiques*, Hermann, Paris, 1961.

[Se1] R. T. Seeley, Complex powers of elliptic operators, in: *Proc. Symp. Pure Math.*, vol. 10, Amer. Math. Soc., Providence, R. I., 1967, pp. 288–307.

[Se2] R. T. Seeley, The resolvent of an elliptic boundary problem, *Amer. J. Math.* **91** (1969), 889–920.

[Se3] R. T. Seeley, Analytic extensions of the trace associated with elliptic boundary problems, *Amer. J. Math.* **91** (1969), 963–983.

[Se4] R. T. Seeley, A sharp asymptotic remainder estimate for the eigenvalues of the Laplacian in a domain of $\mathbb{R}^3$, *Adv. in Math.* **29** (1978), 244–269.

[Se5] R. T. Seeley, An estimate near the boundary for the spectral counting function of the Laplace operator, *Amer. J. Math.* **102** (1980), 869–902.

[Sen] M. Senechal, *Quasicrystals and Geometry*, Cambridge Univ. Press, Cambridge, 1995.

[Ser] J.-P. Serre, *A Course in Arithmetic*, English translation, Springer-Verlag, Berlin, 1973.

[ShlW] M. F. Shlesinger and B. West, Complex fractal dimensions of the bronchial tree, *Phys. Rev. Lett.* **67** (1991), 2106–2109.

[Shu] M. A. Shubin, *Pseudodifferential Operators and Spectral Theory*, Springer-Verlag, Berlin, 1987.

[Sim] B. Simon, *Functional Integration and Quantum Physics*, Academic Press, New York, 1979.

[Sin] Y. G. Sinai, The asymptotic behaviour of the number of closed geodesics on a compact manifold of negative curvature, *Transl. Amer. Math. Soc.* **73** (1968), 227–250.

[Sma] S. Smale, Differentiable dynamical systems, *Bull. Amer. Math. Soc.* **73** (1967), 747–817.

[SmiFoSp] L. A. Smith, J.-D. Fournier and E. A. Spiegel, Lacunarity and intermittency in fluid turbulence, *Phys. Lett. A* **114** (1986), 465–468.

[Sor] D. Sornette, Discrete scale invariance and complex dimensions, *Phys. Rep.* **297** (1998), 239–270.

[Stein] J. Steiner, Über parallele Flächen, *Monatsb. preuss. Akad. Wiss.*, Berlin, 1840, pp. 114–118. (Reprinted in: *Gesamm. Werke* vol. II, pp. 173–176.)

[Step] S. A. Stepanov, On the number of points of a hyperelliptic curve over a finite prime field, *Izv. Akad. Nauk SSSR, Ser. Mat.* **33** (1969), 1103–1114.

[Sto-c] C. J. Stone, On moment generating functions and renewal theory, *Ann. Math. Stat.* **36** (1965), 1298–1301.

[Sto-r] R. Stone, Operators and divergent series, *Pacific J. of Math.* **217** No. 2 (2004), 331–374.

[Str1] R. S. Strichartz, Fourier asymptotics of fractal measures, *J. Functional Anal.* **89** (1990), 154–187.

[Str2] R. S. Strichartz, Self-similar measures and their Fourier transforms, I, *Indiana Univ. Math. J.* **39** (1990), 797–817; II, *Trans. Amer. Math. Soc.* **336** (1993), 335–361; III, *Indiana Univ. Math. J.* **42** (1993), 367–411.

[Su] D. Sullivan, Entropy, Hausdorff measures old and new, and limit sets of geometrically finite Kleinian groups, *Acta Math.* **153** (1984), 259–277.

[Ta] J. T. Tate, *Fourier Analysis in Number Fields and Hecke's Zeta-Functions*, Ph.D. Dissertation, Princeton University, Princeton, N. J., 1950. (Reprinted in: *Algebraic Number Theory*, J. W. S. Cassels and A. Fröhlich (eds.), Academic Press, New York, 1967, pp. 305–347.)

[Tep1] A. Teplyaev, Spectral zeta functions of symmetric fractals, in: *Progress in Probability*, vol. 57, Birkhäuser-Verlag, Basel, 2004, pp. 245–262.

[Tep2] A. Teplyaev, Spectral zeta functions of fractals and the complex dynamics of polynomials, *Trans. Amer. Math. Soc.*, in press. (Also: e-print, `arXiv:math.SP/0505546`, 2005.)

[Ter] A. Terras, *Harmonic Analysis on Symmetric Spaces and Applications*, vol. I, Springer-Verlag, New York, 1985.

[Ti] E. C. Titchmarsh, *The Theory of the Riemann Zeta-Function*, 2nd ed. (revised by D. R. Heath-Brown), Oxford Univ. Press, Oxford, 1986.

[Tr1] C. Tricot, Douze définitions de la densité logarithmique, *C. R. Acad. Sci. Paris Sér. I Math.* **293** (1981), 549–552.

[Tr2] C. Tricot, Two definitions of fractional dimension, *Math. Proc. Cambridge Philos. Soc.* **91** (1982), 57–74.

[Tr3] C. Tricot, *Curves and Fractal Dimensions*, Springer-Verlag, New York, 1995.

[Va] V. S. Varadarajan, Some remarks on the analytic proof of the Prime Number Theorem, *Nieuw Archief voor Wiskunde* **16** (1998), 153–160.

[vB] M. van den Berg, Heat content and Brownian motion for some regions with a fractal boundary, *Probab. Theory and Related Fields* **100** (1994), 439–456.

[vB-Gi] M. van den Berg and P. B. Gilkey, A comparison estimate for the heat equation with an application to the heat of the s-adic von Koch snowflake, *Bull. London Math. Soc.* **30** (1998), 404–412.

[vB-Le] M. van den Berg and M. Levitin, Functions of Weierstrass type and spectral asymptotics for iterated sets, *Quart. J. Math. Oxford* (2) **47** (1996), 493–509.

[vF1] M. van Frankenhuijsen, Counting the number of points on an algebraic curve, in: *Algoritmen in de algebra, a seminar on algebraic algorithms* (A. H. M. Levelt, ed.), Nijmegen, 1993.

[vF2] M. van Frankenhuijsen, *Hyperbolic Spaces and the abc Conjecture*, Ph.D. Dissertation (Proefschrift), Katholieke Universiteit Nijmegen, Netherlands, 1995. at ⟨http://www.research.uvsc.edu/~machiel/programs.html⟩.

[vF3] M. van Frankenhuijsen, Arithmetic progressions of zeros of the Riemann zeta function, *J. of Number Theory*, **115** (2005), 360–370.

[vFWatk] M. van Frankenhuijsen and M. Watkins, Finite arithmetic progressions of zeros of L-series, preprint, 2004.

[vL-vdG] J. H. van Lint and G. van der Geer, *Introduction to Coding Theory and Algebraic Geometry*, DMV Seminar, Band 12, Birkhäuser, Basel, 1988.

[vL-tR-W] J. van de Lune, H. J. J. te Riele and D. Winter, On the zeros of the Riemann zeta function in the critical strip, IV, *Math. Comp.* **46** (1986), 667–681.

[Voi] D. V. Voiculescu, Lectures on free probability theory, in: *Lectures on Probability and Statistics* (P. Bernard, ed.), Ecole d'Eté de Probabilités de Saint-Flour XXVIII-1998, Lecture Notes in Math., vol. 1738, Springer-Verlag, Berlin, pp. 279–349.

[vK1] H. von Koch, Sur une courbe continue sans tangente obtenue par une construction géométrique élémentaire, *Ark. Mat., Astron. Fys.* **1** (1904), 681–702. (Engl. translation, with annotations, reprinted in [Ed, pp. 25–45].)

[vK2] Une méthode géométrique élémentaire pour l'étude des courbes planes, *Acta Math.* **30** (1906), 145–174.

[vM1] H. von Mangoldt, *Auszug aus einer Arbeit unter dem Titel: Zu Riemann's Abhandlung 'Über die Anzahl der Primzahlen unter einer gegebenen Grösse'*, Sitzungsberichte preuss. Akad. Wiss., Berlin, 1894, pp. 883–896.

[vM2] H. von Mangoldt, Zu Riemann's Abhandlung 'Über die Anzahl der Primzahlen unter einer gegebenen Grösse', *J. Reine Angew. Math.* **114** (1895), 255–305.

[Vor] A. Voros, Spectral functions, special functions and the Selberg zeta function, *Commun. Math. Phys.* **110** (1987), 439–465.

[Watk] M. Watkins, Arithmetic progressions of zeros of Dirichlet L-functions, preprint, 1998.

[Wats] G. N. Watson, *A Treatise on the Theory of Bessel Functions*, 2nd ed., Cambridge Mathematical Library, Cambridge Univ. Press, Cambridge, 1995.

[Wei1] A. Weil, On the Riemann hypothesis in function-fields, *Proc. Nat. Acad. Sci. U.S.A.* **27** (1941), 345–347. (Reprinted in [Wei7, vol. I, pp. 277–279].)

[Wei2] A. Weil, Sur les courbes algébriques et les variétés qui s'en déduisent, *Pub. Inst. Math. Strasbourg* VII (1948), pp. 1–85. (Reprinted in: *Courbes algébriques et variétés abéliennes*, Hermann, Paris, 1971.)

[Wei3] A. Weil, Number of solutions of equations in finite fields, *Bull. Amer. Math. Soc.* **55** (1949), 497–508. (Reprinted in [Wei7, vol. I, pp. 399–410].)

[Wei4] A. Weil, Sur les "formules explicites" de la théorie des nombres premiers, *Comm. Sém. Math. Lund*, Université de Lund,

Tome supplémentaire (dédié à Marcel Riesz), (1952), pp. 252–265. (Reprinted in [Wei7, vol. II, pp. 48–61].)

[Wei5] A. Weil, Fonction zêta et distributions, *Séminaire Bourbaki*, 18ème année, 1965/66, no. 312, Juin 1966, pp. 1–9. (Reprinted in [Wei7, vol. III, pp. 158–163].)

[Wei6] A. Weil, Sur les formules explicites de la théorie des nombres, *Izv. Mat. Nauk (Ser. Mat.)* **36** (1972), 3–18; English translation in: *Math. USSR, Izv.* **6** (1973), 1–17. (Reprinted in [Wei7, vol. III, pp. 249–264].)

[Wei7] A. Weil, *André Weil*: *Oeuvres Scientifiques* (Collected Papers), vols. I, II and III, 2nd ed. (with corrected printing), Springer-Verlag, Berlin and New York, 1980.

[Wey1] H. Weyl, Über die Abhängigkeit der Eigenschwingungen einer Membran von deren Begrenzung, *J. Reine Angew. Math.* **141** (1912), 1–11. (Reprinted in [Wey4, vol. I, pp. 431–441].)

[Wey2] H. Weyl, Das asymptotische Verteilungsgesetz der Eigenwerte linearer partieller Differentialgleichungen, *Math. Ann.* **71** (1912), 441–479. (Reprinted in [Wey4, vol. I, pp. 393–430].)

[Wey3] H. Weyl, On the volume of tubes, *Amer. J. Math.* **61** (1939), 461–472. (Reprinted in [Wey4, vol. III, pp. 658–669].)

[Wey4] H. Weyl, *Hermann Weyl*: *Gesammelte Abhandlungen* (Collected Works), vols. I and III, Springer-Verlag, Berlin and New York, 1968.

[WhW] E. T. Whittaker, G. N. Watson, *A Course of Modern Analysis*, 4th ed., Cambridge Mathematical Library, Cambridge Univ. Press, Cambridge, 1996.

[Wid] D. Widder, *The Laplace Transform*, Princeton Univ. Press, Princeton, 1946.

[Zyg] A. Zygmund, *Trigonometric Series*, vols. I and II, 2nd ed., Cambridge Univ. Press, Cambridge, 1959.

Acknowledgements

We would like to thank the Institut des Hautes Etudes Scientifiques (IHES), of which we were members in 1998–2000 when some of the main ideas for [Lap-vF5] were conceived. The work of Michel L. Lapidus was supported by the National Science Foundation under grants DMS-9207098, DMS-9623002 and DMS-0070497, and that of Machiel van Frankenhuijsen was supported by the Marie Curie Fellowship ERBFMBICT960829 of the European Community and by a summer stipend from Utah Valley State College. The second author would also like to thank the Department of Mathematics and the School of Science and Health at Utah Valley State College for their continuing and generous support.

The first author would also like to thank Alain Connes, Rudolf H. Riedi and Christophe Soulé for helpful conversations and/or references during the preliminary phase of this work. In addition, the authors are grateful to Gabor Elek and Jim Stafney as well as to several anonymous referees, for their helpful comments on the preliminary versions of [Lap-vF5].

We are indebted to Mark Watkins for sharing with us his beautiful proof of the finiteness of shifted arithmetic progressions of zeros of L-series. We also wish to thank Ben Hambly for his comments on a preliminary version of Section 12.4.1.

We are very grateful to Erin Pearse, one of the first author's current Ph.D. students, for many helpful comments on [Lap-vF5] after its publication and on several preliminary versions of the present book. We are also grateful to him for making his Ph.D. oral examination (based on Chapter 6 of [Lap-vF5], Chapter 8 of this book) available to us, along with Figure 8.1. This has enabled us to improve the pedagogical presentation of parts of Chapter 8. Also, Figures 12.7–12.10 (from [LapPe1], some adapted from this paper) and Figure 12.11 were created by him. Moreover, we wish to acknowledge his help in combining two figures from the book in order to obtain the design appearing on the front cover. Finally, we would like to thank him for having suggested to us the term "languid" to refer to the growth conditions **L1** and **L2**, which were called hypotheses ($\mathbf{H}_1$) and ($\mathbf{H}_2$) in [Lap-vF5].

In addition to Erin Pearse, the first author would like to thank his other current Ph.D. students and members of his "Fractal Research Group", Vicente Alvarez, Scot Childress, Britta Daudert, Tim (Hung) Lu, John Rock and Jun Tanaka, for helpful comments and feedback. He would also like to thank the many participants in his Seminar on "Mathematical Physics and Dynamical Systems", graduate students, visitors, postdocs and faculty members alike.

We are very grateful to Ann Kostant, Executive Editor for the Mathematical Sciences at Birkhäuser Boston, and now also Editorial Director for Mathematics at Springer, for her enthusiasm and her constant encouragement, as well as for her guidance in preparing the manuscripts of [Lap-vF5] and the present book for publication. She went well beyond the call of duty in helping us bring this project to fruition. In particular, her wonderful attention both to the big picture and many important details certainly played a significant role in the succesful completion of this book.

We would also like to thank several anonymous referees for their efforts in reviewing this book at various stages of its preparation, as well as for their expertise, enthusiasm, and many useful suggestions and constructive criticisms which definitely helped us improve the readability and overall quality of this research monograph.

Last but not least, the first author would like to thank his wife, Odile, and his children, Julie and Michaël, for supporting him through long periods of sleepless nights, either at the Résidence de l'Ormaille of the IHES in Bures-sur-Yvette or at their home in Riverside, California, while the theory presented in the research monograph [Lap-vF5] was being developed and when that book and the present one were in the process of being written.

Part of this work was presented by the authors at the Special Session on "Analysis, Diffusions and PDEs on Fractals" held during the Annual Meeting of the American Mathematical Society (San Diego, January 1997), at the Special Session on "Dynamical, Spectral and Arithmetic Zeta-Functions" held during the Annual Meeting of the American Mathematical Society (San Antonio, January 1999),[1] the Special Session on "Fractal Geometry, Number Theory and Dynamical Systems" held during the "First Joint Mathematical International Meeting of the American Mathematical

[1] Abstracts #918-35-537 and 539, Abstracts Amer. Math. Soc. **18** No. 1 (1997), 82–83, and Abstracts #939-58-84 and 85, *ibid.* **20** No. 1 (1999), 126–127.

Society and the Société Mathématique de France" at the Ecole Normale Supérieure (Lyon, France, July 2001), and at the Special Session on "Fractal Geometry and Applications: A Jubilee of Benoît Mandelbrot" held during the Annual Meeting of the American Mathematical Society (San Diego, January 2002). It was also presented by the first author in invited talks at the CBMS-NSF Conference on "Spectral Problems in Geometry and Arithmetic" (Iowa City, August 1997) and at the Conference on "Recent Progress in Noncommutative Geometry" (Lisbon, Portugal, September 1997), as well as in the Workshops and Programs on "Spectral Geometry" (June–July 1998) and on "Number Theory and Physics" (September 1998), both held at the Erwin Schroedinger International Institute for Mathematical Physics in Vienna, Austria.

In addition, it was presented by the first author at the Basic Research Institute in the Mathematical Sciences (BRIMS) in Bristol, UK, in April 1999 and in the Program on "Mathematics and Applications of Fractals" (March–April 1999) held at the Isaac Newton Institute for Mathematical Sciences of the University of Cambridge, England, the Workshop on "Fractals and Dynamical Systems" held at the Chinese University (Hong Kong, China, December 2000), the Special Session on "Dynamical Systems, with Emphasis on Geometric Aspects and Symbolic Dynamics" held during the Fifth Joint Meeting of the American Mathematical Society and the Sociedad Matematica Mexicana (Morelia, Mexico, May 2001), the Conference "Fractals in Gratz 2001" (Gratz, Austria, June 2001), the Special Session on "Fractals" held at Ohio State University during the Sectional Meeting of the American Mathematical Society (Columbus, September 2001), the First and Second Conferences on "Analysis and Probability on Fractals" held at Cornell University (Ithaca, June 2002 and June 2005), the Workshop and Program on "Topology in Condensed Matter Physics" held at the Max Planck Institute for the Physics of Complex Systems (Dresden, Germany, July 2002), the Satellite Conference on "Fractal Geometry and Applications" (Nanjing, China, September 2002) held in conjunction with the International Congress of Mathematicians in Beijing (August 2002), the Special Session on "Analytic Number Theory" held during the Sectional Meeting of the American Mathematical Society (Salt Lake City, October 2002), the Conference on "New Geometric Methods in Modern Science" (Paris, France, March 2003), the Special Session on "Noncommutative Geometry and Geometric Analysis" held at the University of Colorado during the First Joint Meeting of the Central and Western Sections of the American Mathematical Society (Boulder, October 2003), the Special Session on "Iterated Function Systems and Analysis on Fractals" held at Northwestern University during the Sectional Meeting of the American Mathematical Society (Evanston, October 2004), the Workshop on "Fractal Analysis" (Eisenach, Germany, September 2005) and the Workshop on "Traces in Geometry, Number Theory and Quantum Fields" held at the Max Planck Institute for Mathematics (Bonn, Germany, October 2005).

The second author presented the work on finite arithmetic progressions of zeros of the Riemann zeta function in the Special Session on "Inverse Spectral Geometry" held during the Annual Meeting of the American Mathematical Society (Atlanta, 2005).

Part of this work will also be presented by the authors at the Special Session on "Fractal Geometry: Connections to Dynamics, Geometric Measure Theory, Mathematical Physics and Number Theory" to be held at San Francisco State University during the Sectional Meeting of the American Mathematical Society (San Francisco, April 2006). It will also be presented by the first author at the International Conference on "Contemporary Problems of Mathematical Analysis and Mathematical Physics" (Taormina, Sicily, Italy, June–July 2006), as well as in a series of lectures at the International Summer School on "Arithmetic and Geometry around Quantization" (AGAQ 2006) to be held at Galatasaray University (Istanbul, Turkey, June 2006), and by the second author in the Special Session on "Number Theory" to be held during the 2006 Fall Western Section Meeting of the AMS in Salt Lake City, Utah, October 7–8, 2006.

Part of this research was conducted and this book was written at a number of institutions in the US and abroad, including the authors' own universities. In addition to those already mentioned above, the first author would like to acknowledge the University of Rome (Tor Vergata), Italy, the University of Copenhagen, Denmark, and the Université Paris VII, France, for a number of visits during the past few years, as well as the Mathematical Sciences Research Institute in Berkeley, of which he was a member during the Programs on "Random Matrix Theory: Models and Applications" from May through June 1999 and on "Spectral Invariants" from April through June 2001, as well as the Centre Emile Borel of the Institut Henri Poincaré (IHP) of which he was a member of the Program on "Noncommutative Geometry and K-Theory" from March through July 2004 while living in the Résidence de l'Ormaille of the IHES from March through September 2004. The material and/or financial support, during this latter period, of the IHP, IHES and the Clay Mathematics Institute is gratefully ackowledged by the first author.

Conventions

$f(x) = O(g(x))$	$f(x)/g(x)$ is bounded
$f(x) = o(g(x))$	$f(x)/g(x)$ tends to 0
$f(x) \sim g(x)$	$f(x)/g(x)$ tends to 1
$f(x) \ll g(x)$	same meaning as $f(x) = O(g(x))$
$\approx$	approximately equal to
$d \mid n$	d divides n
$A \backslash B$	the set of points in A that do not lie in B
$\#A$	the cardinality of the finite set A
$\mathbb{N}$	the set of nonnegative integers $0, 1, 2, 3, \ldots$
$\mathbb{N}^* = \mathbb{N} \backslash \{0\}$	the set of positive integers $1, 2, 3, \ldots$
$\mathbb{Z}$ and $\mathbb{Q}$	the sets of integers and rational numbers, respectively
$\mathbb{R}$ and $\mathbb{C}$	the sets of real and complex numbers, respectively
$\mathbb{R}_+^*$	the multiplicative group of positive real numbers
i	the square root of -1
$s = \sigma + it$	s is a complex number with $\sigma = \operatorname{Re} s$ and $t = \operatorname{Im} s$
$\log x$	the natural logarithm of x
$\log_a x$	$\log x / \log a$, the logarithm of x with base a
$[x]$	the greatest integer less than or equal to x
$\{x\} = x - [x]$	the fractional part of x

Index of Symbols

Author Index

Subject Index